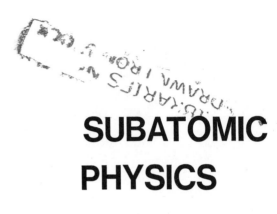

SUBATOMIC
PHYSICS

SCIENCE

SUBATOMIC

PHYSICS

Hans Frauenfelder

University of Illinois

Ernest M. Henley

University of Washington

PRENTICE-HALL, INC., Englewood Cliffs, N.J.

SCIENCE

Library of Congress Cataloging in Publication Data

FRAUENFELDER, HANS. 1922–
   Subatomic physics.

   (Prentice-Hall physics series)
   Bibliography: p.
   1.–Nuclear physics. 2.–Particles (Nuclear
physics) I.–Henley, Ernest M., joint author.
II.–Title.
QC776.F723   539.7   73-12966
ISBN 0-13-859082-6

PRENTICE-HALL PHYSICS SERIES

Consulting Editors

Francis M. Pipkin

George A. Snow

SUBATOMIC PHYSICS

Hans Frauenfelder
Ernest M. Henley

10 9 8 7 6 5 4 3 2 1

Printed in the United States of America

PRENTICE-HALL INTERNATIONAL, INC., *London*
PRENTICE-HALL OF AUSTRALIA, PTY. LTD., *Sydney*
PRENTICE-HALL OF CANADA, LTD., *Toronto*
PRENTICE-HALL OF INDIA PRIVATE LIMITED, *New Delhi*
PRENTICE-HALL OF JAPAN, INC., *Tokyo*

To Vreneli and Elaine

SCIENCE

Contents

PART **IV**

INTERACTIONS

10. The Electromagnetic Interaction 229

11. The Weak Interaction 273

12. Hadronic Interactions 319

PART **V**

MODELS

13. Quarks and Regge Poles 353

Preface

Subatomic Physics, the physics of nuclei and particles has been one of the frontiers of science since its birth in 1896. From the study of the radiations emitted by radioactive nuclei to the scattering experiments that point to the presence of subunits in nucleons, from the discovery of the hadronic interactions to the realization that the photon possesses hadronic (strong) attributes, and that weak and electromagnetic forces may be intimately related, subatomic physics has enriched science with new concepts and deeper insights into the laws of nature.

Subatomic Physics does not stand isolated, it bears on many aspects of life. Ideas and facts emerging from studies of the subatomic world change our picture of the macrocosmos. Concepts discovered in subatomic physics are needed to understand the creation and abundance of the elements, and the energy production in the sun stars. Nuclear power may provide most of the future energy sources. Nuclear bombs affect national and international decisions. Pion beams may become a tool to treat cancer. Tracer and Mössbauer techniques give information about structure and reactions in solid state physics, chemistry, biology, metallurgy, and geology.

Subatomic Physics, because it reaches into so many areas, should not only be accessible to physicists, but also to other scientists and to engineers. The chemist observing the Mössbauer effect, the geologist using a radioactive dating method, the physician injecting a radioactive isotope, or the nuclear engineer designing a power plant have no immediate need to understand isospin or inelastic electron scattering. Nevertheless, their work may be more satisfying and they may be able to find new connections if they have a grasp of the basic principles of subatomic physics. While the present book is mainly intended as an introduction for physicists, we hope that it will also be useful to other scientists and to engineers.

Subatomic Physics deals with all entities smaller than the atom; it combines nuclear and particle physics. The two fields have many concepts and features in common. Consequently, we treat them together and attempt to stress unifying ideas, concepts, and currently unsolved problems. We also show how subatomic physics is involved in power production, in astrophysics, and in chemistry. The level of presentation is aimed at the senior undergraduate or first-year graduate student who has some understanding of electromagnetism, special relativity, and quantum theory. While many aspects of subatomic physics can be elucidated by hand-waving and analogies, a proper understanding requires equations. One of the most infuriating sentences in textbooks is "It can be shown . . . ." We would like to avoid this sentence but it is just not possible. We include most derivations but use equations without proof in two situations. Many of the equations from other fields will be quoted without derivation in order to save space and time. The second situation arises when the proper tools, for instance Dirac theory or field quantization, are too advanced. We justify omission of the derivations in both situations by an analogy. Mountain climbers usually like to reach the unexplored parts of a climb quickly rather than spend days walking through familiar terrain. Quoting equations from quantum theory and electrodynamics corresponds to reaching the starting point of an adventure by car or cable car. Some peaks can only be reached by difficult routes. An inexperienced climber, not yet capable of mastering such a route, can still learn by watching from a safe place. Similarly, some equations can only be reached by difficult derivations, but the reader can still learn by exploring the equations without following their derivations. Therefore, we will quote some relations without proof, but we will try to make the result plausible and to explore the physical consequences. Some more difficult parts will be denoted with bullets ($\bullet$); these parts can be omitted on first reading.

Acknowledgments

In writing the present book, we have had generous help from many friends who have either cleared up questions, provided us with data, or have read through parts of the manuscript. We should particularly like to thank G. A. Baym, J. M. Bardeen, J. S. Blair, D. Bodansky, M. K. Brussel, S. J. Chang, P. G. Debrunner, H. Drickamer, W. A. Fowler, E. Greenbaum, J. L. Groves, I. Halpern, L. M. Jones, B. W. Lee, E. Münck, F. M. Pipkin, H. Primakoff, G. D. Ravenhall, G. A. Snow, R. L. Schult, C. P. Slichter, C. H. Llewellyn Smith, J. D. Sullivan and L. Wilets. Valuable criticism and constructive suggestions have come from many students who have read through the notes that formed the preliminary editions of the present book. We also should like to thank Mrs. D. Johnson and Mrs. N. Garman who have cheerfully typed the many versions of the manuscript, Mrs. J. Spaeth who did some of the typing, Mr. George Morris who did the artwork, and W. H. Grimshaw and E. Thompson who guided the book to completion.

Some of the joint work of the normally widely separated authors has taken place at each other's home institution. We are grateful for the support we received from the University of Illinois and the University of Washington, which made these visits possible. We also met at Los Alamos, at the Aspen Center of Physics, and CERN, and we express our appreciation for the stimulating hospitality of these centers. Further, we gratefully acknowledge support by the John Simon Guggenheim Memorial Foundation.

Our approach to teaching subatomic physics has been profoundly shaped and influenced by some of our teachers and friends. In particular one of us (HF) is deeply indebted to his teacher, Paul Scherrer, who instilled in most of his students the enthusiasm for the beauty of physics and the need to explain the observed phenomena in the simplest and most physical terms.

The effort required to bring this text to completion would have been impossible without the support of our wives, who helped with kind understanding for the endless working hours when we labored alone, and with gracious hospitality, when we collaborated in Seattle or Urbana.

General Bibliography

The reader of the present book is expected to have some understanding of electromagnetism, special relativity, and quantum theory. We shall quote many equations from these fields without proof, but shall indicate where derivations can be found. The books listed here will be referred to in the text by the name of the author.

Electrodynamics

J. D. Jackson, *Classical Electrodynamics*, Wiley, New York, 1962. Jackson's book is not exactly an undergraduate text, but it is beautifully written and provides an exceptionally lucid treatment of classical electrodynamics.

Modern Physics

R. M. Eisberg, *Fundamentals of Modern Physics*, Wiley, New York, 1961. Quoted as Eisberg. This book gives most of the needed background in special relativity, quantum mechanics, and atomic theory.

Quantum Mechanics

The number of books on quantum mechanics is large. If the information contained in Eisberg or a similar undergraduate text is not sufficient, it is best to proceed to a more advanced text. Experience has shown that one text alone is unlikely to answer all queries with equal lucidity. Therefore, it is best to glance at the treatment for a given problem in two or three texts and then study the one that is most appealing. We will refer to the following books:

R. P. Feynman, R. B. Leighton, and M. Sands, *The Feynman Lectures on Physics*, Addison-Wesley, Reading, Mass., 1965. Referred to as Feynman Lectures.

A. Messiah, *Quantum Mechanics*, Wiley, New York, 1961.

D. Park, *Introduction to the Quantum Theory*, McGraw Hill, New York, 1964.

E. Merzbacher, *Quantum Mechanics*, Wiley, New York, 1970.

Mathematical Physics

J. Mathews and R. L. Walker, *Mathematical Methods of Physics*, Benjamin, Reading, Mass., 1964, 1970, is a concise and easy-to-read book that covers the mathematical tools needed. If the information cannot be found here, reach for the encyclopedic work

P. M. Morse and H. Feshbach, *Methods of Theoretical Physics*, 2 Vols, McGraw Hill, New York, 1953.

Finally we should like to say that physics, despite its cold appearance, is an intensely human field. Its progress depends on hard-working people. Behind each new idea lie countless sleepless nights and long struggles for clarity. Each major experiment involves strong emotions, often bitter competition, and nearly always dedicated collaboration. Each new step is bought with disappointments; each new advance hides failures. Many concepts are connected to interesting stories and sometimes funny anecdotes. A book like ours cannot dwell on these aspects, but we add a list of books related to subatomic physics that we have read with enjoyment.

L. Fermi, *Atoms in the Family*, University of Chicago Press, Chicago 1954.

L. Lamont, *Day of Trinity*, Atheneum, New York, 1965.

R. Moore, *Niels Bohr*, A. A. Knopf, New York, 1966

V. F. Weisskopf, *Physics in the Twentieth Century: Selected Essays*, MIT Press, Cambridge, 1972.

G. Gamow, *My World Line*, Viking, New York, 1970

E. Segrè, *Enrico Fermi, Physicist*, University of Chicago Press, Chicago, 1970.

M. Oliphant, *Rutherford Recollections of the Cambridge Days*, Elsevier, Amsterdam, 1972.

W. Heisenberg, *Physics and Beyond; Encounters and Conversations*, Allen and Unwin, London, 1971.

Robert Jungk, *The Big Machine*, Scribner, New York, 1968.

SUBATOMIC PHYSICS

1
Background and Language

The exploration of subatomic physics started in 1896 with Becquerel's discovery of radioactivity; since then it has been a constant source of surprises, unexpected phenomena, and fresh insights into the laws of nature. In a science fiction story, which was later made into a movie, a team of astronauts flies through hyperspace to a distant galaxy. After a hazardous journey they fight giant animals inside frightening caves. They finally discover that they have traveled through hyperspace into the interior of a caterpillar on earth and are attacked by microbes. In subatomic physics the inverse has happened. The nucleus is the heavy, dense, and small center of the atom. As physicists penetrated deeper into this small object, they encountered the universe. They learned about thermonuclear reactions in the sun, produced new and strange forms of matter, and came face to face with problems related to the creation of the elements in the universe.

In this first chapter we shall describe the orders of magnitude encountered in subatomic physics, define our units, and introduce the language needed for studying subatomic phenomena.

1.1 Orders of Magnitude

Subatomic physics is distinguished from all other sciences by one feature: It is the playground of three different interactions, and two of them act only when the objects are very close together. Biology, chemistry, and atomic and solid-state physics are dominated by the long-range electromagnetic force. Phenomena in the universe are ruled by two long-range forces, gravity and electromagnetism. Subatomic physics, however, is a subtle interplay of three interactions—the hadronic, the electromagnetic, and the weak—and the hadronic and the weak vanish at atomic and larger distances. (It is possible that at least one more interaction, a superweak one, exists, but evidence is not yet conclusive.) The hadronic (or strong, or nuclear) force holds nuclei together; its range is very short, but it is strong.

DISTANCES

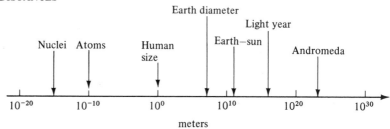

Fig. 1.1. Typical distances. The region below about 10^{-17} m is unexplored. It is unknown if new forces and new phenomena appear.

EXCITATION ENERGIES

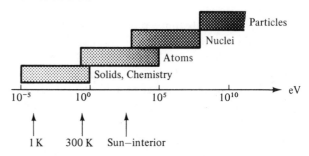

Fig. 1.2. Range of excitation energies. The temperatures corresponding to the energies are also given.

DENSITY

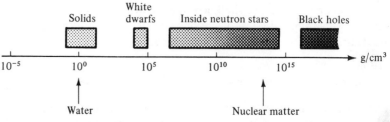

Fig. 1.3. Range of densities.

The weak interaction has an even shorter range. At this point *hadronic*, *weak*, and *short range* are just names, but we shall become familiar with the forces as we go along.

Figures 1.1, 1.2, and 1.3 give an idea of the order of magnitude involved in the various phenomena. We present them here without discussion; they speak for themselves.

1.2 Units

The *basic units* to be used are given in Table 1.1. The prefixes defined in Table 1.2 give the decimal fractions or multiples of the basic units. As

Table 1.1 BASIC UNITS. c is the velocity of light.

Quantity	Unit	Abbreviation
Length	Meter	m
Time	Second	sec
Energy	Electron volt	eV
Mass		eV/c^2
Momentum		eV/c

Table 1.2 PREFIXES FOR POWERS OF 10

Power	Name	Symbol	Power	Name	Symbol
10^1	Deca	da	10^{-1}	Deci	d
10^2	Hecto	h	10^{-2}	Centi	c
10^3	Kilo	k	10^{-3}	Milli	m
10^6	Mega	M	10^{-6}	Micro	μ
10^9	Giga	G	10^{-9}	Nano	n
10^{12}	Tera	T	10^{-12}	Pico	p
			10^{-15}	Femto	f
			10^{-18}	Atto	a

examples, 10^6 eV = MeV, 10^{-12} sec = psec, and 10^{-15} m = fm. The last unit, femtometer, is often also called *Fermi*, and it is extensively used in particle physics. The introduction of the electron volt as an energy unit requires a few words of justification. One eV is the energy gained by an electron if it is accelerated by a potential difference of 1 V(volt):

$$1 \text{ eV} = 1.60 \times 10^{-19} \text{ C (coulomb)} \times 1 \text{ V} = 1.60 \times 10^{-19} \text{ J (joule)}$$
$$= 1.60 \times 10^{-12} \text{ erg.} \tag{1.1}$$

The electron volt (or any decimal multiple thereof) is a convenient energy unit because particles of a given energy are usually produced by acceleration in electromagnetic fields. To explain the units for mass and momentum we require one of the most important equations of special relativity, connecting total energy E, mass m, and momentum $\mathbf{p}$ of a free particle[1]:

$$E^2 = p^2c^2 + m^2c^4. \tag{1.2}$$

This equation states that the total energy of a particle consists of a part independent of the motion, the rest energy mc^2, and a part that depends on the momentum. For a *particle without mass*, Eq. (1.2) reads

$$E = pc; \tag{1.3}$$

on the other hand, for a *particle at rest*, the famous relation

$$E = mc^2 \tag{1.4}$$

follows. These equations make it clear why the units eV/c^2 for mass and eV/c for momentum are convenient. For instance, if the mass and energy of a particle are known, then the momentum in eV/c follows immediately from Eq. (1.2). In the previous equations, we have denoted a vector by $\mathbf{p}$ and its magnitude by p.

In equations where we require electromagnetic quantities we shall use *Gaussian units*. Gaussian units are used by Jackson and his Appendix 4 (p. 621) gives clear prescriptions for the conversion from Gaussian to mks units.

1.3 The Language—Feynman Diagrams

In our discussions we shall use concepts and equations from electrodynamics, special relativity, and quantum mechanics. The fact that we need some *electrodynamics* is not surprising. After all, most particles and nuclei are charged; their mutual interaction and their behavior in external electric and magnetic fields are governed by Maxwell's laws.

The fact that the theory of *special relativity* is essential can be seen most clearly from two features. First, subatomic physics involves the creation and destruction of particles, or, in other words, the change of energy into matter and vice versa. If the matter is at rest, the relation between energy and matter is given by Eq. (1.4); if it is moving, Eq. (1.2) must be used. Second, the particles produced by modern accelerators move with velocities that are close to the velocity of light, and nonrelativistic (Newtonian) mechanics does not apply. Consider two coordinate systems, K and K'. System K' has its axes parallel to those of K but is moving with a velocity v in the positive z direction relative to K. The connection between the coordinates (x', y', z', t') of system K' and (x, y, z, t) of K is given by the *Lorentz transformation*,[2]

1. Eisberg, Eq. (1.25), or Jackson, Eq. (12.11).
2. Eisberg, Eq. (1.13); Jackson, Eq. (11.19).

$$x' = x, \qquad y' = y$$
$$z' = \gamma(z - vt)$$
$$t' = \gamma\left(t - \frac{\beta}{c}z\right)$$

(1.5)

where

$$\gamma = \frac{1}{(1 - \beta^2)^{1/2}}, \qquad \beta = \frac{v}{c}.$$

(1.6)

Momentum and velocity are connected by the relation

$$\mathbf{p} = m\gamma\mathbf{v}.$$

(1.7)

Squaring this expression and using Eqs. (1.2) and (1.6) yield

$$\beta \equiv \frac{v}{c} = \frac{pc}{E}.$$

(1.8)

As one application of the Lorentz transformation to subatomic physics, consider the muon, a particle that we shall encounter often. It is basically a heavy electron with a mass of 106 MeV/c^2. While the electron is stable, the muon decays with a mean life τ:

$$N(t) = N(0)e^{-t/\tau}.$$

Here, $N(t)$ is the number of muons present at time t. If $N(t_1)$ muons are present at time t_1, only $N(t_1)/e$ are still around at time $t_2 = t_1 + \tau$. The mean life of a muon *at rest* has been measured as 2.2 μsec. Now consider a muon produced at the NAL (National Accelerator Laboratory) accelerator with an energy of 100 GeV. If we observe this muon in the laboratory, what mean life τ_{lab} do we measure? Nonrelativistic mechanics would say 2.2 μsec. To obtain the correct answer, the Lorentz transformation must be used. In the muon's rest frame (unprimed), the mean life is the time interval between the two times t_2 and t_1 introduced above, $\tau = t_2 - t_1$. The corresponding times, t'_2 and t'_1, in the laboratory (primed) system are obtained with Eq. (1.5) and the *observed* mean life $\tau_{\text{lab}} = t'_2 - t'_1$ becomes

$$\tau_{\text{lab}} = \gamma\tau.$$

With Eqs. (1.6) and (1.8), the ratio of mean lifes becomes

$$\frac{\tau_{\text{lab}}}{\tau} = \gamma = \frac{E}{mc^2}.$$

(1.9)

With $E = 100$ GeV, $mc^2 = 106$ MeV, $\tau_{\text{lab}}/\tau \approx 10^3$. The mean life of the muon observed in the laboratory is about 1000 times longer than the one in the rest frame (called proper mean life).

Quantum mechanics was forced on physics because of otherwise unexplained properties of atoms and solids. It is therefore not surprising that subatomic physics also requires quantum mechanics for its description. Indeed the existence of quantum levels and the occurrence of interference phenomena in subatomic physics make it clear that quantum phenomena occur. But will the knowledge gained from atomic physics be sufficient?

The dominant features of atoms can be understood without recourse to relativity, and nonrelativistic quantum mechanics describes nearly all atomic phenomena well. In contradistinction, subatomic physics cannot be explained without relativity, as outlined above. It is therefore to be expected that nonrelativistic quantum mechanics is inadequate. An example of its failure can be explained simply: Assume a particle described by a wave function $\psi(\mathbf{x}, t)$. The normalization condition[3]

$$\int_{-\infty}^{+\infty} \psi^*(\mathbf{x}, t)\psi(\mathbf{x}, t)d^3x = 1 \qquad (1.10)$$

states that the particle must be found somewhere at all times. However, the *creation* and *destruction* of particles is a phenomenon that occurs frequently in subatomic physics. A spectacular example is shown in Fig. 1.4. On the left-hand side, a bubble chamber picture is reproduced. (Bubble

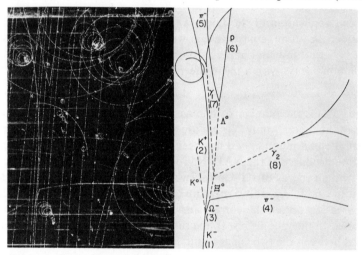

Fig. 1.4. Liquid hydrogen bubble chamber picture. This photograph and the tracing at right show the production and the decay of many particles. Part of the story is told in the text. (Courtesy Brookhaven National Laboratory, where the photograph was taken in 1964.)

chambers will be discussed in Section 4.4). On the right-hand side, the important tracks in the bubble chamber are redrawn and identified. We shall describe the various particles in Chapter 5. Here we just assume that particles with the names indicated in Fig. 1.4 exist and do not worry about their properties. The figure then tells the following story: A K^-, or negative kaon, enters the bubble chamber from below. The bubble chamber is filled with hydrogen and the only particle with which the kaon can collide with appreciable probability is the nucleus of the hydrogen atom, namely the proton. The negative kaon indeed collides with a proton and produces a positive kaon, a neutral kaon, and an *omega minus*. The Ω^- decays into a

---

3. The integral should properly be written as $\iiint d^3x$. Following custom, we write only one of the three integrals.

Ξ^0 and a π^-, and so forth. The events shown in Fig. 1.4 make the essential point forcefully: Particles are created and destroyed in physical processes. Without special relativity, these observations cannot be understood. Equally strongly, Eq. (1.10) cannot be valid since it states that the total probability of finding the particle described by ψ must be independent of time. Nonrelativistic quantum mechanics cannot describe the creation and destruction of particles.[4]

A detailed discussion of the hadronic and weak processes, including creation and destruction of particles, is beyond the scope of this book. However, we need at least a language to describe these phenomena. Such a language exists and is used universally. It is the method of Feynman diagrams or graphs. We shall use a pidgin variant and warn that the diagrams can do more and have a more sophisticated use than would appear from the way we describe them here. The Feynman graphs for two of the processes contained in Fig. 1.4 are given in Fig. 1.5. The first one describes the decay of a lambda (Λ^0) into a proton and a negative pion, and the second one the collision of a negative kaon and a proton, giving rise to a neutral and a positive kaon and an omega minus. In both diagrams, the interaction is drawn as a "blob" to indicate that the exact mechanism remains to be explored. In the following chapters we shall use Feynman diagrams often and explain more details as we need them.

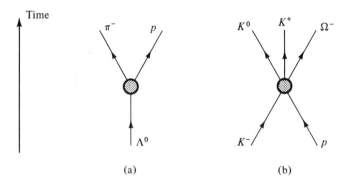

Fig. 1.5. Feynman diagrams for (a) the decay $\Lambda^\circ \longrightarrow p\pi^-$ and (b) the reaction $K^-p \longrightarrow K^\circ K^+\Omega^-$.

1.4 References

Special relativity is treated in many books, and every teacher and reader has his favorites. Good first introductions can be found in the Feynman Lectures, Vol. I, Chapters 15–17, and in J. H. Smith, *Introduction to Special*

4. The theorem that nonrelativistic quantum mechanics cannot describe unstable elementary particles was proved by Bargmann. The proof can be found in Appendix 7 of F. Kaempffer, *Concepts in Quantum Mechanics*, Academic Press, New York, 1965. The appendix is entitled "If Galileo Had Known Quantum Mechanics."

Relativity, Benjamin, Reading, Mass., 1967. A concise and complete exposition is given in Jackson, Chapters 11 and 12. These two chapters form an excellent base for all applications to subatomic physics. A detailed account with particular reference to subatomic physics is also given in R. D. Sard, *Relativistic Mechanics*, Benjamin, Reading, Mass., 1970. An exciting and unconventional book is E. F. Taylor and J. A. Wheeler, *Spacetime Physics*, W. H. Freeman, San Francisco, 1963, 1966.

Books on quantum mechanics have already been listed at the end of the Preface. However, a few additional remarks concerning Feynman diagrams are in order here. There is no place where Feynman diagrams can be learned painlessly. Relatively gentle introductions can be found in

R. P. Feynman, *Theory of Fundamental Processes*, Benjamin, Reading, Mass., 1962.

F. Mandl, *Introduction to Quantum Field Theory*, Wiley-Interscience, New York, 1959.

J. M. Ziman, *Elements of Advanced Quantum Theory*, Cambridge University Press, Cambridge, 1969.

PROBLEMS

1.1. Use what information you can find to get a number characterizing the strength of each of the four basic interactions. Justify your numbers.

1.2. Discuss the range of each of the four basic interactions.

1.3. List a few important processes for which the electromagnetic interaction is essential.

1.4. For what cosmological and astrophysical phenomena is the weak interaction essential?

1.5. It is known that the muon (the heavy electron, with a mass of about 100 MeV/c^2) has a radius that is smaller than 0.1 fm. Compute the minimum density of the muon. Where would the muon lie in Fig. 1.3? What problems does this crude calculation raise?

1.6. Verify Eq. (1.8).

1.7. Verify Eq. (1.9).

1.8. Consider a pion with a kinetic energy of 200 MeV. Find its momentum in MeV/c.

1.9. A proton is observed to have a momentum of 5 MeV/c. Compute its kinetic energy in MeV.

1.10. For a certain experiment, kaons with a kinetic energy of 1 GeV are needed. They are selected with a magnet. What momentum does the magnet have to select?

1.11. Find two examples where special relativity is essential in subatomic physics.

1.12. How far does a beam of muons with kinetic energy of
(a) 1 MeV,
(b) 100 GeV
travel in empty space before its intensity is reduced to one half of its initial value?

1.13. Repeat Problem 1.12 for charged and for neutral pions.

1.14. Which subatomic phenomena exhibit quantum mechanical interference effects?

1.15. If the strong and weak forces are assumed to be approximately constant over 1 fm, find the order of magnitudes for

$$F_h : F_{em} : F_{weak} : F_{gravit}$$

for two protons that are 1 fm apart. Use any physical knowledge or arguments at your disposal to obtain the desired ratios.

PART I

Tools

One of the most frustrating experiences in life is to be stranded without proper tools. The situation can be as simple as being in the wilderness with a broken shoe strap but no wire or knife. It can be as simple as having a leaking radiator hose in Death Valley and no tape to fix it. In these instances we at least know what we miss and what we need. Confronted with the mysteries of subatomic physics, we also need tools and we often do not know what is required. However, during the past 75 years, we have learned a great deal, and many beautiful tools have been invented and constructed. We have accelerators to produce particles, detectors to see them and to study their interactions, and instruments to quantify what we observe. In the following three chapters we sketch some important tools.

2

Accelerators

2.1 Why Accelerators?

Accelerators cost a lot of money. What can they do? Why are they crucial for studying subatomic physics? As we proceed through various fields of subatomic physics, these questions will be answered. Here we shall simply point out a few of the important aspects.

Accelerators produce beams of charged particles with energies ranging from a few MeV to a few hundred GeV. Intensities can be as high as 10^{16} particles/sec, and the beams can be concentrated onto targets of only a few mm^2 in area. The particles that are most often used as primary projectiles are protons and electrons.

Two tasks can be performed well only by accelerators, namely the production of new particles and new states, and the investigation of the detailed structure of subatomic systems. Consider, first, particles and nuclei. Only very few stable particles exist in nature—the proton, the electron, the neutrino, and the photon. Only a limited number of nuclides are available in terrestrial matter, and they are usually in the ground state. To escape the narrow limitations of what is naturally available, new states must be produced artificially. To create a state of mass m, we need at least the energy $E = mc^2$. Very often, considerably more energy is required, as we shall find out. So far, no limit on the mass of new particle states has been found, and we do not know if one exists. Clearly, higher energies are a prerequisite to finding out.

High energies are not only needed to produce new states; they are also essential in finding out details concerning the structure of subatomic systems. It is easy to see that the particle energy has to be higher as the dimension to be looked at becomes smaller. The de Broglie wavelength of a particle with momentum p is given by

$$\lambda = \frac{h}{p}, \qquad (2.1)$$

where h is Planck's constant. In most expressions, we shall use the *reduced* de Broglie wavelength,

$$\lambdabar = \frac{\lambda}{2\pi} = \frac{\hbar}{p}, \qquad (2.2)$$

where h-bar, or Dirac's $\hbar$, is

$$\hbar = \frac{h}{2\pi} = 6.5820 \times 10^{-22} \text{ MeV-sec.} \qquad (2.3)$$

As is known from optics, in order to see structural details of linear dimensions d, a wavelength comparable to, or smaller than, d must be used:

$$\lambdabar \leq d. \qquad (2.4)$$

The momentum required then is

$$p \geq \frac{\hbar}{d}. \qquad (2.5)$$

To see small dimensions, high momenta and thus high energies are needed. As an example, we consider $d = 1$ fm and protons as a probe. We shall see that a nonrelativistic approximation is permitted here; the minimum kinetic energy of the protons then becomes, with Eq. (2.5),

$$E_{\text{kin}} = \frac{p^2}{2m_p} = \frac{\hbar^2}{2m_p d^2}. \qquad (2.6)$$

It is straightforward to insert the constants $\hbar$ and m_p from Table Al in the Appendix. However, we shall use this example to compute E_{kin} in a more roundabout but also more convenient way: Express as many quantities as possible as dimensionless ratios. E_{kin} has the dimension of an energy, as does $m_p c^2 = 938$ MeV. The kinetic energy is consequently rewritten as a ratio:

$$\frac{E_{\text{kin}}}{m_p c^2} = \frac{1}{2d^2} \left(\frac{\hbar}{m_p c} \right)^2.$$

A glance at Table Al shows that the quantity in parentheses is just the

Compton wavelength of the proton,

$$\lambda_p = \frac{h}{m_p c} = 0.210 \text{ fm}, \qquad (2.7)$$

so that the kinetic energy is given by

$$\frac{E_{\text{kin}}}{m_p c^2} = \frac{1}{2}\left(\frac{\lambda_p}{d}\right)^2 = 0.02. \qquad (2.8)$$

The kinetic energy required to see linear dimensions of the order of 1 fm is about 20 MeV. Since this kinetic energy is much smaller than the rest energy of the nucleon, the nonrelativistic approximation is justified. Nature does not provide us with intense particle beams of such energies; they must be produced artificially. (Cosmic rays contain particles with much higher energies, but the intensity is so low that only very few problems can be attacked in a systematic way.)

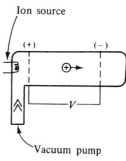

Fig. 2.1. Prototype of the simplest accelerator.

The common way to produce a particle beam of high energy is to accelerate charged particles in an electric field. The force exerted on a particle of charge q by an electric field $\boldsymbol{\varepsilon}$ is

$$\mathbf{F} = q\boldsymbol{\varepsilon}. \qquad (2.9)$$

In the simplest accelerator, two grids with a potential difference V at a distance d (Fig. 2.1), the average field is given by $|\boldsymbol{\varepsilon}| = V/d$, and the energy gained by the particle is

$$E = Fd = qV. \qquad (2.10)$$

Of course, the system must be placed in a vacuum; otherwise the accelerated particles will collide with air molecules and continuously lose much of the acquired energy. Figure 2.1. therefore shows a vacuum pump. Moreover, an ion source is also indicated—it produces the charged particles. These elements—particle source, accelerating structure, and vacuum pump—appear in every accelerator.

Can particle beams of 20 MeV be reached with simple machines as sketched in Fig. 2.1? Anyone who has played with high voltages knows that such an approach is not easy. At a few kV, voltage breakdowns can occur and it requires experience to exceed even 100 kV. Indeed, it has taken considerable ingenuity and work to bring *electrostatic generators* to the point where they can produce particles with energies of the order of 10 MeV. However, it is impossible to achieve energies that are orders of magnitude higher, no matter how sophisticated the electrostatic generator. A new idea is needed, and such an idea was found—successive application of a given voltage to the same particle. Actually, a few times during the long road to the giant accelerators of today it looked as though the maximum accelerator energy had been reached. However, every apparently unsurmountable difficulty was overcome by an ingenious new approach.

We shall discuss only three types of accelerators, the electrostatic generator, the linear accelerator, and the synchrotron.

2.2 Electrostatic Generators (Van de Graaff)

It is difficult to produce a very high voltage directly, for instance, by a combination of transformer and rectifier. In the Van de Graaff generator,[1] the problem is circumvented by transporting a charge Q to one terminal of a condenser C; the resulting voltage,

$$V = \frac{Q}{C},$$ (2.11)

is used to accelerate the ions. The main elements of a Van de Graaff generator are shown in Fig. 2.2. Positive charges are sprayed onto an insulating belt by using a voltage of about 20–30 kV. The positive charge is carried to the terminal by the motor-driven belt; it is removed there by a set of

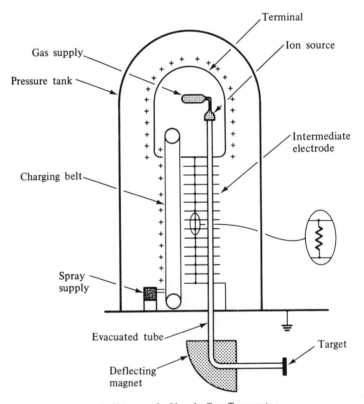

Fig. 2.2. Schematic diagram of a Van de Graaff generator.

1. R. J. Van de Graaff, *Phys. Rev.* **38**, 1919A (1931). R. J. Van de Graaff, J. G. Trump, and W. W. Buechner, *Rept. Progr. Phys.* **11**, 1 (1948).

needles and travels to the terminal surface. Positive ions (protons, deuterons, etc.) are produced in the ion source and are accelerated in the evacuated accelerating column. The beam emerging from the column is usually deflected by a magnet onto the target. If the entire system is placed in air, voltages of up to about a few MeV can be reached before artificial lightning discharges the terminal. If the system is placed in a pressure tank filled with an inert gas, for instance, nitrogen and carbon dioxide at 15 atm, voltages of up to 12 MV can be obtained.

Twice the maximum voltage can be utilized in *tandem* machines, sketched in Fig. 2.3. Here, the terminal is in the middle of a long high-pressure tank; the ion source is at one end and it produces negative ions, for instance H^-. These ions are accelerated toward the central terminal where they are stripped of their two electrons by passage through a foil or a gas-containing canal. The positive ions now fall toward the target and again acquire energy. The total energy gain is therefore twice that of a single-stage machine.

Van de Graaff generators in various energy and price ranges can be obtained commercially, and they are ubiquitous. They have a high beam intensity (up to 100 μA); their beam can be continuous and well collimated; and the output energy is well stabilized (± 10 keV). They are the current workhorses of nuclear structure research. However, their present maximum energy is limited to about 30–40 MeV for protons, and they can therefore not be used in elementary particle research.

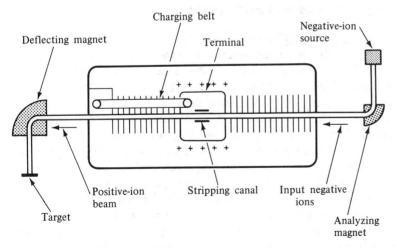

Fig. 2.3. Tandem Van de Graaff. Negative ions are first accelerated to the central position terminal. There they are stripped of their electrons and accelerated as positive ions to the target.

Photo 1: The Tandem Van de Graaff accelerator at the University of Washington, Seattle, Wash. After the beam leaves the accelerator, it enters a beam transport system: A quadrupole magnet at the exit focuses the beam. It is then deflected by 90° and finally split into a number of beam lines that go to the various experimental areas. (Courtesy University of Washington.)

2.3 Linear Accelerators (Linacs)

To reach very high energies, particles must be accelerated many times over. Conceptually the simplest system is the linear accelerator,[2] sketched in Fig. 2.4. A series of cylindrical tubes are connected to a high-frequency oscillator. Successive tubes are arranged to have opposite polarity. The beam of particles is injected along the axis. Inside the cylinders the electric field is always zero; in the gaps it alternates with the generator frequency. Consider now a particle of charge e that crosses the first gap at a time when the accelerating field is at its maximum. The length L of the next cylinder is so chosen that the particle arrives at the next gap when the field has changed sign. It therefore again experiences the maximum accelerating voltage and has already gained an energy $2\,eV_0$. To achieve this feat, L must be equal to $\frac{1}{2}vT$, where v is the particle velocity and T the period of the oscillator. Since the velocity increases at each gap, the cylinder lengths must increase also. For electron linacs, the electron velocity

2. R. Wideröe, *Arch. Elektrotech.* **21**, 387 (1928). D. H. Sloan and E. O. Lawrence, *Phys. Rev.* **38**, 2021 (1931).

Photo 2: Early stages in the proton linear accelerator at Los Alamos. The design here is essentially that shown in Fig. 2.4; the drift tubes are clearly recognizable. (Courtesy Los Alamos Scientific Laboratory.)

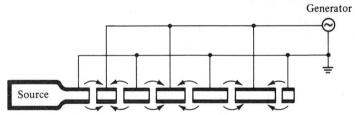

Fig. 2.4. Drift tube linac. The arrows at the gaps indicate the direction of the electric field at a given time.

soon approaches c and L tends to $\frac{1}{2}cT$. The drift-tube arrangement is not the only possible one; electromagnetic waves propagating inside cavities can also be used to accelerate the particles. In both cases large rf power sources are required for the acceleration, and enormous technical problems had to be solved before linacs became useful machines. At present, Stanford has an *electron linac* that is 3 km ("2 miles") long and produces electrons of more than 20-GeV energy. A *proton linac* of 800-MeV energy with

Photo 3: One accelerating cavity of the 800 MHz part of the accelerator at Los Alamos. The "side-coupling" of the cavities shown here provides for greater efficiency. (Courtesy Los Alamos Scientific Laboratory.)

a beam current of 1 mA, a so-called meson factory, has been constructed at Los Alamos.

2.4 Beam Optics

In the description of linacs we have swept many problems under the rug, and we shall leave most of them there. However, one question must occur to anyone thinking about a machine that is a few km in length: How can the beam be kept well collimated? The beam of a flashlight, for instance, diverges, but it can be refocused with lenses. Do lenses for charged particle beams exist? Indeed they do, and we shall discuss here some of the elementary considerations, using the analogy to ordinary optical lenses.

In light optics, the path of a monochromatic light ray through a system of thin lenses and prisms can be found easily by using geometrical optics.[3] Consider, for instance, the combination of a positive and a negative thin lens, with equal focal lengths f and separated by a distance d (Fig. 2.5). This combination is *always* focusing, with an overall focal length given by

$$f_{\text{comb}} = \frac{f^2}{d}. \tag{2.12}$$

Fig. 2.5. The combination of a focusing and a defocusing thin lens with equal focal lengths is always focusing.

In principle one could use electric or magnetic lenses for the guidance of charged particle beams. The electric field strength required for the effective focusing of high-energy particles is, however, impossibly high, and only magnetic elements are used. The deflection of a monochromatic (monoenergetic) beam by a desired angle, or the selection of a beam of desired momentum, is performed with a *dipole magnet*, as shown in Fig.

3. See, for instance, H. D. Young, *Fundamentals of Optics and Modern Physics*, McGraw-Hill, New York, 1968, or any other introductory physics text.

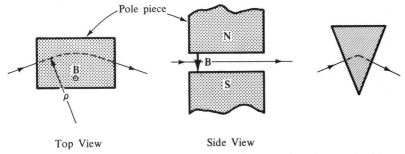

Pole piece

N

B

S

Top View   Side View

Fig. 2.6. Rectangular dipole magnet. The optical analog is a prism, shown at the right.

2.6. The radius of curvature, ρ, can be coumpted from the *Lorentz equation*,[4] which gives the force **F** exerted on a particle with charge q and velocity **v** in an electric field $\boldsymbol{\varepsilon}$ and a magnetic field **B**:

$$\mathbf{F} = q\left\{\boldsymbol{\varepsilon} + \frac{1}{c}\, \mathbf{v} \times \mathbf{B}\right\}. \tag{2.13}$$

The force is normal to the trajectory. For the normal component of the force, Newton's law, $\mathbf{F} = d\mathbf{p}/dt$, and Eq. (1.7) give

$$F_n = \frac{pv}{\rho}, \tag{2.14}$$

so that with Eq. (2.13) the radius of curvature becomes[5]

$$\rho = \frac{pc}{|q|\,B}. \tag{2.15}$$

Problems arise when a beam should be focused. Figure 2.6 makes it clear that an ordinary (dipole) magnet bends particles only in one plane and that focusing can be achieved only in this plane. No magnetic lens with properties analogous to that of an optical focusing lens can be designed, and this fact stymied physicists for many years. A solution was finally found in 1950 by Christofilos and independently by Courant, Livingston, and Snyder in 1952.[6] The basic idea of the so-called strong focusing can be explained simply by referring to Fig. 2.5: If focusing and defocusing ele-

4. Jackson, Eq. (6.87).
5. Equation (2.15) is given in Gaussian units, where the unit for B is 1 G, and the unit of potential is 1 stat $V = 300$ V. To compute ρ for a particle with unit charge ($|q| = e$), express pc in eV; then Eq. (2.15) yields

$$B \text{ (Gauss)} \times \rho \text{ (cm)} = \frac{V}{300}. \tag{2.15a}$$

As an example, consider an electron with a kinetic energy of 1 MeV; pc follows from Eq. (1.2) as $pc = (E_{kin}^2 + 2E_{kin}mc^2)^{1/2} = 1.42 \times 10^6$ eV. V then is 1.42×10^6 V and $B\rho = 4.7 \times 10^3$ G-cm. Equation (2.15) can also be rewritten in mks units, where the unit of B is 1 T (Tesla) $= 10$ Wb (Webers) m$^{-2} = 10^4$ G.
6. E. D. Courant, M. S. Livingston, and H. S. Snyder, *Phys. Rev.* **88**, 1190 (1952).

ments of equal focal lengths are alternated, a net focusing effect occurs. In beam transport systems, strong focusing is most often achieved with *quadrupole magnets*. A cross section through such a magnet is shown in Fig. 2.7. It consists of four poles; the field in the center vanishes and the

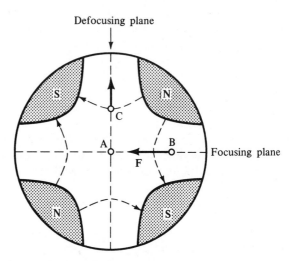

Fig. 2.7. Cross section through a quadrupole magnet. Three positive particles enter the magnet parallel to the central symmetry axis at points A, B, and C. The particle at A is not deflected, B is pushed toward the center, and C is deflected outward.

magnitude of the field increases from the center in all directions. To understand the operation of a quadrupole magnet, consider three positive particles, going into the magnet at the points denoted by A, B, and C. Particle A in the center is not deflected; the Lorentz force, Eq. (2.13), pushes particle B toward and particle C away from the central symmetry axis. The magnet therefore behaves as a focusing element in one plane and a defocusing element in the other plane. A combination of two quadrupole magnets focuses in both planes if the second magnet is rotated around the central axis by 90° with respect to the first one. Such *quadrupole doublets* form essential elements of all modern particle accelerators and also of the beam lines that lead from the machines to the experiments. With these focusing devices, a beam can be transported over distances of many km with small intensity loss.

2.5 Synchrotrons

Why do we need another accelerator type? The linac obviously can produce particles of arbitrary energy. However, consider the price: Since the 20-GeV Stanford linac is already 3 km long, a 500-GeV accelerator would have to be about 75 km long; construction and power problems would be enormous. It makes more sense to let the particles run around a smaller track repeatedly. The first circular accelerator, the *cyclotron*, was proposed by Lawrence in 1930.[7] Cyclotrons have been of enormous importance in the development of subatomic physics, and some very modern and sophisticated ones will come into operation within the next few years. We omit discussion of the cyclotron here because its cousin, the *synchrotron*, has many similar features and achieves higher energies.

The synchrotron was proposed independently by McMillan and by Veksler in 1945.[8] Its essential elements are shown in Fig. 2.8. The injector sends particles of an initial energy E_i into the ring. Dipole magnets with a radius of magnetic curvature ρ bend the particles around the ring while quadrupole systems maintain the collimation. The particles are accelerated in a number of rf cavities which are supplied with a circular frequency ω. The actual path of the particles consists of straight segments in the accelerating cavities, the focusing elements, and some other elements and of circular segments in the bending magnets. The radius of the ring, R, is therefore larger than the radius of curvature, ρ.

Now consider the situation just after injection of the particles with energy E_i and momentum p_i, where energy and momentum are connected by

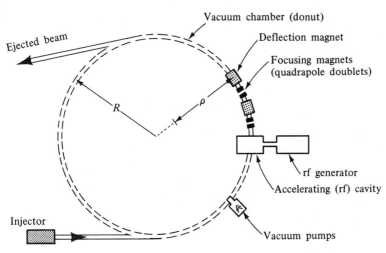

Fig. 2.8. Essential elements of a synchrotron. Only a few of the repetitive elements are shown.

7. E. O. Lawrence and N. E. Edlefsen, *Science* **72**, 376 (1930). E. O. Lawrence and M. S. Livingston, *Phys. Rev.* **40**, 19 (1932).

8. E. M. McMillan, *Phys. Rev.* **68**, 143 (1945); V. Veksler, *J. Phys.* (U.S.S.R.) **9**, 153 (1945).

Eq. (1.2). Assume that the rf power has not yet been turned on. The particles will then coast around the ring with a velocity v, and the time T for one full turn is given with Eq. (1.8) as

$$T = \frac{2\pi R}{v} = \frac{2\pi R E_i}{p_i c^2}. \tag{2.16}$$

The corresponding circular frequency, Ω, is

$$\Omega = \frac{2\pi}{T} = \frac{p_i c^2}{R E_i}, \tag{2.17}$$

and the magnetic field required to keep them on the track follows from Eq. (2.15) as

$$B = \frac{p_i c}{|q| \rho}. \tag{2.18}$$

Once the rf power is turned on, the situation changes. First, the radio frequency ω, must be an integer multiple, k, of Ω in order to always give the circulating particles the push at the right time. Equation (2.17) then shows that the applied rf must increase with increasing energy up to the point where the particles are fully relativistic so that $pc = E$. The magnetic field also must increase:

$$\omega = k\Omega = \frac{kc}{R}\frac{pc}{E} \longrightarrow \frac{kc}{R}; \qquad B = \frac{pc}{|q|\rho}. \tag{2.19}$$

If these two conditions are satisfied, then the particles are properly accelerated. The procedure is as follows: A burst of particles of energy E_i is injected at the time $t = 0$. The magnetic field and the rf are then increased from their initial values B_i and ω_i to final values B_f and ω_f, always maintaining the relations (2.19). The energy of the bunch of particles is increased during this process from the injection energy E_i to the final energy E_f. The time required for bringing the particles up to the final energy depends on the size of the machine; for very big machines, a pulse per sec is about par.

Equation (2.19) shows another feature of these big accelerators: Particles cannot be accelerated *from start* to the final energy in one ring. The range over which the rf and the magnetic field would have to vary is too big. The particles are therefore preaccelerated in smaller machines and then injected. Consider, for instance, the 300-GeV synchrotron at NAL (Fig. 2.9): An electrostatic generator (Cockcroft-Walton) produces a beam of 750-keV protons; a linac then brings the energy up to 200 MeV; a *booster* synchrotron takes over and raises it to 8 GeV, after which the big machine finally starts. The enormous dimensions of the entire enterprise are evident from Fig. 2.9.

Synchrotrons can accelerate protons or electrons. Electron synchrotrons share one property with other circular electron accelerators: They are an intense source of short-wavelength light. The origin of *synchrotron radiation* can be explained on the basis of classical electrodynamics. Maxwell's equations predict that any accelerated charged particle radiates. A

particle that is forced to remain in a circular orbit is continuously accelerated in the direction toward the center, and it emits electromagnetic
radiation. The power radiated by a particle with charge e moving with
velocity $v = \beta c$ on a circular path of radius R is given by[9]

$$P = \frac{2e^2c}{3R^2} \frac{\beta^4}{(1 - \beta^2)^2}. \tag{2.20}$$

The velocity of a relativistic particle is close to c; with Eqs. (1.6) and (1.9)
and with $\beta \approx 1$, Eq. (2.20) becomes

$$P \approx \frac{2e^2c}{3R^2} \gamma^4 = \frac{2e^2c}{3R^2} \left(\frac{E}{mc^2}\right)^4. \tag{2.21}$$

The time T for one revolution is given by Eq. (2.16), and the energy lost
in one revolution is

$$-\delta E = PT \approx \frac{4\pi e^2}{3R} \left(\frac{E}{mc^2}\right)^4. \tag{2.22}$$

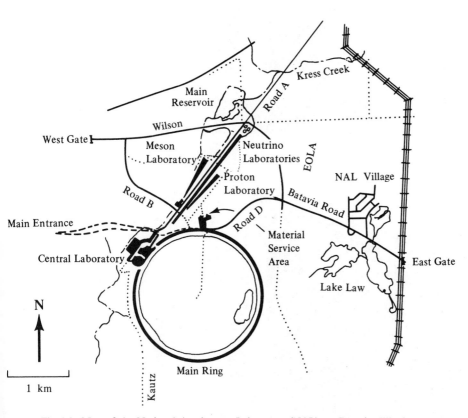

Fig. 2.9. Map of the National Accelerator Laboratory (NAL), at Batavia, Illinois.
(Courtesy National Accelerator Laboratory.)

9. Jackson, Eqs. (14.31) and (11.76).

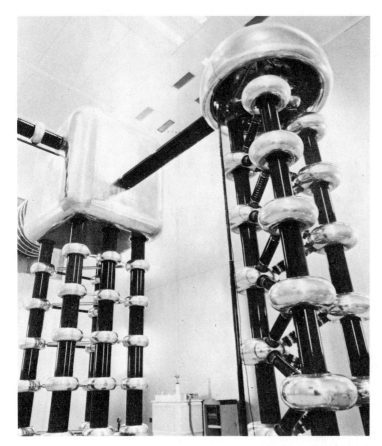

Photo 4

Photo 5

Photo 6

Photo 7

Photos 4–7: The photographs 4–7 show
the essential parts of the 300 GeV proton
synchrotron at the National Accelerator
Laboratory. Protons are accelerated to
750 keV in an electrostatic accelerator
(Cockcroft-Walton); a linear accelerator
then brings the energy up to 200 MeV and
injects the protons into a booster synchrotron. The booster synchrotron raises the
energy to about 8 GeV; the final energy is
reached in the main ring. (Courtesy
National Accelerator Laboratory.)

The difference between the proton and electron synchrotron is obvious from Eq. (2.22). For equal radii and equal total energies E, the ratio of energy losses is

$$\frac{\delta E(e^-)}{\delta E(p)} = \left(\frac{m_p}{m_e}\right)^4 \approx 10^{13}. \tag{2.23}$$

The energy loss must be taken into account in the design of electron synchrotrons. Fortunately, the emitted radiation is useful for solid-state studies, and high energy's loss is solid-state's gain.[10]

2.6 Laboratory and Center-of-Momentum Frames

Trying to achieve higher energies with ordinary accelerators is somewhat like trying to earn more money—you do not keep all you earn. In the second case, the tax collector takes an increasing bite, and in the first case, an increasing fraction of the total energy in a collision goes into center-of-mass motion and is not available for exciting internal degrees of freedom. To discuss this fact, we briefly describe the laboratory (lab) and center-of-momentum (c.m.) coordinates. Consider the following two-body reaction,

$$a + b \longrightarrow c + d, \tag{2.24}$$

and call a the projectile and b the target particle. In the *laboratory frame*, the target is at rest and the projectile strikes it with an energy E^{lab} and a momentum $\mathbf{p}^{\text{lab}}$. After the collision both particles in the final state, c and d, are usually moving. In the center-of-mass frame or, more correctly, the *center-of-momentum frame*, both particles approach each other with equal but opposite momenta. The two frames are defined by

$$\text{lab frame:} \quad \mathbf{p}_b^{\text{lab}} = 0, \qquad E_b^{\text{lab}} = m_b c^2 \tag{2.25}$$

$$\text{c.m. frame:} \quad \mathbf{p}_a^{\text{c. m.}} + \mathbf{p}_b^{\text{c. m.}} = 0. \tag{2.26}$$

It is only the energy of one particle relative to the other one that is available for producing particles or for exciting internal degrees of freedom. The uniform motion of the center of momentum of the whole system is irrelevant. The energies and momenta in the c.m. system are thus the important ones.

A simple example can provide an understanding of how much one is robbed in the laboratory system. New particles can, for instance, be produced by bombarding protons with pions,

$$\pi p \longrightarrow \pi N^*,$$

where N^* is a particle of high mass ($m_{N^*} > m_p \gg m_\pi$). In the c.m. frame, the pion and proton collide with opposite momenta; the total momentum in

10. R. P. Godwin, *Springer Tracts Modern Phys.* **51**, 1 (1969).

the initial and hence also in the final state is zero. The highest mass can be reached if the pion and the N^* in the final state are produced at rest because then no energy is wasted to produce motion. This collision in the c.m. frame is shown in Fig. 2.10. The total energy in the final state is

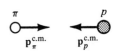

Before collision

$$W^{\text{c. m.}} = (m_\pi + m_{N^*})c^2 \approx m_{N^*}c^2. \qquad (2.27)$$

The total energy is conserved in the collision so that

$$W^{\text{c. m.}} = E_\pi^{\text{c. m.}} + E_p^{\text{c. m.}}. \qquad (2.28)$$

After collision

The pion energy, E_π^{lab}, required in the laboratory system to produce the N^*, can be computed by using the Lorentz transformation. We shall use a different approach here in order to introduce the idea of a *relativistic invariant*. Consider a system of i particles with energies E_i and momenta $\mathbf{p}_i$. In a derivation similar to the one that leads to Eq. (1.2) it is possible to show that one can write

Fig. 2.10. Production of a new particle, N^*, in collision $\pi p \longrightarrow \pi N^*$, seen in the c.m. frame.

$$\left(\sum_i E_i\right)^2 - \left(\sum_i \mathbf{p}_i\right)^2 c^2 = M^2 c^4. \qquad (2.29)$$

Here, M is called the *total mass* or *invariant mass* of the system of i particles; it is equal to the sum of the rest masses of the i particles only if they are all at rest in their common c.m. frame. The important fact expressed by Eq. (2.29) is the Lorentz invariance: The right-hand side (RHS) is a constant and must therefore be the same in *all* coordinate systems. It then follows that the left-hand side (LHS) is also a relativistic invariant (sometimes called a relativistic scalar) that has the same value in all coordinate systems. We apply this invariance to the collision equation (2.24) as seen in the c.m. and the lab systems,

$$(E_a^{\text{c. m.}} + E_b^{\text{c. m.}})^2 - (\mathbf{p}_a^{\text{c. m.}} + \mathbf{p}_b^{\text{c. m.}})^2 c^2 = (E_a^{\text{lab}} + E_b^{\text{lab}})^2 - (\mathbf{p}_a^{\text{lab}} + \mathbf{p}_b^{\text{lab}})^2 c^2,$$
$$(2.30)$$

or with Eqs. (2.25) and (2.26),

$$W^2 = (E_a^{\text{c. m.}} + E_b^{\text{c. m.}})^2 = (E_a^{\text{lab}} + m_b c^2)^2 - (\mathbf{p}_a^{\text{lab}} c)^2$$
$$= 2E_a^{\text{lab}} m_b c^2 + (m_a^2 + m_b^2)c^4. \qquad (2.31)$$

Equation (2.31) connects W^2, the square of the total c.m. energy, to the laboratory energy. With $E_a^{\text{lab}} \gg m_a c^2$, $m_b c^2$, the energy W becomes

$$W \approx [2E_a^{\text{lab}} m_b c^2]^{1/2}. \qquad (2.32)$$

Only the energy available in the c.m. frame is useful for producing new particles or exploring internal structure. Equation (2.32) shows that this energy, W, increases only as the square root of the laboratory energy at *high* energies.

2.7 Colliding Beams

The price for working in the laboratory system is high, as is stated plainly by Eq. (2.32). If the machine energy is increased by a factor of

100, the effective gain is only a factor of 10. In 1956, Kerst and his colleagues and O'Neill therefore suggested the use of colliding beams to attain very high useful energies.[11] Two proton beams of 21.6 GeV colliding head-on would be equivalent to one 1000-GeV accelerator. The main technical obstacle is intensity; both beams must be much more intense than the ones available in normal accelerators in order to produce sufficient events in the regions where they collide. In the past few years, this problem has been solved, owing to strong focusing and progress in vacuum technology. At present, a number of colliding beam machines are already in operation. The proton storage ring at CERN, shown in Fig. 2.11 and called ISR (for *intersecting storage rings*), is the largest presently working. The mode of operation can be explained easily: Protons are accelerated in the 28-GeV proton synchrotron. They are then extracted on a tangent and guided to a switchyard. There, a bending magnet sends alternate bunches into alternate branches of a fork. From each branch, the protons are injected into a storage ring (ISR). The two storage rings overlap closely.

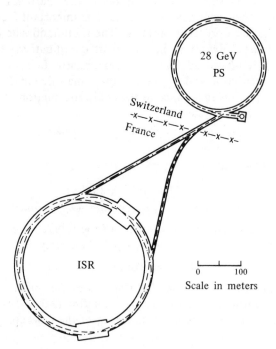

Fig. 2.11. 28-GeV proton synchrotron (PS) and intersecting storage rings (ISR) at CERN. (From *CERN Courier*, **6**, 127 (1966)).

11. D. W. Kerst et al., *Phys. Rev.* **102**, 590 (1956). G. K. O'Neill, *Phys. Rev.* **102**, 1418 (1956).

Photo 8: CERN proton synchrotron and intersecting storage ring (ISR) line. The main ring branches off to the left, the protons that fill the ISR are guided straight ahead, as indicated in Fig. 2.11. (Photo courtesy of CERN.)

Both rings have rf acceleration cavities that increase the energy of each particle bunch just enough to move it away from the injection orbit. The next pulse can then again be accommodated, and many pulses can be "stacked" in the ring. Each ring will accept about 400 pulses and carry a proton current of about 20 A. Collisions between the two 28-GeV beams can be observed at a number of intersections. The maximum total c.m. energy is 56 GeV, corresponding to a conventional accelerator energy of about 1700 GeV.

2.8 References

An exciting and very readable introduction to accelerators has been written by R. R. Wilson and R. Litttauer: *Accelerators, Machines of Nuclear Physics,* Anchor Books, Doubleday, Garden City, N.Y, 1960.

On a more advanced level, a careful and lucid treatment of the various particle accelerators is given by E. Persico, E. Ferrari, and S. E. Segre, *Principles of Particle Accelerators,* Benjamin, Reading, Mass., 1968.

Photo 9: An intersection region of the ISR before experimental equipment was installed. To the right in the lower foreground is one of the beam lines coming up from the proton synchrotron. (Photo courtesy of CERN.)

Considerably more details and concise stories of the development of each type can be found in M.S. Livingston and J. P. Blewett, *Particle Accelerators.*, McGraw-Hill, New York, 1962.

R. B. Neal, ed; *The Stanford Two-Mile Accelerator*, Benjamin, Reading, Mass., 1968, contains a collection of detailed articles describing all aspects of the 20-GeV electron linac. Cyclic accelerators are clearly described in J. J. Livingood. *Principles of Cyclic Particle Accelerators*, Van Nostrand Reinhold, New York, 1961.

A review of recent advances is found in J. P. Blewett, *Advan. Electron Phys.* **29**, 233 (1970).

Relativistic kinematics, which we have only touched upon briefly, is treated in detail in R. Hagedorn, *Relativistic Kinematics.*, Benjamin, Reading, Mass., 1963, and in E. Byckling and K. Kajantie, *Particle Kinematics*, J. Wiley, New York, 1973.

Beam optics is discussed in K. G. Steffen, *High Energy Beam Optics*, Wiley-Interscience, New York, 1965, and in A. Septier, ed., *Focusing of Charged Particles*, Academic Press, New York, 1967 (two volumes).

A guide to the literature on accelerators is J. P. Blewett, "Research Letter PA-1 on Particle Accelerators," *Am. J. Phys.* **34**, 9 (1966).

PROBLEMS

2.1. An electron accelerator is to be designed to study properties of linear dimensions of 1 fm. What kinetic energy is required?

2.2. Estimate the capacity of a typical Van de Graaff terminal with respect to the ground (order of magnitude only). Assume that the terminal is to be charged to 1 MV. Compute the charge on the terminal. How long does it take to reach this voltage if the belt carries a current of 0.1 mA?

2.3. Consider a proton linac, working with a frequency of $f = 200$ MHz. How long must the drift tubes be at the point where the proton energy is
(a) 1 MeV?
(b) 100 MeV?
What is approximately the smallest energy with which the protons can be injected, and what determines the lower limit? Why does the frequency at the Los Alamos linac change from 200 to 800 MHz at a proton energy of about 200 MeV?

2.4. A proton beam of kinetic energy of 10 MeV enters a dipole magnet of 2 m length. It should be deflected by 10°. Compute the field that is necessary.

2.5. A proton beam of kinetic energy 200 GeV enters a 2-m-long dipole magnet with a magnetic field of 20 kG. Compute the deflection of the beam.

2.6. The maximum magnetic field that can be obtained in a conventional magnet is about 20 kG. Assume an accelerator that follows the earth's equator. What is the maximum energy to which protons can be accelerated in such a machine?

2.7. Use Fig. 2.9 and the data given in Section 2.5 to estimate over what range the frequency and the magnetic field must be changed in the main ring of the NAL machine during one accelerating cycle.

2.8. Verify Eq. (2.29).

2.9. Assume collisions of protons from the accelerator described in Problem 2.6 with stationary protons. Compute the total energy, W, in GeV in the c.m. frame. Compare W with the corresponding quantity obtained in a colliding beam experiment, with each beam having a maximum energy E_0. How big must E_0 be in order to get the same W?

2.10.
(a) Verify Eq. (2.20).
(b) Compute the energy loss per turn for a 10-GeV electron accelerator if the radius R is 100 m.
(c) Repeat part (b) for a radius of 1 km.

2.11. Describe a typical ion source. What are the physical processes involved? How is one constructed?

2.12. In what way is a conventional cyclotron different from a synchrotron? What limits the maximum energy obtainable in a cyclotron? Why are high-energy accelerators predominantly synchrotrons?

2.13. What is meant by *phase stability*? Discuss this concept for linacs and for synchrotrons.

2.14. What is the duty cycle of an accelerator? Discuss the duty cycle for the Van de Graaff generator, the linac, and the synchrotron. Sketch the *beam structure*, i.e., the intensity of the ejected beam as a function of time for these three machines.

2.15. How is the beam ejected in a synchrotron?

2.16. How and why will superconductivity be important in the field of accelerator physics?

2.17. Why is it expensive to build very-high-energy electron synchrotrons or very-high-energy proton linacs?

2.18. Modern cyclotrons are being constructed in various places, for instance, at Indiana University and at the Swiss Institute for Nuclear Research (SIN). Sketch the principles on which these two cyclotrons are designed. In what way do they differ from the classic cyclotrons?

2.19. Discuss the direction of emission and the polarization of synchrotron radiation. Why is it useful in solid-state studies?

2.20. Compare the ratio of the appropriate (kinetic or total) c.m. energy to the laboratory energy for

(a) Nonrelativistic energies.

(b) Extreme relativistic energies.

2.21. Assume a beam current of 10 A in each storage ring of the CERN ISR, with the beams focused in the intersection region into an area of 1 cm² and intersecting over a length of 10 cm. Compare the number of collisions per sec with those of the NAL accelerator assuming a beam current of 10^{-7} A incident on a 10-cm-long liquid hydrogen target. Assume equal cross sections and equal beam diameters.

3

Passage of Radiation Through Matter

In everyday life we constantly use our understanding of the passage of matter through matter. We do not try to walk through a closed steel door, but we brush through if the passage is only barred by a curtain. We stroll through a meadow full of tall grass but carefully avoid a field of cacti. Difficulties arise if we do not realize the appropriate laws; for example, driving on the right-hand side of a road in England or Japan can lead to disaster. Similarly, a knowledge of the passage of radiation through matter is a crucial part in the design and the evaluation of experiments. The present understanding has not come without surprises and accidents. The early X-ray pioneers burned their hands and their bodies; many of the early cyclotron physicists have cataracts. It took many years before the exceedingly small interaction of the neutrino with matter was experimentally observed because it can pass through a light year of matter with only small attenuation. Then there was the old cosmotron beam at Brookhaven which was accidentally found a few km away from the accelerator, merrily traveling down Long Island.

The passage of charged particles and of photons through matter is governed by *atomic* physics. True, some interactions with nuclei occur. However, the main energy loss and the main scattering effects come from the interaction with the atomic electrons. We shall therefore give few details and no theoretical derivations in the present chapter but shall summarize the important concepts and equations.

3.1 Concepts

Consider a well-collimated beam of monoenergetic particles passing through a slab of matter. The properties of the beam after passage depend on the nature of the particles and of the slab, and we first consider two extreme cases, both of great interest. In the first case, shown in Fig. 3.1(a), a particle undergoes many interactions. In each interaction, it loses a small amount of energy and suffers a small-angle scattering. In the second,

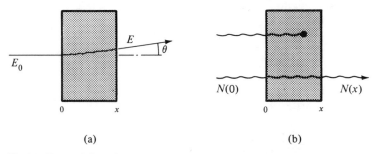

(a)   (b)

Fig. 3.1. Passage of a well-collimated beam through a slab. In (a), each particle suffers many interactions; in (b), a particle is either unharmed or eliminated.

shown in Fig. 3.1(b), the particle either passes unscathed through the slab or it is eliminated from the beam in one "deadly" encounter. The first case applies, for instance, to heavy charged particles, and the second one approximates the behavior of photons. (Electrons form an intermediate case.) We shall now discuss the two cases in more detail.

Many Small Interactions. Each interaction produces an energy loss and a deflection. Losses and deflections add up statistically. After passing through an absorber the beam will be degraded in energy, will no longer be monoenergetic, and will show an angular spread. Characteristics of the beam before and after passage are shown in Fig. 3.2. The number of particles left in the beam can be observed as a function of the absorber thickness x. Up to a certain thickness, essentially all particles will be transmitted. At some thickness, some of the particles will no longer emerge; at a thickness R_0, called the mean range, half of the particles will be stopped, and finally, at sufficiently large thickness, no particles will emerge. The behavior of the number of transmitted particles versus absorber thickness is shown in Fig. 3.3. The fluctuation in range is called straggling.

"All-or-Nothing" Interactions. If an interaction eliminates the particle from the beam, the characteristics of the transmitted beam are different

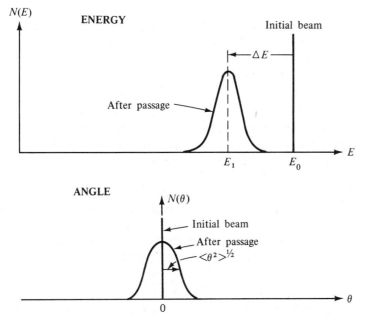

Fig. 3.2. Energy and angular distribution of a beam of heavy charged particles before and after passing through an absorber.

from the one just discussed. Since the transmitted particles have not undergone an interaction, the transmitted beam has the same energy and angular spread as the incident one. In each elementary slab of thickness dx the number of particles undergoing interactions is proportional to the number of incident particles, and the constant of proportionality is called the absorption coefficient μ:

$$dN = -N(x)\mu \, dx.$$

Integration gives

$$N(x) = N(0)e^{-\mu x}. \tag{3.1}$$

The number of transmitted particles decreases exponentially, as indicated in Fig. 3.4. No range can be defined, but the *average distance* traveled by a particle before undergoing a collision is called the *mean free path*, and it is equal to $1/\mu$.

3.2 Heavy Charged Particles

Heavy charged particles lose energy mainly through collisions with bound electrons via Coulomb interactions. The electrons can be lifted to higher discrete energy levels (excitation), or they can be ejected from the atom (ionization). Ionization dominates if the particle has an energy large

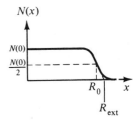

Fig. 3.3. Range of heavy charged particles. $N(x)$ is the number of particles passing through an absorber of thickness x. R_0 is the mean range; R_{ext} is called the extrapolated range.

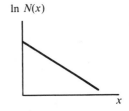

Fig. 3.4. In all-or-nothing interactions, the number of transmitted particles, $N(x)$, decreases exponentially with the absorber thickness x.

35

compared to atomic binding energies. The rate of energy loss due to collisions with electrons has been calculated classically by Bohr and quantum mechanically by Bethe and by Bloch.[1] The result, usually called the Bethe-Bloch equation, is

$$-\frac{dE}{dx} = \frac{4\pi n z^2 e^4}{m_e v^2}\left\{\ln\frac{2m_e v^2}{I[1-(v/c)^2]} - \left(\frac{v}{c}\right)^2\right\}. \tag{3.2}$$

Here, $-dE$ is the energy lost in a distance dx; n is the number of electrons per cm³ in the stopping substance; m_e is the electron mass; ze and v are, respectively, charge and velocity of the particle; and I is the mean excitation potential of the atoms of the stopping substance. [Eq. (3.2) is an approximation, but it suffices for our purpose.]

In practical applications, the thickness of an absorber is not measured in length units but in terms of ρx, where ρ is the density of the absorber. ρx is usually given in g/cm², and it can be found experimentally by determining the mass and the area of the absorber and taking the ratio of the two. The specific energy loss tabulated or plotted is then

$$\frac{dE}{d(\rho x)} = \frac{1}{\rho}\frac{dE}{dx}.$$

Figure 3.5 gives the specific energy loss of protons in hydrogen and lead as

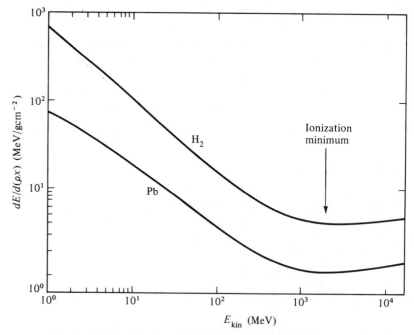

Fig. 3.5. Specific energy loss, $dE/d(\rho x)$, for protons in hydrogen and lead.

1. N. Bohr, *Phil. Mag.* **25**, 10 (1913); H. A. Bethe, *Ann. Physik* **5**, 325 (1930); F. Bloch, *Ann. Physik* **16**, 285 (1933).

a function of the kinetic energy E_{kin}. Figure 3.5 and Eq. (3.2) show the salient features of the energy loss of heavy particles in matter clearly. The specific energy loss is proportional to the number of electrons in the absorber and proportional to the *square* of the particle charge. At a certain energy, for protons about 1 GeV, an *ionization minimum* occurs. Below the minimum, $dE/d(\rho x)$ is proportional to $1/v^2$. Consequently, as a nonrelativistic particle slows down in matter, its energy loss increases. However, Eq. (3.2) breaks down when the particle velocity becomes comparable to, or less than, the velocity of the electrons in the atoms. The energy loss then decreases again, and the curves in Fig. 3.5 turn down below about 1 MeV. Above the ionization minimum, $dE/d(\rho x)$ increases slowly. It is often useful to remember that the energy loss at the minimum and for at least two decades above is about the same for all materials and that it is of the order

$$-\frac{dE}{d(\rho x)}(\text{at minimum}) \approx 2 \text{ MeV/g-cm}^{-2}. \qquad (3.3)$$

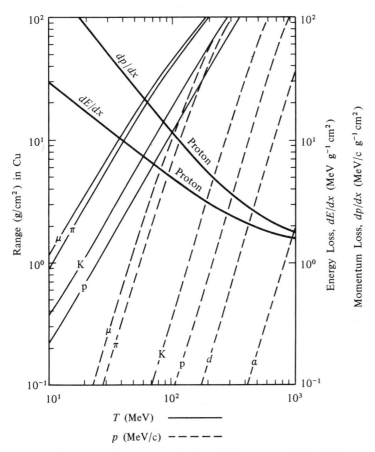

Fig. 3.6. Specific energy loss and range of heavy particles in copper. [From M. Roos et al., *Phys. Letters* **33B**, 18 (1970).] The quantity x is identical to ρx in the text.

Equation (3.2) also shows that the specific energy loss does not depend on the mass of the particle (provided it is much heavier than the electron) but only on its charge and velocity. The curves in Fig. 3.5 therefore are valid also for heavy particles other than the protons if the energy scale is appropriately shifted.

The *range* of a particle in a given substance is obtained from Eq. (3.2) by integration:

$$R = \int_{T_o}^{0} \frac{dT}{(dT/dx)}. \tag{3.4}$$

Here T is the kinetic energy and the subscript o refers to the initial value. Some useful information concerning range and specific energy loss is summarized in Fig. 3.6.

Two more quantities shown in Fig. 3.2, the spread in energy and the spread in angle, are important in experiments, but they are not essential for a first view of the subatomic world. We shall therefore not discuss them here; the relevant information can be found in the references given in Section 3.5.

3.3 Photons

Photons interact with matter chiefly by three processes:
1. Photoelectric effect.
2. Compton effect.
3. Pair production.
A complete treatment of the three processes is rather complicated and requires the tools of quantum electrodynamics. The essential facts, however, are simple. In the photoelectric effect, the photon is absorbed by an atom, and an electron from one of the shells is ejected. In the Compton effect, the photon scatters from an atomic electron. In pair production, the photon is converted into an electron-positron pair. This process is impossible in free space because energy and momentum cannot be conserved simultaneously when a photon decays into two massive particles. It occurs in the Coulomb field of a nucleus which helps to balance energy and momentum.

The energy dependences of processes 1–3 are very different. At low energies, below a few keV, the photo effect dominates, the Compton effect is small, and pair production is energetically impossible. At an energy of $2m_ec^2$, pair production becomes possible, and it soon dominates completely.

Two of the three processes, photoeffect and pair production, eliminate the photons undergoing interaction. In Compton scattering, the scattered photon is degraded in energy. The all-or-nothing situation described in Section 3.1 and depicted in Fig. 3.1(b) is therefore a good approximation, and the transmitted beam should show an exponential behavior, as described by Eq. (3.1). The absorption coefficient μ is a sum of three terms,

$$\mu = \mu_{\text{photo}} + \mu_{\text{Compton}} + \mu_{\text{pair}} \qquad (3.5)$$

and each term can be computed accurately. The behavior of the three terms, and of the total absorption coefficient, is shown in Fig. 3.7. As is customary, the photon energy $\hbar\omega$ is expressed in terms of $m_e c^2 = 0.511$ MeV.

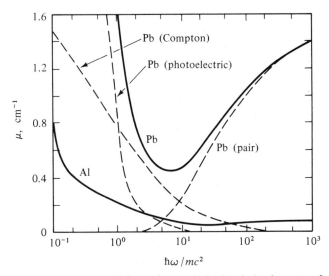

Fig. 3.7. Total absorption coefficients of γ rays by lead and aluminum as a function of energy (solid lines). Photoelectric absorption of aluminum is negligible at the energies considered here. Dashed lines show separately the contributions of photoelectric effect, Compton scattering, and pair production for Pb. Abscissa, logarithmic energy scale; $\hbar\omega/mc^2 = 1$ corresponds to 511 keV. (From W. Heitler, *The Quantum Theory of Radiation*, The Clarendon Press, Oxford, 1936, p. 216.)

3.4 Electrons

The energy-loss mechanism of electrons differs from that of heavier charged particles for several reasons. The most important difference is energy loss by radiation; this mechanism is unimportant for heavy particles but dominant for high-energy electrons. Radiation makes it necessary to consider two energy regions separately. At energies well below the *critical energy* E_c, given approximately by

$$E_c \approx \frac{600 \text{ MeV}}{Z}, \qquad (3.6)$$

excitation and ionization of the bound absorber electrons dominate. [In Eq. (3.6), Z is the charge number of the stopping atoms.] Above the critical energy, radiation loss takes over. We shall treat the two regions separately.

Ionization Region $(E < E_c)$. In this region, the energy loss of an electron and a proton of equal velocity are nearly the same and Eq. (3.2) can be taken over with some small modifications. There is, however, one major difference, as sketched in Fig. 3.8. The path of the heavy particle is straight and the $N(x)$ versus x curve is as given in Fig. 3.3. The electron, owing to its small mass, suffers many scatterings with considerable angles. The behavior of the number of transmitted electrons versus absorber thickness is sketched in Fig. 3.8. An extrapolated range R_p is defined as shown in Fig. 3.8. Between about 0.6 and 12 MeV the extrapolated range in aluminum is well represented by the linear relation

$$R_p \text{ (in g/cm}^2) = 0.526\, E_{\text{kin}} \text{ (in MeV)} - 0.094. \tag{3.7}$$

Radiation Region $(E > E_c)$. A charged particle passing by a nucleus of charge Ze experiences the Coulomb force and it is deflected [Fig. 3.9(a)]. The process is called *Coulomb scattering*. The deflection accelerates (decelerates) the passing particle. As pointed out in Section 2.5, acceleration

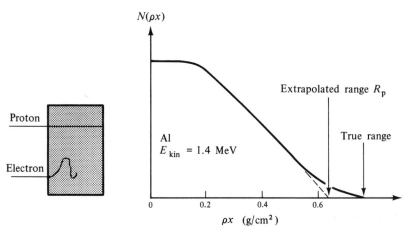

Fig. 3.8. Passage of a proton and an electron with equal total pathlength through an absorber. The $N(x)$ vs. x behavior for electrons is given at right.

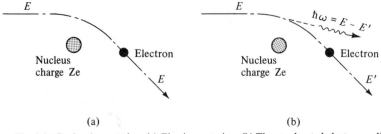

Fig. 3.9. Coulomb scattering. (a) Elastic scattering. (b) The accelerated electron radiates and loses energy in the form of a photon (Bremsstrahlung).

produces radiation. In the case of electrons in a synchrotron, it is called *synchrotron radiation;* in the case of charged particles scattered in the Coulomb field of nuclei, it is called *Bremsstrahlung* (braking radiation). Equations (2.21) and (2.22) show that, for equal acceleration, the energy carried away by photons will be proportional to $(E/mc^2)^4$. Bremsstrahlung is thus an important energy-loss mechanism for electrons, but it is very small for heavier particles, such as muons, pions, and protons.

Actually, Eq. (2.21) has been calculated by using classical electrodynamics. Bremsstrahlung, however, must be treated quantum mechanically. Bethe and Heitler have done so, and the essential results are as follows.[2] The number of photons with energies between $\hbar\omega$ and $\hbar(\omega + d\omega)$ produced by an electron of energy E in the field of a nucleus with charge Ze is proportional to Z^2/ω:

$$N(\omega)d\omega \propto Z^2\frac{d\omega}{\omega}. \tag{3.8}$$

Owing to the emission of these photons, the electron loses energy, and the distance over which its energy is reduced by a factor e is called the *radiation length* and conventionally denoted by X_0. In terms of X_0, the radiative energy loss for large electron energies is

$$-\left(\frac{dE}{dx}\right)_{\text{rad}} \approx \frac{E}{X_0} \quad \text{or} \quad E = E_0 e^{-x/X_0}. \tag{3.9}$$

The radiation length is given either in g/cm² or in cm; a few values of X_0 and of the critical energy E_c are given in Table 3.1.

Table 3.1 VALUES OF THE CRITICAL ENERGY E_c AND THE
RADIATION LENGTH X_0 FOR VARIOUS SUBSTANCES

Material	Z	Density (g/cm³)	Critical energy (MeV)	Radiation length g/cm²	Radiation length cm
H₂(liquid)	1	0.071	340	62.8	887
He (liquid)	2	0.125	220	93.1	745
C	6	1.5	103	43.3	28
Al	13	2.70	47	24.3	9.00
Fe	26	7.87	24	13.9	1.77
Pb	82	11.35	6.9	6.4	0.56
Air		0.0012	83	37.2	30,870
Water		1	93	36.4	36.4

According to Eq. (3.9), a highly energetic electron loses its energy exponentially and after about seven radiation lengths has only 10^{-3} of its initial energy left. However, concentrating on the primary electron is misleading. Many of the Bremsstrahlung photons have energies greatly in

2. H. A. Bethe and W. Heitler, *Proc. Roy. Soc.*(London) **A146**, 83 (1934).

excess of 1 MeV and can produce electron-positron pairs (Section 3.3). In fact, the mean free path, that is, the average distance, X_p, traveled by a photon before it produces a pair, is also related to the radiation length:

$$X_p = \tfrac{9}{7} X_0. \tag{3.10}$$

In successive steps, a high-energy electron creates a *shower*. (Of course a shower can also be initiated by a photon.) The detailed theory of such a shower is very complicated and in practice computer calculations are performed. Figure 3.10 shows the number n of electrons in a shower as a function of the thickness of the absorber. The energy E_0 of the incident

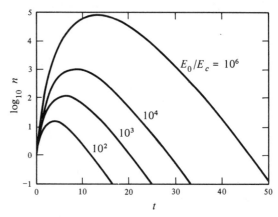

Fig. 3.10. Number n of electrons in a shower as a function of the thickness traversed, t, in radiation lengths. [These curves were taken from the work of B. Rossi and K. Greisen, *Rev. Modern Phys.* **13**, 240 (1941).]

electron is measured in units of the critical energy; the thickness is expressed in units of the radiation length X_0. Figure 3.10 expresses the development and death of a shower: The increase in the number of electrons is very rapid at the beginning. As the cascade progresses, the average energy per electron (or per photon) becomes smaller. At some point it becomes so small that the photons can no longer produce pairs, and the shower dies.

3.5 References

The basic ideas underlying the computation of the energy loss of charged particles in matter are described lucidly in N. Bohr, "Penetration of Atomic Particles Through Matter," *Kgl. Danske Videnskab. Selskab Mat-fys Medd.* XVIII, No. 8 (1948), and in E. Fermi, *Nuclear Physics*, notes compiled by J. Orear, A. H. Rosenfeld, and R. A. Schluter, University of Chicago Press, Chicago, 1950.

Details concerning the passage of radiation through matter, tables, figures, and additional references can be found in the following articles:

H. A. Bethe and J. Ashkin, in *Experimental Nuclear Physics*, Vol. 1 (E. Segrè, ed.), Wiley, New York, 1953. This article is the standard work and most later papers refer to it.

R. M. Sternheimer, in *Methods of Experimental Physics*, Vol. 5: *Nuclear Physics* (L. C. L. Yuan and C. S. Wu, eds.), Academic Press, New York, 1961.

G. Knop and W. Paul; also C. M. Davisson, in *Alpha-, Beta- and Gamma-Ray Spectroscopy* (K. Siegbahn, ed.), North-Holland, Amsterdam, 1965.

Additional references are listed and described in W. P. Trower, "Resource Letter PD-1 on Particle Detectors," *Am. J. Phys.* **38**, 795 (1970).

Curves, tables, and equations relating to the passage of radiation through matter are also given in the *American Institute of Physics Handbook*, 3rd ed., McGraw-Hill, New York, 1972, Section 8.

PROBLEMS

3.1. An accelerator produces a beam of protons with kinetic energy of 100 MeV. For a particular experiment, a proton energy of 50 MeV is required. Compute the thickness of
(a) a carbon and
(b) a lead absorber,
both in cm and in g/cm^2, necessary to reduce the beam energy from 100 to 50 MeV. Which absorber would be preferable? Why?

3.2. A counter has to be placed in a muon beam of 100-MeV kinetic energy. No muons should reach the counter. How much copper is needed to stop all muons?

3.3. We have stated that the transmission of charged particles through matter is dominated by atomic, and not nuclear, interactions. When is this statement no longer true; i.e., when do nuclear interactions become important?

3.4. A beam stop is required at the end of accelerators to prevent the particles from running wild. How many *m* of solid dirt would be required at NAL to completely stop the 200-GeV protons, assuming only electromagnetic interactions? Why is the actual beam stop length less?

3.5. Cosmic-ray muons are still observed in mines that are more than 1-km underground. What is the minimum initial energy of these muons? Why are no cosmic-ray protons or pions observed in these underground laboratories?

3.6. Discuss and understand the simplest derivation of Eq. (3.2).

3.7. Show that the mean free path of a particle undergoing exponential absorption as described by Eq. (3.1) is given by $1/\mu$.

3.8. A beam of 1-mA protons of kinetic energy of 800 MeV passes through a 1-cm^3 copper cube. Compute the maximum energy deposited per sec in the copper. Assume the cube to be thermally insulated, and compute the temperature rise per sec.

3.9. Compare the energy loss of nonrelativistic π^+, K^+, d, $^3He^{2+}$, $^4He^{2+} \equiv \alpha$ to that of protons of the same energy in the same material.

3.10. In an experiment, alpha particles of 20-MeV energy enter a scattering chamber through a copper foil that is 1 mm thick.
(a) Use the form of Eq. (3.2) to find the energy

of the proton beam that has the same energy loss as the α beam.

(b) Compute the energy loss.

3.11. Use Eq. (3.2) and Fig. 3.5 to sketch the ionization along the path of a heavy charged particle (Bragg curve). What happens at very low energies, i.e., toward the end of the particle track? [The behavior for very small energies is not contained in Eq. (3.2) or Fig. 3.5—you have to find it in the references given in Section 3.5.]

3.12. Use Eq. (3.2) to calculate numerically the energy loss of a 20-MeV proton in aluminum ($I = 150$ eV).

3.13. A radioactive source emits gamma rays of 1.1-MeV energy. The intensity of these gamma rays must be reduced by a factor 10^4 by a lead container. How thick (in cm) must the container walls be?

3.14. ^{57}Fe has a gamma ray of 14-keV energy. A source is contained in a metal cylinder. It is desired that 99% of the gamma rays escape the cylinder. How thin must the walls be made if the cylinder is

(a) Aluminum?

(b) Lead?

3.15. A source emits gamma rays of 14 and 6 keV. The 6-keV gamma rays are 10 times more intense than the 14-keV rays. Select an absorber that cuts the intensity of the 6-keV rays by a factor of 10^3 but affects the 14-keV rays as little as possible. What is your choice? By what factor is the 14-keV intensity reduced?

3.16. The three processes discussed in Section 3.3 are not the only interactions of photons. List and briefly discuss other types of photon interactions.

3.17. A radioactive source contains two gamma rays of equal intensity with energies of 85 and 90 keV, respectively. Compute the intensity of the two gamma lines after passing through a 1-mm lead absorber. Explain your result.

3.18. Electrons of 1-MeV kinetic energy should be stopped in an aluminum absorber. How thick, in cm, must the absorber be?

3.19. What is the energy of an electron that has approximately the same total (true) pathlength as a 10-MeV proton?

3.20. An electron of 10^3-GeV energy strikes the surface of the ocean. Describe the fate of the electron. What is the maximum number of electrons in the resulting shower? At which depth, in m, does the maximum occur?

3.21. A 10-GeV electron from the Stanford Linear Accelerator (SLAC) passes through a 1-cm aluminum plate. How much energy is lost?

3.22. Show that pair production is not possible without the presence of a nucleus to take up momentum.

3.23. Show that the maximum energy that can be transferred to an electron in a single collision by a particle of kinetic energy T and mass M ($M \gg m_e$) is $(4m_e/M)T$.

4

Detectors

What would a physicist do if he were asked to study ghosts and telepathy? We can guess. He would probably (1) perform a literature search and (2) try to design detectors to observe ghosts and to receive telepathy signals. The first step is of doubtful value because it could easily lead him away from the truth. The second step, however, would be essential. Without a detector that allows the physicist to *quantify* his observations, his announcement of the discovery of ghosts would be rejected by *Physical Review Letters*. In experimental subatomic physics, detectors are just as important and the history of progress is to a large extent the history of increasingly more sophisticated detectors. Even without accelerators and using only the rare cosmic-ray particles, a great deal can be learned by making the detectors bigger and better. In the following sections, we shall discuss four different types of detectors. Because of the arbitrary restriction to four, many beautiful and elegant tools are not treated here. However, once the ideas behind typical instruments are understood, it is easy to pick up more details concerning others. We also add a brief section about electronics because it is an integral part of any detection system.

4.1 Scintillation Counters

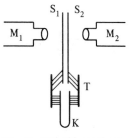

The first scintillation counter, called spinthariscope, was constructed in 1903 by Sir William Crookes. It consisted of a ZnS screen and a microscope; when alpha particles hit the screen, a light flash could be seen. In 1910, Geiger and Marsden performed the first coincidence experiment. As Fig. 4.1 shows, they used two screens, S_1 and S_2, and two observers with microscopes M_1 and M_2. If the radioactive gas between the two screens emitted two alpha particles within a "short" time and if each hit one screen, each observer would see a flash. They probably shouted to indicate the time of arrival.

The human eye is slow and unreliable and the scintillation counter was abandoned for many years. It was reintroduced in 1944 with a photomultiplier replacing the eye. The basic arrangement for a modern scintillation counter is shown in Fig. 4.2. A scintillator is joined to one (or more) photomultipliers through a light pipe. A particle passing through the scintillator produces excitations; deexcitation occurs through emission of photons. These photons are transmitted through a shaped light pipe to the photocathode of a photomultiplier. There, photons release electrons which are accelerated and focused onto the first dynode. For each primary elec-

Fig. 4.1. Coincidence observation "by eye." (From E. Rutherford, *Handbuch der Radiologie*, Vol. II, Akademische Verlagsgesellschaft, Leipzig, 1913.)

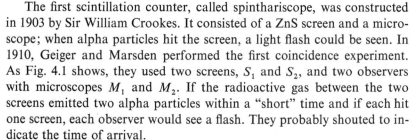

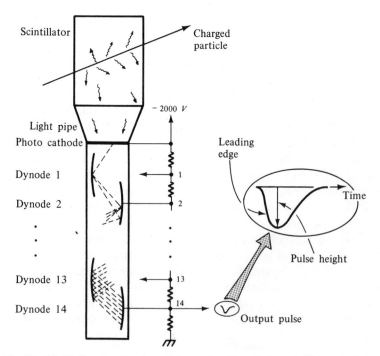

Fig. 4.2. Scintillation counter. A particle passing through the scintillator produces light which is transmitted through a light pipe onto a photomultiplier.

tron hitting a dynode, two to five secondary electrons are released. Up to 14 multiplying stages are used, and overall multiplying factors of up to 10^9 can be achieved. The few incident photons therefore produce a measurable pulse at the output of the multiplier. The shape of the pulse is shown schematically in the insert of Fig. 4.2. The pulse height is proportional to the total energy deposited in the scintillator.

Two types of scintillators are widely used, sodium iodide and plastics. *Sodium iodide* crystals are usually doped with a small amount of thallium and denoted by NaI(Tl). The Tl atoms act as luminescence centers. The efficiency of these inorganic crystals for gamma rays is high, but the decay of each pulse is slow, about 0.25 μsec. Moreover, NaI(Tl) is hygroscopic and large crystals are very expensive. *Plastic scintillators*, for instance polystyrene with terphenyl added, are cheap; they can be bought in large sheets and can be machined in nearly any desired shape. Scintillator and light pipes together can look like a work of art (Fig. 4.3). The decay time is only a few nsec, but the efficiency for photons is low. They are therefore mainly used for the detection of charged particles.

A few remarks are in order concerning the mechanism of observation of gamma rays in NaI(Tl) crystals. For a gamma ray of less than 1 MeV, only photoeffect and Compton effect have to be considered. Photoeffect results in an electron with an energy $E_e = E_\gamma - E_b$, where E_b is the binding energy of the electron before it was ejected by the photon. The electron will, as a rule, be completely absorbed in the crystal. The energy deposited in the crystal gives rise to a number of light quanta that are then seen by the photomultiplier. In turn, these photons result in a pulse proportional to E_e and with a certain width ΔE. This photo or full-energy peak is shown in Fig. 4.4. The energy of the electrons produced by the Compton effect

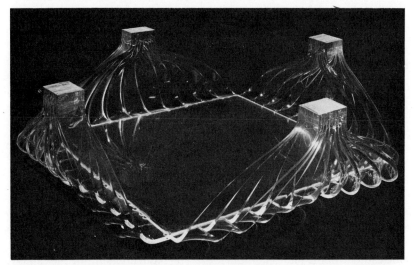

Fig. 4.3. Light pipes for transporting the light from scintillators to photo multipliers. (Courtesy New England Nuclear, Pilot Chemicals Division.)

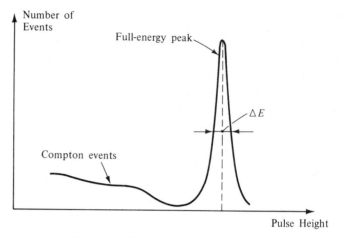

Fig. 4.4. Scintillation spectrum, NaI(Tl) crystal.

depends on the angle at which they are scattered. The Compton effect therefore gives rise to a spectrum, as indicated in Fig. 4.4. The width of the full-energy peak, measured at half-height, depends on the number of light quanta produced by the incident gamma ray; typically $\Delta E/E_\gamma$ is of the order of 20% at $E_\gamma = 100$ keV and 6–8% at 1 MeV. At energies above 1 MeV, the incident gamma ray can produce an electron-positron pair; the electron is absorbed, and the positron annihilates into two 0.51-MeV photons. These two photons can escape from the crystal. The energy deposited is E_γ if no photon escapes, $E_\gamma - m_e c^2$ if one escapes, and $E_\gamma - 2m_e c^2$ if both annihilation photons escape.

The energy resolution $\Delta E/E$ deserves some additional consideration. Is a resolution of about 10% sufficient to study the gamma rays emitted by nuclei? In some cases, it is. In many instances, however, gamma rays have energies so close together that a scintillation counter cannot separate them. Before discussing a counter with better resolution, it is necessary to understand the sources contributing to the width. The chain of events in a scintillation counter is as follows: The incident gamma ray produces a photoelectron with energy $E_e \approx E_\gamma$. The photoelectron, via excitation and ionization, produces n_{lq} light quanta, each with an energy of $E_{\mathrm{lq}} \approx 3$ eV ($\lambda \approx 400$ nm). (For clarity we call the incident photon *the gamma ray* and the optical photon *the light quantum*.) The number of light quanta is given by

$$n_{\mathrm{lq}} \approx \frac{E_\gamma}{E_{\mathrm{lq}}} \epsilon_{\mathrm{light}},$$

where $\epsilon_{\mathrm{light}}$ is the efficiency for the conversion of the excitation energy into light quanta. Of the n_{lq} light quanta, only a fraction ϵ_{coll} are collected at the cathode of the photomultiplier. Each light quantum hitting the cathode has a probability $\epsilon_{\mathrm{cathode}}$ of ejecting an electron. The number n_e of electrons produced at the input of the photomultiplier is therefore

$$n_e = \frac{E_\gamma}{E_{lq}} \epsilon_{\text{lighs}} \epsilon_{\text{coll}} \epsilon_{\text{cathode}}. \qquad (4.1)$$

Typical values for the efficiencies are

$$\epsilon_{\text{light}} \approx 0.1, \qquad \epsilon_{\text{coll}} \approx 0.4, \qquad \epsilon_{\text{cathode}} \approx 0.2,$$

so that the number of electrons released at the photocathode after absorption of a 1-MeV gamma ray is $n_e \approx 3 \times 10^4$. (The value $\epsilon_{\text{light}} \approx 0.1$ is appropriate for a NaI(Tl) crystal; the corresponding value for a plastic scintillator is about 0.03.) Since all processes in Eq. (4.1) are statistical, n_e will be subject to fluctuations, and these produce most of the observed line width. An additional broadening comes from the multiplication in the photomultiplier which is also statistical. To discuss the line width, we digress to present some of the fundamental statistical concepts.

4.2 Statistical Considerations

Random processes play an important part in subatomic physics. The standard example is a collection of radioactive atoms, each atom decaying independently of all the others. We shall consider here an equivalent problem that came up in the previous section, the production of electrons at the photocathode of a multiplier. The question to be answered is illustrated in Fig. 4.5. Each incident photon produces n photoelectrons as output. We can repeat the measurement of the number of output electrons N times, where N is very large. In each of these N identical measurements, we shall find a number $n_i, i = 1, \ldots, N$. The *average* number of output electrons is then given by

$$\bar{n} = \frac{1}{N} \sum_{i=1}^{N} n_i. \qquad (4.2)$$

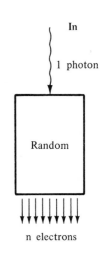

Fig. 4.5. Production of photoelectrons as a random process.

The question of interest can be stated: How are the various values n_i distributed around $\bar{n}$? Another way of phrasing the same question is: What is the probability $P(n)$ of finding a particular value n in a given measurement if the average number is $\bar{n}$? Or, to make it more specific, consider a process where the average number of output electrons is small, say $\bar{n} = 3.5$. What is the probability of finding the value $n = 2$? This problem has occupied mathematicians for a long time, and the answer is well known[1]: The probability $P(n)$ of observing n events is given by the *Poisson distribution*,

$$P(n) = \frac{(\bar{n})^n}{n!} e^{-\bar{n}}, \qquad (4.3)$$

1. A derivation can, for instance, be found in H. D. Young, *Statistical Treatment of Experimental Data*, McGraw-Hill, New York, 1962, Eq. (8.5). A fascinating collection of excerpts from works on statistics is contained in Vols. II and III of J. R. Newman, *The World of Mathematics*, Simon and Schuster, New York, 1956.

where $\bar{n}$ is the average defined by Eq. (4.2). As behooves a probability, the sum over all possible values n is 1, $\sum_{n=0}^{\infty} P(n) = 1$. With Eq. (4.3), the previous questions can now be answered, and we first turn to the most specific one. With $\bar{n} = 3.5$, $n = 2$, Eq. (4.3) gives $P(2) = 0.185$. It is straightforward to compute the probabilities for all interesting values of n. The corresponding histogram is shown in Fig. 4.6. It shows that the distribution is very wide. There is a nonnegligible probability of measuring values as small as zero or as large as 9. If we perform only one experiment and find, for instance, a value of $n = 7$, we have no idea what the average value would be.

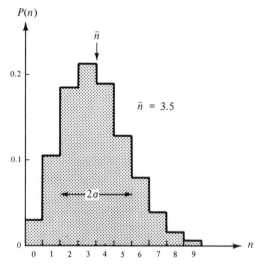

Fig. 4.6. Histogram of the Poisson distribution for $\bar{n} = 3.5$. The distribution is not symmetric about $\bar{n}$.

A glance at Fig. 4.6 shows that it is not enough to measure and record the average, $\bar{n}$. A measure of the *width* of the distribution is also needed. It is customary to characterize the width of a distribution by the *variance* σ^2:

$$\sigma^2 = \sum_{n=0}^{\infty} (\bar{n} - n)^2 P(n), \tag{4.4}$$

or by the square root of the variance, called the *standard deviation*. For the Poisson distribution, Eq. (4.3), variance and standard distribution are easy to compute, and they are given by

$$\sigma^2 = \bar{n}, \qquad \sigma = \sqrt{\bar{n}}. \tag{4.5}$$

It is customary to quote the result of the measurement of a quantity limited by statistical deviations in the following way:

$$\text{result} = \bar{n} \pm \sigma. \tag{4.6}$$

For small values of $\bar{n}$, the distribution is not symmetric about $\bar{n}$, as is evident from Fig. 4.6. Equation (4.6) then must be understood with this property in mind.

So far we have discussed the Poisson distribution for *small* values of n. Experimentally, such a situation arises, for instance, at the first dynode of a photomultiplier, where each incident electron produces two to five secondary electrons. Data are then given in the form of histograms, as in Fig. 4.6. In many instances, $\bar{n}$ can be very *large*. In the case of the scintillation counter discussed in the previous section, the number of photoelectrons at the photomultiplier is on the average $\bar{n} = 3 \times 10^3$. For $\bar{n} \gg 1$, Eq. (4.3) is cumbersome to evaluate. However, for large n, $\bar{n}$ can be considered a continuous variable, and Eq. (4.3) can be approximated by

$$P(n) = \frac{1}{(2\pi n)^{1/2}} e^{-(\bar{n}-n)^2/2n},\qquad(4.7)$$

which is easier to evaluate. Moreover, the behavior of $P(n)$ is now dominated by the factor $(\bar{n} - n)^2$ in the exponent. Particularly near the center of the distribution, n can be replaced by $\bar{n}$ except in the factor $(\bar{n} - n)^2$, and the result is

$$P(n) = \frac{1}{(2\pi\bar{n})^{1/2}} e^{-(\bar{n}-n)^2/2\bar{n}},\qquad(4.8)$$

This expression is symmetric about $\bar{n}$ and is called a *normal* or *Gaussian distribution*. The standard deviation and the variance are still given by Eq. (4.5). As an example of the limiting case where the Poisson distribution can be represented by the normal one, we show in Fig. 4.7 $P(n)$ for $\bar{n} = 3 \times 10^3$, the number of photoelectrons of our example in the previous section. The standard deviation is equal to $(3 \times 10^3)^{1/2} = 55$, resulting in a fractional deviation $\sigma/\bar{n} \approx 2\%$. To compare this value to $\Delta E/E_\gamma$ we note that ΔE is the *full* width at half maximum (FWHM). With Eqs. (4.5) and (4.8) it is straightforward to see that Δn, the full width at half maximum,

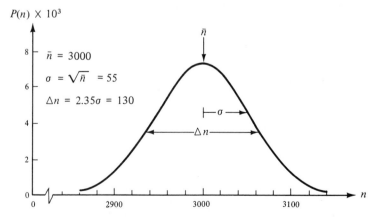

Fig. 4.7. Poisson distribution for $\bar{n} \gg 1$ where it becomes a normal distribution.

is related to the standard deviation by

$$\Delta n = 2.35\sigma. \qquad (4.9)$$

With $\Delta E/E_\gamma = \Delta n/\bar{n}$, the expected fractional energy resolution becomes about 5%. Since the value must still be corrected for additional fluctuations, for instance, in the multiplier, the agreement with the experimentally observed resolution of 6–8% is satisfactory.

4.3 Semiconductor Detectors

Scintillation counters started a revolution in the detection of nuclear radiations, and they reigned unchallenged from 1944 to the late 1950s. They are still essential for many experiments, but in many areas they have been replaced by semiconductor detectors. Before discussing these, we compare in Fig. 4.8 a complex gamma-ray spectrum as seen by a semiconductor and by a scintillation detector. The superior energy resolution of the solid-state counter is obvious. How is it achieved? In the scintillation counter, the efficiences in Eq. (4.1) reduce the number of photoelectrons counted; it is difficult to imagine how each of the efficiency factors in Eq. (4.1) could be improved to about 1. A different approach is therefore needed and the solid-state (semiconductor) detector offers one.

The idea underlying the semiconductor counter is old and it is used in ionization chambers: A charged particle moving through a gas or a solid

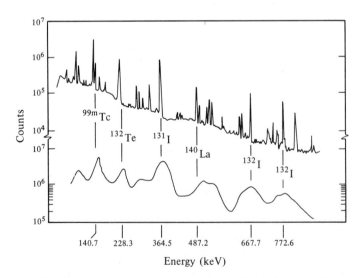

Fig. 4.8. Complex gamma-ray spectrum, due to gross fission products, observed by a germanium detector (upper curve) and a scintillation detector (lower curve). [From F. S. Goulding and Y. Stone, *Science* **170**, 280 (1970). Copyright 1970 by the American Association for the Advancement of Science.]

produces ion pairs, and the number of these pairs is given by

$$n_{\text{ion}} = \frac{E_e}{W}, \qquad (4.10)$$

where W is the energy needed to produce one ion pair. If the ion pairs are separated in an electric field and if the total charge is collected and measured, the energy of the electron can be found. A gas-filled ionization chamber uses this principle, but it has two disadvantages: (1) The density of a gas is low so that the energy deposited by a particle is small. (2) The energy needed for the production of an ion pair is large ($W = 42$ eV for He, 22 eV for Xe, and 34 eV for air). Both disadvantages are avoided in a semiconductor detector, as sketched in Fig. 4.9. If a charged particle passes through a semiconductor, ion pairs will be created. The energy W is about 2.9 eV for germanium and 3.5 eV for silicon. The energies are so low because ionization does not occur from an atomic level to the continuum but from the valence band to the conduction band.[2] The electric field will sweep the negative charges toward the positive and the positive charges toward the negative surface. The resulting current pulse is fed to a low-noise amplifier. At room temperature, thermal excitation can produce an unwanted current, and many semiconductor detectors are therefore cooled to liquid nitrogen temperature.

The low value of W and the collection of all ions explains the high resolution of semiconductor detectors shown in Fig. 4.8. Figure 4.10 pre-

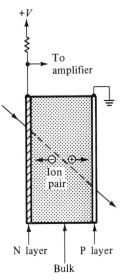

Fig. 4.9. Ideal, fully depleted semiconductor detector with heavily doped surface layers of opposite types.

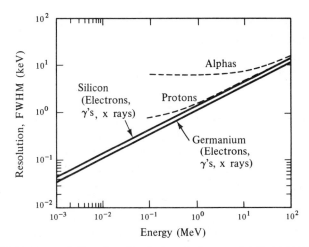

Fig. 4.10. Energy resolution of semiconductor counters as a function of energy. The amplifier noise contribution is not taken into account. [From F. S. Goulding and Y. Stone, *Science* **170**, 280 (1970). Copyright 1970 by the American Association for the Advancement of Science.] FWHM means full width at half maximum.

2. A simple description of the band structure of semiconductors can be found in H. D. Young, *Fundamentals of Optics and Modern Physics*, McGraw-Hill, New York, 1968, Section 11.6, or in the *Feynman Lectures*, Vol. III, Chapter 14.

sents the energy resolution as a function of particle energy for germanium and silicon detectors.

While semiconductor detectors have a much higher density than gas-filled ionization chambers, they cannot be made as large as scintillation counters. Large semiconductor counters have volumes of up to 100 cm^3, and it is hoped that improved technology will allow production of detectors with volumes of up to 1000 cm^3. Scintillation counters can be made orders of magnitude larger, and they do not have to be cooled. For any given application one must therefore consider which type of counter will be more suitable and more convenient.

4.4 Bubble Chambers

The two types of counters discussed so far record the passage or the stopping of a particle and give information about the energy deposited in the counter, but they are blind to complexity. If, for instance, two particles pass through at the same time, the counter gives only the total energy. Of course, we can arrange a number of counters to observe more details, but even then the system will give answers only to the question asked, and it will usually not reveal unsuspected phenomena. Clearly, scintillation and semiconductor counters must be supplemented by detectors that look at all processes as unbiasedly as possible. The bubble chamber is one such detector. Since its invention by Glaser in 1952, it has played a crucial role in the elucidation of the properties of subatomic particles.[3]

The physical phenomenon underlying the bubble chamber is best described in Glaser's own words[4]: "A bubble chamber is a vessel filled with a transparent liquid which is so highly superheated that an ionizing particle moving through it starts violent boiling by initiating the growth of a string of bubbles along its path." A superheated liquid is at a temperature and pressure such that the actual pressure is lower than the equilibrium vapor pressure. The condition is unstable, and the passage of a single charged particle initiates bubble formation. To achieve the superheated condition, the liquid in the chamber (Fig. 4.11) is first kept at the equilibrium pressure; the pressure is then rapidly dropped by moving a piston. A few msec after the chamber has become sensitive, the process is reversed and the chamber pressure is brought back to its equilibrium value. The time during which the chamber is sensitive is synchronized with the arrival time of a pulse of particles from an accelerator. The bubbles are illuminated with an electronic photoflash and recorded by stereoscopic photographs.

Glaser's first chambers contained only a few cm^3 of liquid. Develop-

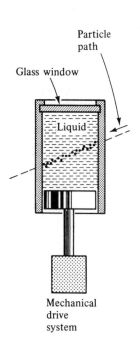

Particle path

Glass window

Liquid

Mechanical drive system

Fig. 4.11. Bubble chamber—schematic diagram.

3. L. W. Alvarez, *Science* **165**, 1071 (1969).
4. D. A. Glaser and D. C. Rahm, *Phys. Rev.* **97**, 474 (1955).

Photo 10. Bubble chamber. While the original bubble chamber was small and simple, modern versions have become very large and sophisticated. (Courtesy Argonne National Laboratory.)

ment was rapid, however; in less than 20 years, the volume increased by more than 10^6. The modern bubble chambers are monsters and cost millions of dollars. They require enormous magnets to curve the paths of the charged particles. The superheated liquid, often hydrogen, is explosive if it comes in contact with oxygen, and accidents have happened despite stringent safety regulations. The existing bubble chambers produce about 35 million photographs/yr, and data evaluation is complex.

Two examples demonstrate the beautiful and exciting events that have already been seen. Figure 1.4 shows the production and the decay of the omega minus, a most remarkable particle that we shall encounter later. Figure 4.12 represents the first neutrino interaction observed in pure hydrogen. It was found on November 13, 1970, in the 3.6-m (12-ft) hydrogen bubble chamber of the Argonne National Laboratory. This chamber is the largest existing one, and it contains about 20,000 liters of hydrogen. A superconducting magnet produces a field of about 18 kG in the chamber volume of 25 m³.

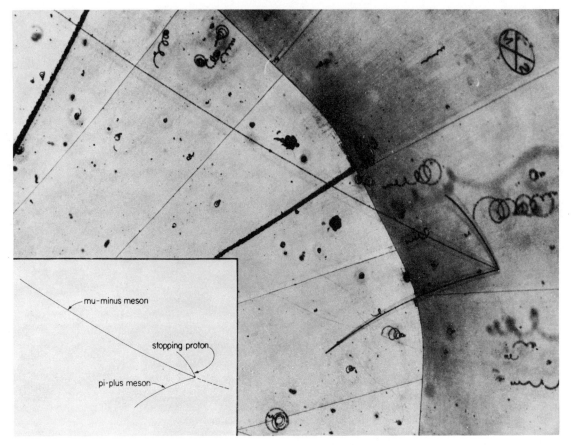

Fig. 4.12. Neutrino interaction in the hydrogen bubble chamber of the Argonne National Laboratory. A neutrino enters from the right as shown in the inset and interacts with the proton of a hydrogen atom to yield a positive pion, a proton, and a muon. (Courtesy Argonne National Laboratory.)

4.5 Spark Chambers

Bubble chambers are beautiful tools, but they have one shortcoming: They are not selective because they cannot be triggered. A bubble chamber is like a surveillance camera at a bank that grinds away and photographs every visitor. To find the picture of a robber, every frame has to be developed and scanned. Clearly, it is more efficient to install a camera that is ready all the time but takes pictures only if it is alerted, for instance, by a bank teller or by a magnetometer detecting a gun. The bubble chamber cannot be triggered because the liquid must be superheated *before* the particle passes through the liquid in order for boiling to occur. *Spark* chambers have many of the advantages of bubble chambers, and they can be triggered.

Spark chambers are based on a simple fact. If the voltage across two metal plates, spaced by a distance of the order of cm, is increased beyond a certain value, a breakdown occurs. If an ionizing particle passes through the volume between the plates, it produces ion pairs, and the breakdown takes the form of a spark that follows the track of the particle. Since the ions remain between the plates for a few μsec, the voltage can be applied *after* passage of the particle: A spark chamber is a triggerable detector.

The elements of a spark chamber system are shown in Fig. 4.13. The problem to be studied in this simplified arrangement is the reaction of an incoming charged particle with a nucleus in the chamber, giving rise to at least two charged products. Thus the *signature* of the desired events is "one charged in, two charged out." Three scintillation counters, A, B, and C, detect the three charged particles. If the particles pass through the three counters, the LOGIC circuit activates the high-voltage supply, and a high-voltage pulse (10–20 kV) is applied to the plates within less than 50 nsec. The resulting sparks are recorded on stereophotographs.

The standard spark chamber arrangement of the type just discussed has been used in many experiments, and chambers have been designed to solve many problems. Thin plates are employed if only the direction of charged particles is desired; thick lead plates are used if gamma rays are to be observed or if electrons have to be distinguished from muons. The electrons produce showers in the lead plates and can thus be recognized. Very small chambers have been successful in nuclear physics problems; enormous ones have helped to detect neutrinos.

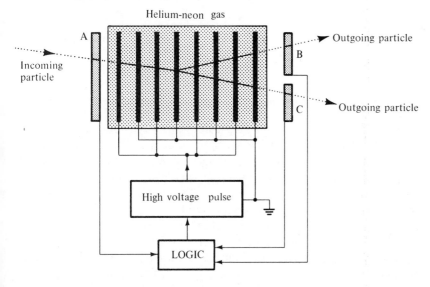

Fig. 4.13. Spark chamber arrangement. The spark chamber consists of an array of metal plates in a helium-neon mixture. If the counter-and-logic system has decided that a wanted event has occurred, a high-voltage pulse is sent to alternate plates, and sparks are produced along the ionization trails.

The optical spark chambers share one disadvantage with the bubble chambers: Data have to be extracted from tracks on photographs. *Automated* spark chambers have been introduced to alleviate the data reduction problem. In these, the plates are replaced by wires that form a coordinate system. Sparks give rise to pulses in the wires, and these pulses are recorded electromagnetically, for instance, by small ferrite cores. The information is then fed directly into computers.

4.6 Counter Electronics

The original scintillation counter, and even the original coincidence arrangement (Fig. 4.1), needed no electronics; the human eye and the human brain provided the necessary elements, and recording was achieved with paper and pen. Nearly all modern detectors, however, contain electronic components as integral elements. A typical example is the circuitry associated with the scintillation counter (Fig. 4.14).

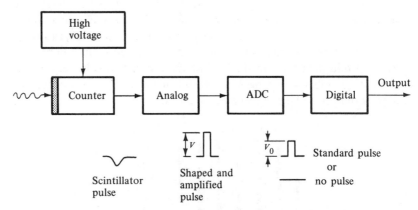

Fig. 4.14. Schematic representation of the main components of counter electronics.

A well-regulated *power supply* provides the voltage for the photomultiplier. The output pulse of the multiplier is shaped and amplified in the *analog* part. The height V of the final pulse is proportional to the height of the original pulse. In the ADC, the *analog-to-digital converter*, the information is transformed into digital form. In the simplest case, only pulses are accepted that have a height between V_0 and $V_0 + \Delta V$. If a pulse is in this window, the output of the ADC is a standard pulse; if the input pulse lies outside the window, no output pulse appears. The *digital part* works with the standard pulse. It can, for instance, be a scaler; for every 10 (or 10^n, n integer) input pulses, one output pulse appears. The output is then a number which gives, in units of 10^n, the number of input pulses within a certain energy interval.

The example here is only one of many possibilities. Setting up a detector electronics system is straightforward because standardized building blocks can be bought; the physicist selects and matches the proper components.[5] We shall not discuss the building blocks here in detail but only make two additional remarks concerning ADC and data processing. In the example above, the ADC consisted of a single-channel analyzer. An output resulted if and only if the input pulse possessed a height within a predetermined range. Clearly, such a procedure is wasteful because the information in all other pulses is thrown away. Usually none of the interesting analog pulses are suppressed. All analog pulses are digitized: A digital signal is associated with every analog pulse. The digital signal can, for instance, be a number proportional to the height of the analog pulse. Usually, this number will be expressed in binary units.

The second remark concerns data processing. In the example above, data processing is simple: The output is connected to a register and the number of counts in a given time is recorded. If all analog pulses are digitized, data processing becomes more involved. One straightforward way is to use many scalers and many registers. This approach is unwieldy, and multichannel analyzers are employed where the digital information is stored in a two-dimensional memory. A pulse with a given pulse height, i.e., a given digital signal, is always stored in the same part of the memory. The memory can then be read out onto an oscilloscope or onto magnetic tape. The most versatile data-processing system is an on-line computer.

4.7 Electronics: Logic

In the previous section we discussed the electronics associated with a single detector. Electronic units do considerably more, however, than just process the data from one counter. A simple example, shown in Fig. 4.15, is the stopping of muons in matter. Muons from an accelerator pass through two counters and enter an absorber where they slow down and finally decay into an electron and two neutrinos:

$$\mu \longrightarrow e\nu\bar{\nu}.$$

We have already mentioned in Section 1.3 that the mean life for the decay of a muon at rest is 2.2 μsec. The specific question to be asked in the experiment sketched in Fig. 4.15 is now as follows: The muon should pass through counters A and B but should stop in the absorber and therefore *not* traverse counter C. After a delay of about 1 μsec, an electron should be observed in counter D. The *logic* must record a muon only if these events happen as described. In shorthand, the requirement can be written as AB$\bar{\text{C}}$D(del), where the AB$\bar{\text{C}}$D means a coincidence between ABD and

5. Of course, he also has to find the necessary funds. To do so, he usually prepares a detailed proposal in which he outlines his research ideas, why they are important, and what they will presumably yield in new information. He then submits his proposal to various agencies and waits.

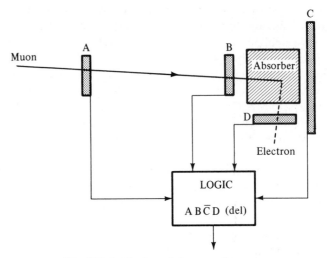

Fig. 4.15. Logic elements in a counting system.

an anticoincidence of this threefold coincidence with C. Furthermore, D must respond at least 1 μsec later than A and B. Such problems can be solved in a straightforward way with logic circuits.

Four *logic elements* are particularly important and useful: AND, OR, NAND, and NOR. The function of these four types can be explained with the aid of Fig. 4.16. The general logic element shown has three inputs and one output. Input and output pulses are of standard size (called 1); 0 denotes no pulse. An AND element produces no output (0) if only one or two pulses arrive. If, however, three pulses arrive within the resolving time (a few nsec), a standard output pulse (1) results. OR produces an output pulse if one or more input pulses arrive. NAND (NOT AND) and NOR (NOT OR) are the logical complements; they produce pulses whenever AND, respectively OR, would *not* produce a pulse. The functions of the four elements are summarized in Table 4.1. The element NOR

Input

A   B   C

Logic
element

Output

Fig. 4.16. Logic element.

Table 4.1 FUNCTION OF THE FOUR LOGIC ELEMENTS AND, OR, NAND, AND NOR. 1 denotes a standard pulse, 0 no pulse. The elements are symmetric in A, B, and C. Only typical cases are shown.

| *Input* | | | *Output* | | | |
A	B	C	AND	NAND	OR	NOR
1	1	1	1	0	1	0
1	1	0	0	1	1	0
1	0	0	0	1	1	0
0	0	0	0	1	0	1

requires one remark: It puts out a steady signal as long as there is no input pulse present; the signal disappears if at least one pulse arrives.

With the four elements shown in Table 4.1, the electronic systems for even highly sophisticated experiments can be assembled. In general, AND elements serve as coincidence circuits, NAND and NOR provide anti-coincidence signals, and OR is used as a buffer.

4.8 References

The literature on particle detectors is listed and reviewed in W. P. Trower, "Resource Letter PD-1 on Particle Detectors," *Am. J. Phys.* **38**, 7 (1970).

Particle detectors are discussed in the following books:

L. C. L. Yuan and C. S. Wu, eds., *Methods of Experimental Physics*, Vol. 5-A: *Nuclear Physics*, Academic Press, New York, 1961.

D. M. Ritson, ed., *Techniques of High Energy Physics*, Wiley-Interscience, New York, 1961.

K. Siegbahn, ed., *Alpha-, Beta- and Gamma-Ray Spectroscopy*, North-Holland, Amsterdam, 1965.

W. J. Price, *Nuclear Radiation Detection*, McGraw-Hill, New York, 1968.

Experiments in Modern Physics, Academic Press, New York, 1966, by A. Melissinos is a well-written textbook; it discusses experiments with detectors that can be used or constructed in an undergraduate laboratory.

More specialized books and articles include

Scintillation counters: J. B. Birks, *The Theory and Practice of Scintillation Counting*, Pergamon, Elmsford N.Y., 1964.

Semiconductor counters: A. J. Tavendale, *Ann. Rev. Nucl. Sci.* **17**, 73 (1967).

Bubble and spark chambers: R. P. Shutt, ed., *Bubble and Spark Chambers*, Academic Press, New York, 1967 (two volumes); O. C. Allkofer, *Spark Chambers*, Karl Thiemig, Munich, 1969, and G. Charpak, *Ann. Rev. Nucl. Sci.* **20**, 195 (1970).

There exist many good books on the application of statistics to experiments. An easy-to-read introduction is H. D. Young, *Statistical Treatment of Experimental Data*, McGraw-Hill, New York, 1962. The application of statistical considerations to subatomic physics is discussed in detail in R. D. Evans, *The Atomic Nucleus*, McGraw-Hill, New York, 1955, Chapters 26–28. A clear and very brief discussion of probability and statistics can be found in Jon Mathews and R. L. Walker, *Mathematical Methods of Physics*, Benjamin, Reading, Mass., 1964. Tables of the various distributions are collected in R. S. Burington and D. C. May Jr., *Handbook of Proba-*

bility and Statistics, Handbook Publishers McGraw-Hill, New York, 1969, and in W. H. Beyer, *CRC Handbook of Tables for Probability and Statistcs*, Chemical Rubber Co., Cleveland, 1968. A detailed treatment of statistical methods and their application to subatomic physics at an advanced level are given in D. Drijard, W. T. Eadie, F. E. James, M. G. W. Roos, and B. Sadoulet, *Statistical Methods in Experimental Physics*, North-Holland, Amsterdam, 1971.

Recent treatments of nuclear electronics are E. Kowalski, *Nuclear Electronics*, Springer, Berlin, 1970, and H. H. Chiang, *Basic Nuclear Electronics*, Wiley-Interscience, New York, 1969.

The digital aspects are also discussed in H. V. Malmstadt and C. G. Enke, *Digital Electronics for Scientists*, Benjamin, Reading, Mass., 1969. The general aspects of electronics are covered in H. V. Malmstadt, C. G. Enke, and E. C. Toren, *Electronics for Scientists*, Benjamin, Reading, Mass., 1962. A considerable amount of information can be gathered from the publications of the various manufacturers. The use of computers-on-line as data acquisition and evaluation systems is explained by S. J. Lindenbaum, *Ann. Rev. Nucl. Sci.* **16**, 619 (1966).

PROBLEMS

4.1. Find the circuit diagram for a photomultiplier. Discuss the importance and the choice of the components.

4.2. A proton with kinetic energy E_k impinges on a 5-cm-thick plastic scintillator. Sketch the light output as a function of E_k.

4.3. Three-MeV photons are counted by a 3 × 3 in. NaI(Tl) counter.
(a) Sketch the spectrum.
(b) Find the probability of observing the photon in the full-energy peak.

4.4. The 14-keV gamma rays from ^{57}Fe must be counted with a NaI(Tl) counter. Higher-energy gamma rays are a nuisance. Find the optimum thickness of the NaI(Tl) crystal.

4.5. Compute and draw the Poisson distribution for $\bar{n} = 1$ and $\bar{n} = 100$.

4.6. Sketch the derivation of Eq. (4.3). Verify Eq. (4.5).

4.7. Compute the variance of $P(n)$ in Eq. (4.8).

4.8. Verify that Eq. (4.8) is the limiting case of a Poisson distribution.

4.9. Prove Eq. (4.9).

4.10. For the Poisson distribution, compare

$$\frac{P(2\bar{n})}{P(\bar{n})} \quad \text{for } \bar{n} = 1, 3, 10, 100.$$

4.11. A scintillation counter used underground counts, on the average, eight muons/hr. An experiment is run for 10^3 hr, and counts are recorded every hr. How often do you expect to find $n = 2, 4, 7, 8, 16$ counts in the records?

4.12. Consider a germanium counter. Discuss the processes in more detail than in the text. In particular, answer the questions
(a) Why does the major part of the counter have to be depleted?
(b) Why is it not possible to simply use metal foils on both sides to collect the charge?
(c) How big a current pulse can be expected for a 100-keV photon?
(d) What limits the low-energy range of such a counter?

4.13. Compute the efficiency of a 1-cm-thick germanium counter for photons of

(a) 100 keV.

(b) 1.3 MeV.

4.14. Sketch the construction of a large bubble chamber.

4.15. Consider the 12-ft Argonne bubble chamber. What is the highest-energy proton that will stop in the chamber? Assume that the same chamber is filled with propane. Compute the range of the proton in this chamber. What energy proton can now be stopped?

4.16. Compute the magnetic energy stored in the Argonne 12-ft bubble chamber. From what height (in m) would an average car have to be dropped to equal this energy?

4.17. Discuss the principle of a streamer chamber. How is the voltage produced that is necessary to cause streamers?

4.18. What limits the speed with which a spark chamber can be triggered? Find typical delay times in the various components of the logical chain.

4.19. Use the elements listed in Table 4.1 to sketch the logic for the experiment of Fig. 4.15.

4.20. Sketch electronic circuits with which the four logic elements AND, OR, NAND, and NOR can be realized.

Particles and Nuclei

The situation is familiar. At a meeting we are introduced to some stranger. A few minutes later we realize with embarrassment that we have already forgotten his name. Only after being reintroduced a few times do we begin to fit the stranger into our catalog of people. The same phenomenon takes place when we encounter new concepts and new facts. At first they slip away rapidly, and only after grappling with them a number of times do we become familiar with them. The situation is particularly true with particles and nuclei. There are so many that at first they seem not to have sharp identities. So what is the difference between a muon and a pion?

In Part II we shall introduce many subatomic particles and describe some of their properties. Such a first introduction is not sufficient to give a clear picture, and we shall therefore return again to particle and nuclear characteristics in later chapters. Hopefully they will lose their "look-alike" status, and it will become clear, for instance, that muons and pions have less in common than man and microbe.

The first and most obvious questions are: What are particles? Can composite and elementary particles be distinguished? We shall try to explain why it is difficult to respond unambiguously to the apparently simple questions. Consider first the Franck-Hertz experiment[1] in which a gas, for instance helium or mercury, is studied by the passage of electrons through it. Below an energy of 4.9 eV in mercury vapor, the Hg atom behaves like an elementary particle. At an electron energy of 4.9 eV the first excited state of Hg is reached, and the mercury atom begins to reveal its structure. At 10.4 eV, an electron is knocked out; at 18.7 eV, a second electron is removed and it is apparent that electrons are atomic constituents. A similar situation exists with nuclei. At low electron energies, the electron cannot excite the nuclear levels, and the nucleus appears as an elementary particle. At higher electron energies, the nuclear levels become apparent, and it is possible to knock out nuclear constituents, protons and neutrons. The question is now shifted to the new actors, proton and neutron. Are they elementary? Protons and neutrons can also be probed with electrons. At energies of a few hundred MeV it becomes apparent that the nucleons, neutron and proton, are not point particles but have a "size" of the order of 1 fm. It also turns out that the nucleons have excited states, just as atoms and nuclei do. These excited states decay very rapidly, usually with the emission of pions. Where do these pions come from? Did they exist inside the nucleon, or were they created in the process of emission? Moreover, recent experiments with electrons of about 10-GeV energy indicate that subunits may exist inside nucleons.[2] We do not yet know if such subunits (sometimes called *partons*) truly exist, what they are, and if they have an independent existence outside their host particle. Hopefully, experiments with the NAL accelerator and with storage rings will elucidate these questions, but these preliminary remarks should make it clear that the problem of elementary particles is more involved than it appears at first. We shall return to these fundamental questions in Chapter 13, but in the following sections, we shall be more modest and present the experimental facts concerning subatomic particles. To describe and classify the particles, conserved quantities, such as energy, angular momentum, and charge are used. Conserved quantities and the corresponding symmetries will be treated in detail in Part III.

1. Eisberg, Section 5.5.
2. W. Kendall and W. Panofsky, *Sci. Amer.* **224**, 60 (June 1971).

5

The Subatomic Zoo

A conventional zoo is a collection of various animals, some familiar and some strange. The subatomic zoo also contains a great variety of inhabitants, and a number of questions concerning the catching, care, and feeding of these come to mind: (1) How can the particles be produced? (2) How can they be characterized and identified? (3) Can they be grouped in families? In the present chapter, we concentrate on the second question. In the first two sections, the properties that are essential for the characterization of the particles are introduced. Some members of the zoo already appear in these two sections as examples. In the later sections, the various families are described in more detail. Since there are so many animals in the subatomic zoo, some initial confusion in the mind of the reader is unavoidable. We hope, however, that the confusion will give way to order as the same particles appear again and again.

5.1 Mass and Spin. Fermions and Bosons

A first identification of a particle is usually made by measuring its *mass, m*. In principle, the mass can be found from Newton's law by observing the acceleration, **a**, in a force field, **F**:

$$m = \frac{\mathbf{F}}{\mathbf{a}}.$$ (5.1)

Equation (5.1) is not valid relativistically, but the correct generalization poses no problems. We only note that with mass we always mean *rest mass*. The actual determination of masses will be discussed in Section 5.3. The rest masses of subatomic particles vary over a wide range. Some, such as the photon and the neutrinos, have zero rest mass. The lightest massive particle is the electron with a mass, m_e, of about 10^{-27} g. Its rest energy, $E = mc^2$, is 0.51 MeV. The next heavier particle is the muon with a mass of about $200m_e$. From there on, the situation gets more complex, and many particles with strange and wonderful properties have masses that lie between about 270 times the electron mass to one or two orders of magnitude higher. Nuclei, which of course are also subatomic particles, start with the proton, the nucleus of the hydrogen atom, with a mass of about $2000m_e$. The heaviest known nucleus is about 260 times more massive than the proton. The masses (not counting zero) consequently vary by a factor of nearly a million. We shall return to the masses a few more times, and details will become clearer as more specific examples appear. However, just as it is impossible to understand chemistry without a thorough knowledge of the periodic table, it is difficult to obtain a clear picture of the subatomic world without an acquaintance with the main occupants of the subatomic zoo. It is therefore a good idea to look frequently at Tables A3–A5 in the Appendix.

A second property that is essential in classifying particles is the *spin* or *intrinsic angular momentum*. Spin is a purely quantum mechanical property, and it is not easy to grasp this concept at first. As introduction we therefore begin to discuss the *orbital* angular momentum which has a classical meaning. Classically, the orbital angular momentum of a particle with momentum **p** is defined by

$$\mathbf{L} = \mathbf{r} \times \mathbf{p},$$ (5.2)

where **r** is the radius vector connecting the center of mass of the particle to the point to which the angular momentum is referred. Classically, orbital angular momentum can take on any value. Quantum mechanically, the magnitude of **L** is restricted to certain values. Moreover, the angular momentum vector can assume only certain orientations with respect to a given direction. The fact that such a *spatial quantization* exists appears to violate intuition. However, the existence of spatial quantization is beautifully demonstrated in the Stern-Gerlach experiment,[1] and it follows logically from the postulates of quantum mechanics. In quantum mechanics, **p** is replaced by the operator $-i\hbar(\partial/\partial x, \partial/\partial y, \partial/\partial z) \equiv -i\hbar \nabla$ and the orbital angular momentum consequently also becomes an operator[2]

1. Eisberg, Section 11.3; Feynman Lectures, II-35-3.
2. Eisberg, Eqs. (10.51) and (10.52); Merzbacher, Chapter 9.

whose z component, for instance, is given by

$$L_z = -i\hbar\left(x\frac{\partial}{\partial y} - y\frac{\partial}{\partial x}\right) = -i\hbar\frac{\partial}{\partial \varphi}. \qquad (5.3)$$

Here φ is the azimuthal angle in polar coordinates. The wave function of a particle with definite angular momentum can then be chosen to be an eigenfunction of $\mathbf{L}^2$ and L_z:[3]

$$\mathbf{L}^2\psi_{\ell m} = \ell(\ell + 1)\hbar^2\psi_{\ell m}$$
$$L_z\psi_{\ell m} = m\hbar\psi_{\ell m}. \qquad (5.4)$$

The first equation states that the magnitude of the angular momentum is quantized and restricted to values $[\ell(\ell + 1)]^{1/2}\hbar$. The second equation states that the component of the angular momentum in a given direction, called z by general agreement, can assume only values $m\hbar$. The quantum numbers ℓ and m must be *integers*, and for a given value of ℓ, m can assume the $2\ell + 1$ values from $-\ell$ to $+\ell$. The spatial quantization is expressed in a *vector diagram*, shown in Fig. 5.1 for $\ell = 2$. The component along the arbitrarily chosen z direction can assume only the values shown.

We repeat again that the quantization of the orbital angular momentum Eq. (5.2) leads to integral values of ℓ and hence to *odd* values of $2\ell + 1$, the number of possible orientations. It was therefore a surprise when the alkali spectra showed unmistakable doublets. Two orientations demand $2\ell + 1 = 2$ or $\ell = \frac{1}{2}$. Many attempts were made before 1924 to explain this half-integer number. The first half of the correct solution was found by Pauli in 1924; he suggested that the electron possesses a classically nondescribable two-valuedness, but he did not associate a physical picture with this property. The second half of the solution was provided by Uhlenbeck and Goudsmit, who postulated a spinning electron. The two-valuedness then arises from the two different directions of rotation.

Of course, a way has to be found to incorporate the value $\frac{1}{2}$ into quantum mechanics. It is easy to see that the quantum mechanical operators that correspond to $\mathbf{L}$, Eq. (5.2), satisfy the *commutation relations*

$$L_xL_y - L_yL_x = i\hbar L_z$$
$$L_yL_z - L_zL_y = i\hbar L_x \qquad (5.5)$$
$$L_zL_x - L_xL_z = i\hbar L_y.$$

It is postulated that the commutation relations, Eq. (5.5), are more fundamental than the classical definition, Eq. (5.2). To express this fact, the symbol $\mathbf{L}$ is reserved for the orbital angular momentum, and a symbol $\mathbf{J}$ is introduced that stands for any angular momentum. $\mathbf{J}$ is assumed to

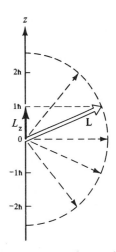

Fig. 5.1. Vector diagram for an angular momentum with quantum number $\ell = 2$, $m = 1$. The other possible orientations are indicated by dashed lines.

---

3. Some confusion can arise from the usual convention that classical quantities (e.g., $\mathbf{L}$) and the corresponding quantum mechanical operators (e.g., $\mathbf{L}$) are denoted by the same symbol. Moreover, the quantum numbers are often also denoted by similar symbols (ℓ or L). We follow this convention because most books and papers use it. After some initial bewilderment, the meaning of all symbols should become clear from the context. Occasionally we use the subscript *op* for quantum mechanical operators.

satisfy the commutation relations

$$J_x J_y - J_y J_x = i\hbar J_z$$
$$J_y J_z - J_z J_y = i\hbar J_x \qquad (5.6)$$
$$J_z J_x - J_x J_z = i\hbar J_y.$$

The consequences of Eq. (5.6) can be explored by using algebraic techniques.[4] The result is a vindication of Pauli's and of Goudsmit and Uhlenbeck's proposals. The operator **J** satisfies eigenvalue equations analogous to the ones for the orbital operator, Eq. (5.4):

$$J^2 \psi_{JM} = J(J + 1)\hbar^2 \psi_{JM} \qquad (5.7)$$
$$J_z \psi_{JM} = M\hbar \psi_{JM}. \qquad (5.8)$$

However, the allowed values of J are not only integers but also half-integers:

$$J = 0, \tfrac{1}{2}, 1, \tfrac{3}{2}, 2, \ldots. \qquad (5.9)$$

For each value of J, M can assume the $2J + 1$ values from $-J$ to $+J$.

Equations (5.7)–(5.9) are valid for any quantum mechanical system. As for any angular momentum, the particular value of J depends not only on the system but also on the reference point to which the angular momentum is referred. Now we return to *particles*. It turns out that each particle has an *intrinsic angular momentum*, usually called *spin*. Spin cannot be expressed in terms of the classical position and momenta coordinates, as in Eq. (5.2), and it has no analog in classical mechanics. Spin is often pictured by assuming the particle to be a small fast-spinning top. However, for any acceptable radius of the particle the velocity at the surface of the particle then exceeds the velocity of light, and the picture therefore is not really tenable. In addition, even particles with zero rest mass, such as the photon and the neutrino, possess a spin. The existence of spin has to be accepted as a fact. In the rest frame of the particle, any orbital contribution to the total angular momentum disappears, and the spin is the angular momentum in the rest frame. It is an immutable characteristic of a particle. The spin operator is denoted by **J** or by **S**;[5] it satisfies the eigenvalue equations (5.7) and (5.8). The quantum number J is a constant and characterizes the particle, while the quantum number M describes the orientation of the particle in space and depends on the choice of the reference axis.

How can J be determined experimentally? For a macroscopic system, the classical angular momentum can be measured. For a particle such a measurement is not feasible. However, if we succeed in determining the number of possible orientations in space, the spin quantum number J, usually just called *the spin*, follows because there are $2J + 1$ possible orientations.

---

4. A clear and concise derivation is given in Messiah, Chapter XIII.

5. *S* will later also be used for *strangeness*, and therefore *S* does not always denote the spin quantum number.

We have noted above that integer J values occur in connection with orbital angular momentum, which has a classical limit, but that half-integral values have no classical counterpart. As we shall see soon, particles with integer and half-integer spins exist. Examples for the integer class are the photon and the pion, whereas electrons, neutrinos, muons, and nucleons have spin $\frac{1}{2}$. Does the difference between integer and half-integer values express itself in some profound way? It indeed does, and the two classes of particles behave very differently. The difference becomes apparent when the properties of wave functions are studied. Consider a system of two *identical* particles, denoted by 1 and 2. The particles have the same spin J, but their orientation, given by $J_z^{(i)}$, can be different. The wave function of the system is written as

$$\psi(\mathbf{x}^{(1)}, J_z^{(1)}; \mathbf{x}^{(2)}, J_z^{(2)}) \equiv \psi(1, 2).$$

If the two particles are interchanged, the wave function becomes $\psi(2,1)$. It is a remarkable fact of nature that all wave functions for identical particles are either symmetric or antisymmetric under the interchange $1 \rightleftharpoons 2$:

$$\begin{aligned} \psi(1, 2) &= +\psi(2, 1), \quad \text{symmetric} \\ \psi(1, 2) &= -\psi(2, 1), \quad \text{antisymmetric.} \end{aligned} \tag{5.10}$$

Complete symmetry or antisymmetry under interchange of any two particles is easily extended to n identical particles.[6]

There exists a profound connection between *spin and symmetry* that was first noted by Pauli and that was proved by him using relativistic quantum field theory: The wave function of a system of n identical particles with half-integer spin, called *fermions*, changes sign if any two particles are interchanged. The wave function of a system of n identical particles with integer spin, called *bosons*, remains unchanged under the interchange of any two particles. The spin-symmetry relation is summarized in Table 5.1.

Table 5.1 BOSONS AND FERMIONS

Spin J	Particles	Behavior of wave function under interchange of any two identical particles
Integer	Bosons	Symmetric
Half-integer	Fermions	Antisymmetric

The connection between spin and symmetry leads to the *Pauli exclusion principle*. Assume that two particles have exactly the same quantum numbers. The two particles are then said to be in the same *state*. An interchange $1 \rightleftharpoons 2$ will leave the wave function unchanged. However, if the two particles are fermions, the wave function changes sign, and it consequently

6. Park, Chapter 11.

must vanish. The exclusion principle hence states that one quantum mechanical state can be occupied by only one *fermion*.[7] The principle is extremely important in all of subatomic physics.

5.2 Electric Charge and Magnetic Dipole Moment

Many particles possess *electric charges*. In an external electromagnetic field, the force on a particle of charge q will be given by Eq. (2.13),

$$\mathbf{F} = q\left(\boldsymbol{\varepsilon} + \frac{1}{c}\mathbf{v} \times \mathbf{B}\right). \tag{5.11}$$

The deflection of the particle in a purely electric field $\boldsymbol{\varepsilon}$ determines q/m. If m is known, q can be determined. Historically, progress went the inverse way: The electron charge was determined by Millikan in his oil drop experiment. With q and q/m known, the electron mass was found.

The *total charge* of a subatomic particle determines its interaction with $\boldsymbol{\varepsilon}$ and $\mathbf{B}$, as expressed by the Lorentz equation (5.11). It is a remarkable and not understood observation that the charge always appears in integer multiples of the elementary quantum e. Because of this fact, the total charge gives little information about the structure of a subatomic system. Other electromagnetic properties, however, do so, and the most prominent is the *magnetic dipole moment*. With a bad conscience (because we know that it is not really correct), we picture an elementary particle as a spinning body (Fig. 5.2).

If electric charges are distributed throughout the particle, they will spin also and give rise to current loops, which produce a magnetic dipole moment, $\boldsymbol{\mu}$. How does such a current distribution interact with an external magnetic field $\mathbf{B}$? Classical electrodynamics shows that a current loop as in Fig. 5.3. leads to an energy

$$E_{\text{mag}} = -\boldsymbol{\mu}\cdot\mathbf{B}, \tag{5.12}$$

where the magnitude of the magnetic dipole moment $\boldsymbol{\mu}$ is, in Gaussian units, given by

$$\mu = \frac{1}{c}\,\text{current} \times \text{area}. \tag{5.13}$$

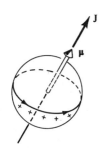

Fig. 5.2. Magnetic dipole moment. In a classical picture the spinning particle gives rise to electric current loops, which, in turn, produce a magnetic dipole moment.

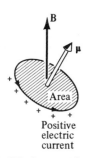

Fig. 5.3. A current loop gives rise to a magnetic moment $\boldsymbol{\mu}$. The direction of the magnetic moment is perpendicular to the plane bounded by the current.

7. Pauli describes the situation in the following words:

"If one pictures by boxes the nondegenerate states of an electron in an atom, the exclusion principle maintains that a box can contain no more than one electron. This, for example, makes the atoms much larger than if many electrons could be contained in the innermost shell. Quantum theory maintains that other particles such as photons or light particles show opposite behavior; that is, as many as possible fill the same box. One can call particles obeying the exclusion principle the 'antisocial' particles, while photons are 'social.' However, in both cases sociologists will envy the physicists on account of the simplifying assumption that all particles of the same type are exactly alike."

From W. Pauli, *Science* **103**, 213 (1946). Reprinted in *Collected Scientific Papers by Wolfgang Pauli* (R. Kronig and V. F. Weisskopf, eds.), Wiley-Interscience, New York, 1964.

The direction of $\boldsymbol{\mu}$ is perpendicular to the plane of the current loop; positive current and $\boldsymbol{\mu}$ form a right-handed screw.[8] A connection between magnetic moment and angular momentum is established by considering a particle of charge q moving with velocity $\mathbf{v}$ in a circular orbit of radius r (Fig. 5.4). The particle revolves $v/(2\pi r)$ times/sec and hence produces a current $qv/2\pi r$. With Eqs. (5.2) and (5.13), $\boldsymbol{\mu}$ and $\mathbf{L}$ are related by

$$\boldsymbol{\mu} = \frac{q}{2mc}\mathbf{L}. \tag{5.14}$$

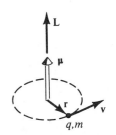

Fig. 5.4. A particle of mass m and charge q on a circular orbit produces a magnetic moment $\boldsymbol{\mu}$ and an orbital angular moment $\mathbf{L}$.

This result suffers from two defects. It has been derived by using classical physics, and it applies to a point particle moving in a circular orbit. Nevertheless, Eq. (5.14) exhibits two significant facts: $\boldsymbol{\mu}$ points in the direction of $\mathbf{L}$, and the ratio μ/L is given by $q/2mc$. These two facts indicate a way to define a quantum mechanical operator $\boldsymbol{\mu}$ for a particle with mass m and spin J. Even in this case, $\boldsymbol{\mu}$ should be parallel to $\mathbf{J}$ because there is no other preferred direction; the operators $\boldsymbol{\mu}$ and $\mathbf{J}$ are consequently related by

$$\boldsymbol{\mu} = \text{const. } \mathbf{J}.$$

According to Eq. (5.14), the constant has the dimension e/mc, and it is convenient to write const. $= g(e/2mc)$. The new constant g is then dimension-less, and the relation between μ and J becomes

$$\boldsymbol{\mu} = g\frac{e}{2mc}\mathbf{J}. \tag{5.15}$$

The constant g measures the deviation of the actual magnetic moment from the simple value $e/2mc$. Note that e and not q is used in Eq. (5.15). While q can be positive or negative, e is defined to be positive, and the sign of μ is given by the sign of the g factor. $\mathbf{J}$ has the same units as $\hbar$ so that $\mathbf{J}/\hbar$ is dimensionless. Equation (5.15) is therefore rewritten as

$$\boldsymbol{\mu} = g\mu_0 \frac{\mathbf{J}}{\hbar} \tag{5.16}$$

$$\mu_0 = \frac{e\hbar}{2mc}. \tag{5.17}$$

The constant μ_0 is called a *magneton*, and it is the unit in which magnetic moments are measured. Its value depends on the mass that is used. In atomic physics and in all problems involving electrons, m in Eq. (5.17) is taken to be the electron mass, and the unit is called the *Bohr magneton* (μ_B):

$$\mu_B = \frac{e\hbar}{2m_e c} = 0.5788 \times 10^{-14} \text{ MeV/G}. \tag{5.18}$$

In subatomic physics, magnetic moments are expressed in terms of *nuclear magnetons*, obtained from Eq. (5.17) with $m = m_p$:

$$\mu_N = \frac{e\hbar}{2m_p c} = 3.1525 \times 10^{-18} \text{ MeV/G}. \tag{5.19}$$

---

8. Jackson, Eqs. (5.60) and (5.73).

The nuclear magneton is about 2000 times smaller than the Bohr magneton.

Information about the structure of a particle is contained in the g factor. For a very large number of nuclear states and for a small number of particles, the g factor has been measured. It is the problem of theory to account for the observed values.

The energy levels of a particle with magnetic moment $\boldsymbol{\mu}$ in a magnetic field $\mathbf{B}$ are obtained from the Schrödinger equation,

$$H\psi = E\psi,$$

where the Hamiltonian H is assumed to have the form

$$H = H_0 + H_{\mathrm{mag}} = H_0 - \boldsymbol{\mu}\cdot\mathbf{B},$$

or, with Eq. (5.16),

$$H = H_0 - \frac{g\mu_0}{\hbar}\mathbf{J}\cdot\mathbf{B}. \tag{5.20}$$

The spin-independent Hamiltonian H_0 gives rise to an energy E_0: $H_0\psi = E_0\psi$. To find the energy values corresponding to the complete Hamiltonian, the z axis is conveniently chosen along the magnetic field so that $\mathbf{J}\cdot\mathbf{B} = J_z B_z \equiv J_z B$. With Eq. (5.8), the eigenvalues E of the Hamiltonian H are

$$E = E_0 - g\mu_0 MB. \tag{5.21}$$

Here, M assumes the $2J + 1$ values from $-J$ to $+J$. The corresponding *Zeeman* splitting is shown in Fig. 5.5 for a spin $J = \frac{3}{2}$.

Experimentally the splitting $\Delta E = g\mu_0 B$ between two Zeeman levels is determined. If B is known, g follows. Nevertheless the value quoted in the literature is usually not g but a quantity μ, defined by

$$\mu = g\mu_0 J. \tag{5.22}$$

Here J is the quantum number defined in Eq. (5.7). As can be seen from Fig. 5.5, $2\mu B$ is the total splitting of the Zeeman levels. [Quantum mechanically, μ is the expectation value of the operator Eq. (5.16) in the state $M = J$.] To determine μ, g and J have to be known. J can in principle be found from the Zeeman effect because the total number of levels is equal to $2J + 1$.

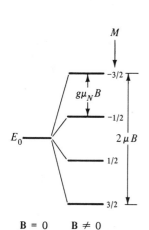

B = 0   **B ≠ 0**

Fig. 5.5. Zeeman splitting of the energy levels of a subatomic particle with spin J and g factor g in an external magnetic field **B**. **B** is along the z axis, $g > 0$.

5.3 Mass Measurements

The mass is the home address of a particle or nucleus, and it is therefore no surprise that there exist many methods for its measurement. We shall discuss only three here, and we have selected three that are different in character and apply to very different situations.

Subatomic particles are quantum systems, and nearly all of them possess excited states. Schematically the level diagrams appear as shown in Fig. 5.6. Even though the basic aspects are similar for nuclei and particles, units and notation differ. In the case of *nuclei*, the mass of the ground state is quoted not for the nucleus alone but for the neutral atom, including all electrons. The international unit for the *atomic mass* is one twelfth of the atomic mass of ^{12}C. This unit is called the *relative nuclidic mass* unit and is abbreviated *u*, for unified mass unit. In terms of grams and MeV, it is

$$1u = 1.66043 \times 10^{-24} \text{ g (mass)}$$
$$= 931.481 \text{ MeV (energy).} \qquad (5.23)$$

The masses of nuclear ground states are given in *u*. The excited nuclear states are not characterized by their masses but by their excitation energies (MeV above ground state). In the case of *particles*, rest energies are given, and they are quoted in MeV or GeV. This procedure is arbitrary but makes sense because in the nuclear case excitation energies are small compared to the rest energy of the ground state, whereas in the particle case excitation energies and ground-state energies are comparable.

After these preliminary remarks we turn to *mass spectroscopy*, the determination of nuclear masses. The first mass spectrometer was built in 1910 by J. J. Thomson, but the main advances are due to F. W. Aston. The components of Aston's mass spectrometer are shown in Fig. 5.7. Atoms are ionized in an ion source. The ions are accelerated by a voltage of 20–50 kV. The beam is collimated by slits and passes through an electric and a magnetic field. These fields are so chosen that ions of different velocity but with the same charge-to-mass ratio are focused on the photographic plate. The positions of the various ions on the photographic plate permit a determination of the relative masses to a very high degree of accuracy.

NUCLEUS

$(mc^2)_1$

$(mc^2)_0$

PARTICLE

Fig. 5.6. Level diagrams of nuclei and particles. The notation is explained in the text.

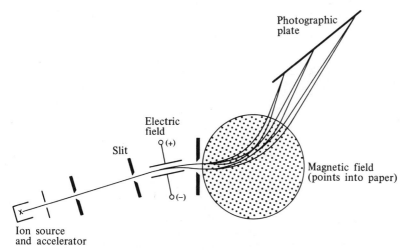

Fig. 5.7. Aston's mass spectrometer.

Mass spectroscopy works well for nuclei, but it is difficult (or impossible) to apply to most particles. In the mass spectrometer, all ions start with a very small (thermal) velocity and are accelerated in the same field. Their relative masses can therefore be determined very accurately. However, particles are produced in reactions, and their initial velocities are not accurately known. Moreover, some of the particles are neutral and cannot be deflected. Different approaches are necessary, and they are based on Eqs. (1.2) and (1.7):

$$E^2 = p^2c^2 + m^2c^4 \qquad (1.2)$$

$$\mathbf{p} = m\gamma\mathbf{v} \qquad (1.7)$$

$$\gamma = \frac{1}{(1-(v/c)^2)^{1/2}}. \qquad (1.6)$$

These relations show that the mass of a particle can be computed if momentum and energy or momentum and velocity are known. Many techniques are based on this fact, and the arrangement shown in Fig. 5.8 provides an example. A magnet selects particles with momentum **p**. Two scintillation counters, S_1 and S_2, record the passage of a particle. The signals from these two counters are fed to an oscilloscope, and the time delay between pulses S_2 and S_1 can be measured on the screen. With the distance between S_1 and S_2 known, the velocity can be computed. Together, momentum and velocity give the mass.

The method just discussed fails if the particle is neutral or if its lifetime is so short that neither momentum nor velocity can be measured. As an example of how it is even then possible to obtain a mass, we discuss the *invariant mass plot*. Consider the reaction

$$p\pi^- \longrightarrow n\pi^+\pi^-, \qquad (5.24)$$

taking place in a hydrogen bubble chamber. The reaction can proceed in two different ways, shown in Fig. 5.9. If it proceeds as in Fig. 5.9(a), the

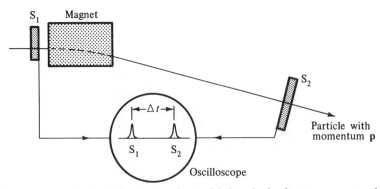

Fig. 5.8. Determination of the mass of a particle by selecting its momentum **p** and measuring its velocity v.

three particles in the final state will be created incoherently. It is, however, also possible that a neutron and a new particle, called a *neutral rho*, will be produced [Fig. 5.9(b)]. The neutral rho then decays into two pions. Is it possible to distinguish between the two cases? Yes, as we see now.

If the rho lives for a sufficiently long time, there will be a gap between the proton and the pion tracks. We shall see in Section 5.7 that the lifetime of the ρ^0 is about 6×10^{-24} sec. Even if the ρ^0 moves with the velocity of light, it will travel only about 1.5 fm during one mean life, about a factor 10^{10} less than needed for observation. How can the ρ^0 be detected and its mass be determined? To see how the trick is done, consider the energies and momenta involved (Fig. 5.10). Earlier, in Eq. (2.29), we defined the total or invariant mass of a system of particles. Applying this definition to the two pions and using the notation defined in Fig. 5.10, the invariant mass m_{12} of the two pions is

$$m_{12} = \frac{1}{c^2}[(E_1 + E_2)^2 - (\mathbf{p}_1 + \mathbf{p}_2)^2 c^2]^{1/2}. \qquad (5.25)$$

If a magnetic field is applied to the bubble chamber, the momenta of the two charged pions can be determined. The energy can be found from their range (Fig. 3.6) or their ionization. For every observed pion pair, the invariant mass m_{12} can then be computed from Eq. (5.25). If the reaction proceeds according to Fig. 5.9(a), with no correlation between the two pions and the neutron, they will share energy and momentum statistically. The number of pion pairs with a certain invariant mass, $N(m_{12})$, can be calculated in a straightforward way, and the result is called a *phase-space spectrum*. (Phase space will be discussed in Section 10.2.) It is sketched in Fig. 5.11. If, on the other hand, the reaction proceeds via the production of a ρ, energy and momentum conservation demand

$$E_\rho = E_1 + E_2$$
$$\mathbf{p}_\rho = \mathbf{p}_1 + \mathbf{p}_2. \qquad (5.26)$$

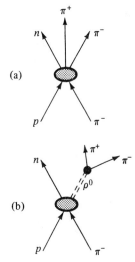

(a)

(b)

Fig. 5.9. The reaction $p\pi^- \longrightarrow n\pi^+\pi^-$ can proceed in two different ways: (a) The three particles in the final state can all be produced in one step, or (b) in the first step, two particles, n and ρ^0, are created. ρ^0 then decays into two pions.

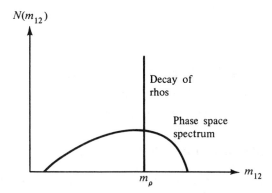

Fig. 5.11. Invariant mass spectrum if pion pairs are produced independently (phase space) or if they result from the decay of a rho of small decay width.

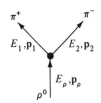

Fig. 5.10. Energies and momenta involved in the decay of the ρ^0.

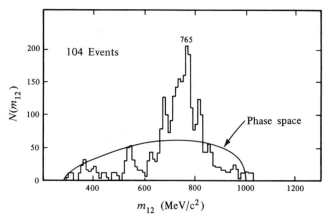

Fig. 5.12. Invariant mass spectrum of the two pions produced in the reaction $p\pi^{-} \longrightarrow n\pi^{+}\pi^{-}$. [After A. R. Erwin, R. March, W. D. Walker, and E. West, *Phys. Rev. Letters* **6**, 628 (1961).]

The mass of the rho is given by Eq. (1.2) as

$$m_{\rho} = \frac{1}{c^2}[E_{\rho}^2 - \mathbf{p}_{\rho}^2 c^2]^{1/2}$$

or, with Eqs. (5.25) and (5.26), as

$$m_{\rho} = m_{12}. \tag{5.27}$$

If the pions result from the decay of a particle, their invariant mass will be a constant and will be equal to the mass of the decaying particle. Figure 5.12 shows an early result, the invariant mass spectrum of pion pairs produced in the reaction Eq. (5.24) with pions of momentum 1.89 GeV/c. A broad peak at an invariant mass of 765 MeV/c^2 is unmistakable. The particle giving rise to this peak is called the rho. Even though it lives only about 6×10^{-24} sec, its existence is well established and its mass known.

The invariant mass spectrum is not restricted to particle physics; it has also been used in nuclear physics. Consider, for instance, the reaction

$$p + {}^{11}\text{B} \begin{cases} \nearrow 3\alpha \\ \searrow {}^{8}\text{Be} + \alpha \\ \phantom{\searrow {}^{8}\text{Be}} \hookrightarrow 2\alpha \end{cases}$$

Since ^{8}Be lives only for 2×10^{-16} sec before decaying into two alpha particles, three alphas are observed in either case. Nevertheless, the formation of ^{8}Be can be studied with the invariant mass spectrum.

5.4 A First Glance at the Subatomic Zoo

The techniques discussed so far have led to the discovery of well over 100 particles and a much larger number of nuclei. How can these be or-

dered in a meaningful way? A first separation is achieved by considering
the interactions that act on each particle. Four interactions are known to
exist, as pointed out in Section 1.1. In order of increasing strength they are
the gravitational, the weak, the electromagnetic, and the hadronic interaction. In principle, then, the four interactions can be used to classify
subatomic particles. However, the gravitational interaction is so weak that
it plays no role in present-day subatomic physics. For this reason we shall
restrict our attention to the three other interactions.

How can we discover which interactions govern the behavior of a
particular particle? First consider the electron. It clearly is subject to the
electromagnetic interaction because it carries an electric charge and is
deflected in electromagnetic fields. Does it participate in the weak interaction? The prototype of a weak process is the neutron decay,

$$n \longrightarrow pe^-\bar{\nu}.$$

Table A.3 in the Appendix indicates that this decay is very slow; the neutron lives on the average for about 15 min before decaying into a proton,
an electron, and a neutrino. If we call the neutron decay a weak decay, then
the electron participates in it. Does the electron interact hadronically? To
find out, nuclei are bombarded with electrons, and the behavior of the
scattered electrons is investigated. It turns out that the scattering can be
explained by invoking the electromagnetic force alone; the electron does
not interact hadronically. Decay and collision processes are also used to
investigate the interactions of all other particles. The result is summarized
in Table 5.2.

Subatomic particles are divided into three groups, the photon, leptons, and hadrons. The photon takes part in the electromagnetic interaction,
despite the fact that it has no electric charge. This fact follows, for instance,

Table 5.2 INTERACTIONS AND SUBATOMIC PARTICLES. The entries in the table are not always unambiguous. The photon,
for instance, behaves at high energies as if it could interact
both hadronically and weakly. Some aspects of this ambiguity
will be discussed in Section 10.8.

Particle	Type	Weak	Electromagnetic	Hadronic
Photon	Boson	No	Yes	No
Leptons				
Neutrino	Fermion	Yes	No	No
Electron	Fermion	Yes	Yes	No
Muon	Fermion	Yes	Yes	No
Hadrons				
Mesons	Bosons	Yes	Yes	Yes
Baryons	Fermions	Yes	Yes	Yes

from the emission of photons by accelerated charges [Eq. (2.20)]. Neutrino, electron, and muon are grouped together under the name *leptons*. All leptons have a weak interaction. The charged leptons, in addition, are also subject to the electromagnetic force. All other particles, including all nuclei, are hadrons, and their behavior is governed by the hadronic, the electromagnetic, and the weak interactions. The subdivision of hadrons into mesons and baryons will be discussed in detail in later chapters. One remarkable fact emerges when the masses of all particles are considered: There appears to exist a connection between mass and interaction. The lightest of the hadrons, the pion, is heavier than the heaviest of the leptons, the muon. In the following sections, we shall discuss the particles in more detail.

5.5 Photons

The particle properties of light invariably lead to some confusion. It is not possible to eliminate all confusion at an elementary level because a satisfactory treatment of photons requires quantum electrodynamics. However, a few remarks may at least make some of the important physical properties clearer. Consider an electromagnetic wave with circular frequency ω and with a reduced wavelength $\lambdabar = \lambda/2\pi$ moving in a direction given by the unit vector $\hat{k}$ (Fig. 5.13). Instead of giving $\hat{k}$ and $\lambdabar$ separately, a wave vector $\mathbf{k} = \hat{k}/\lambdabar$ is introduced. It points in the direction $\hat{k}$ and has a magnitude $1/\lambdabar$. According to Einstein, a monochromatic electromagnetic wave is composed of N monoenergetic photons, each with energy E and momentum $\mathbf{p}$, where

$$E = \hbar\omega, \qquad \mathbf{p} = \hbar\mathbf{k}. \tag{5.28}$$

The number of photons in the wave is such that the total energy $W = NE = N\hbar\omega$ is equal to the total energy in the electromagnetic wave.

Equation (5.28) shows that photons are endowed with energy and momentum. How about angular momentum? In 1909, Poynting predicted that

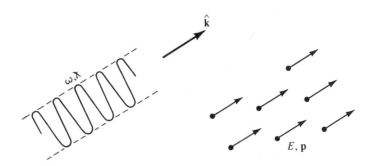

Fig. 5.13. An electromagnetic wave is composed of photons with energy E and momentum $\mathbf{p}$.

a circularly polarized electromagnetic wave carries angular momentum, and he proposed an experiment to verify this prediction: If a circularly polarized wave is absorbed, the angular momentum contained in the electromagnetic field is transferred to the absorber, which should then rotate. The first successful experiment was performed by Beth in 1935.[9] A modern variant, a *microwave motor*, is shown in Fig. 5.14. A circularly polarized microwave impinges on a suspended dipole at the end of a circular wave guide. Some energy and some angular momentum are absorbed by the dipole and it begins to rotate. The ratio of absorbed energy to absorbed angular momentum can easily be calculated,[10] and it is

$$\frac{\Delta E}{\Delta J_z} = \omega. \qquad (5.29)$$

This relation shows that the torque experiment is easier with microwaves than with optical light because the angular momentum transfer for a given energy transfer increases as $1/\omega$.

Equation (5.29) has been computed on the basis of classical electromagnetism. It can be translated into quantum mechanics by assuming that n photons, moving along the z axis with energy $\Delta E = n\hbar\omega$ and with angular momentum $\Delta J_z = nJ_z$, are absorbed. Equation (5.29) then yields

$$J_z = \hbar. \qquad (5.30)$$

The angular momentum carried by one photon is $\hbar$. This result can be restated by saying that the photon has *spin 1*.

Spin 1 for the photon is not surprising. Remember that a spin-1 particle has three independent orientations. To describe the three orientations, a quantity with three independent components is needed. A vector fills the bill, since it has three independent components. The electromagnetic field is a vector field: It is described by vectors $\mathcal{E}$ and $\mathbf{B}$ and corresponds to a vector particle—a particle with spin 1.[11] There is, however, a fly in the ointment. It is well known from classical optics that an electromagnetic wave has only *two* independent polarization states. Could it be that the photon has spin $\frac{1}{2}$? This possibility can be ruled out quickly. The connection between spin and symmetry, discussed in Section 5.1, would make a spin-$\frac{1}{2}$ photon a fermion, and it would obey the exclusion principle. Not more than one photon could be in one state; classical electromagnetic waves and television would be impossible. The solution to the apparent paradox comes not from quantum theory but from relativity. The photon has zero mass;

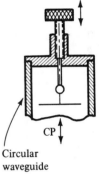

Circular waveguide

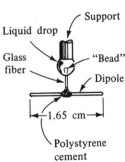

Fig. 5.14. A drop-suspended dipole exposed to a circularly polarized microwave rotates because the angular momentum of the electromagnetic field exerts a torque. [From P. J. Allen, *Am. J. Phys.* **34**, 1185 (1964).]

9. R. A. Beth, *Phys. Rev.* **50**, 115 (1936). Reprinted in *Quantum and Statistical Aspects of Light*, American Institute of Physics, New York, 1963.

10. See, for instance, R. T. Weidner and R. L. Sells, *Elementary Classical Physics*, Allyn and Bacon, Boston, 1965, Eq. (47.5).

11. The situation is actually somewhat more complicated. The correct description of the electromagnetic field is through the potential; the scalar and the vector potential together form a four vector, (A^0, A). It therefore appears at first as if this four vector corresponded to four degrees of freedom. However, the Lorentz subsidiary condition removes one degree and we are back to three.

it is light and moves with the velocity of light. There is no coordinate system in which the photon is at rest. The argument leading to Eq. (5.8) and to the $2J + 1$ possible orientations is, however, made in the rest system, and it breaks down for the photon. In fact, any massless particle can at most have two spin orientations, parallel or antiparallel to its momentum, regardless of its spin.[12] We can summarize the result of the previous arguments by saying that the free photon is a spin-1 particle that can have its spin either parallel or antiparallel to the direction of motion.[13] The two states are called right- and left-circularly polarized or states of positive and negative helicity, respectively.

5.6 Leptons

Electrons, muons, and neutrinos are called leptons. The leptons are at the same time the best studied and the most mysterious of all particles. On the one hand, their properties are extremely well measured and explored; the theoretical description of their behavior and, in particular, of their *g* factors is incredibly successful. On the other hand, they do not seem to fit into any known scheme. The muon appears to be completely out of place in this world: No evidence for a hadronic interaction of the muon has yet been found, but its mass is nearly as large as that of the pion. If mass and interaction are related, as we have stated earlier, why, then, is the muon so heavy? The sole raison d'être of the muon appears at present to be a reminder to physicists that they still do not comprehend subatomic physics.

Half of all leptons are listed in Table 5.3. The word *half* requires preliminary explanation. One of the best-documented facts of subatomic

Table 5.3 LEPTONS

Lepton	Spin	$(Mass)c^2$	Magnetic moment	Lifetime
ν_e	$\frac{1}{2}$	< 60 eV	0	Stable
ν_μ	$\frac{1}{2}$	< 1.6 MeV	0	Stable
e^-	$\frac{1}{2}$	0.5110041	$-1.001\ 159\ 6577$	Stable
		$\pm .0000016$ MeV	$\pm .000\ 000\ 0035 e\hbar/2m_e c$	$(> 2 \times 10^{21} y)$
μ^-	$\frac{1}{2}$	105.6599	$-1.001\ 166\ 16$	2.1994
		$\pm .0014$ MeV	$\pm .000\ 000\ 31\ e\hbar/2m_\mu c$	$\pm .0006$
				$\times 10^{-6}$ sec

12. E. P. Wigner, *Rev. Modern Phys.* **29**, 255 (1957).
13. Two words of warning are in order here. Single photons do not have to be eigenstates of momentum and angular momentum. It is possible to form linear combinations of eigenstates that correspond to single photons but do not have well-defined momentum and angular momentum. The second remark concerns the term *polarization vector*. In electromagnetism it is conventional to call the direction of the electric vector the polarization direction. *A photon with its spin along the momentum has its electric vector perpendicular to the momentum.*

physics is that each particle has an antiparticle, with opposite charge, but otherwise very similar properties. Each of the four leptons in Table 5.3 has an antilepton, and these antileptons have been observed. A more careful explanation of the idea of antiparticles will follow in Chapters 7 and 9.

The manner in which we have introduced the neutrino and the muon here is really terrible. It can be compared to introducing a master criminal, such as Professor Moriarty,[14] by listing his weight, height, and hair color rather than by telling of his feats. In reality, the neutrino behaved like a master criminal, and it escaped suspicion at first and then detection for a long time. The muon arrived disguised as a hadron and managed to confuse physicists for a considerable period before it was unmasked as an imposter. The introduction, as we have performed it, can be excused only by noting that excellent accounts of the histories of the neutrino and muon exist.[15]

5.7 Decays

Two facts compel us to digress and talk about decays before attacking the hadrons. The first is the comparison of muon and electron. The electron is stable, whereas the muon decays with a lifetime of 2.2 μsec. Does this fact indicate that the electron is more fundamental than the muon? The second fact emerges from comparing Figs. 5.11 and 5.12. In Fig. 5.11, the rho is indicated as a sharp line with mass m_ρ; the actually observed rho displays a wide *resonance* with a width of over 100 MeV/c^2. Is this width of experimental origin, or does it have fundamental significance? To answer the questions raised by the two observations we turn to a discussion of *decays*.

Consider an assembly of independent particles, each having a probability λ of decaying per unit time. The number decaying in a time dt is given by

$$dN = -\lambda N(t)\, dt, \tag{5.31}$$

where $N(t)$ is the number of particles present at time t. Integration yields the exponential decay law,

$$N(t) = N(0)e^{-\lambda t}. \tag{5.32}$$

Figure 5.15 shows $\log N(t)$ versus t. Half-life and mean life are indicated. In one half-life, one half of all atoms present decay. The mean life is the

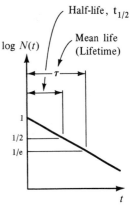

Fig. 5.15. Exponential decay.

14. A. C. Doyle, *The Complete Sherlock Holmes*, Doubleday, New York, 1953.

15. C. S. Wu, "The Neutrino," in *Theoretical Physics in the Twentieth Century* (M. Fierz and V. F. Weisskopf, eds.), Wiley-Interscience, New York, 1960; C. D. Anderson, *Am. J. Phys.* **29**, 825 (1961); C. N. Yang, *Elementary Particles*, Princeton University Press, Princeton, N.J., 1961.

average time a particle exists before it decays; it is connected to λ and $t_{1/2}$ by

$$\tau = \frac{1}{\lambda} = \frac{t_{1/2}}{\ln 2} = 1.44 t_{1/2}. \tag{5.33}$$

To relate the exponential decay to properties of the decaying state, the time dependence of the wave function of a particle at rest ($\mathbf{p} = 0$) is shown explicitly as

$$\psi(t) = \psi(0)e^{-iEt/\hbar}. \tag{5.34}$$

If the energy E of this state is *real*, the probability of finding the particle is *not* a function of time because

$$|\psi(t)|^2 = |\psi(0)|^2.$$

A particle described as a wave function of the type of Eq. (5.34) with real E does not decay. To introduce an exponential decay of a state described by $\psi(t)$, a small imaginary part is added to the energy,

$$E = E_0 - \tfrac{1}{2}i\Gamma, \tag{5.35}$$

where E_0 and Γ are real and where the factor $\frac{1}{2}$ is chosen for convenience. With Eq. (5.35), the probability becomes

$$|\psi(t)|^2 = |\psi(0)|^2 e^{-\Gamma t/\hbar}. \tag{5.36}$$

It agrees with the decay law [Eq. (5.32)] if

$$\Gamma = \lambda \hbar. \tag{5.37}$$

With Eqs. (5.34) and (5.35) the wave function of a decaying state is

$$\psi(t) = \psi(0)e^{-iE_0 t/\hbar}e^{-\Gamma t/2\hbar}. \tag{5.38}$$

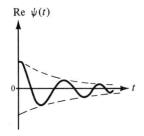

Re $\psi(t)$

Fig. 5.16. Real part of the wave function of a decaying state. It is assumed that the decaying state is formed at $t = 0$.

The real part of $\psi(t)$ is shown in Fig. 5.16 for positive times. The addition of a small imaginary part to the energy permits a description of an exponentially decaying state, but what does it mean? The energy is an observable; does an imaginary component make sense? To find out we note that $\psi(t)$ in Eq. (5.38) is a function of time. What is the probability that the emitted particle has an energy E? In other words, we would like to have the wave function as a function of energy rather than time. A change from $\psi(t)$ to $\psi(E)$ is effected by a Fourier transformation, a generalization of the ordinary Fourier expansion. A short and readable introduction is given by Mathews and Walker[16]; here we present only the essential equations. Consider a function $f(t)$. Under rather general conditions it can be expressed as an integral,

$$f(t) = (2\pi)^{-1/2} \int_{-\infty}^{+\infty} d\omega\, g(\omega)e^{-i\omega t}. \tag{5.39}$$

16. Mathews and Walker, Chapter 4. Short tables of Fourier transforms are given in the *Standard Mathematical Tables*, Chemical Rubber Co., Cleveland, Ohio. Extensive tables can be found in A. Erdelyi, W. Magnus, F. Oberhettinger, and F. G. Tricomi, *Tables of Integral Transforms*, McGraw-Hill, New York, 1954.

The expansion coefficient in the ordinary Fourier series has become a function $g(\omega)$. Inversion of Eq. (5.39) gives

$$g(\omega) = (2\pi)^{-1/2} \int_{-\infty}^{+\infty} dt\, f(t) e^{+i\omega t}. \tag{5.40}$$

The variables t and ω are chosen so that the product ωt is dimensionless; otherwise exp $i\omega t$ does not make sense. Thus t and ω can be time and frequency or coordinate and wave number. We now set $f(t)$ in Eq. (5.40) equal to $\psi(t)$, Eq. (5.38). If the decay starts at the time $t = 0$, the lower limit on the integral can be set equal to zero, and $g(\omega)$ becomes

$$g(\omega) = (2\pi)^{-1/2}\psi(0) \int_{0}^{\infty} dt\, e^{+i(\omega - E_0/\hbar)t} e^{-\Gamma t/2\hbar} \tag{5.41}$$

or

$$g(\omega) = \frac{\psi(0)}{(2\pi)^{1/2}} \frac{i\hbar}{(\hbar\omega - E_0) + i\Gamma/2}. \tag{5.42}$$

The function $g(\omega)$ is proportional to the probability amplitude that the frequency ω occurs in the Fourier expansion of $\psi(t)$. Since $E = \hbar\omega$, the probability density $P(E)$ of finding an energy E is also proportional to $|g(\omega)|^2 = g^*(\omega)g(\omega)$[17]:

$$P(E) = \text{const.}\; g^*(\omega)g(\omega) = \text{const.}\; \frac{\hbar^2}{2\pi} \frac{|\psi(0)|^2}{(E - E_0)^2 + \Gamma^2/4}.$$

The condition

$$\int_{-\infty}^{+\infty} P(E)\, dE = 1 \tag{5.43}$$

yields

$$\text{const.} = \frac{\Gamma}{\hbar^2\, |\psi(0)|^2},$$

and $P(E)$ finally becomes

$$P(E) = \frac{\Gamma}{2\pi} \frac{1}{(E - E_0)^2 + (\Gamma/2)^2}. \tag{5.44}$$

The energy of a decaying state is not sharp. The small imaginary part in Eq. (5.35) leads to a decay *and* it introduces a broadening of the state. The width acquired by the state because of its decay is called *natural line width*. The shape is called a Lorentzian or Breit-Wigner curve; it is sketched in Fig. 5.17. Γ turns out to be the full width at half maximum. With Eqs. (5.33) and (5.37), the product of lifetime and width becomes

$$\tau\Gamma = \hbar. \tag{5.45}$$

This relation can be interpreted as a Heisenberg uncertainty relation, $\Delta t\, \Delta E \geq \hbar$. To measure the energy of the state or particle to within an

17. For photons, the relation $E = \hbar\omega$ connects the energy to the frequency of the electromagnetic wave. For massive particles, it *defines* the frequency ω; the derivation leading to Eq. (5.44) remains correct because it is independent of the actual form of ω.

uncertainty $\Delta E = \Gamma$, a time $\Delta t = \tau$ is needed. Even if a longer time is used, the energy cannot be measured more accurately.

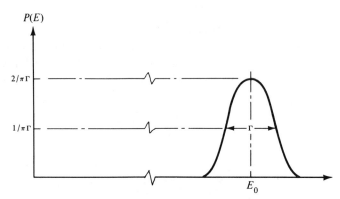

Fig. 5.17. Natural line shape of a decaying state. Γ is the full width at half maximum.

We can now answer the second question posed at the beginning of this section: The width observed in the decay of the rho is caused by decay; the instrumental width is much smaller. Table A 4 in the Appendix gives $\Gamma_\rho = 125$ MeV, and the lifetime becomes

$$\tau_\rho = \frac{\hbar}{\Gamma_\rho} = 6 \times 10^{-24} \text{ sec.}$$

We still have not answered the first question: Are decaying particles less fundamental than stable ones? To answer it, a few examples of unstable particles are listed in Table 5.4.

A number of facts emerge from Table 5.4:

1. No connection between simplicity and decay appears. Electron and

Table 5.4 SELECTED DECAYS. The entry under Class indicates the type of decay. W means weak, EM electromagnetic, and H hadronic.

Particle	Mass (MeV/c^2)	Main decays	Decay energy (MeV)	Lifetime (sec)	Class
μ	106	$e\nu\bar{\nu}$	105	2.2×10^{-6}	W
$\pi^\pm$	140	$\mu\nu$	34	2.6×10^{-8}	W
π^0	135	$\gamma\gamma$	135	7.6×10^{-17}	EM
η	549	$\gamma\gamma, \pi\pi\pi$	549	3×10^{-19}	EM
ρ	765	$\pi\pi$	485	6×10^{-24}	H
n	940	$pe^-\bar{\nu}$	0.8	0.93×10^3	W
Λ	1116	$p\pi^-, n\pi^0$	39	2.5×10^{-10}	W
Δ	1236	$N\pi$	159	6×10^{-24}	H
^{8}Be*	—	2α	3	6×10^{-22}	H

muon differ only in mass, yet the muon decays. The deuteron, a composite of neutron and proton, is not listed because it is stable, but the free neutron decays. The charged pions decay slowly, but the neutral one decays rapidly. The data suggest that a particle decays if it can and that it is stable only if there is no state of lower energy (mass) to which it is allowed to decay. Stability does not appear to be a criterion for *elementarity*.

2. Comparsion of particles with about the same decay energy shows that classes occur. We know that hadronic, electromagnetic, and weak forces exist and thus expect corresponding decays. Indeed, all three types show up. Detailed calculations are required to justify that the three interactions can give rise to decays with the listed lifetimes. Nevertheless, a very crude idea of typical lifetimes can be gained by comparing the delta (Δ), the neutral pion, and the lambda. These have decay energies between 40 and 160 MeV and decay into two particles. Approximate values for the corresponding lifetimes are

$$
\begin{array}{lll}
\text{hadronic decay } (\Delta) & 10^{-23} \text{ sec} & \\
\text{electromagnetic decay } (\pi^0) & 10^{-18} \text{ sec} & \textbf{(5.46)} \\
\text{weak decay } (\Lambda) & 10^{-10} \text{ sec.} &
\end{array}
$$

The ratios of these lifetimes give *very* approximately the ratios of strengths of the three forces. To obtain better measures of the relative strengths, the interactions must be studied in more detail, as will be done in Part IV.

3. The type of particle or quantum emitted is not always an indication of the interaction at work. Lambda and delta both decay into proton and pion, yet the delta decays about 10^{14} times faster. *Selection rules* must be involved, and it will be one of the tasks of later chapters to find these rules.

5.8 Mesons

In Table 5.2, hadrons are separated into mesons and baryons. We shall explain the difference between these two types of hadrons in more detail in Chapter 7, where a new quantum number, the *baryon number*, will be introduced. It is similar to the electric charge: Particles can have baryon numbers 0, ± 1, ± 2, ... The prototype of a baryon-number-1 particle is the nucleon. Like the electric charge, baryon number is "conserved," and a state with baryon number 1 can decay only to another state with baryon number 1. Mesons are hadrons with baryon number 0. All mesons have a transient existence and decay through one of the three interactions discussed in the previous section.

The first meson to appear in the zoo was the *pion*. Since its existence was predicted more than 10 years before it was found experimentally, it is worth explaining the basis of the prophecy. To do so, it is necessary to return to the photon and the electromagnetic interaction. Because of relativity, it is generally assumed that no interactions at a distance exist.[18] The electromagnetic force between two electrons, for instance, is assumed to be mediated by photons. Figure 5.18 explains the idea. One electron emits a photon which is absorbed by the other electron. The exchange of photons or *field quanta* gives rise to the electromagnetic interaction between the two charged particles, whether it occurs in a collision or in a bound state, such as positronium (e^-e^+ atom). The exchange process is best considered in the c.m. of the two colliding electrons. Since the collision is elastic, the energies of the electrons are unchanged so that $E'_1 = E_1$, $E'_2 = E_2$. Before the emission of the photon, the total energy is $E = E_1 + E_2$. After emission but before reabsorption of the quantum the total energy is given by $E = E_1 + E_2 + E_\gamma$, and energy is not conserved. Is such a violation allowed? Energy conservation can indeed be broken for a time Δt because of the Heisenberg uncertainty relation

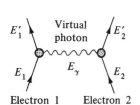

Fig. 5.18. Exchange of a photon between two electrons, 1 and 2. The virtual photon is emitted by one and absorbed by the other electron.

$$\Delta E \Delta t \geq \hbar. \tag{5.47}$$

Equation (5.47) states that the time Δt required to observe an energy to within the uncertainty ΔE must be greater than $\hbar/\Delta E$. Nonconservation of energy within an amount ΔE is therefore unobservable if it occurs within a time T given by

$$T \leq \frac{\hbar}{\Delta E}. \tag{5.48}$$

A photon of energy $\Delta E = \hbar\omega$ consequently cannot be observed if it exists for less than a time

$$T = \frac{\hbar}{\hbar\omega} = \frac{1}{\omega}. \tag{5.49}$$

Since the unobserved photon exists for less than the time T, it can travel at most a distance

$$r = cT = \frac{c}{\omega}. \tag{5.50}$$

The frequency ω can be arbitrarily small, and the distance over which a photon can transmit the electromagnetic interaction is arbitrarily large. Indeed, the Coulomb force has a distance dependence $1/r^2$ and presumably extends to infinity. Since the exchanged photon is not observed, it is called a *virtual photon*.

By 1934, it was known that the hadronic force is very strong and that it has a range of about 2 fm, but there was total ignorance as to what

---

18. In Newton's theory of gravitation it is assumed that the interaction between two bodies is instantaneous. A rapid acceleration of the sun, for instance, would affect the earth immediately and not after 8 min. This basic tenet is in conflict with the special theory of relativity which assumes that no signal can travel faster than the speed of light. This inconsistency led Einstein to his general theory of relativity. [S. Chandrasekhar, *Am. J. Phys.* **40**, 224 (1972).] In quantum theory a force that is transmitted with at most the speed of light is pictured as being caused by the exchange of quanta. Even the possible existence of particles with speed exceeding that of light (tachyons) does not change the argument. [O. M. Bilaniuk and E. C. G. Sudarshan, *Phys. Today*, **22**, 43 (May 1969); G. Feinberg, *Phys. Rev.* **159**, 1089 (1967).]

caused it. Yukawa, a Japanese theoretical physicist, then suggested in a brilliant paper that a "new sort of quantum" could be responsible.[19] Yukawa's arguments are more mathematical than we can present here, but the analogy to the virtual photon exchange permits an estimate of the mass m of the "new quantum," the pion. In Yukawa's approach, the force between two hadrons, for instance two neutrons, is mediated by an unobserved pion, as sketched in Fig. 5.19. The minimum energy of the virtual pion is given by $E = m_\pi c^2$ and its maximum velocity by c. With Eq. (5.48), the maximum distance that the virtual pion is allowed to travel by the uncertainty relation is given by

$$R \leq cT = \frac{\hbar}{m_\pi c} \approx 1.4 \text{ fm.} \tag{5.51}$$

The range is therefore at most equal to the Compton wavelength of the pion. Originally, of course, the argument was turned around, and the mass of the postulated hadronic quantum was estimated by Yukawa as 100 MeV/c^2.

Physicists were delighted when a particle with a mass of about 100 MeV/c^2 was found in 1938. Delight turned to dismay when it was realized that the newcomer, the muon, did not interact strongly with matter and hence could not be held responsible for the hadronic force. In 1947, the true Yukawa particle, the pion, was finally discovered in nuclear emulsions.[20] After 1947, more mesons kept turning up, and at present the list is long. Some of these new mesons live long enough to be studied by conventional techniques. Some decay so rapidly that the invariant-mass-spectra method, discussed in Section 5.3, had to be invented. A complete list of the known mesons is given in the Appendix; in Table 5.5, the hadronically stable mesons are listed.

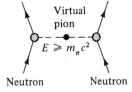

Fig. 5.19. Exchange of a virtual pion between two neutrons.

Table 5.5 HADRONICALLY STABLE MESONS. The mesons listed here decay either by weak or by electromagnetic processes.

Particle	Mass (MeV/c^2)	Charge (e)	Mean life (sec)
π^0	135.0	0	0.89×10^{-16}
$\pi^\pm$	139.6	$+, -$	2.60×10^{-8}
$K^\pm$	493.8	$+, -$	1.24×10^{-8}
K^0	497.8	0	Complicated
η	548.8	0	2.5×10^{-19}

19. H. Yukawa, *Proc. Math. Soc. Japan* **17**, 48 (1935). Reprinted in D. M. Brink, *Nuclear Forces*, Pergamon, Elmsford, N.Y., 1965. This book also contains a reprint of the articles by G. C. Wick on which our discussion of the connection between force range and quantum mass is based.

20. C. M. G. Lattes, H. Muirhead, G. P. S. Occhialini, and C. F. Powell, *Nature* **159**, 694 (1947).

5.9 Baryon Ground States

The spectrum of baryons is even richer than that of mesons. We begin the survey by considering *nuclear ground states*. By about 1920 it was well established that the electric charge Q and the mass M of a particular nuclear species are characterized by two integers, Z and A:

$$Q = Ze \tag{5.52}$$

$$M \approx A m_p. \tag{5.53}$$

The first relation was found to hold accurately, and the second one approximately. The nuclear charge number Z was determined by Rutherford's alpha-particle scattering, by X-ray scattering, and by the measurement of the energy of characteristic X rays. It was also found that Z is identical to the chemically determined *atomic number* of the corresponding element. The *mass number* A was extracted from mass spectroscopy, where it turned out that a given element can have nuclei with different values of A. The ground state of any nuclear species can, according to Eqs. (5.52) and (5.53), be characterized by two integers, A and Z. Before the discovery of the neutron, the interpretation of these facts was rather unclear. When the neutron was finally found by Chadwick in 1932,[21] everything fell into place: A nucleus (A, Z) is composed of Z protons and $N = A - Z$ neutrons; since neutrons and protons are about equally heavy, the total mass is approximately given by Eq. (5.53). The charge is entirely due to the protons so that Eq. (5.52) is also satisfied.

At this point, we can get some definitions out of the way: A *nuclide* is a particular nuclear species with a given number of protons and neutrons: *isotopes* are nuclides with the same number of protons, Z; *isotones* are nuclides with the same neutron number, N; and *isobars* are nuclides with the same total number of nucleons, A. A particular nuclide is written as (A, Z) or ^A_Z element. The alpha particle, for instance, is characterized by $(4, 2)$ or $^4_2$He or simply ^{4}He.

Stable nuclides, characterized by $N = A - Z$ and Z, are represented as small squares in an N-Z plot in Fig. 5.20. The plot indicates that stable nuclides exist only in a small band in the N-Z plane. The band starts off at 45° (equal proton and neutron numbers) and slowly veers toward neutron-rich nuclides. This behavior will provide a clue to an understanding of properties of the nuclear force.

Figure 5.20 contains only stable nuclides. In Section 5.7 we have pointed out that stability is not an essential criterion in considering hadrons. Unstable nuclear ground states therefore can also be added to the N-Z plot. We shall explore some properties of such an extended plot in Chapter 14.

21. J. Chadwick, *Nature* **129**, 312 (1932); *Proc. Roy. Soc.* (London) **A136**, 692 (1932).

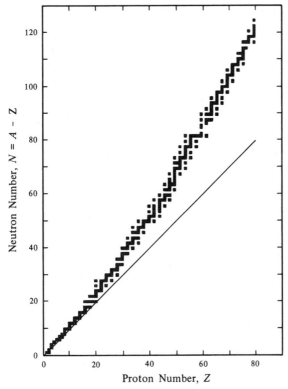

Fig. 5.20. Plot of the stable nuclides. Each stable nuclide is indicated as a square in this *N-Z* plot. The solid line would correspond to nuclides with equal proton and neutron numbers. (After D. L. Livesey, *Atomic and Nuclear Physics*, Blaisdell, 1966.)

At the mass number $A = 1$, nuclear and particle physics meet. The proton and the neutron, the two building blocks of all heavier nuclides, can either be considered the simplest nuclei or they can be called particles. It is a surprising fact that the two nucleons are not the only $A = 1$ hadrons. Other baryons with the mass number $A = 1$ exist; they are called hyperons.

As an example of the investigation of *hyperons*, we consider the production of the lambda. If negative pions of a few GeV of energy pass through a hydrogen bubble chamber, events such as the one shown in Fig. 5.21 are observed: The negative pion "disappears," and further downstream two *V*-like events appear. At first, the two *V*s seem to be very similar. However, when the energies and momenta of the four particles are determined (Section 5.3) it turns out that one *V* consists of two pions, and the other of a pion and a proton. Invariant mass plots, such as explained in Section 5.3, show that the particle giving rise to the two pions has a mass of about 500 MeV/c^2, while the particle decaying into proton and pion has a mass of 1116 MeV/c^2. The first particle, the neutral kaon, has mass number 0, and it has already been discussed in the previous section. The second particle is called *lambda*. (The name, of course, refers to the characteristic appearance of the tracks of the proton and the pion.)

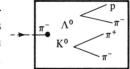

Fig. 5.21. Observation of the process $p\pi^- \longrightarrow \Lambda^0 K^0$ in a hydrogen bubble chamber.

The lifetime of each particle can be computed from the distance traveled in the bubble chamber and from its momentum. The complete reaction reads

$$p\pi^- \longrightarrow \Lambda^0 K^0$$

$$\begin{array}{l} \quad\quad\quad\quad \lfloor\!\rightarrow\!\pi^+\pi^- \\ \quad\quad\quad \lfloor\!\rightarrow\! p\pi^- \end{array} \quad\quad (5.54)$$

The lambda is not the only hyperon; a number of other hadronically stable particles of similar character have been found. These earn the designation *hadronically stable* because their lifetimes are much longer than 10^{-22} sec, and they are called baryons because they all ultimately decay to *one* proton or neutron. The hadronically stable baryons are listed in Table 5.6.

Table 5.6 HADRONICALLY STABLE BARYONS. More details concerning these particles are given in the Appendix.

Particle	Charge (e)	Mass (MeV/c²)	Mean life (sec)
N	+	938.3	$> 2 \times 10^{28}\,y$
	0	939.6	0.93×10^3
Λ	0	1115.6	2.5×10^{-10}
Σ	+	1189.4	0.80×10^{-10}
	0	1192.5	$< 1.0 \times 10^{-14}$
	−	1197.3	1.5×10^{-10}
Ξ	0	1314.7	3.0×10^{-10}
	−	1321.2	1.7×10^{-10}
Ω	−	1672.5	1×10^{-10}

5.10 Excited States and Resonances

In atomic physics, the development of concepts and theories is intimately linked with the exploration of excited states, in particular those of the hydrogen atom. The Balmer series, the Ritz combination principle, the Bohr theory, the Schrödinger equation, the Dirac equation, and the Lamb shift are all connected with the hydrogen spectrum. Without the simplicity *and* the richness of the hydrogen spectrum, progress would have been slower. In subatomic physics, the situation is more complex. The nuclear system that most closely resembles the hydrogen atom is the deuteron, a bound system consisting of a proton and a neutron. This system has only one bound state and consequently does not provide the richness of information that the hydrogen atom yielded. It is necessary to consider the excited states of more complicated systems, such as heavier nuclides. Moreover, excited states of baryons and mesons exist, and they must be studied in detail in the hope that they will provide clues to an understanding of hadronic physics.

An understanding of the features of excited hadronic states requires a knowledge of some results of quantum mechanics, and these can be discussed most easily by treating the square well. Consider a particle with mass m in a square well as shown in Fig. 5.22. It is straightforward to solve the Schrödinger equation for this problem and to find the allowed energy levels. First consider the case $E < 0$, where the numerical or graphical solution of the Schrödinger equation produces a number of bound states. *Bound* indicates that a particle in one of these levels will remain attached to the force center.

The Schrödinger equation for the square well is an eigenvalue equation, $H\psi = E_i\psi$, and the eigenvalues E_i represent sharp energy states. In reality, however, all states but the lowest one usually decay, for instance by photon emission. We have seen in Section 5.7 that decaying states possess a finite width and that the energy is composed of a large real and a small imaginary part, as in Eq. (5.35). For a bound state, the large real component is negative if the zero point of the energy is taken to be the value of the potential at infinity, as in Fig. 5.22.

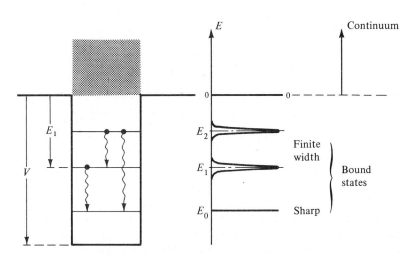

Fig. 5.22. Energy levels in a square well. The ground state is sharp. The excited states can decay to the ground state by photon emission, and they display a natural line width. States with positive energy form a continuum.

For positive energies, E can have any value. In other words, the spectrum forms a *continuum*. One would therefore guess that nothing interesting can happen in this region. This guess is false. To study the situation, scattering events have to be considered. In the one-dimensional case, as in Fig. 5.23, scattering is simple: A particle beam is assumed to impinge on the potential well from the left (Fig. 5.23). Classically, such a particle will pass unhindered over the well. In quantum mechanics the situation is more complicated. The Schrödinger equation can easily be solved, and it

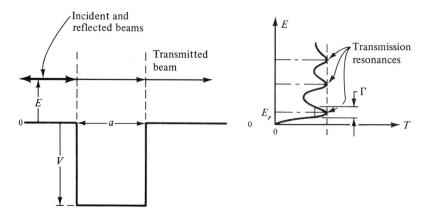

Fig. 5.23. Scattering of a particle with energy E from a one-dimensional potential well. Classically, all incident particles will be transmitted. Quantum mechanically, at small energies, the transmission coefficient T is unity only at certain energies. The appearance of *transmission resonances* in the behavior of the transmission as a function of particle energy E is shown at the right.

turns out that only a fraction of the incident beam is transmitted; another fraction is reflected at the barrier. The transmitted fraction, T, is given by[22]

$$\frac{1}{T} = 1 + \frac{V^2}{4E(E + |V|)} \sin^2 ka, \qquad (5.55)$$

where E is the kinetic energy of the incident particles, $V \, (< 0)$ the depth, and a the width of the potential well. The wave number k is given by

$$k^2 = \frac{2m}{\hbar^2}(E + |V|). \qquad (5.56)$$

Equations (5.55) and (5.56) demonstrate that the transmission coefficient T is unity only at certain energies. The behavior of T as a function of E is sketched in Fig. 5.23, where the appearance of *transmission resonances* is evident. The behavior of a particle with an energy E_r corresponding to maximum transmission can be investigated by using wave packets rather than plane waves to describe the incident beam. It turns out that the incident particle remains in the well region for a time that is much longer than that expected from classical mechanics.[23] The mean time spent in the well region, τ, and the width of the corresponding resonance, Γ, satisfy Eq. (5.45). Mathematically, the existence of a resonance at the energy E_r can again be described, in analogy to Eq. (5.35), by introducing a complex energy,

$$E = E_r - \tfrac{1}{2}i\Gamma.$$

Here E_r is positive, and Γ can be comparable to E_r.

22. Eisberg, Eq. (8.55); Park, Eq. (4.38).
23. Detailed discussions can be found in Merzbacher, Chapter 6, and in D. Bohm, *Quantum Theory*, Prentice-Hall, Englewood Cliffs, N.J., 1951, Chapters 11 and 12.

The appearance of a resonance in the continuum is not restricted to the simple one-dimensional case just discussed but is a more general phenomenon. To treat the problem with more relevance to actual situations, scattering of particles from a three-dimensional potential has to be studied. The basic ideas, however, are already contained in our simple example: Resonances can appear in the continuous energy spectrum, and they are characterized by the energy of their maximum, E_r, and by their width, Γ. Width and position together can be described by introducing a complex energy, $E = E_r - \frac{1}{2}i\Gamma$.

The use of a complex energy allows a classification of the energy levels of a quantum system. The classification is illustrated in Fig. 5.24. A point in the complex energy plane represents energy and width of a particular state. In addition to resonances, every positive energy corresponds to a permissible solution of the scattering problem. This fact is expressed in Fig. 5.24 by drawing the continuum along the positive energy axis.[24]

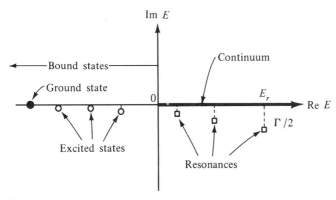

Fig. 5.24. Classification of the energy levels of a quantum system in the complex energy plane. Re $E = 0$ is determined by the potential at infinity. The widths Γ in actual resonances are usually much smaller than indicated here.

Resonances are characterized by unique quantum numbers; energy, width, and quantum numbers of the states appearing in a particular system depend on the constituents of the system and on the forces acting among them. It is the task of experimental subatomic physics to find the levels and determine their quantum numbers, and it is the goal of theoretical subatomic physics to explain the properties of the observed bound states and resonances in terms of models and forces.

5.11 Excited States of Baryons

The problem of finding all excited states of the baryons is probably hopeless. It is crucial, however, to find enough states to be able to discover

24. In a more advanced treatment of scattering, the bound states and the resonances appear as poles, and the continuum as a cut of the scattering matrix in the complex energy plane.

regularities, get clues to the construction of theories, and test the theories. Even this more restricted requirement is very difficult to fulfill in subatomic physics. A great deal of ingenuity and effort is expended on *nuclear and particle spectroscopy*, the study of nuclear and particle states. In the present section we shall give some examples of how excited states and resonances are found.

As a first example, we consider the nuclide ^{58}Fe, with a natural abundance of 0.31 %. Two ways in which the energy levels of ^{58}Fe have been investigated are sketched in Fig. 5.25. An accelerator, for instance, a Van de Graaff, produces a proton beam of well-defined energy. The beam is

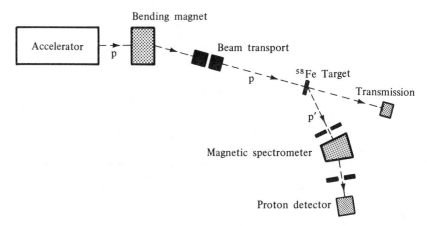

Fig. 5.25. Investigation of the energy levels of ^{58}Fe by transmission and by inelastic scattering.

momentum-analyzed and transported to a scattering chamber where it hits a thin target. The target consists of an iron foil that has been enriched in ^{58}Fe. The transmission through the foil can be studied as a function of the energy of the incident proton, or the scattered protons can be momentum-analyzed. Consider the second case, denoted by (p, p'). The notation (p, p') indicates that incoming and scattered particles are protons but that the scattered particle has a different energy in the c.m. The momentum and hence the energy of the scattered proton p' are determined in a magnetic spectrometer, i.e., a combination of bending magnet, slits, and detectors. If the kinetic energy of the incident proton is E_p and that of the scattered one is E'_p, the nucleus received an energy $E_p - E'_p$, and a level at this energy was excited. The experiment constitutes a nuclear Franck-Hertz effect. (A correction has to be applied because the ^{58}Fe* nucleus recoils, and the recoil energy must be subtracted from $E_p - E'_p$ in order to find the correct excitation energy.) A typical result of such an experiment is shown in Fig. 5.26. The appearance of many excited levels is unmistakable. The reaction (p, p') is only one of many that are used to excite and study nuclear levels. Other possibilities are (e, e'), (γ, γ'), (γ, n), (p, n), (p, γ), $(p, 2p)$, (d, p),

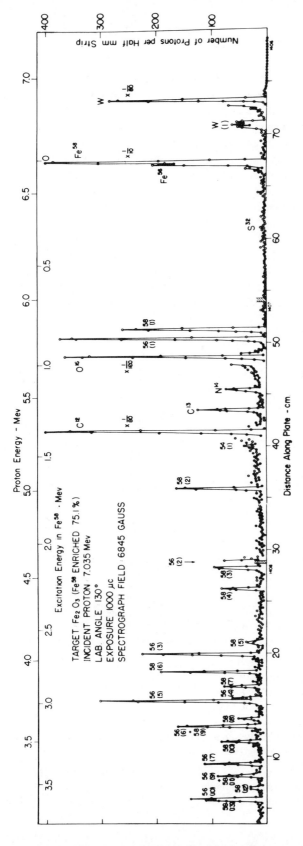

Fig. 5.26. Spectrum of protons scattered from enriched ^{58}Fe (75.1 %) target. The detector consists of photographic plates so that many lines can be observed simultaneously. [From A. Sperduto and W. W. Buechner, *Phys. Rev.* **134**, B142 (1964).] Since the target still contains some isotopes other than ^{58}Fe, additional lines appear. The iron lines are labeled by the mass number A.

(d, n), and so forth. Decays are also sources of information, and Fig. 4.8 gives an example of a partial gamma-ray spectrum. Data from a large variety of experiments are used to piece together a level diagram of a particular nuclide. For ^{58}Fe, the level diagram is shown in Fig. 5.30.

As the excitation energy is increased, the situation becomes more complex. In a simplified picture it can be discussed by referring to Fig. 5.23 with the essential aspects shown in Fig. 5.27. At an excitation energy of about 8 MeV, the top of the well is reached, and it becomes possible to eject a nucleon from the nucleus, for instance, by a reaction (γ, n), (γ, p), (e, ep), or (e, en). Just above the well, such processes are still not very likely, and most excited states will return to the nuclear ground state by the emission of one or more photons, because particle emission is inhibited by reflections from the nuclear surface (Fig. 5.23), angular momentum effects, and the small number of states available per unit energy (small phase space). Nevertheless, the states are no longer bound but are now classed

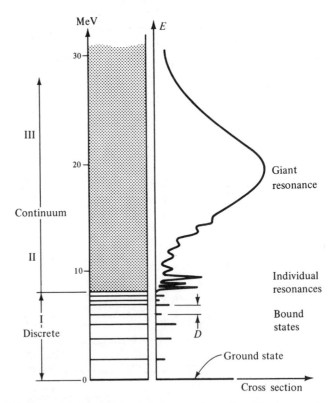

Fig. 5.27. Typical features of the excited states of a nucleus. The cross-section curve is idealized; it can be approximated by inelastic electron scattering or by studying the absorption of gamma rays as a function of gamma-ray energy. Three regions are distinguished: I, bound (discrete) states; II, individual resonances; and III, statistical region (overlapping resonances).

as resonances. In the idealized cross-section curve in Fig. 5.27, the individual resonances are shown in region II. As the energy is further increased, the resonances become more numerous and their widths increase. They begin to overlap, and the individual structure averages out. In region III, called the statistical region, the envelope of the overlapping individual resonances is measured, and it displays a prominent feature, called the giant resonance: At around 20-MeV excitation energy, the total cross section goes through a pronounced maximum. At much higher energies, the continuum loses all features.

The three regions shown in Fig. 5.27 are characterized by three numbers, the average level width, $\bar{\Gamma}$; the average distance between levels, $\bar{D}$; and the excitation energy, E. Typical values of these three quantities for the three regions are given in Table 5.7. Details vary widely from nuclide to nuclide, but the gross features remain.

Table 5.7 NUCLEAR ENERGY LEVEL CHARACTERISTICS FOR THE THREE REGIONS SHOWN IN FIG. 5.27. E is the excitation energy, $\bar{\Gamma}$ the average level width, and $\bar{D}$ the average level spacing.

Region	Characteristics	E (MeV)	Typical values $\bar{\Gamma}$ (eV)	$\bar{D}$ (eV)
I, bound states	$\bar{\Gamma} \ll \bar{D} \approx E$	1	10^{-3}	10^5
II, resonance region	$\bar{\Gamma} < \bar{D} \ll E$	8	1	10^2
III, statistical region	$\bar{D} \ll \bar{\Gamma} \ll E$	20	10^4	1

Exploration of the excited states of baryons with $A = 1$ is more difficult for three reasons: (1) No bound states exist and resonances are harder to study than bound states. (2) Most of the resonances decay by hadronic processes, their widths are large, and it is difficult to separate individual levels. (3) The only stable baryon that can be used as a target is the proton; liquid hydrogen targets are standard equipment in all high-energy laboratories. No isolated neutron targets exist. All other baryons (Table 5.6) have such a short lifetime that experiments of the type shown in Fig. 5.25 are not possible, and indirect methods must be used.

The first excited proton state was discovered by Fermi and collaborators in 1951. They measured the scattering of pions from protons and found that the cross-section increased rapidly with energy up to about 200-MeV pion kinetic energy and then leveled off or decreased again.[25] Brueckner suggested that this behavior could be interpreted as being due to a nucleon isobar (excited nucleon state) with spin $\frac{3}{2}$.[26] It took some

25. H. L. Anderson, E. Fermi, E. A. Long, and D. E. Nagle, *Phys. Rev.* **85**, 936 (1952).
26. K. A. Brueckner, *Phys. Rev.* **86**, 106 (1952).

more time and many more experiments before it became clear that the *Fermi resonance* is only the first of many excited states of the nucleon.

The investigation of excited proton states proceeds similarly to the study of excited nuclear states. High-energy particles, mainly electrons or pions, impinge on a hydrogen target, and the transmitted and the scattered beams are detected and analyzed. The behavior of the total cross section for pions on protons is given in Fig. 5.28. The appearance of resonances is evident. Since 1951, a great deal of effort has been expended to find such resonances and determine their quantum numbers. The pres-

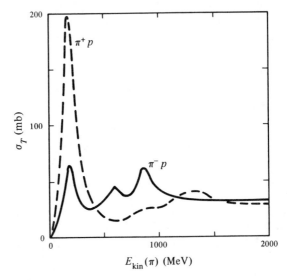

Fig. 5.28. Total cross section as a function of pion kinetic energy for the scattering of positive and negative pions from protons. (1 mb = 1 millibarn = 10^{-27} cm^2.)

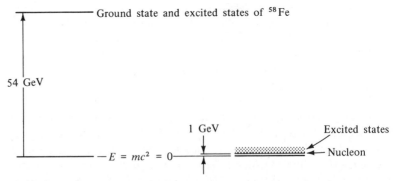

Fig. 5.29. Total rest energies of the states in ^{58}Fe and of the nucleon and its excited states. On the scale shown here, the excited states of the nuclide ^{58}Fe are so close to the ground state that they cannot be distinguished without magnification. A magnified spectrum is provided in Fig. 5.30.

ently known ones are listed in Table A5 in the Appendix. The Fermi resonance discussed above and shown as the first peak in Fig. 5.28 is called $\Delta(1236)$, where the number denotes the rest energy of the resonance in MeV.

In Figs. 5.29 and 5.30, we compare the energy spectra of the nuclide ^{58}Fe and of the nucleon. Figure 5.29 depicts the total masses (rest energies), while Fig. 5.30 presents the excitation spectra, namely the energies above

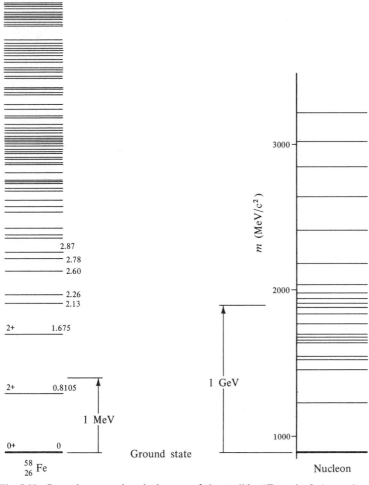

Fig. 5.30. Ground state and excited states of the nuclide ^{58}Fe and of the nucleon (neutron and proton). The region above the nuclear ground state in Fig. 5.29 has been enlarged by a factor of 10^4. The spectrum of the nucleon in Fig. 5.29 has been magnified 25 times. The nuclear states have widths of the order of eV or less and consequently can be observed separately. The excited particle states or resonances, on the other hand, have widths of the order of a few hundred MeV; they overlap and are often very difficult to find. It is likely that many additional levels exist.

the ground states. The figures make it clear that the nuclear excitation energies are very small compared to the rest energy of the ground state, whereas the particle excitation energies can be large compared to the rest energy of the ground state. The particle excitation energies are 2–3 orders of magnitude larger than nuclear excitation energies. Another difference exists between nuclear and particle excited states: Nuclei possess bound states *and* resonances, as indicated in Fig. 5.27. The excited particle states, on the other hand, are all resonances.

Finally, we note that we have treated nuclear and particle spectroscopy here extremely briefly; we have sketched only one way of finding the excited states. Many other ones exist. Moreover, the determination of the various quantum numbers of a state (spin, parity, charge, isospin, magnetic moment, quadrupole moment) can be an exceedingly difficult business. In fact, some of these quantum numbers can be measured only for very few states. The references in Section 5.12 describe most of the techniques and ideas of subatomic spectroscopy, but we shall not treat this topic further.

5.12 References

Information concerning the members of the subatomic zoo increases rapidly in scope, and any compilation becomes outdated nearly as soon as it is printed. The properties of elementary particles are periodically reviewed by the Particle Data Group in "Review of Particle Properties", and the compilation is published yearly, alternately in *Reviews of Modern Physics* and *Physics Letters B*.

The properties of nuclear levels are summarized, for instance, in C. M. Lederer, J. M. Hollander, and I. Perlman, *Table of Isotopes*, Wiley, New York, 1968.

More recent information can be found in the journals *Nuclear Data Tables* and *Nuclear Data Sheets* published by Academic Press.

Nuclear spectroscopy is reviewed in many places, and the following books provide additional information on most of the problems treated in the present chapter:

F. Ajzenberg-Selove, ed., *Nuclear Spectroscopy*, Academic Press, New York, 1960 (two volumes).

E. Segrè, ed., *Experimental Nuclear Physics*, Wiley, New York, 1953, 1959 (three volumes).

L. C. L. Yuan and C. S. Wu, eds., *Methods of Experimental Physics*, Vol. 5, Academic Press, New York, 1963.

K. Siegbahn, ed., *Alpha-, Beta-, and Gamma-Ray Spectroscopy*, North-Holland, Amsterdam, 1965 (two volumes).

G. Giacomelli, *Progr. Nucl. Phys.* **12**, Part 2, 77 (1970).

Particle spectroscopy is much less well covered in textbooks. The field moves so fast that nobody has yet had time to write and edit books that are the equivalent to the ones listed above for nuclear spectroscopy. The information must be collected from the original papers, reviews, and conference proceedings. Two recent reviews and two recent conference proceedings with relevant information are R. D. Tripp, "Spin and Parity Determination of Elementary Particles," *Ann. Rev. Nucl. Sci.* **15**, 325 (1965); I. Butterworth, "Boson Resonances," *Ann. Rev. Nucl. Sci.* **19**, 179 (1969); H. Filthuth, ed., *Proceedings of the Heidelberg International Conference on Elementary Particles*, North-Holland, Amsterdam, 1968; and J. Prentki and J. Steinberger, eds., *Proceedings of the 14th International Conference on High-Energy Physics*, CERN, Geneva, 1968.

The *photon concept*, treated very briefly in Section 5.5, often leads to long and heated arguments. An interesting brief discussion is given in M. O. Scully and M. Sargent III, "The Concept of the Photon," *Phys. Today*, **25**, 38 (March 1972). A more complete exposition can be found in M. Sargent III, M. O. Scully, and W. E. Lamb, Jr., *Quantum Electronics*, to be published.

PROBLEMS

5.1. Does a vanishing mass indicate that the corresponding particle has no gravitational interaction? If not, how can the force in a gravitational field be defined?

5.2. Discuss the Mössbauer experiment that indicates that photons falling in the earth's gravitational field gain energy. Why can such an experiment not be performed with optical photons? [R. V. Pound and J. L. Snider, *Phys. Rev.* **140B**, 788 (1965).]

5.3. Use Eq. (5.4) and the corresponding complete expressions for the operators L^2 and L_z to find the eigenvalues l and m for the functions

$$Y_0^0(\theta, \varphi) = (4\pi)^{-1/2}$$

$$Y_1^0(\theta, \varphi) = \frac{1}{2}\left(\frac{3}{\pi}\right)^{1/2}\cos\theta$$

$$Y_1^{\pm 1}(\theta, \varphi) = \mp\frac{1}{2}\left(\frac{3}{2\pi}\right)^{1/2}\sin\theta e^{\pm i\varphi}.$$

Here θ and φ are the angles defining spherical coordinates.

5.4. Verify Eq. (5.5)

5.5. Assume that electron and muon are uniform spheres with a radius of 0.1 fm. Compute the velocity at the surface caused by the rotation with spin $(\frac{3}{4})^{1/2}\hbar$.

5.6. Consider a system consisting of two identical particles and assume that the total wave function is of the form

$$\psi(\mathbf{x}_1, \mathbf{x}_2) = A\psi(\mathbf{x}_1)\varphi(\mathbf{x}_2) + B\psi(\mathbf{x}_2)\varphi(\mathbf{x}_1).$$

Find the values of A and B that make the total wave function normalized to unity and (a) symmetric, (b) antisymmetric, or (c) neither under interchange $1 \rightleftharpoons 2$.

5.7. Does a particle with zero electric charge necessarily have no interaction with an external electromagnetic field? Give an example of a neutral particle that does interact with an external electromagnetic field. Find an example for a particle that does not. Does a particle with electric charge necessarily interact with an external electromagnetic field?

5.8. A nucleus with a spin $J = 2$ and a g factor

of $g = -2$ is placed in a magnetic field of 1 MG.

(a) Where can such a field be found?

(b) Sketch the corresponding splitting of the energy levels. Label the levels with magnetic quantum numbers M. Find the value of the splitting between two adjacent levels in eV and in K.

5.9. Show that the magnetic dipole moment of a particle with spin $J = 0$ must vanish.

5.10. The discussion of the mass determination of nuclides in the text is greatly simplified. In actual experiments, the so-called *doublet method* is used. Discuss the basic idea underlying this method.

5.11. The determination of the mass of a particle often requires knowledge of its velocity. Discuss the principle of the Cerenkov counter. Show that the Cerenkov counter is a velocity-dependent detector.

5.12. How were the masses of the following particles determined:

(a) Muon.

(b) Charged pion.

(c) Neutral pion.

(d) Charged kaon.

(e) Charged sigma.

(f) Cascade particle (Ξ).

5.13. In Eq. (5.24), $\pi^- p \longrightarrow n\pi^+\pi^-$, the neutron in the final state escapes unobserved. The fact that the "missing" particle is a neutron is verified by using a *missing mass plot*: Assume a reaction of the form $a + b \longrightarrow 1 + 2 + 3 + \cdots$. Denote the total energy by $E_\alpha = E_a + E_b$ and the total momentum of the two colliding particles by $p_\alpha = p_a + p_b$. Similarly, denote the corresponding sums for all *observed* particles in the final state by E_β and p_β. The unobserved (neutral) particles then carry away the "missing" energy $E_m = E_\alpha - E_\beta$ and the "missing" momentum $p_m = p_\alpha - p_\beta$. The "missing mass" is defined by

$$m_m^2 c^4 = E_m^2 - p_m^2 c^2.$$

(a) Sketch a missing mass plot, i.e., a plot of the number of events expected with mass m_m versus m_m, if the only unobserved particle is a neutron.

(b) Repeat part (a) for the case where a neutron and a neutral pion escape.

(c) Find a missing mass plot in the literature.

5.14. Discuss the reaction $d\pi^+ \longrightarrow pp\pi^+\pi^-\pi^0$. The invariant mass spectrum of the three pions in the final state provides evidence for two short-lived mesons. Read the relevant literature and discuss how these mesons have been found.

5.15. Consider Eq. (5.24). Assume that the two pions do not form a resonant state (rho) but are emitted independently. Compute the upper and lower limit on the phase-space spectrum in Fig. 5.11.

5.16. Verify Eq. (5.29).

5.17. Discuss the determination of the present limit on the mass of

(a) The electron neutrino and

(b) The muon neutrino.

(c) How can the limit on the mass of the muon neutrino be improved?

5.18. How can the stability of electrons be measured? Try to design a simple experiment and estimate the limit on the lifetime that you expect to get from your experiment.

5.19. What was Professor Moriarty's profession? Where did he finally disappear?

5.20. Describe the experimental facts that led Pauli to postulate the existence of the neutrino.

5.21. ^{64}Cu decays with a branching ratio of 62% to ^{64}Ni and with a branching ratio of 38% by electron emission to ^{64}Zn. The overall half-life of ^{64}Cu is 12.8 hr. A spectrometer (magnet and scintillation counter) is adjusted so that only the electron decay to ^{64}Zn is observed. How long does it take until the intensity of this decay mode is reduced by a factor of 2?

5.22. Verify Eq. (5.33).

5.23. Find the Fourier transform of the function

$$f(x) = \begin{cases} 1, & |x| < a, \\ 0, & |x| > a. \end{cases}$$

5.24. Find the Fourier transform of

$$f(x) = \begin{cases} 0, & x < -1, \\ \frac{1}{2}, & -1 < x < 1, \\ 0, & x > 1. \end{cases}$$

5.25. Verify Eq. (5.42).

5.26. The level giving rise to the 14.4-keV gamma ray in ^{57}Fe decays with a half-life of 98 nsec. Compute Γ, the full width at half-height, in eV.

5.27. Verify Eq. (5.44).

5.28. Discuss methods to measure lifetimes of the order of
(a) $10^6 y$.
(b) 1 sec.
(c) 10^{-8} sec.
(d) 10^{-12} sec.
(e) 10^{-20} sec.

5.29. The rho is believed to contribute to the hadronic force between hadrons. Compute the range of this force.

5.30. What experiments would you perform to check if the muon is the quantum predicted by Yukawa? Compare your proposal to the actual evidence that led to the conclusion that the muon is not the Yukawa particle. [M. Conversi, E. Pancini, and O. Piccioni, *Phys. Rev.* **71**, 209 (1947); E. Fermi, E. Teller, and V.F. Weisskopf, *Phys. Rev.* **71**, 314 (1947).]

5.31. Does an electron bound in an atom satisfy Eq. (1.2)?

5.32. Discuss the following methods for determining the nuclear charge Z:
(a) X-ray scattering.
(b) Observation of characteristic X rays.

5.33. Before the discovery of the neutron, the nucleus was pictured as consisting of A protons and $A - Z$ electrons. Discuss arguments against this hypothesis.

5.34. At which pion kinetic energy does the process $p\pi^- \longrightarrow \Lambda^\circ K^\circ$ begin to occur? (i.e., determine the threshold for the reaction).

5.35. List two reactions that lead to the production of the Ξ^-; compute the corresponding threshold energies.

5.36.
(a) Derive Eq. (5.55).
(b) Sketch the transmission T as a function of E for a one-dimensional square well with the parameters $(2mV_0)^{1/2}a/\hbar = 100$.

5.37. Consider a one-dimensional potential well with parameters $a = 10$ fm and $V_0 = -100$ MeV. Find (numerically or graphically) the lowest two energy levels of a proton in this well.

5.38. Consider a well as shown in Fig. P.5.38.

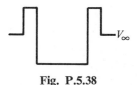

Fig. P.5.38

(a) Indicate the energy region where bound states exist.
(b) How will particles behave in the region above V_∞?

5.39. The experiment discussed in Section 5.11 demands the use of enriched ^{58}Fe.
(a) How is enriched iron prepared?
(b) What is the price of 1 mg of enriched ^{58}Fe?

5.40. In elastic and in inelastic scattering, some energy is given to the target particle in the form of recoil.
(a) Consider the reaction ^{58}Fe(p, p') ^{58}Fe*. Assume that the incident protons have an energy of 7 MeV, that the scattered proton is observed at 130° in the laboratory, and that excitation to the first excited state of ^{58}Fe is studied. What is the energy of the scattered proton?
(b) Assume that you try to excite the first nucleon resonance, $N^*(1236)$, by inelastic proton-proton scattering and that the primary proton kinetic energy is 1 GeV. What is the maximum scattering angle at which the scattered proton can be observed? At which energy will the peak in the inelastically scattered protons occur at this angle?

5.41. Discuss resonance fluorescence:
(a) What is the process?
(b) How can resonance fluorescence be observed in nuclei?
(c) What information can be obtained from it?

6

Structure of
Subatomic Particles

In Chapter 5 the members of the subatomic zoo have been classified according to interaction, symmetry, and mass. In the present chapter, we shall investigate some particles in more detail; in particular, we shall study the ground-state structure of some nuclides, of the charged leptons, and of the nucleons. What do we mean by *ground-state structure*? For *atoms*, the answer is familiar: Structure denotes the spatial distribution of the electrons, and it is described by the ground-state wave function. For the hydrogen atom, neglecting spin, the probability density $\rho(\mathbf{x})$ at point $\mathbf{x}$ is given by

$$\rho(\mathbf{x}) = \psi^*(\mathbf{x})\psi(\mathbf{x}), \tag{6.1}$$

where $\psi(\mathbf{x})$ is the electron wave function at $\mathbf{x}$. The electric charge density is given by $e\rho(\mathbf{x})$; the charge and the electron probability density are proportional to each other. Actually, the structure includes the excited states, and only if the wave functions of all possible atomic states are known is the structure completely determined. We shall, however, restrict the discussion to the ground state.

For *nuclei*, the concept of a charge distribution still makes sense, but charge and matter distribution are not identical. For *nucleons*, a new problem arises. The momenta needed to investigate the structure are so high that the nucleons, which are initially at rest, recoil with velocities that are close to the velocity of light. It is then very difficult to compute the nucleon charge distribution from the observed cross section. To avoid this problem, the nucleon structure is described in terms of *form factors*. While it takes some time to get used to this concept, it is closer to the experimental information than the charge distribution. For *leptons*, no structure is found at all, even at the smallest distances studied, less than 0.1 fm. They appear to be true pointlike Dirac particles.

6.1 The Approach: Elastic Scattering

Elastic scattering experiments have provided a great deal of insight into the structure of subatomic particles. How do such studies differ from the spectroscopic experiments discussed in Chapter 5? There is no sharp boundary, but the essential aspects can be described as follows. Both kinds of studies use an arrangement of the type shown in Fig. 5.25. In spectroscopy one angle is selected, and the spectrum of the scattered particles is explored at this angle. The energy levels of the nuclide under investigation can be taken from data similar to the ones given in Fig. 5.26. In structure (elastic form factor) experiments the detector looks only at the elastic peak. The intensity of the elastic peak is then determined as a function of the scattering angle. (Note that the energy at the elastic peak changes with scattering angle because of the recoil of the target particle; the detector must be adjusted correspondingly at each new angle.) The observed intensity is translated into a differential cross section, a quantity that we shall define in Section 6.2. From the cross section, the information concerning the structure of the target particle can be obtained.

In 1911, Rutherford observed the elastic scattering of alpha particles from nuclei; he found a small deviation from the scattering law derived for point nuclei and thus got a good idea concerning the size of the nucleus.[1] Many of the later investigations were also done with hadrons, mainly alpha particles or protons. These experiments, however, have one serious drawback: Nuclear size effects are intertwined with nuclear force effects, and the two must be disentangled. *Leptonic* probes do not suffer from this handicap, and the most detailed information concerning the nuclear charge distribution has been obtained with electrons and muons.

6.2 Cross Sections

Collisions are the most important processes used to study structure in subatomic physics. The behavior of a collision is usually expressed in terms of a cross section. To define cross section, a monoenergetic particle beam of well-defined energy is assumed to impinge on a target (Fig. 6.1). The flux F of the incident beam is *defined* as the number of particles crossing

1. E. Rutherford, *Phil. Mag.* **21**, 669 (1911).

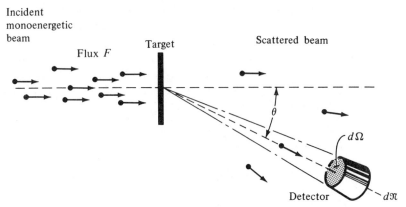

Incident
monoenergetic
beam

Flux F

Target

Scattered beam

$d\Omega$

θ

Detector

$d\mathfrak{N}$

Fig. 6.1. An incident monoenergetic beam is scattered by a target; the counter observing the scattered particles makes an angle θ with respect to the incident beam direction, subtends a solid angle $d\Omega$, and records $d\mathfrak{N}$ particles per unit time.

a unit area perpendicular to the beam per unit time. If the beam is uniform and contains n_i particles per unit volume, moving with velocity v with respect to the stationary target, the flux is given by

$$F = n_i v. \tag{6.2}$$

In most calculations, the number of incident particles is normalized to *one particle per volume V*. The number n_i is then equal to $1/V$. Particles scattered by the target are observed with a counter that detects all particles scattered by an angle θ into the solid angle $d\Omega$. The number $d\mathfrak{N}$ recorded per unit time is proportional to the incident flux F, the solid angle $d\Omega$, and the number N of independent scattering centers in the target that are intercepted by the beam[2]:

$$d\mathfrak{N} = FN\sigma(\theta)d\Omega. \tag{6.3}$$

The constant of proportionality is designated by $\sigma(\theta)$; it is called the *differential scattering cross section*, and we also write

$$\sigma(\theta) \, d\Omega = d\sigma(\theta) \quad \text{or} \quad \sigma(\theta) = \frac{d\sigma(\theta)}{d\Omega}. \tag{6.4}$$

The total number of particles scattered per unit time is obtained by integrating over all solid angles,

$$\mathfrak{N}_s = FN\sigma_{\text{tot}}, \tag{6.5}$$

where

$$\sigma_{\text{tot}} = \int \sigma(\theta) \, d\Omega \tag{6.6}$$

is called the *total scattering cross section*. Equation (6.5) shows that the

2. It is assumed here that each particle scatters at most once in the target and that each scattering center acts independently of each other one.

total cross section has the dimension of an area, and it is customary to quote subatomic cross sections in barns (b) or decimal fractions of barns, where

$$1 \text{ b} = 10^{-24} \text{ cm}^2 = 100 \text{ fm}^2. \tag{6.7}$$

The significance of σ_{tot} can be understood by computing the fraction of particles that are scattered. Figure 6.2 represents the target seen in the beam direction. The area a intercepted by the beam contains N scattering centers. The total number of incident particles per unit time is given by

$$\mathfrak{N}_{in} = Fa;$$

the total number of scattered particles is given by Eq. (6.5) so that the ratio of scattered to incident particle numbers is

$$\frac{\mathfrak{N}_s}{\mathfrak{N}_{in}} = \frac{N\sigma_{tot}}{a}. \tag{6.8}$$

The interpretation of this relation is straightforward: If no multiple scattering events occur, then the fraction of particles scattered is equal to the effective fraction of the total area occupied by scattering centers. $N\sigma_{tot}$ consequently must be the total area of all scattering centers and σ_{tot} the area of one scattering center. We stress that σ_{tot} is the area effective in scattering. It depends on the type and energy of the particles and is only occasionally equal to the actual geometrical area of the scattering center.

Finally, we note that if n is the number of scattering centers per unit volume, d the target thickness, and a the area intercepted by the beam, N is given by

$$N = and.$$

If the target consists of nuclei with atomic weight A and has a density ρ, n is given by

$$n = \frac{N_0\rho}{A}, \tag{6.9}$$

where $N_0 = 6.0222 \times 10^{23}$ mole^{-1} is Avogadro's number.

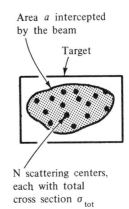

Area a intercepted by the beam

Target

N scattering centers, each with total cross section σ_{tot}

Fig. 6.2. An area a of the target is struck by the incident beam. The area a contains N scattering centers, each with cross section σ_{tot}.

6.3 Rutherford and Mott Scattering

The classical picture of elastic scattering of an alpha particle by the Coulomb field of a nucleus of charge Ze is shown in Fig. 6.3. This event is called *Rutherford scattering* if the nucleus is spinless; the alpha particle also has spin 0. The cross section for scattering of a spin-0 particle by a spinless nucleus can be computed classically or quantum mechanically, with the same result. The *Rutherford scattering formula* is one of the few equations that can be taken over into quantum mechanics without change, and this fact was a source of great pride to Rutherford.[3]

3. Rutherford scorned complicated theories and used to say that a theory is good only if it could be understood by a barmaid. (G. Gamow, *My World Line*, Viking, New York, 1970.)

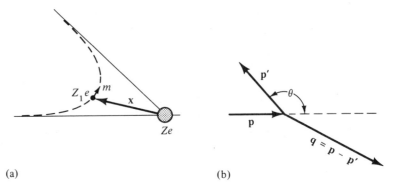

Fig. 6.3. Rutherford scattering. (a) Classical trajectory of a particle with charge $Z_1 e$ in the field of a heavy nucleus with charge Ze. (b) Representation of the collision in momentum space.

A fast way to derive the differential cross section for Rutherford scattering is based on the first Born approximation. In general, the differential cross section is written as

$$\frac{d\sigma}{d\Omega} = |f(\mathbf{q})|^2, \tag{6.10}$$

where $f(\mathbf{q})$ is called the scattering amplitude and where $\mathbf{q}$ is the momentum transfer,

$$\mathbf{q} = \mathbf{p} - \mathbf{p}'. \tag{6.11}$$

$\mathbf{p}$ is the momentum of the incident and $\mathbf{p}'$ that of the scattered particle. For elastic scattering, Fig. 6.3(b) shows that the magnitude of the momentum transfer is connected to the scattering angle θ by

$$q = 2p \sin \tfrac{1}{2}\theta. \tag{6.12}$$

In the first Born approximation it is assumed that the incident and the scattered particle can be described by plane waves. The scattering amplitude can then be written as[4]

$$f(\mathbf{q}) = -\frac{m}{2\pi\hbar^2} \int V(\mathbf{x}) e^{i\mathbf{q}\cdot\mathbf{x}/\hbar}\, d^3x. \tag{6.13}$$

$V(\mathbf{x})$ is the scattering potential. If it is spherically symmetric, integration over angles can be performed, and the scattering amplitude becomes, with $x = |\mathbf{x}|$,

$$f(\mathbf{q}^2) = -\frac{2m}{\hbar q} \int_0^\infty dx\, x \sin\left(\frac{qx}{\hbar}\right) V(x). \tag{6.14}$$

---

4. We introduce Eq. (6.10) and the Born approximation here without derivation. This omission will be rectified later, in Section 6.8 and, with a different approach, in Problem 10.3. The student who has not yet encountered Eqs. (6.10) and (6.13) should simply use them as a tool here and then study their derivation later. Derivations are also given in Eisberg, Section 15.3; Merzbacher, Section 11.4; and Park, Section 9.3.

Since f no longer depends on the direction of $\mathbf{q}$ but only on its magnitude, it is now written as $f(\mathbf{q}^2)$.

For Rutherford scattering, the potential $V(x)$ is the Coulomb potential.[5] Ordinarily, the Coulomb interaction between two charges $q_1 q_2$ at a distance x is written as

$$V(x) = \frac{q_1 q_2}{x}.$$

In the scattering experiment shown in Fig. 6.3, the nucleus is surrounded by its electron cloud, and the nuclear charge Ze is shielded. Shielding is taken into account by writing

$$V(x) = \frac{Z_1 Z e^2}{x} e^{-x/a}, \tag{6.15}$$

where a is a length characteristic of *atomic* dimension. With Eq. (6.15), the integral in Eq. (6.14) can be done, and the scattering amplitude becomes

$$f(\mathbf{q}^2) = -\frac{2m Z_1 Z e^2}{q^2 + (\hbar/a)^2}. \tag{6.16}$$

In all collisions exploring the structure of nuclei, the momentum transfer q is at least of the order of a few MeV/c, and the term $(\hbar/a)^2$ can be neglected completely. With Eqs. (6.16) and (6.10) the Rutherford differential cross section becomes

$$\left(\frac{d\sigma}{d\Omega}\right)_R = \frac{4m^2 (Z_1 Z e^2)^2}{q^4}. \tag{6.17}$$

The Rutherford scattering formula, Eq. (6.17), is based on a number of assumptions. The four most important ones are

1. The Born approximation.
2. The target particle is very heavy and does not take up energy (no recoil).
3. The incident and target particle have spin 0.
4. The incident and target particle have no structure; they are assumed to be point particles.

These four restrictions have to be justified or removed. We shall retain and justify the first two and partially remove the second two.

1. The Born approximation assumes that the incident and the outgoing particle can be described by plane waves. Such an assumption is allowed as long as

$$\frac{Z_1 Z e^2}{\hbar c} \ll 1. \tag{6.18}$$

If condition (6.18) is not satisfied, a more detailed calculation is necessary

---

5. In the original Rutherford experiments, the probing particles were α particles. These are hadrons, and if they get close to the nucleus, the hadronic force must also be taken into account. The experiments discussed here are performed with electrons, and no problems from hadronic forces arise.

(phase-shift analysis or higher Born approximations).[6] The essential physical aspects can, however, be understood by using the first Born approximation, and we shall not go beyond it.

2. Only elastic scattering is considered here. The target particle remains in its ground state, and it does not accept excitation energy. Moreover, it is assumed to be so heavy that its recoil energy can be neglected. However, as Fig. 6.3(b) shows, a very large momentum can be transferred to the target particle. At first the idea of a collision with large momentum transfer but with negligible energy transfer seems unrealistic. A simple experiment will convince an unbeliever that such a process is possible: Take a car or motorcycle and race straight into a concrete wall. If well constructed, the wall will take up the entire momentum but will accept very little energy. Most of the later discussion will be concerned with the scattering of electrons from nuclei and nucleons. In this case, restriction 2 is satisfied as long as the ratio of incident electron energy to target rest energy is small. At higher energy, the cross section can be corrected for nucleon or nuclear recoil in a straightforward manner. Essential results remain unaffected, and we shall therefore not treat the recoil corrections.

3. As just pointed out, most experiments to be discussed concern the scattering of electrons. In this case, the spin has to be taken into account. Scattering of spin-$\frac{1}{2}$ particles with charge $Z_1 = 1$ from spinless target particles has been treated by Mott, and the cross section for Mott scattering is[7]

$$\left(\frac{d\sigma}{d\Omega}\right)_{\text{Mott}} = 4(Ze^2)^2 \frac{E^2}{(qc)^4}\left(1 - \beta^2 \sin^2 \frac{\theta}{2}\right). \tag{6.19}$$

E is the energy of the incident electron and $v = \beta c$ its velocity. The term $\beta^2 \sin^2 \theta/2$ comes from the interaction of the electron's magnetic moment with the magnetic field of the target. In the rest frame of the target, this field vanishes, but in the electron's rest frame, it is present. The term is peculiar to spin $\frac{1}{2}$, it disappears as $\beta \longrightarrow 0$, and it is as important as the ordinary electric interaction as $\beta \longrightarrow 1$ since the magnetic and electric forces are then of equal strength. In the limit $\beta \longrightarrow 0$ ($E \longrightarrow mc^2$), the Mott cross section reduces to the Rutherford formula, Eq. (6.17).

4. The aim of the present chapter is the exploration of the structure of subatomic particles, and restriction 4 must consequently be removed. This task will be performed in the following section.

6.4 Form Factors

How is the cross section modified if the colliding particles possess extended structures? We shall treat leptons in Section 6.6 and find that they

6. D. R. Yennie, D. G. Ravenhall, and R. N. Wilson, *Phys. Rev.* **95**, 500 (1954).

7. A relatively easy-to-read derivation of Eq. (6.19) can be found in R. Hofstadter, *Ann. Rev. Nucl. Sci.* **7**, 231 (1958). A more sophisticated proof is given in J. D. Bjorken and S. D. Drell, *Relativistic Quantum Mechanics*, McGraw-Hill, New York, 1964, p. 106, or in J. J. Sakurai, *Advanced Quantum Mechanics*, Addison-Wesley, Reading, Mass., 1967, p. 193.

behave like point particles. This fact renders them ideal as probes, and the modification of Eq. (6.19) must take only the spatial distribution of the target particle into account. For simplicity, we shall assume here that the target particle possesses a spherically symmetric density distribution. It will then be shown below that the cross section for scattering of electrons from such a target is of the form

$$\frac{d\sigma}{d\Omega} = \left(\frac{d\sigma}{d\Omega}\right)_{\text{Mott}} |F(\mathbf{q}^2)|^2. \tag{6.20}$$

The multiplicative factor $F(\mathbf{q}^2)$ is called the *form factor*, and

$$\mathbf{q}^2 = (\mathbf{p} - \mathbf{p}')^2 \tag{6.21}$$

is the square of the momentum transfer.

Form factors play an increasingly important role in subatomic physics because they are the most convenient link between experimental observation and theoretical analysis. Equation (6.20) expresses the fact that the form factor is the direct result of a measurement. To discuss the theoretical side, consider a system that can be described by a wave function $\psi(\mathbf{r})$, which in turn can be found as the solution of a Schrödinger equation. For an object of charge Q, the charge density can be written as $Q\rho(\mathbf{r})$, where $\rho(\mathbf{r})$ is a normalized probability density, $\int d^3r\, \rho(\mathbf{r}) = 1$. It will be shown below that the form factor can be written as the Fourier transform of the probability density

$$F(\mathbf{q}^2) = \int d^3r\, \rho(\mathbf{r})\, e^{i\mathbf{q}\cdot\mathbf{r}/\hbar}. \tag{6.22}$$

The form factor at zero momentum transfer, $F(0)$, is usually normalized to be 1 for a charged particle; however for a neutral one $F(0) = 0$. The chain linking the experimentally observed cross section to the theoretical point of departure can thus be sketched as follows:

Experiment   Comparison   Theory

$\dfrac{d\sigma}{d\Omega} \longrightarrow |F(q^2)| \Longleftrightarrow F(q^2) \longleftarrow \rho(\mathbf{r}) \longleftarrow \psi(\mathbf{r}) \longleftarrow$ Schrödinger equation

In reality, individual steps can be more complicated than shown here, but the essential aspects of the chain remain.

We verify these introductory remarks by computing the scattering of a spinless electron from a finite spherically symmetric nucleus in the first Born approximation (Fig. 6.4). The scattering potential $V(x)$ in Eq. (6.13) at the position of the electron consists of contributions from the entire nucleus. Each volume element d^3r contains a charge $Ze\rho(r)\, d^3r$ and gives a contribution

$$dV(x) = -\frac{Ze^2}{z}\, e^{-z/a}\rho(r)\, d^3r,$$

so that

$$V(x) = -Ze^2 \int d^3r\, \rho(r)\, \frac{e^{-z/a}}{z}. \tag{6.23}$$

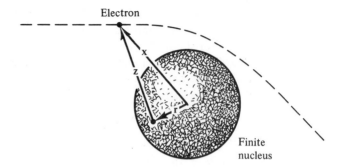

Fig. 6.4. Scattering of a point electron by a spinless nucleus with extended charge distribution.

The vector **z** from the volume element d^3r to the electron is shown in Fig. 6.4. Introducing $V(x)$ into Eq. (6.13) and using $\mathbf{x} = \mathbf{r} + \mathbf{z}$ yields

$$f(\mathbf{q}^2) = \frac{mZe^2}{2\pi\hbar^2} \int d^3r \, e^{i\mathbf{q}\cdot\mathbf{r}/\hbar} \, \rho(r) \int d^3x \, \frac{e^{-z/a}}{z} \, e^{i\mathbf{q}\cdot\mathbf{z}/\hbar}.$$

For fixed **r**, d^3x can be replaced by d^3z. The integral over d^3z is then the same as encountered in the evaluation of Eq. (6.16), and it gives

$$\int d^3z \, \frac{e^{-z/a}}{z} \, e^{i\mathbf{q}\cdot\mathbf{z}/\hbar} = \frac{4\pi\hbar}{\mathbf{q}^2 + (\hbar/a)^2} \longrightarrow \frac{4\pi\hbar}{\mathbf{q}^2}. \tag{6.24}$$

The integral over d^3r is the form factor, defined in Eq. (6.22), and the cross section $d\sigma/d\Omega = |f|^2$ becomes

$$\frac{d\sigma}{d\Omega} = \left(\frac{d\sigma}{d\Omega}\right)_R |F(\mathbf{q}^2)|^2. \tag{6.25}$$

The computation for electrons with spin follows the same lines; Eq. (6.20) is the correct generalization of Eq. (6.25). One remark is in order concerning the density $\rho(r)$. By Eq. (6.22), the density $\rho(r)$ has been defined in such a way that

$$\int \rho(r) \, d^3r = 1. \tag{6.26}$$

Equation (6.20) indicates how the form factor $|F(\mathbf{q}^2)|$ can be determined experimentally: The differential cross section is measured at a number of angles, the Mott cross section is computed, and the ratio gives $|F(\mathbf{q}^2)|$. The step from $F(\mathbf{q}^2)$ to $\rho(r)$ is less easy. In principle, Eq. (6.22) can be inverted and then reads

$$\rho(r) = \frac{1}{(2\pi)^3} \int d^3q \, F(\mathbf{q}^2) e^{-i\mathbf{q}\cdot\mathbf{r}/\hbar}. \tag{6.27}$$

Equations (6.22) and (6.27) are the three-dimensional generalization of Eqs. (5.39) and (5.40). The expression for $\rho(r)$ shows that the probability distribution is determined completely if $F(q^2)$ is known for all values of q^2. Experimentally, however, the maximum momentum transfer is limited by the available particle momentum. Moreover, as we shall see soon, the cross section becomes very small at large values of q^2, and it is then extremely difficult to determine $F(q^2)$. The practical approach is therefore different: Forms for $\rho(\mathbf{r})$ with a number of free parameters are assumed. The parameters are determined by computing $F(\mathbf{q}^2)$ with Eq. (6.22) and fitting the expression to the measured form factors.[8]

To provide some insight into the meaning of form factors and probability distributions, we shall connect $F(\mathbf{q}^2)$ to the nuclear radius and give examples of the relation between form factor and probability distribution. For $qR \ll 1$, where R is approximately the nuclear radius, the exponential in Eq. (6.22) can be expanded, and $F(q^2)$ becomes

$$F(\mathbf{q}^2) = 1 - \frac{1}{6\hbar^2} \mathbf{q}^2 \langle r^2 \rangle + \cdots \qquad (6.28)$$

where $\langle r^2 \rangle$ is defined by

$$\langle r^2 \rangle = \int d^3r \, r^2 \rho(r) \qquad (6.29)$$

and is called the mean-square radius. For small values of the momentum transfer, only the zeroth and second moments of the charge distribution are measured, and further details cannot be obtained.

If the probability density is Gaussian,

$$\rho(r) = \rho_0 e^{-(r/b)^2}, \qquad (6.30)$$

then the form factor can be computed easily, and it becomes

$$F(\mathbf{q}^2) = e^{-\mathbf{q}^2 b^2/4\hbar^2}, \qquad \langle r^2 \rangle = \tfrac{3}{2} b^2. \qquad (6.31)$$

If b becomes very small, the distribution approaches a point charge and the

Table 6.1 Probability Densities and Form Factors for Some One-Parameter Charge Distributions. (After R. Herman and R. Hofstadter, *High-Energy Electron Scattering Tables*, Stanford University Press, Stanford, Calif., 1960.)

Probability density, $\rho(r)$	Form factor, $F(\mathbf{q}^2)$
$\delta(r)$	1
$\rho_0 e^{-r/a}$	$(1 + \mathbf{q}^2 a^2/\hbar^2)^{-2}$
$\rho_0 e^{-(r/b)^2}$	$e^{-\mathbf{q}^2 b^2/4\hbar^2}$
$\left.\begin{array}{l}\rho_0, \ r \leq R \\ 0, \ \ r > R\end{array}\right\}$	$\dfrac{3[\sin(\lvert \mathbf{q} \rvert R/\hbar) - (\lvert \mathbf{q} \rvert R/\hbar) \cos(\lvert \mathbf{q} \rvert R/\hbar)]}{(\lvert \mathbf{q} \rvert R/\hbar)^3}$

8. One famous problem is apparent from the chain shown after Eq. (6.22). Experimentally, the absolute square of the form factor is obtained and not the form factor. The same problem appears in X-ray structure determinations. To get more information on the form factor, interference effects must be studied. In X-ray investigations of large molecules, interference is produced by substituting a heavy atom, for instance, gold, into the large molecule, and the resultant change of the X-ray pattern is observed. What can be used in subatomic physics?

form factor tends toward unity. This limiting case is the point from which we started. A few probability densities and form factors are given in Table 6.1.

A final word concerns the dependence of the form factor on experimental quantities. Equation (6.22) shows that $F(\mathbf{q}^2)$ depends only on the square of the momentum transferred to the target particle and not on the energy of the incident particle. $F(\mathbf{q}^2)$, for a specific value of $\mathbf{q}^2$, can therefore be determined with projectiles of different energies. Equation (6.12) indicates that it is only necessary to change the scattering angle correspondingly, and the same value of $F(\mathbf{q}^2)$ should result. Incidentally, the fact that $F(\mathbf{q}^2)$ depends only on $\mathbf{q}^2$ is true only in the first Born approximation; it is not valid in higher order. It can therefore be used to test the validity of the first Born approximation.

6.5 The Charge Distribution of Spherical Nuclei

The investigation of nuclear structure by electron scattering has been pioneered by Hofstadter and his collaborators.[9] The basic arrangement is similar to the one shown in Fig. 5.25: An electron accelerator produces an intense beam of electrons with energies between 250 MeV and a few GeV. The electrons are transported to a scattering chamber where they strike the target. The intensity of the elastically scattered electrons is determined as a function of the scattering angle. The differential cross section for the scattering of 750-MeV electrons from ^{40}Ca and ^{48}Ca is shown in Fig. 6.5. The solid curve is the best fit to the data. The fit yields values of $|F(\mathbf{q}^2)|$, and from these values information about the charge distribution is obtained.

The crudest approximation to the nuclear charge distribution is a one-parameter function, for instance, a uniform or a Gaussian distribution. Such distributions give poor fits, and the simplest useful approximation is the two-parameter *Fermi distribution*

$$\rho(r) = \frac{N}{1 + e^{(r-c)/a}}. \tag{6.32}$$

N is a normalization constant and c and a are the parameters describing the nucleus. The Fermi distribution is shown in Fig. 6.6; c is called the half-density radius and t the surface thickness. The parameter a in Eq. (6.32) and t are related by

$$t = (4 \ln 3)a. \tag{6.33}$$

The results of many experiments can be summarized in terms of the parameters defined in Eqs. (6.29) and (6.32):

1. For medium- and heavyweight nuclei the root-mean-square charge

9. R. Hofstadter, H. R. Fechter, and J. A. McIntyre, *Phys. Rev.* **92**, 978 (1953).

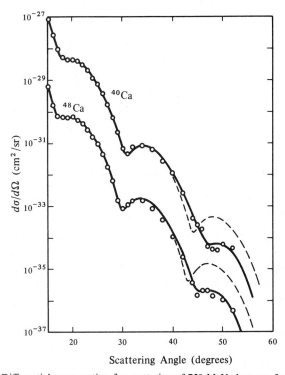

Fig. 6.5. Differential cross section for scattering of 750-MeV electrons from calcium isotopes. The cross section for ^{40}Ca has been multiplied by a factor of 10, and that for ^{48}Ca by 10^{-1}. [From J. B. Bellicard et al., *Phys. Rev. Letters* **19**, 527 (1967).]

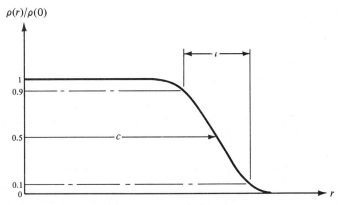

Fig. 6.6. Fermi distribution for the nuclear charge density. c is the half-density radius and t the surface thickness.

radius can be approximated by the relation

$$\langle r^2 \rangle^{1/2} = r_0 A^{1/3}, \qquad r_0 = 0.94 \text{ fm}, \tag{6.34}$$

where A is the mass number (number of nucleons). The nuclear volume consequently is proportional to the number of nucleons. The nuclear density is approximately constant; nuclei behave more like solids or liquids than atoms.

2. The half-density radius and the skin thickness satisfy approximately

$$c \text{ (in fm)} = 1.18\, A^{1/3} - 0.48$$
$$t \approx 2.4 \text{ fm.} \tag{6.35}$$

From these values, the density of nucleons at the center follows as

$$\rho_n \approx 0.17 \text{ nucleon/fm}^3. \tag{6.36}$$

This value approaches the density of nuclear matter, namely the density that an infinitely large nucleus, without surface effects, is presumed to have.

3. The charge distribution is more complex than the two-parameter Fermi distribution. In particular the density in the interior of nuclei is not constant as assumed in Eq. (6.32); it can decrease or increase toward the center. Two specific cases are shown in Fig. 6.7.

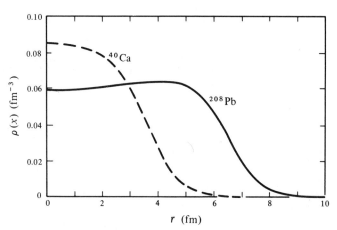

Fig. 6.7. Probability distribution for ^{40}Ca and ^{208}Pb, obtained by electron scattering. (Courtesy of D. G. Ravenhall.)

4. In the older literature, written at a time when the shape of nuclei was not yet well known, it was customary to describe the nuclear radius differently. A nucleus of uniform density and radius R was assumed. From Eq. (6.29) it follows that R^2 and $\langle r^2 \rangle$ are connected by

$$\langle r^2 \rangle = 4\pi \int_0^R \frac{3r^4\, dr}{4\pi R^3} = \frac{3}{5} R^2. \tag{6.37}$$

R approximately satisfies the relation

$$R = R_0 A^{1/3}, \qquad R_0 = 1.2 \text{ fm.} \qquad (6.38)$$

The information given so far in this section provides a glimpse into the structure of nuclei. Considerably more is known already—finer details have been investigated and data have been gathered on many nuclides. It is likely that the near future will bring more improvements. Finally, we must remember that the information provided by charged lepton scattering concerns the nuclear charge and current distribution and that corresponding data on the hadronic structure (matter distribution) can be obtained only by using hadrons as probes.[10]

6.6 Leptons Are Point Particles

We return now to the g factor of the electron. By 1926, the idea of the spinning electron and its magnetic moment was generally accepted,[11] but the value of the g factor [Eq. (5.16)],

$$g(1926) = -2,$$

had to be taken from experiment. (The minus sign indicates that the magnetic moment points in the direction opposite to the spin for a negative electron.) It was exactly twice as large as the g factor for orbital motion, Eq. (5.14). In other words, even though the electron has spin $\frac{1}{2}$, it carries *one* Bohr magneton. In 1928, Dirac introduced his famous equation; the existence of a magnetic moment and the value $g = -2$ turned out to be natural consequences.[12]

In 1947, Kusch and Foley measured the g factor carefully by using the then-new microwave technique and discovered that it showed a small deviation from -2.[13] Within a very short time, Schwinger could explain the deviation. The experiment was accurate to about 5 parts in 10^5, and the theory was somewhat better. Since then, theoretical and experimental physicists have been in a race to improve the numbers. The winner has consistently been physics, because everybody has learned more. Since the com-

---

10. "International Congress on Nuclear Sizes and Density Distributions" *Rev. Modern Phys.* **30**, 414–569 (1958); R. Wilson, "What Is the Radius of a Nucleus?" *Comments Nucl. Particle Phys.* **4**, 116 (1970).

11. A fascinating description of the history of the spin is presented by B. L. Van der Waerden, in *Theoretical Physics of the Twentieth Century* (M. Fierz and V. F. Weisskopf, eds.), Wiley-Interscience, New York, 1960. See also S. A. Goudsmit, *Phys. Today* **14**, 18 (June 1961) and P. Kusch, *Phys. Today* **19**, 23 (February 1966).

12. For a derivation of the magnetic moment of the electron in Dirac theory, see, for instance, Merzbacher, Eq. (24.37), or Messiah, Section XX, 29. Actually, the magnetic moment can already be derived as a nonrelativistic phenomenon, as, for instance, in A. Galindo and C. Sanchez del Rio, *Am. J. Phys.* **29**, 582 (1961), or R. P. Feynman, *Quantum Electrodynamics*, Benjamin, Reading, Mass., 1961, p. 37.

13. P. Kusch and H. M. Foley, *Phys. Rev.* **72**, 1256 (1947); **74**, 250 (1948).

parison between theory and experiment is very important, a few words on both are in order here.

The theoretical explanation invokes virtual photons, a concept already discussed in Section 5.8. A physical electron does not always exist as a Dirac electron. Part of the time it emits a virtual photon which it then reabsorbs. (Classically, this process corresponds to the electron's interaction with its own electromagnetic field.) The measurement of the g factor involves the interaction of the electron with photons; the presence of virtual photons changes the interaction and consequently also the g factor. Figure 6.8 shows how the simple interaction of a photon with a Dirac

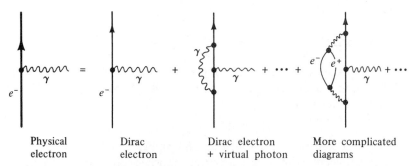

| Physical electron | Dirac electron | Dirac electron + virtual photon | More complicated diagrams |

Fig. 6.8. A physical electron is not just a pure Dirac electron. The presence of virtual photons affects the properties of the electron; in particular it changes the g factor by an amount that can be calculated and measured.

electron is altered and complicated by the electron's own electromagnetic field. The net effect is to add an *anomalous* magnetic moment. An enormous amount of labor has been put into calculating the magnetic moment of a Dirac particle taking into account corrections of the type shown in Fig. 6.8. The result is expressed in terms of the number

$$a = \frac{|g| - 2}{2}. \tag{6.39}$$

A pure *Dirac particle*, that is, a particle with properties as predicted by the Dirac equation alone, would have a value $a = 0$. The value of a for a *physical electron* has been computed by many people, and the present best theoretical value is[14]

$$a_e^{\text{th}} = \frac{1}{2}\left(\frac{\alpha}{\pi}\right) - 0.328479\left(\frac{\alpha}{\pi}\right)^2 + 1.29\left(\frac{\alpha}{\pi}\right)^3, \tag{6.40}$$

where α is the fine structure constant, $\alpha = e^2/hc$.

The early experimental results for a_e were based on an approach that can be explained with Fig. 5.5: If an electron is placed in an external magnetic field, Zeeman splitting results. A precise determination of the energy

14. S. J. Brodsky and S. D. Drell, *Ann. Rev. Nucl. Sci.* **20**, 147 (1970); M. J. Levine and J. Wright, *Phys. Rev. Letters* **26**, 1351 (1971).

difference between levels and of the externally applied field yields g. Indeed, the discovery of a nonvanishing parameter a_e occurred with such a technique. Present experiments are based on a different approach in which $|g| - 2$, and not g, is determined[15]: In a uniform magnetic field, the spin and the momentum of a particle with spin $\frac{1}{2}$ and $|g| = 2$ retain a constant angle between them. Now consider an experimental arrangement as in Fig. 6.9. Longitudinally polarized electrons, i.e., electrons with spin and

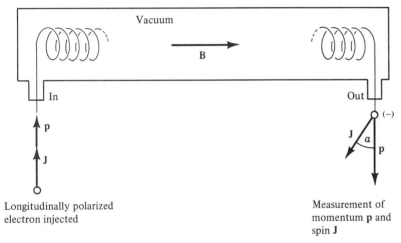

Longitudinally polarized electron injected

Measurement of momentum **p** and spin **J**

Fig. 6.9. Basic approach underlying the direct determination of $a = (|g| - 2)/2$. For details see the text.

momentum pointing in the same direction, are injected into a solenoidal magnetic field. In this field, the electrons move in circular orbits, and their spins and momenta are observed after a large number of revolutions. If the g factor were exactly 2, spin and magnetic moment of the outcoming electrons would still be parallel, regardless of the time spent in the field **B**. The small anomalous part a, however, causes a slightly different rotation for spin and magnetic moment. After a time t in the field **B**, the angle α between **p** and **J** becomes

$$\alpha = a\omega_c t, \tag{6.41}$$

where

$$\omega_c = \frac{eB}{mc} \tag{6.42}$$

is the cyclotron frequency. If the product Bt is very large, α also becomes very large and a can be measured very accurately. This method has been applied to electrons and muons of both signs. The most accurately measured value is

$$a_{e^-}^{\text{exp}} = 0.001\ 159\ 658(4)$$

15. A more detailed description of the ideas underlying the $|g| - 2$ experiments is given in R. D. Sard, *Relativistic Mechanics*, Benjamin, Reading, Mass., 1971.

where (4) indicates an error of ± 4 in the last place.[16] Less accurate values of a have been obtained for the other charged leptons.[17] A comparison of the theoretical with the experimental data is given in Table 6.2.

Table 6.2 COMPARISON OF THE THEORETICAL AND EXPERIMENTAL VALUES OF $a = (|g| - 2)/2$

Particle	$(a^{\text{th}} - a^{\text{exp}})/a^{\text{th}}$
e^-	$(2 \pm 5) \times 10^{-6}$
e^+	$(5 \pm 10) \times 10^{-4}$
$\mu^\pm$	$(2.5 \pm 2.7) \times 10^{-4}$

Table 6.1 states that the theoretical and experimental values for a_{e^-} agree to better than 1 part in 10^5; the theoretical values for the g factor hence agree to better than 1 part in 10^8. Quantum electrodynamics, the quantum theory of the interactions of charged leptons and photons, is an unbelievably successful theory.

The theoretical computations are performed under the assumption that the leptons are point particles. Any deviation from this assumption would appear as a discrepancy in Table 6.1. The agreement can be translated into a limit on the radius of the leptons; leptons must be smaller than 0.1 fm. The assumption of a structureless electron, made in the previous sections, is verified.

Experiments performed with high-energy charged leptons also demonstrate that quantum electrodynamics predicts all observed phenomena correctly.[14] The incredible success of quantum electrodynamics raises the question, is there a limit at which it breaks down? Does the electron ultimately have a finite extension? Why is the muon heavier than the electron? If its larger mass were caused by some hadronic interaction, it should also show up in the g factor. Present-day physics cannot answer these questions, and one can only bet as to whether the next major advance will come from a new experimental surprise or from a theoretical breakthrough.

6.7 Nucleon Elastic Form Factors

By 1932 it was well known that electrons have a spin $\frac{1}{2}$ and a magnetic moment of 1 μ_B, (Bohr magneton), as predicted by the Dirac equation. Two other spin-$\frac{1}{2}$ particles were also known to exist, the proton and the

16. J. C. Wesley and A. Rich, *Phys. Rev.* **A4**, 1341 (1971).

17. J. R. Gilleland and A. Rich, *Phys. Rev.* **A5**, 38 (1972); J. Bailey, W. Bartl, G. Von Bochmann, R. C. A. Brown, F. J. M. Farley, H. Jöstlein, E. Picasso, and R. W. Williams, *Phys. Letters* **B28**, 287 (1968); *Nuovo Cimento* **9A**, 369 (1972).

neutron. It was firmly believed that these would also have magnetic moments as predicted by the Dirac equation, one nuclear magneton for the proton and zero moment for the neutron. Enter Otto Stern. Stern had principles in selecting his experiments: "Try only crucial experiments. Crucial experiments are those that test universally accepted principles." When he started setting up equipment to measure the magnetic moment of the proton, his friends teased him and told him that he should not waste his time on an experiment whose outcome was foreordained. The surprise was great when Stern and his collaborators found a magnetic moment of about 2.5 μ_N for the proton and about -2 μ_N for the neutron.[18]

How can the departure of the magnetic moments of the proton and the neutron from the "Dirac values" be understood? The current explanation of the anomalous nucleon moments assumes that these particles would have moments predicted by the Dirac theory if there were no hadronic interaction. However, the hadronic interaction exists and it gives rise to virtual particles that surround ("clothe") the Dirac ("bare") nucleon. In terms of Feynman diagrams, a real nucleon consists of a superposition of many states, as indicated in Fig. 6.10. An actual proton, for instance, has

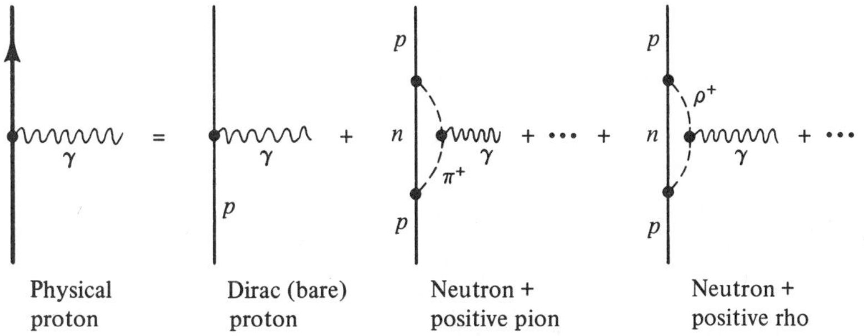

Physical proton   Dirac (bare) proton   Neutron + positive pion   Neutron + positive rho

Fig. 6.10. A physical proton is pictured as a superposition of many states, for instance a bare proton, a bare neutron plus a pion, and so forth. Measurement of the magnetic moment involves the interaction of the electromagnetic field (photon) with the proton, and the photon probes not only the bare proton but also the charged particles in the meson cloud.

a certain probability amplitude of being a bare proton, a certain amplitude of being a neutron plus a positive pion, and so forth.

In the previous section we learned that leptons obey the Dirac equation and that they behave like point particles. It is tempting to assume that bare nucleons also are point particles and that the finite extension of real nucleons is due to the virtual particle cloud. Pions are the lightest of these, and they account for the "outermost" part. Since the pions have to return, they can go out only to about half the pion Compton wavelength [Eq. (5.51)]. The maximum radius of the nucleons consequently is expected to

18. I. Estermann, R. Frisch, and O. Stern, *Nature* **132**, 169 (1933); R. Frisch and O. Stern, *Z. Physik* **85**, 4 (1933).

be about $\hbar/2m_\pi c$ or about 0.7 fm. The pion cloud also contributes to the magnetic moment, and the anomalous moments can be explained at least approximately. The word *approximately* raises a question: Fig. 6.10 and 6.8 look very similar. Why can the anomalous g factor for the electron be computed so accurately, whereas the one for the nucleons cannot? The answer lies in the forces involved. The anomalous g factor of leptons is caused by the electromagnetic force which is relatively weak and can be treated by perturbation theory. The anomalous nucleon moments are due to hadronic effects; these are strong and no satisfactory way to perform the calculations has yet been found.

The best way to explore the charge and current distributions of nucleons is again electron scattering. Experimentally, the problem is straightforward for protons. A liquid hydrogen target is placed in an electron beam, and the differential cross section of the elastically scattered electrons is determined. For neutrons, the situation is not so easy. No neutron targets exist, and it is necessary to use deuteron targets and subtract the effect of the proton. The subtraction procedure introduces uncertainties. The e^-n cross section is consequently less well known than the e^-p cross section.

For spinless target particles, the form factor can be extracted from the cross section by using Eq. (6.20). Nucleons have spin $\frac{1}{2}$, and Eq. (6.20) must be generalized. Without calculation, we can guess some features of the result. $F(q^2)$ in Eq. (6.20) describes the distribution of the electric charge, and it can be called an *electric* form factor. The proton also possesses, in addition to its charge, a magnetic moment. It is unlikely that it behaves like a point moment and sits at the center of the proton. It is to be expected that the magnetization is also distributed over the volume of the nucleon, and this distribution will be described by a magnetic form factor.[19] The detailed computation indeed proves that elastic scattering from a spin-$\frac{1}{2}$ particle with structure must be described by two form factors; the laboratory cross section can be written as

$$\frac{d\sigma}{d\Omega} = \left(\frac{d\sigma}{d\Omega}\right)_{\text{Mott}} \left\{ \frac{G_E^2 + bG_M^2}{1+b} + 2bG_M^2 \tan^2\left(\frac{\theta}{2}\right) \right\}, \qquad (6.43)$$

where

$$b = \frac{-q^2}{4m^2c^2}. \qquad (6.44)$$

Equation (6.43) is called the Rosenbluth formula;[20] m is the mass of the nucleon, θ the scattering angle, and q the four-momentum transferred to the nucleon.[21] The Mott cross section is given by Eq. (6.19). G_E and G_M

---

19. Nuclei with spin $J \geq \frac{1}{2}$ also possess magnetic moments, and the magnetization is also distributed over the volume of the nucleus. For such nuclei, the discussion given in Section 6.5 must be generalized.

20. M. N. Rosenbluth, *Phys. Rev.* **79**, 615 (1950).

21. Here a word of explanation is in order: The variable q is the four-momentum transfer. It is defined as

$$q = \left\{ \frac{E}{c} - \frac{E'}{c}, \mathbf{p} - \mathbf{p}' \right\}$$

are the electric and magnetic form factors, respectively, and they are both functions of q^2. The designations *electric* and *magnetic* stem from the fact that for $q^2 = 0$, the static limit, they are given by

$$G_E(q^2 = 0) = \frac{Q}{e}$$

$$G_M(q^2 = 0) = \frac{\mu}{\mu_N}, \tag{6.45}$$

where Q and μ are the charge and magnetic moment, respectively, of the nucleon. Specifically, $G_E(0)$ and $G_M(0)$ for the proton and the neutron are

$$G_E^p(0) = 1, \qquad G_E^n(0) = 0$$

$$G_M^p(0) = 2.79, \qquad G_M^n(0) = -1.91. \tag{6.46}$$

Early electron-proton scattering experiments,[22] performed with an electron energy of 188 MeV, were analyzed by fitting the observed differential cross section with an expression of the form of Eq. (6.43) with fixed values of the parameters G. An example is shown in Fig. 6.11. Comparison of the various theoretical curves with the experimental one indicates that the proton is not a point particle. The conclusion based on the discussion of the anomalous magnetic moment is consequently verified by a direct measurement. However, an electron energy of about 200 MeV is too small to permit studies at significant values of the momentum transfer and to get information on the q^2 dependence of G_E and G_M. Since 1956, many experiments have been performed at accelerators with electron energies up to 20 GeV. To extract the form factors from the measured cross sections, the cross section for a fixed value of q^2 is normalized by division by the Mott cross section and plotted versus $\tan^2 \theta/2$, as shown in Fig. 6.12. Such a plot should yield a straight line; from the slope, the value of G_M^2 is obtained. The intersection with the y axis then yields G_E^2.

Figure 6.13 gives the magnetic form factor of the proton. For convenience, $G_M/(\mu/\mu_N)$ is plotted, where μ is the proton magnetic moment. In addition, a particularly simple fit to the data is shown: It has been found empirically that a *dipole fit*, of the form

$$G_D(q^2) = \frac{1}{(1 + |q|^2/q_0^2)^2}, \tag{6.47}$$

with $q_0^2 = 0.71 \ (\text{GeV}/c)^2$, fits the shape of the form factor curve well. The dipole fit has no theoretical basis, but like the Balmer formula, it is simple

---

Its square,

$$q^2 = \frac{1}{c^2}(E - E')^2 - (\mathbf{p} - \mathbf{p}')^2 = \frac{1}{c^2}(E - E')^2 - \mathbf{q}^2,$$

is a Lorentz-invariant quantity. [Jackson, Eq. (12.5).] Since q^2 is a Lorentz scalar, its use is preferred in high-energy physics. For elastic scattering in the c.m. or at low energies, $q^2 = -\mathbf{q}^2$.

22. R. W. McAllister and R. Hofstadter, *Phys. Rev.* **102**, 851 (1956).

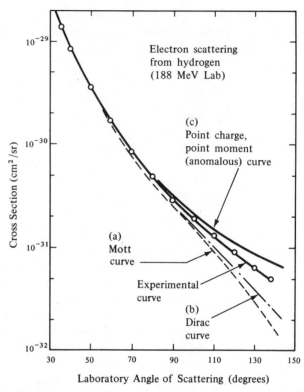

Fig. 6.11. Electron-proton scattering with 188-MeV electrons. [R. W. McAllister and R. Hofstadter, *Phys. Rev.* **102**, 851 (1956).] The theoretical curves correspond to the following values of G_E and G_M: Mott (1; 0), Dirac (1; 1), anomalous (1; 2.79).

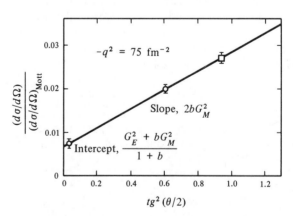

Fig. 6.12. Rosenbluth plot. See the text for description.

and probably tries to tell us something. However, the fit is not perfect, and deviations from it are shown in the inset to Fig. 6.13.

Figure 6.13 gives the magnetic form factor of the proton. Similar data, but with larger errors, are obtained for the other nucleon form factors.

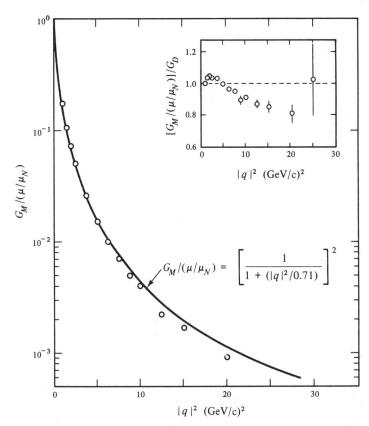

Fig. 6.13. Values of the proton magnetic form factor G_M, normalized by division with the proton magnetic moment, plotted versus the momentum transfer q^2. An empirical *dipole fit* to the data is shown as a solid line. The insert shows the ratio of the measured values of G_M to the dipole fit. From P. N. Kirk et al., *Phys. Rev.* **D8**, 63 (1973).

In a good approximation, the form factors satisfy

$$G_E^p(q^2) \approx \frac{G_M^p(q^2)}{(\mu_p/\mu_N)} \approx \frac{G_M^n(q^2)}{(\mu_n/\mu_N)} = G_D(q^2)$$

$$G_E^n \approx 0.$$

(6.48)

Some features of the nucleon structure emerge from these relations:

1. Nucleons are not point particles. For point particles, the form factors are constant. The charge distribution corresponding to the observed

dipole form factor, Eq. (6.47), follows from Table 6.1 as

$$\rho(r) = \rho(0)e^{-r/a}$$

$$a = \frac{\hbar}{q_0} = 0.23 \text{ fm.}$$

(6.49)

Nucleons are extended systems but do not have well-defined surfaces. One remark must be added: The Fourier transform used here is valid only for small values of $|q|^2$. For large values of $|q|^2$, the proton that was initially at rest recoils with a velocity approaching that of light, and $\exp(-r/a)$ no longer represents the charge distribution.

2. All the form factors except the charge form factor of the neutron have about the same q^2 dependence and they satisfy the *scaling law* expressed by Eq. (6.48).

3. If a certain property, for instance, the charge, is described by a form factor G, with $G(0) = 1$, then Eq. (6.28) shows that the mean-square radius for this property can be found from the slope of $G(q^2)$ at the origin:

$$\langle r^2 \rangle = -6\hbar^2 \left(\frac{dG(q^2)}{dq^2} \right)_{q^2=0}.$$

(6.50)

With the dipole fit, Eq. (6.48), the mean-square radii become

$$\langle r_E^2(\text{proton}) \rangle \approx \langle r_M^2(\text{proton}) \rangle \approx \langle r_M^2(\text{neutron}) \rangle \approx 0.7 \text{ fm}^2.$$

(6.51)

The earlier estimate for the proton radius, obtained by considering virtual pions, qualitatively agrees with these values. The assumption that the deviation of the magnetic moments from the Dirac values is caused by the hadronic structure is therefore verified.

4. Determination of the mean square charge radius of the neutron is made difficult by uncertainties that arise from the use of a deuterium target. Fortunately, there is another way to determine $\langle r_E^2(\text{neutron}) \rangle$, namely by scattering low-energy neutrons from electrons bound in atoms. The largest contribution to the interaction between neutrons and electrons comes from the dipole-dipole force between the magnetic moments of the electron and the neutron. In closed shell atoms, all electron spins are paired, and this term does not contribute. The two terms of interest then are the interaction of the neutron magnetic moment with the Coulomb field of the electrons (Foldy term) and the contribution due to a possible electric charge "inside" the neutron giving rise to a nonvanishing $\langle r_E^2(\text{neutron}) \rangle$.[23] The generalization of Eq. (6.50) includes both terms:

$$-6\hbar^2 \left(\frac{dG_E^n(q^2)}{dq^2} \right)_{q^2=0} = \frac{3}{2} \frac{\mu_n}{\mu_N} \left(\frac{\hbar}{m_n c} \right)^2 + \langle r_E^2(\text{neutron}) \rangle.$$

(6.52)

The contribution from the Foldy term gives

$$\frac{3}{2} \frac{\mu_n}{\mu_N} \left(\frac{\hbar}{m_n c} \right)^2 = 0.126 \text{ fm}^2.$$

(6.53)

23. L. L. Foldy, *Phys. Rev.* **83**, 688 (1955); *Rev. Modern Phys.* **30**, 471 (1958).

The scattering cross section for slow neutrons from electrons determines $(dG_E^n/dq^2)_{q^2=0}$. The experiments are not easy, and at the present time, the results from various groups still disagree.[24] It is therefore impossible to quote a reliable value, but the average

$$6\hbar^2\left(\frac{dG_E^n}{dq^2}\right)_{q^2=0} = 0.121 \pm 0.006 \text{ fm}^2 \qquad (6.54)$$

already demonstrates two salient facts:

a. The slope of the form factor G_E^n at the origin is not zero. Even though Eq. (6.48) indicates that $G_E^n(q^2)$ is close to zero, it cannot vanish entirely.

b. The mean-square charge radius of the neutron is given by

$$\langle r_E^2(\text{neutron})\rangle \approx 0.008 \pm 0.006 \text{ fm}^2. \qquad (6.55)$$

At present the very small or vanishing neutron charge radius is a mystery. The neutron, crudely speaking, is all magnetism and "contains" very little electric charge.

6.8 Deep Inelastic Electron Scattering

The Thomson model of the atom, in vogue before 1911, assumed that the positive and the negative charges were distributed uniformly through-out the atom. Rutherford's scattering experiment[1] proved that the positive charge is concentrated in the nucleus; this discovery profoundly affected atomic physics and founded nuclear physics. It is possible that recent experiments on highly inelastic electron scattering will have a similar impact on particle physics and we consequently discuss the most surprising results of these experiments here.[25]

In inelastic electron scattering, the differential cross section is measured for electrons that have lost a certain amount of energy to the proton. The diagrams for elastic and inelastic electron scattering are shown in Fig. 6.14. The interaction between electron and proton is mediated by a photon, as in Fig. 5.18. In elastic scattering, no new particles are created, and only an electron and a proton are present in the final state. In inelastic scattering, additional particles are produced. In the experiment to be discussed here, these particles are not observed, and only energies and momenta of the incident and the scattered electrons are measured (Fig. 6.14). From the values E, $\mathbf{p}$, E', and $\mathbf{p}'$ one determines two quantities, ν, and q^2, that characterize the scattering event and that have definite physical interpretations. Thus, ν is the energy lost by the electron,

$$\nu = E - E', \qquad (6.56)$$

ELASTIC

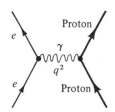

INELASTIC

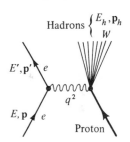

Fig. 6.14. Elastic and inelastic electron scattering.

24. V. E. Krohn and G. R. Ringo, *Phys. Rev.* **148**, 1303 (1966), *Phys Rev.* **8**, 1305 (1973); L. L. Koester, private communication and to be published.

25. H. W. Kendall and W. K. H. Panofsky, *Sci. Amer.* **224**, 60 (June 1971); S. D. Drell, in "Subnuclear Phenomena, International School of Physiscs" *Ettore Majorana*, 2 (1970); J. I. Friedman and H. W. Kendall, *Ann. Rev. Nucl. Sci.* **22**, 203 (1972).

and q^2 is the four momentum transferred from the electron to the proton[21]

$$q^2 = \frac{v^2}{c^2} - (\mathbf{p} - \mathbf{p}')^2. \tag{6.57}$$

For the energy and the momentum of the hadrons in the final state, energy and momentum conservation give

$$E_h = v + mc^2$$
$$\mathbf{p}_h = \mathbf{p} - \mathbf{p}'.$$

Here, m is the mass of the proton, and all quantities are measured in the laboratory. In terms of E_h and $\mathbf{p}_h$, or q and v, one can define a further dynamical variable, W,

$$W^2 = E_h^2 - (\mathbf{p}_h c)^2 = m^2 c^4 + q^2 c^2 + 2vmc^2. \tag{6.58}$$

In the c.m. of the hadrons in the final state, where $\mathbf{p}_h = 0$, $W = E_h$. W thus is the total energy of the final hadrons in their c.m. All three quantities, v, q^2, and W, only two of which are independent, are Lorentz invariants and therefore have the same values in any frame of reference; the invariant q^2 is usually denoted by t, $t \equiv q^2$. (In high-energy physics, t does not always stand for time.)

A typical scattering spectrum is sketched in Fig. 6.15. The number of particles observed at a fixed angle and fixed incident electron energy (10 GeV) is plotted as a function of the energy E' of the scattered electrons. Three features stand out, the elastic peak, resonances, and a continuum. The resonances correspond to excited states of the nucleon, as shown in Fig. 5.30.

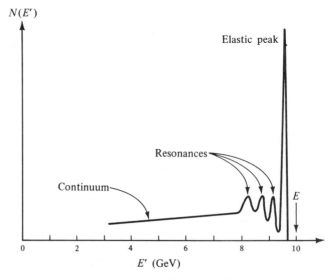

Fig. 6.15. Scattering spectrum: number of scattered electrons observed versus the energy, E', of the scattered electrons. Incident electron energy is 10 GeV.

The elastic cross section has already been discussed in Section 6.7. It is shown in Fig. 6.16 normalized by division through the Mott cross section, Eq. (6.19). Similarly, the differential cross sections for the production of particular resonances can be studied; their angular distributions turn out to be similar to the elastic case. The nucleon in its low-lying excited states consequently has a spatial extension similar to that in its ground state.

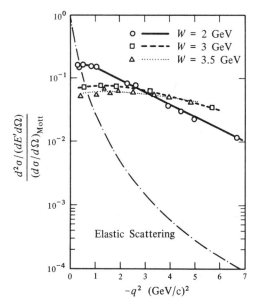

Fig. 6.16. Elastic and double differential cross sections, normalized by division with $\sigma_{\text{Mott}} \equiv (d\sigma/d\Omega)_{\text{Mott}}$. $(d^2\sigma/dE'\,d\Omega)/\sigma_{\text{Mott}}$, in GeV^{-1}, is given for $W = 2, 3,$ and 3.5 GeV. [After M. Breidenbach et al., *Phys. Rev. Letters* **23**, 935 (1969).]

The measurement of the differential cross section in the continuum is somewhat more involved. It is necessary to determine the *double differential cross section* $d^2\sigma/dE'd\Omega$ which is proportional to the probability of a scattering event occurring into a given solid angle $d\Omega$ and into the energy interval between E' and $E' + dE'$. A second problem must be solved: At different scattering angles, what energies E' should be selected? The answer comes from elastic scattering and the resonances: Elastic scattering corresponds to looking at a final state with $W = m_p c^2$; observation of a resonance means selecting a final state with $W = m_{\text{res}} c^2$, where m_{res} is the mass of the resonance. W characterizes the total energy of the hadrons in the final state, and the cross section $d^2\sigma/dE'\,d\Omega$ for the continuum is consequently determined as a function of q^2 for a fixed value of W.

Inelastic electron-proton scattering into the continuum has been studied at SLAC.[26,27] The primary electron energy was varied between 4.5 and 18 GeV; W reached values as high as 5 GeV and $-q^2$ as high as 21 $(\text{GeV}/c)^2$. The large momentum transfers and invariant masses give rise to the name *deep inelastic scattering*. The ratios $(d^2\sigma/dE'\,d\Omega)/(d\sigma/d\Omega)_{\text{Mott}}$ for three values of W are shown in Fig. 6.16. The difference between the elastic and the inelastic continuum scattering is dramatic: The ratio for the elastic cross section decreases rapidly with increasing $|q^2|$, whereas it is nearly independent of $|q^2|$ for the inelastic case. The ratio represents a form factor, and Table 6.1 states that a constant form factor implies a point scatterer. This conclusion is reinforced by looking at the magnitude of the cross-section ratio. The cross section $d^2\sigma/dE'\,d\Omega$ displayed in Fig. 6.16 represents the cross section for scattering into the energy interval between E' and $E' + dE'$, where dE' is 1 GeV. To get the total inelastic cross section from the continuum, $d^2\sigma/dE'\,d\Omega$ must be integrated over all values of E'. To do this integration crudely, we note that the cross-section ratio shown in Fig. 6.16 is nearly independent of q^2 and W over a wide range. Equation (6.58) implies that it is then also independent of E'. Integration over dE' can hence be replaced by multiplication with the total range of E'. E' ranges over nearly 10 GeV. Thus the *total* cross section for inelastic scattering into the continuum is nearly 10 times bigger than $d^2\sigma/dE'\,d\Omega$ in Fig. 6.16, or

$$\left(\frac{d\sigma}{d\Omega}\right)_{\text{cont}} \approx \frac{1}{2}\left(\frac{d\sigma}{d\Omega}\right)_{\text{Mott}}$$

Shades of Rutherford. The Mott cross section applies to a point scatterer, and the deep inelastic scattering thus behaves nearly as if it were produced by point scatterers inside the proton. Further evidence for the existence of point constituents inside the nucleon has come from other experiments. The cross section for the production of muon pairs by 10-GeV photons, for instance, is much larger than expected on the basis of a smooth charge distribution.[28] The nature of these point scatterers and their relation to observed or postulated particles is not yet clear. Feynman has coined the word *partons* to describe them,[29] and attempts have been made to identify them with quarks[30] (Chapter 13) or with the bare nucleons dis-

26. M. Breidenbach, J. I. Friedman, H. W. Kendall, E. D. Bloom, D. H. Coward, H. DeStaebler, J. Drees, L. W. Mo, and R. E. Taylor, *Phys. Rev. Letters* **23**, 935 (1969).

27. G. Miller, E. D. Bloom, G. Buschhorn, D. H. Coward, H. DeStaebler, J. Drees, C. L. Jordan, L. W. Mo, R. E. Taylor, J. I. Friedman, G. C. Hartmann, H. W. Kendall, and R. Verdier, *Phys. Rev.* **D5**, 528 (1972).

28. J. F. Davis, S. Hayes, R. Imlay, P. C. Stein, and P. J. Wanderer, *Phys. Rev. Letters* **29**, 1356 (1972).

29. R. P. Feynman, in *High Energy Collisions*, Third International Conference, State University of New York, Stony Brook, 1969 (C. N. Yang, J. A. Cole, M. Good, R. Hwa, and J. Lee-Franzini, eds.), Gordon & Breach, New York, 1969.

30. J. D. Bjorken and E. A. Paschos, *Phys. Rev.* **185**, 1975 (1969); J. Kuti and V. F. Weisskopf, *Phys. Rev.* **D4**, 3418 (1971).

cussed in Section 6.7.[31] So far no theory has been completely successful and more surprises are likely.[32]

6.9 Scattering and Structure

● The material in Sections 6.4–6.8 demonstrates that much information concerning subatomic structure can be obtained from scattering experiments. Even a glance at a differential cross section, without detailed computation, can reveal gross features. As an example, the information contained in Figs. 6.5, 6.7, 6.11, and 6.13 is reproduced schematically in Fig. 6.17. It highlights one difference between heavy nuclei and nucleons:

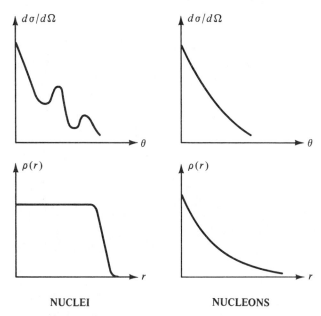

NUCLEI   NUCLEONS

Fig. 6.17. Cross section and charge distribution: The appearance of diffraction minima in the cross section for heavy nuclei implies the existence of a well-defined nuclear surface. Nucleons, in contrast, possess a charge density that decreases smoothly.

Typical heavy nuclei have well-defined surfaces; as in optics, interference effects then produce diffraction minima and maxima in the differential cross section. Nucleons, in contrast, do not have such surfaces; their density decreases smoothly, and they do not show prominent diffraction effects.

The Scattering Amplitude. In the present section, we shall treat scattering in somewhat more detail than we have done before. A glance at any current book on scatter-

31. S. D. Drell and T. D. Lee, *Phys. Rev.* **D5**, 1738 (1972).
32. The present situation is partially described by Mephistopheles' words in Goethe's *Faust*: "Denn eben, wo Begriffe fehlen, da stellt ein Wort zur rechten Zeit sich ein" (When the concepts are missing, a word imposes itself just at the right time).

ing[33] will show that the material presented here constitutes only a minute fraction of what is actually used in research. Even so, it should provide some insight into the connection between scattering and structure.

We begin the discussion with a simple case, nonrelativistic scattering by a fixed potential, $V(\mathbf{x})$, and we approximate the incoming particle by a plane wave moving along the z axis, $\psi = \exp(ikz)$. The solution to the scattering problem is a solution of the time-independent Schrödinger equation,

$$-\frac{\hbar^2}{2m}\nabla^2\psi + V\psi = E\psi$$

or

$$(\nabla^2 + k^2)\psi = \frac{2m}{\hbar^2}V\psi, \tag{6.59}$$

where the wave number k is related to the energy E by

$$k = \frac{p}{\hbar} = \frac{1}{\hbar}\sqrt{2mE}. \tag{6.60}$$

Far away from the scattering center, the scattered wave will be spherical, and it will originate at the scattering center, which is assumed to be at the origin of the coordinate system. The total asymptotic wave function, shown in Fig. 6.18, consequently will be

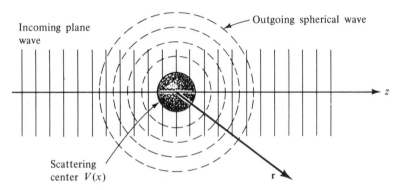

Fig. 6.18. The asymptotic wave function consists of an incoming plane wave and an outgoing spherical wave.

of the form

$$\psi = e^{ikz} + \psi_s, \qquad \psi_s = f(\theta, \varphi)\frac{e^{ikr}}{r}. \tag{6.61}$$

The scattering amplitude f describes the angular dependence of the outgoing spherical wave; its determination is the goal of the scattering experiment.

The connection between differential cross section and scattering amplitude is given by Eq. (6.10). To verify the relation, we note that for the present case of one scattering center ($N = 1$), Eqs. (6.3) and (6.4) give for the differential cross section

$$\frac{d\sigma}{d\Omega} = \frac{(d\mathfrak{N}/d\Omega)}{F_{\text{in}}}.$$

33. M. L. Goldberger and K. M. Watson, *Collision Theory*, Wiley, New York, 1964; R. G. Newton, *Scattering Theory of Waves and Particles*, McGraw-Hill, New York, 1966; L. S. Rodberg and R. M. Thaler, *Introduction to the Quantum Theory of Scattering*, Academic Press, New York, 1967.

The outgoing flux, the number of particles crossing a unit area a at distance r per unit time, is connected to $d\mathfrak{N}/d\Omega$ by

$$F_{\text{out}} = \frac{d\mathfrak{N}}{da} = \frac{d\mathfrak{N}}{r^2 \, d\Omega}$$

so that

$$\frac{d\sigma}{d\Omega} = \frac{r^2 F_{\text{out}}}{F_{\text{in}}}. \tag{6.62}$$

Since the flux is given by the probability density current, the computation of $d\sigma/d\Omega$ is now easy. For the incident wave, $\psi = e^{ikz}$, we find

$$F_{\text{in}} = \frac{\hbar}{2mi} |\psi^* \nabla \psi - \psi \nabla \psi^*| = \frac{\hbar k}{m}.$$

In all directions except forward ($0°$), the scattered wave is given by the second term in Eq. (6.61) so that

$$F_{\text{out}} = \frac{\hbar k}{mr^2} |f(\theta, \varphi)|^2.$$

With Eq. (6.62), the relation (6.10) between scattering amplitude and cross section is verified.[34]

In the *forward direction*, the interference between the incident and the scattered wave can no longer be neglected. It is necessary for the conservation of flux: The scattered particles deplete the incident beam, and the scattering in the forward direction and the total cross section must be related. The relation is called the *optical theorem*: The total cross section and the imaginary part of the forward scattering amplitude are connected by[35]

$$\sigma_{\text{tot}} = \frac{4\pi}{k} \operatorname{Im} f(0°). \tag{6.63}$$

The Scattering Integral Equation. To find the general solution of the Schrödinger equation, Eq. (6.59), we recall that it can be written as the sum of a special solution and of the appropriate solution of the corresponding homogeneous equation, where $V = 0$. To find a special solution of Eq. (6.59), it is convenient to consider the term $(2m/\hbar^2)V\psi$ on the right-hand side as the given inhomogeneity, even though it contains the unknown wave function ψ. As a first step, then, we solve the scattering problem for a point source for which the inhomogeneity becomes a three-dimensional Dirac delta function and Eq. (6.59) takes on the form

$$(\nabla^2 + k^2)G(\mathbf{r}, \mathbf{r}') = \delta(\mathbf{r} - \mathbf{r}'). \tag{6.64}$$

The solution of this equation that corresponds to an outgoing wave is

$$G(\mathbf{r}, \mathbf{r}') = \frac{-1}{4\pi} \frac{e^{ik|\mathbf{r} - \mathbf{r}'|}}{|\mathbf{r} - \mathbf{r}'|}. \tag{6.65}$$

To verify that this *Green's function* indeed satisfies Eq. (6.64), we set, for simplicity, $\mathbf{r}' = 0$, $|\mathbf{r}| = r$, and use the relations[36]

$$\nabla^2 \left(\frac{1}{r}\right) = -4\pi\delta(\mathbf{r}) \tag{6.66}$$

$$\nabla^2(FG) = \nabla^2 F + 2(\nabla F)\cdot(\nabla G) + \nabla^2 G \tag{6.67}$$

34. The derivation given here is superficial. A careful treatment can be found in K. Gottfried, *Quantum Mechanics*. Benjamin, Reading, Mass., 1966, Subsection 12.2.

35. For derivations of the optical theorem, see Park, p. 376; Merzbacher, p. 505; and Messiah, p. 867.

36. For a derivation of Eq. (6.66) see, for instance, Jackson, p. 13.

$$\nabla^2 \text{ (polar coord.)} = \frac{1}{r^2}\frac{\partial}{\partial r}\left(r^2\frac{\partial}{\partial r}\right) + \frac{1}{r^2 \sin\theta}\frac{\partial}{\partial\theta}\left(\sin\theta\frac{\partial}{\partial\theta}\right) + \frac{1}{r^2 \sin^2\theta}\frac{\partial^2}{\partial\phi^2}.$$

$$\text{(6.68)}$$

After some calculations we obtain

$$(\nabla^2 + k^2)\frac{e^{ikr}}{r} = -4\pi\delta(\mathbf{r})e^{ikr} = -4\pi\delta(\mathbf{r}). \tag{6.69}$$

The second step in this identity follows from the fact that $\int d^3r\,\delta(\mathbf{r})f(r)$ and $\int d^3r\,\delta(\mathbf{r})e^{ikr}f(\mathbf{r})$ give the same result, $f(0)$, for any continuous function f. The solution of Eq. (6.59) for a potential $V(\mathbf{r})$ is found by assuming that the inhomogeneity $(2m/\hbar^2)V(\mathbf{r})\psi(\mathbf{r})$ is built up from delta functions, $\delta(\mathbf{r}')$, each with a weight $(2m/\hbar^2)V(\mathbf{r}')\psi(\mathbf{r}')$ so that

$$\psi_s(\mathbf{r}) = \frac{2m}{\hbar^2}\int d^3r'\,G(\mathbf{r}, \mathbf{r}')V(\mathbf{r}')\psi(\mathbf{r}'), \tag{6.70}$$

where $G(\mathbf{r}, \mathbf{r}')$ is the Green's function for a delta function potential, Eq. (6.65). The appropriate solution of the homogeneous Schrödinger equation describes a particle that impinges on the target along the z axis; the general solution is therefore

$$\psi(\mathbf{r}) = e^{ikz} + \frac{2m}{\hbar^2}\int d^3r'\,G(\mathbf{r}, \mathbf{r}')V(\mathbf{r}')\psi(\mathbf{r}'). \tag{6.71}$$

The original Schrödinger differential equation for the wave function ψ has been transformed into an integral equation, called the *scattering integral equation*. For many problems, it is more convenient to start from such an integral equation rather than from the differential equation.

In scattering experiments, the incident beam is prepared far outside the scattering potential, and the scattered particles are also analyzed and detected far away. The detailed form of the wave function inside the scattering region is consequently not investigated, and what is needed is the *asymptotic* form of the scattered wave, $\psi_s(\mathbf{x})$. With $\hat{\mathbf{r}} = \mathbf{r}/r$ and $\mathbf{k} = k\hat{\mathbf{r}}$, as indicated in Fig. 6.19, $|\mathbf{r} - \mathbf{r}'|$ becomes

$$|\mathbf{r} - \mathbf{r}'| = r\left\{1 - \frac{2\mathbf{r}\cdot\mathbf{r}'}{r^2} + \frac{r'^2}{r^2}\right\}^{1/2} \xrightarrow[r\to\infty]{} r - \hat{\mathbf{r}}\cdot\mathbf{r}' \tag{6.72}$$

and the Green's function takes on the asymptotic value

$$G(\mathbf{r}, \mathbf{r}') \underset{r\to\infty}{\sim} \frac{-1}{4\pi}\frac{e^{ikr}}{r}e^{-i\mathbf{k}\cdot\mathbf{r}'}. \tag{6.73}$$

Inserting $G(\mathbf{r}, \mathbf{r}')$ into Eq. (6.70) and comparing with Eq. (6.61) yields the expression for

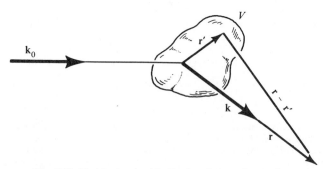

Fig. 6.19. Vectors involved in the description of scattering.

the scattering amplitude,

$$f(\theta, \varphi) = \frac{-m}{2\pi\hbar^2} \int d^3r' e^{-i\mathbf{k}\cdot\mathbf{r}'} V(\mathbf{r}')\psi(\mathbf{r}'). \tag{6.74}$$

The First Born Approximation. The first Born approximation corresponds to the case of a weak interaction. If the interaction were negligible ,the scattering amplitude would vanish and $\psi(\mathbf{r}')$ would be given by $e^{ikz'} \equiv e^{i\mathbf{k}_o\cdot\mathbf{r}'}$. As a first approximation, this value of the wave function is inserted in Eq. (6.74), with the result

$$f(\theta, \varphi) = \frac{-m}{2\pi\hbar^2} \int d^3r' V(\mathbf{r}')e^{i\mathbf{q}\cdot\mathbf{r}'/\hbar}, \tag{6.75}$$

where $\mathbf{q} = \hbar(\mathbf{k}_o - \mathbf{k})$ is the momentum that the scattered particle imparts to the scattering center, as already defined in Eq. (6.11). Equation (6.75) is called the first Born approximation; we quoted this expression in Eq. (6.13) without proof. The scattering of high-energy electrons by nucleons and light nuclei and weak processes can be described adequately by the Born approximation. In Section 6.3, we used it to derive the Rutherford cross section. Next we shall turn to an approximation that is valid under certain conditions even if the force is strong.

Diffraction Scattering—Fraunhofer Approximation. When the wavelength of the incident particle is short compared to the size of the interaction region, a semiclassical approach can be used, even if the force is strong. Such an approach is justified because the average trajectory followed by the particle approaches the classical one. The approximation used for elastic scattering is well known from optics, namely Fraunhofer diffraction. In the scattering of electromagnetic waves, optical or microwaves, the appearance of diffraction patterns has been known for a long time, and their description is well understood.[37] A characteristic example, diffraction from a black disk, is shown in Fig. 6.20. *Black* means that any photon hitting the disk is absorbed. Optical diffraction displays a number of characteristic features of which we stress three:

1. A large forward peak, called diffraction peak.

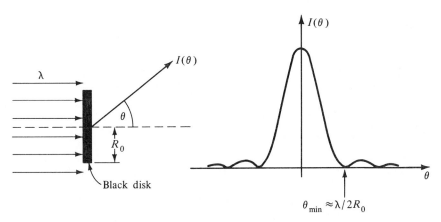

Fig. 6.20. Optical diffraction pattern produced by a black disk.

37. J. R. Meyer-Arendt, *Introduction to Classical and Modern Optics*, Prentice-Hall, Englewood Cliffs, N.J., 1972. Section 2.3; M. V. Klein, *Optics*, Wiley, New York, 1970, Chapter 7; M. Born and E. Wolf, *Principles of Optics*, Pergamon, Elmsford, N.Y., 1959; Jackson, Chapter 9.

2. The appearance of minima and maxima, with the first minimum approximately at an angle

$$\theta_{\min} \approx \frac{\lambda}{2R_0},\tag{6.76}$$

where R_0 is the radius of the disk.

3. At very small wavelengths (corresponding to the energy going to infinity) the total cross section for the scattering of light by the disk tends to a constant value,

$$\sigma \longrightarrow \text{const.} \qquad \text{for } E \longrightarrow \infty.\tag{6.77}$$

It is obvious that a detailed examination of the diffraction pattern for a number of wavelengths permits conclusions to be drawn concerning the shape of the scattering object. Diffraction scattering occurs not only in optics but also in subatomic physics, where it is a useful tool for structure investigations. Diffraction phenomena appear because the wavelength of the incident particles can be chosen to be smaller than the dimension of the target particle. The Fraunhofer approximation applies because the incident and the outgoing wave can be taken to be plane waves. We shall describe the basic ideas underlying the theoretical treatment of diffraction scattering[38] and then present some examples.

We consider a localized scatterer that is hit by a plane wave as in Fig. 6.21. We shall

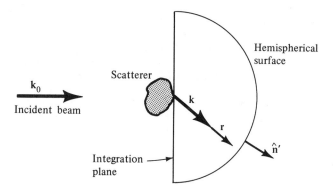

Fig. 6.21. Scattering from a localized target.

show that the scattered wave function, $\psi_s(\mathbf{r})$, is completely specified by the value of the wave function and its derivative on a plane perpendicular to the beam and immediately behind the target. To establish this relation, Eq. (6.64) for the Green's function is multiplied by $\psi_s(\mathbf{r})$ and integrated over some region outside the scatterer:

$$\int d^3r' \psi_s(\mathbf{r'})[\nabla'^2 + k^2]G(\mathbf{r'}, \mathbf{r}) = \int d^3r' \psi_s(\mathbf{r'})\,\delta(\mathbf{r'} - \mathbf{r}) = \psi_s(\mathbf{r}).$$

Outside the target, $\psi_s(\mathbf{r'})$ satisfies the free wave equation, Eq. (6.59) with $V = 0$. Multiplication of this equation with $G(\mathbf{r'}, \mathbf{r})$ and integration over the same region gives

$$\int d^3r' G(\mathbf{r'}, \mathbf{r})\,[\nabla'^2 + k^2]\psi_s(\mathbf{r'}) = 0.$$

Subtraction of the last two equations gives

$$\psi_s(\mathbf{r}) = \int d^3r'\{\psi_s(\mathbf{r'})\nabla'^2 G(\mathbf{r'}, \mathbf{r}) - G(\mathbf{r'}, \mathbf{r})\nabla'^2\psi_s(\mathbf{r'})\}.\tag{6.78}$$

38. J. S. Blair, in *Lectures in Theoretical Physics*, (P. D. Kunz, D. A. Lind, and W. E. Brittin, eds.), Vol. VIII-C, University of Colorado Press, Boulder, 1966. p. 343.

With Green's theorem, this volume integral is changed into the surface integral

$$\psi_s(\mathbf{r}) = \oint ds' \{\psi_s(\mathbf{r}')\hat{\mathbf{n}}' \cdot \nabla' G(\mathbf{r}', \mathbf{r}) - G(\mathbf{r}', \mathbf{r})\hat{\mathbf{n}}' \cdot \nabla' \psi_s(\mathbf{r}')\}. \tag{6.79}$$

Here, $\hat{\mathbf{n}}'$ is a unit vector normal to the surface and pointing outward. If the surface is taken to be a plane behind the target and a hemisphere as shown in Fig. 6.21, then the contribution from the hemisphere at infinity vanishes for finite r. Only the integral over the plane surface behind the target contributes, and the earlier claim has been verified.

Equation (6.79) is exact. To evaluate it, approximations are made. We restrict the discussion to *Fraunhofer scattering*. In optics, Fraunhofer diffraction results if the distances from the scatterer (disk) to the source and to the detector are large compared to the dimension of the scatterer. The relevant condition, $r \gg r'$, can be satisfied here, as can be seen by using Babinet's principle:[37] According to it, the diffraction pattern produced by the target in Fig. 6.18 is the same as that produced by a screen with a hole corresponding to the target. $\psi(\mathbf{r}')$ then is different from zero only in the region of the hole. The integration over ds' in Eq. (6.79) consequently extends only over dimensions comparable to the target size. In subatomic physics, source and detector are always far away from the target and $r \gg r'$ is justified. Inserting the expansion equation (6.72) into $G(\mathbf{r}', \mathbf{r})$ again produces the asymptotic form Eq. (6.73) and (for $r \longrightarrow \infty$)

$$\psi_s(\mathbf{r}) = \frac{e^{ikr}}{r} \left\{ \frac{-1}{4\pi} \int_{plane} ds' [\psi_s(\mathbf{r}')\hat{\mathbf{n}}' \cdot \nabla' e^{-i\mathbf{k} \cdot \mathbf{r}'} - e^{-i\mathbf{k} \cdot \mathbf{r}'}\hat{\mathbf{n}}' \cdot \nabla' \psi_s(\mathbf{r}')] \right\}.$$

Comparison with Eq. (6.61) shows that the quantity in the braces is the scattering amplitude:

$$f(\theta) = \frac{-1}{4\pi} \int_{plane} ds' [\psi_s(\mathbf{r}')\hat{\mathbf{n}}' \cdot \nabla' e^{-i\mathbf{k} \cdot \mathbf{r}'} - e^{-i\mathbf{k} \cdot \mathbf{r}'}\hat{\mathbf{n}}' \cdot \nabla' \psi_s(\mathbf{r}')]. \tag{6.80}$$

As an application, we consider the black disk shown in Fig. 6.20. The integration plane is placed immediately behind the disk. To find the value of $\psi_s(\mathbf{r}')$, we note that it is given by Eq. (6.61) as

$$\psi_s(\mathbf{r}') = \psi(\mathbf{r}') - e^{ikz}. \tag{6.81}$$

For a black scatterer, the total wave $\psi(\mathbf{r}')$ in the integration plane is given everywhere by the plane wave except within the shadow cast by the scatterer, where it is zero. Consequently, the value of $\psi_s(\mathbf{r}')$ is zero everywhere on the plane except behind the disk where it is

$$\psi_s(\mathbf{r}') = -e^{ikz}.$$

The integration plane is set at $z = 0$, and the scattering amplitude becomes

$$f(\theta) = \frac{ik}{4\pi}(1 + \cos\theta) \int_{shadow} ds' e^{-i\mathbf{k} \cdot \mathbf{r}'}.$$

To evaluate the integral, we take the xy plane to be the shadow plane, assume $\mathbf{k}$ to lie in the xz plane, and introduce cylindrical coordinates, z, ρ, and φ. Then $ds' = \rho \, d\rho \, d\varphi$, $\mathbf{k} \cdot \mathbf{r}' = k \sin\theta \rho \cos\varphi$ and

$$f(\theta) = \frac{ik}{4\pi}(1 + \cos\theta) \int_0^{R_0} d\rho \, \rho \int_0^{2\pi} d\varphi \, e^{-ik\rho \sin\theta \cos\varphi}.$$

If $kR_0 \gg 1$, the integral vanishes except for small θ. For small θ, $\cos\theta \approx 1$ and

$$f(\theta) = \frac{ik}{2\pi} \int_0^{R_0} d\rho \, \rho \int_0^{2\pi} d\varphi \, e^{-ik\rho\theta \cos\varphi}.$$

The integral over $d\varphi$ is proportional to the zero-order Bessel function[39],

$$\int_0^{2\pi} d\varphi \, e^{iz \cos \varphi} = 2\pi J_0(z).$$

With

$$\int dz \, z J_0(z) = z J_1(z)$$

we finally obtain

$$f(\theta) = ikR_0^2 \frac{J_1(kR_0\theta)}{kR_0\theta} \tag{6.82}$$

$$\frac{d\sigma}{d\Omega} = (kR_0^2)^2 \left(\frac{J_1(z)}{z}\right)^2, \qquad z = kR_0\theta \tag{6.83}$$

and

$$\sigma = \int d\Omega \frac{d\sigma}{d\Omega} \approx \pi R_0^2. \tag{6.84}$$

The last three relations have been derived for a black disk. However, they are valid also for a black sphere of radius R_0 because sphere and disk have the same shadow.

Equations (6.82)–(6.84) display the three characteristics of Fraunhofer scattering listed earlier:

1. The differential cross section has the shape shown in Fig. 6.20 for $I(\theta)$, with a prominent forward peak.

2. The first zero of $J_1(z)$ occurs at $z = 3.84$; the first minimum of the diffraction pattern therefore appears at

$$\theta_{\min} = \frac{3.84}{kR_0} = 0.61 \frac{\lambda}{R_0}, \tag{6.85}$$

in agreement with estimate Eq. (6.76). The angle $\theta_{\min}$ decreases as $1/k$ for a fixed target size, R_0.

3. The total and the elastic cross sections are independent of energy; they depend only on the size of the interaction region, R_0.

4. In addition to these three characteristics, Eq. (6.83) expresses another fact: Since $J_1(z) \longrightarrow z/2$ in the limit $z \longrightarrow 0$, the differential cross section for elastic forward scattering is given by

$$\frac{d\sigma(0°)}{d\Omega} = \frac{1}{4} k^2 R_0^4. \tag{6.86}$$

At very high energies, the differential cross section at $0°$ increases as k^2 ($k = p/\hbar$).

These observations are often cast in a somewhat different language. In the c.m. the momentum transfer for small-angle elastic scattering is given by $|\mathbf{q}| = 2\hbar k \sin(\theta/2) \approx \hbar k\theta$. In terms of the square of this quantity,

$$-t \equiv \mathbf{q}^2, \tag{6.87}$$

$-dt = 2\hbar^2 k^2 \theta \, d\theta = \hbar^2 k^2 \, d\Omega/\pi$, and Eq. (6.83) is written as

$$\frac{d\sigma}{dt} = \frac{-\pi}{\hbar^2 k^2} \frac{d\sigma}{d\Omega} = \frac{\pi R_0^4}{\hbar^2} \frac{J_1^2(\sqrt{-tR_0^2/\hbar^2})}{tR_0^2/\hbar^2}. \tag{6.88}$$

39. Bessel functions are treated in nearly every text on mathematical physics. See, for instance, Mathews and Walker. For equations, tables, and figures, see M. Abramowitz and I. A. Stegun (eds.), *Handbook of Mathematical Functions*, Government Printing Office, Washington, D.C., 1964 (a best buy).

Instead of Eq. (6.86), the relation

$$\frac{d\sigma}{dt}(t = 0) = \frac{-\pi}{4\hbar^2} R_0^4 \tag{6.89}$$

is obtained. Consequently $d\sigma/dt$ should depend only on t, the square of the momentum transfer and not on the incident energy. The value of $d\sigma/dt$ at $t = 0$ should be independent of the energy of the incident particle.

We shall now present some examples of diffraction scattering in nuclear and particle physics. Consider first *nuclei*.[38] Figure 6.22 shows the differential cross section for

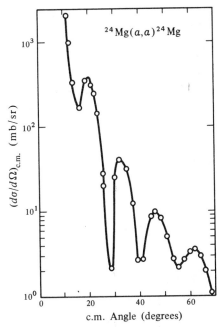

Fig. 6.22. Differential cross section for the elastic scattering of alpha particles from ²⁴Mg. [I. M. Naqib and J. S. Blair, *Phys. Rev.* **165**, 1250 (1968).]

elastic scattering of 42-MeV alpha particles from ²⁴Mg.[40] A sharp forward peak and pronounced diffraction minima and maxima stand out clearly. Equation (6.83) reproduces the position of the minima and maxima well, but with increasing scattering angle, the observed maxima are increasingly smaller than the predicted ones. The reason for the disagreement is obvious: Nuclei do not have sharp edges, as assumed in the derivation of Eq. (6.83). Figure 6.7 indicates that they have a skin of considerable thickness. Moreover, nuclei are not always spherical but may have a permanent deformation, as will be discussed in Section 16.1. Finally, nuclei are partially transparent for low- and medium-energy hadrons. The simple theory can be modified to take these complications into account, and the resulting theory fits the experimental data reasonably well.[41-43]

40. I. M. Naqib and J. S. Blair, *Phys. Rev.* **165**, 1250 (1968).
41. S. Fernbach, R. Serber, and T. B. Taylor, *Phys. Rev.* **75**, 1352 (1949).
42. J. S. Blair, *Phys. Rev.* **115**, 928 (1959).
43. E. V. Inopin and Yu. A. Berezhnoy, *Nucl. Phys.* **63**, 689 (1965).

Diffraction phenomena appear also in *particle* physics.[44-46] We restrict the discussion to elastic proton-proton scattering because it already displays characteristic diffraction features. Differential cross sections, $d\sigma/dt$, for elastic pp scattering for various momenta are shown in Fig. 6.23.[47] The spectacular forward peak stands out clearly,

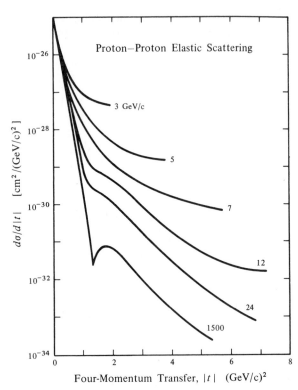

Fig. 6.23. Differential cross section for elastic pp scattering. The parameter assigned to the curves gives the laboratory momentum of the incident protons. The cross sections up to $p_{lab} = 19.3$ GeV/c have been measured at the CERN proton synchrotron; the one for $p_{lab} = 1500$ GeV/c has been obtained with the CERN ISR.

and some other diffraction traits are also evident. In particular, the value of $d\sigma/dt$ at $t = 0$ is approximately independent of the incident momentum, as predicted by Eq. (6.89). With $-d\sigma/dt \approx 10^{-25}$ cm^2/(GeV/c)^2, Eq. (6.89) gives $R_0^2 \approx 0.7$ fm^2. This value is not in disagreement with the value of the electromagnetic radius as given in Eq. (6.51).

44. F. Zachariasen, *Phys. Rept.* **C2**, 1 (1971).

45. B. T. Feld, *Models of Elementary Particles*, Ginn/Blaisdell, Waltham, Mass., 1969, Chapter 11.

46. M. M. Islam, *Phys. Today* **25**, 23 (May 1972).

47. J. V. Allaby et al., *Nuclear Phys.* **B52**, 316 (1973); G. Barbiellini et al., *Phys. Letters* **39B**, 663 (1972); A. Böhm et al., *Phys. Letters* to be published.

Another attribute of diffraction scattering, constancy of the total cross section at high energies, is satisfied over a wide energy range. The total cross section can be obtained in two different ways. One is to measure the decrease in beam intensity after traversing a given thickness of liquid hydrogen. The other is to study the elastic forward scattering and to use the optical theorem, Eq. (6.63). Both methods have been employed, and high-energy measurements have been performed at the 70-GeV proton synchrotron at Serpukhov,[48] at NAL,[49] and at the ISR at CERN.[50] The total cross section is indeed found to be constant within the experimental errors from laboratory energies of about 2 GeV up to 500 GeV; the value is about 39 mb. At ultrahigh energies, the total cross section is no longer constant; ISR[50] and cosmic ray data[51] show a slow increase of σ_{tot} with energy. The total cross section for pp scattering is given in Fig. 6.24 as a function of laboratory momentum. We will briefly return to the increase in Section 12.7.

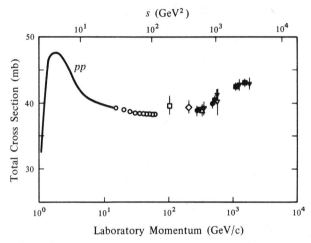

Fig. 6.24. Total proton-proton cross section as a function of the square of the c.m. energy, s, and of the equivalent laboratory momentum. (Data based on experiments carried out at Serpukhov, NAL and at the ISR (CERN). (Courtesy CERN.)

In nuclear physics, the most outstanding diffraction structure is the occurrence of maxima and minima as shown in Fig. 6.22. In particle physics, the smooth distribution of the electric charge and presumably also of nuclear matter washes out the diffraction structure up to momenta of at least 20 GeV/c. At the very highest momenta, however, the first minimum and the following maximum appear as shown in the lowest curve in Fig. 6.23. It is remarkable that the shape of this curve can be explained by using a nuclear matter distribution in the proton that is proportional to the electric distribution described by the dipole form factor, Eq. (6.47).[52]

48. S. P. Denisov et al., *Phys. Letters* **36B**, 415 (1971).

49. V. Bartenev et al., *Phys. Rev. Letters* **29**, 1755 (1972), and *Phys. Rev. Letters* **31**, 1088 (1973).

50. M. Holder et al., *Phys. Letters* **35B**, 361 (1971); S. R. Amendolia et al., *Phys. Letters* **44B**, 119 (1973); U. Amaldi et al., *Phys. Letters* **44B**, 112 (1973).

51. G. B. Yodh, Y. Pal, and J. S. Trefil, *Phys. Rev. Letters* **28**, 1005 (1972).

52. R. Serber, *Rev. Modern Phys.* **36**, 649 (1964); T. T. Chou and C. N. Yang, *Phys. Rev.* **170**, 1591 (1968); L. Durand and R. Lipes, *Phys. Rev. Letters* **20**, 637 (1968); J. N. J. White, *Nuclear Phys.* **B51**, 23 (1973).

The Profile Function.[53-54] The black-disk approximation used so far reproduces the coarse features, but not the finer details, of diffraction scattering. It can be improved by assuming the scatterer to be *gray*. The shadow of a gray scatterer is not uniformly black; its grayness (transmission) is a function of $\boldsymbol{\rho}$, where $\boldsymbol{\rho}$ is the radius vector in the shadow plane (Fig. 6.25). In the black-disk approximation the total wave, $\psi(\mathbf{r}') \equiv$

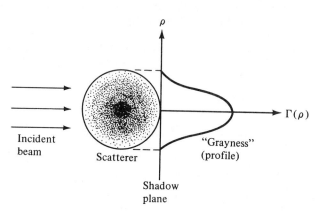

ρ

$\Gamma(\rho)$

Incident
beam

Scatterer

"Grayness"
(profile)

Shadow
plane

Fig. 6.25. Gray scatterer and profile of its shadow. $\Gamma(\rho)$ and ρ are discussed in the text.

$\psi(\boldsymbol{\rho})$, in the shadow plane is zero behind the scatterer, and this fact was used to derive Eq. (6.82). For a gray scatterer it is assumed that the total wave behind the scatterer in the shadow plane is given by

$$\psi(\boldsymbol{\rho}) = e^{i\mathbf{k}_0 \cdot \boldsymbol{\rho}} e^{i\chi(\boldsymbol{\rho})}. \tag{6.90}$$

The total wave is modified by a multiplicative factor. For a black disk, the phase χ is purely imaginary and large. The factor $\exp(i\mathbf{k}_0 \cdot \boldsymbol{\rho})$ is equal to 1, but we keep it because it will turn out to be convenient. With Eq. (6.81) and with $kz = \mathbf{k}_0 \cdot \boldsymbol{\rho}$ in the shadow plane, the scattered wave becomes

$$\psi_s(\boldsymbol{\rho}) = e^{i\mathbf{k}_0 \cdot \boldsymbol{\rho}} \Gamma(\boldsymbol{\rho}), \tag{6.91}$$

where

$$\Gamma(\boldsymbol{\rho}) = 1 - e^{i\chi(\boldsymbol{\rho})} \tag{6.92}$$

is called the profile function.[53] Inserting Eq. (6.91) into Eq. (6.80) produces, with $ds' = d^2\rho$ and $\mathbf{r}' = \boldsymbol{\rho}$, the scattering amplitude at small angles ($\cos\theta \approx 1$):

$$f(\mathbf{q}) = \frac{ik}{2\pi} \int d^2\rho \, e^{i\mathbf{q}\cdot\boldsymbol{\rho}/\hbar} \Gamma(\boldsymbol{\rho}). \tag{6.93}$$

Here $\mathbf{q} = \hbar(\mathbf{k}_0 - \mathbf{k})$ is the momentum transfer. The scattering amplitude is the Fourier transform of the profile function. If the scatterer possesses azimuthal symmetry, integration over the azimuthal angle yields

$$f(\theta) = ik \int d\rho \, \rho\Gamma(\rho) J_0(k\rho\theta). \tag{6.94}$$

53. R. J. Glauber, in *Lectures in Theoretical Physics*, Vol. 1 (W. E. Brittin et al., eds.), Wiley-Interscience, New York, 1959, p. 315; R. J. Glauber, in *High Energy Physics and Nuclear Structure* (G. Alexander, ed.), North-Holland, Amsterdam, 1967, p. 311.
54. W. Czyz, in *The Growth Points of Physics*, Rivista Nuovo Cimento **1**, Special No., 42 1969 (From Conf. European Physical Society).

This expression coincides with $f(\theta)$ for a black scatterer if $\Gamma(\rho) = 1$. The relation connecting $\Gamma(\rho)$ and $f(\theta)$ in Eq. (6.94) is called a Fourier-Bessel (or Hankel) transform.[55] Given a profile function, the scattering amplitude can be calculated. As an example, assume a Gaussian profile function,

$$\Gamma(\rho) = \Gamma(0)\, e^{-(\rho/\rho_0)^2}. \tag{6.95}$$

The Fourier-Bessel transform then becomes[39]

$$f(\theta) = \tfrac{1}{2} ik\Gamma(0)\, \rho_o^2\, e^{-(k\theta\rho_0/2)^2}.$$

With $-t = (\hbar k\theta)^2$, the corresponding differential cross section is

$$-\frac{d\sigma}{dt} = \frac{\pi}{4\hbar^2}\, \Gamma^2(0)\, \rho_0^4\, e^{-(\rho_0^2/2\hbar^2)|t|}. \tag{6.96}$$

A Gaussian profile function leads to an exponentially decreasing cross section $d\sigma/dt$.

The physical interpretation of the profile function becomes clear by considering the total cross section. The optical theorem, Eq. (6.63), with Eq. (6.93) for $\theta = 0°$, yields

$$\sigma_{\text{tot}} = 2\int d^2\rho\, \text{Re}\Gamma(\rho). \tag{6.97}$$

For a black scatterer, $\Gamma(\rho) = 1$ is real, and $f(\theta)$ is purely imaginary. If we assume that in the limit of very high energy the amplitude is imaginary,[56] then Γ is real, and Eq. (6.97) becomes

$$\sigma_{\text{tot}} = 2\int d^2\rho\Gamma(\rho). \tag{6.98}$$

$2\Gamma(\rho)$ can consequently be interpreted as the probability that scattering occurs in the element $d^2\rho$ at the distance ρ from the center. (See Fig. 6.25.) $\Gamma(\rho)$ is the scattering probability density distribution in the shadow plane; hence the name profile function.

As an application of these considerations, we return to elastic pp scattering.[52] Figure 6.23 shows that the diffraction peak drops exponentially for many orders of magnitude. This behavior suggests that the cross section in the region of the forward peak can be approximated by

$$\frac{d\sigma}{dt}(s, t) = \frac{d\sigma}{dt}(s, t = 0)e^{-b(s)|t|}. \tag{6.99}$$

Here, s is the conventional symbol for the square of the total energy of the colliding protons in their c.m. and $b(s)$ is called the slope parameter. It is remarkable that the experimental data over a wide range of s and t can indeed be fitted by such a simple expression. The slope parameter turns out to be a slowly varying logarithmic function of the total energy s, as shown in Fig. 6.26. The exponential drop of $d\sigma/dt$ can be interpreted in terms of a Gaussian profile function, as given in Eq. (6.95). Identification of Eqs. (6.96) and (6.99) leads to the relation

$$\rho_o = \hbar(2b)^{1/2}. \tag{6.100}$$

ρ_o characterizes the width of the Gaussian profile function describing the scattering of two extended protons by hadronic forces. It is therefore not legitimate to compare ρ_o^2, or a corresponding mean-square radius, directly with the mean-square radius of the proton as determined with electromagnetic probes. Nevertheless, it is reassuring that the two measures of the proton size are comparable: The electromagnetic radius is

55. W. Magnus and F. Oberhettinger, *Formulas and Theorems for the Functions of Mathematical Physics*, Chelsea, New York, 1954, pp. 136, 137.

56. The ratio between the real and the imaginary part of the proton-proton forward scattering amplitude has been measured and indeed becomes small at high incident momenta. (G. G. Beznogikh, *Phys. Letters* **39B**, 411 (1972).)

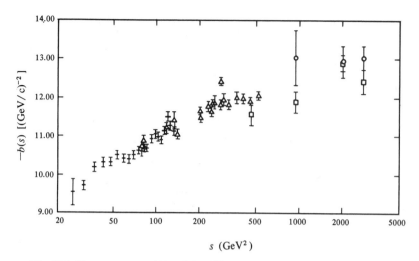

Fig. 6.26. Slope parameter, $b(s)$, of the diffraction peak for elastic pp scattering as a function of the square of the c.m. energy, s. The figure is taken from Bartenev et al., *Phys. Rev. Letters* (to be published) and also contains data from other groups.

given by Eq. (6.51) as $\langle r^2 \rangle^{1/2} \approx 0.8$ fm, whereas a value of $b = 10$ $(GeV/c)^{-2}$, taken from Fig. 6.26, leads to $\rho_0 \approx 0.9$ fm.

The "size" of the proton and the slope parameter $b(s)$ are related through Eq. (6.100); a constant ρ_0 implies a constant $b(s)$. Figure 6.26 shows, however, that $b(s)$ increases logarithmically with the square of the c.m. energy. Since $b(s)$ describes the width of the diffraction peak, increasing $b(s)$ means a shrinking diffraction peak, and it suggests an increase in the size, ρ_0, of the interaction region. This behavior is not fully understood.

The Glauber Approximation.[53-54] So far we have treated diffraction scattering from a single object. We shall now turn to the coherent scattering of a projectile from a target made up of several subunits, for instance, a nucleus built from nucleons. An incoming high-energy particle can collide with a single nucleon, with many in succession, or it can interact strongly with several at once. The treatment of such a multiscattering process is difficult, but diffraction theory makes the problem manageable; it leads to the Glauber approximation.[57]

To arrive at the Glauber approximation, we consider first the optical analog, the passage of a light wave with momentum $p = \hbar k$ through a medium with index of refraction n and thickness d. The electric vector, $\boldsymbol{\varepsilon}_1$, after passage of the wave through the absorber is related to the electric vector of the incident wave, $\boldsymbol{\varepsilon}_0$, by[58]

$$\boldsymbol{\varepsilon}_1 = \boldsymbol{\varepsilon}_0 e^{i\chi_1}, \qquad \chi_1 = k(1-n)d. \tag{6.101}$$

If the index of refraction is complex, then its imaginary part describes the absorption of the wave. If the wave traverses successive absorbers, each characterized by a phase χ_i, the end result is

$$\boldsymbol{\varepsilon}_n = \boldsymbol{\varepsilon}_0 e^{i\chi_1} e^{i\chi_2} \cdots e^{i\chi_n} = \boldsymbol{\varepsilon}_0 e^{i(\chi_1 + \cdots + \chi_n)}. \tag{6.102}$$

The phases of the various absorbers add. The same technique can be applied to the scattering of high-energy particles. Equation (6.90) shows that the wave behind a single

57. R. J. Glauber, *Phys. Rev.* **100**, 252 (1955).
58. The Feynman Lectures 1-31-3.

scatterer is related to the incident wave as the electric waves are related in Eq. (6.101). In the Glauber approximation it is assumed that the phases from the individual scatterers in a compound system, such as a nucleus, also add. To formulate the approximation, we assume that the individual scatterers are arranged as shown in Fig. 6.27. The distance

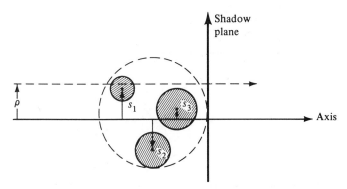

Fig. 6.27. Arrangement of the individual scatterers in a nucleus.

of the center of each scatterer to the axis perpendicular to the shadow plane is denoted by s_i. The distance that determined the profile function for each nucleon is no longer ρ but $\rho - s_i$, and the phase factor for the ith nucleon is given by Eq. (6.92) as

$$e^{i\chi_i} = 1 - \Gamma_i(\rho - s_i).$$

For the total phase factor, additivity of the individual phases gives

$$e^{i\chi} = e^{i\chi_1}e^{i\chi_2}\cdots e^{i\chi_A} = \prod_{i=1}^{A}[1 - \Gamma_i(\rho - s_i)],$$

and for the complete profile function

$$\Gamma(\rho) = 1 - \prod_{i=1}^{A}[1 - \Gamma_i(\rho - s_i)]. \tag{6.103}$$

This relation describes the Glauber approximation. If the profile functions for the individual nucleons are known, the profile function for the entire nucleus can be calculated. One more step is needed to arrive at the Glauber expression for the scattering amplitude. Nucleons are not fixed, as shown in Fig. 6.27; they move around and their probability distribution is given by the relevant wave function. For *elastic scattering*, initial and final wave functions are identical, and $\Gamma(\rho)$ in Eq. (6.93) must be replaced by

$$\int d^3x_1 \cdots d^3x_A \, \psi^*(\mathbf{x}_1, \cdots, \mathbf{x}_A)\Gamma(\rho)\psi(\mathbf{x}_1 \cdots, \mathbf{x}_A) \equiv \langle i|\Gamma(\rho)|i\rangle.$$

The scattering amplitude equation (6.93) thus becomes

$$f(\mathbf{q}) = \frac{ik}{2\pi}\int d^2\rho \, e^{i\mathbf{q}\cdot\boldsymbol{\rho}/\hbar}\langle i|\Gamma(\rho)|i\rangle, \tag{6.104}$$

with an inverse which is

$$\langle i|\Gamma(\rho)|i\rangle = \frac{1}{2\pi ik}\int e^{-i\mathbf{q}\cdot\boldsymbol{\rho}/\hbar} f(\mathbf{q})d^2q.$$

As an example, we consider elastic scattering of a high-energy projectile from the simplest nucleus, the deuteron (Fig. 6.28). When the energy of the incident particle is so high that its wavelength is much smaller than the deuteron radius ($R \approx 4$ fm), one could at first assume that neutron and proton scatter independently and that the total

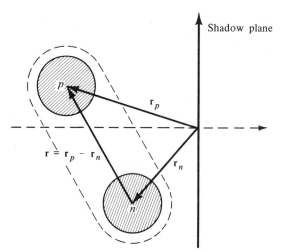

Fig. 6.28. Coordinates used in the description of the scattering from deuterons.

cross section is simply the sum of the individual ones. Use of the Glauber approximation shows that this assumption is wrong, and experiment bears out the calculations. For the deuteron, with $\mathbf{r} = \mathbf{r}_p - \mathbf{r}_n$, Eq. (6.103) becomes

$$\Gamma_d(\boldsymbol{\rho}) = \Gamma_p\left(\boldsymbol{\rho} + \frac{1}{2}\mathbf{r}\right) + \Gamma_n\left(\boldsymbol{\rho} - \frac{1}{2}\mathbf{r}\right) - \Gamma_p\left(\boldsymbol{\rho} + \frac{1}{2}\mathbf{r}\right)\Gamma_n\left(\boldsymbol{\rho} - \frac{1}{2}\mathbf{r}\right). \quad (6.105)$$

Inserting $\Gamma_d(\rho)$ into Eq. (6.104), and using the fact that the deuteron wave function, $\psi_d(\mathbf{r})$, is only a function of the relative coordinate $\mathbf{r}$, gives, for the scattering function of the deuteron,

$$f_d(\mathbf{q}) = f_p(\mathbf{q})F\left(\frac{1}{2}\mathbf{q}\right) + f_n(\mathbf{q})F\left(\frac{1}{2}\mathbf{q}\right) + \frac{i}{2\pi k}\int F(\mathbf{q}')f_p\left(\frac{1}{2}\mathbf{q} - \mathbf{q}'\right)f_n\left(\frac{1}{2}\mathbf{q} + \mathbf{q}'\right)d^2q',$$
$$(6.106)$$

where $F(\mathbf{q})$ is the form factor for the deuteron ground state,

$$F(\mathbf{q}) = \int d^3r\, e^{i\mathbf{q}\cdot\mathbf{r}/\hbar}\, |\psi_d(\mathbf{r})|^2. \quad (6.107)$$

The first two terms in Eq. (6.106) describe the individual scatterings; the last one represents the double scattering correction. For the total cross section, the optical theorem Eq. (6.63) yields

$$\sigma_d = \sigma_p + \sigma_n + \frac{2}{k^2}\int d^2q\, F(\mathbf{q})\, \text{Re}\{f_p(-\mathbf{q})f_n(\mathbf{q})\}. \quad (6.108)$$

The deuteron radius is considerably larger than the range of the hadronic interaction; the form factor $F(\mathbf{q})$ hence is sharply peaked in the forward direction, and the total cross section becomes

$$\sigma_d \approx \sigma_p + \sigma_n + \frac{2}{k^2}\, \text{Re}[f_p(0)f_n(0)]\langle r^{-2}\rangle_d,$$

where $\langle r^{-2}\rangle_d$ is the expectation value of r^{-2} in the deuteron ground state. If the scattering is again assumed to be entirely absorptive so that the forward scattering amplitudes are imaginary, then

$$\sigma_d \approx \sigma_p + \sigma_n - \frac{1}{4\pi}\, \sigma_p\sigma_n\langle r^{-2}\rangle_d. \quad (6.109)$$

The last term here shows the shadow effect of one nucleon on the other one. The shadow or double scattering term has a negative sign: the total cross section is smaller than the sum of that from the individual nucleons. This feature follows already from Eq. (6.105), where the double scattering contribution has the opposite sign from the single scattering one. More generally, expansion of Eq. (6.103) shows that the signs of successive terms alternate. This behavior has been verified experimentally.

The angular distribution of the scattering from deuterons provides considerably more information than the total cross section. With Eqs. (6.10) and (6.88), $d\sigma/dt$ is

$$\frac{d\sigma}{dt} = \frac{-\pi}{\hbar^2 k^2}|f(\mathbf{q})|^2. \tag{6.110}$$

To compute $d\sigma/dt$, $f_d(\mathbf{q})$ from Eq. (6.106) is inserted into Eq. (6.110). Consider specifically proton-deuteron scattering. The scattering amplitudes f_n and f_p can then be obtained from pp and np scattering; the corresponding ideas have already been treated beginning on p. 137. To find the form factor $F(\mathbf{q})$, a specific form of the deuteron wave function must be assumed; for a given ψ_d, $f_d(\mathbf{q})$ and hence $d\sigma/dt$ can be calculated. Figure 6.29 shows $d\sigma/dt$ for scattering of 1- and 2-GeV protons from deuterons. Some

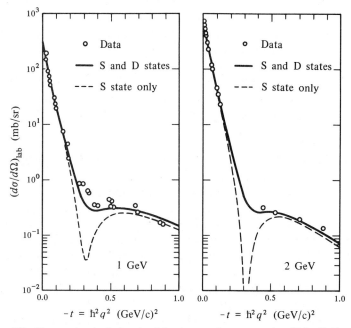

Fig. 6.29. Measured and calculated p-d elastic scattering cross section. [After V. Franco and R. J. Glauber, *Phys. Rev. Letters* **22**, 370 (1969).]

characteristic features stand out: an initial rapid drop, a shallow minimum, and then a slower decrease in $d\sigma/dt$. These features can be understood with Eq. (6.106). The first two terms, corresponding to single scattering, possess diffraction peaks of widths $\propto 1/k$, as indicated in Eq. (6.85). In double scattering, each nucleon absorbs half the momentum transfer; the corresponding diffraction width is larger. The first rapid drop-off is due to single scattering; the double scattering dominates at larger values of t. The explicit calculation of $d\sigma/dt$ shows that scattering indeed explores the structure of a nucleus.[59] As we shall discuss in more detail in Section 12.5, the two nucleons in the deuteron are predominantly in a state with relative orbital angular momentum $L = 0$ (s state), but there is a small admixture of angular momentum $L = 2$ (d state) (Fig. 12.9). If the cross section is calculated with the s-state wave function only, the dashed curves in Fig. 6.29 result, and they disagree with the experimental data. To obtain the good agreement exhibited by the solid lines, a small d-state admixture (6.7%) is required. This small admixture washes out the deep interference minimum between single and double scattering.

The technique described here for the deuteron has been used to explore the structure of other nuclides.[53,54,60] It can also be applied if particles other than the proton, for instance, pion or antiproton, are employed as probes. ●

6.10 References

R. Hofstadter, *Nuclear and Nucleon Structure*, Benjamin, Reading, Mass., 1963. The important papers concerning the structure of subatomic papers are reprinted in this volume. Among the reprints are a few excellent reviews. In searching for pre-1963 information on the structure of subatomic particles, it is a good idea to look here first.

Problems treated in the present chapter are discussed in a number of popular articles. Of particular interest are the following:

H. R. Crane, "The g Factor of the Electron," *Sci. Amer.* **218**, 72 (January 1968).

H. W. Kendall and W. K. H. Panofsky, *Sci. Amer.* **224**, 61 (June 1971).

M. J. Perl, "How Does the Muon Differ from the Electron?," *Phys. Today*, **24** 34 (July 1971).

The lepton g factors are reviewed in A. Rich and J. C. Wesley, *Rev. Modern Phys.* **44**, 250 (1972).

A good first introduction to the ideas and the various techniques underlying the study of nuclear sizes is L. R. B. Elton, *Nuclear Sizes*, Oxford University Press, London, 1961. A comprehensive discussion of structure studies using electrons is H. Überall, *Electron Scattering from Complex Nuclei*, Academic Press, New York, 1971 (two volumes).

In the present chapter, only one technique for determining the nuclear charge distribution has been treated, namely elastic electron scattering. However, many other approaches exist. Of particular importance is the observation of muonic X rays. This topic is reviewed in the following publications:

S. Devons and I. Dueroth, *Advan. Nucl. Phys.* **2**, 295 (1969).

C. S. Wu and L. Wilets, *Ann. Rev. Nucl. Sci.* **19**, 527 (1969).

Y. N. Kim, *Mesic Atoms and Nuclear Structure*, American Elsevier–North-Holland, New York, 1971.

59. V. Franco and R. J. Glauber, *Phys. Rev. Letters* **22**, 370 (1969).
60. W. Czyz, *Advan. Nucl. Phys*, **4**, 61 (1971).

The experimental data on nuclear radii are summarized in R. Hof-stadter and H. R. Collard, in Landolt-Börnstein, *Numerical Data and Functional Relationships in Science and Technology*, Group I, Vol. 2: *Nuclear Radii* (H. Schopper, ed.), Springer, Berlin, 1967.

No really elementary treatment of the form factors of elementary particles is known to us. A diligent student may be able to extract some information from the following reviews, written at a higher level than the present book:

S. D. Drell, "Form Factors of Elementary Particles," in *Selected Topics on Elementary Particle Physics*," Academic Press, New York, 1963.

T. A. Griffy and L. I. Schiff, "Electromagnetic Form Factors," in *High-Energy Physics*, Vol. 1, Academic Press, New York, 1967.

D. Bartoli, F. Felicetti, and V. Silvestrini, "Electromagnetic Structure of Hadrons," Rivista Nuovo Cimento **2**, 241 (1972).

PROBLEMS

6.1. An electron beam of 10-GeV energy and a current of 10^{-8} A is focused onto an area of 0.5 cm². What is the flux F?

6.2. Assume that a beam pulse at a 100-GeV accelerator contains 10^{13} protons, is focused onto a 2-cm² area, and is extracted uniformly over a time of 0.5 sec. Compute the flux.

6.3. A copper target of thickness 0.1 cm intercepts a particle beam of 4-cm² area. Nuclear scattering is observed.

(a) Compute the number of scattering centers intercepted by the beam.

(b) Assume a total cross section of 10 mb for an interaction. What fraction of the incident beam is scattered?

6.4. Positive pions of kinetic energy of 190 MeV impinge on a 50-cm-long liquid hydrogen target. What fraction of the pions undergoes pion-proton scattering? (See Fig. 5.28.)

6.5. Consider the collision of an alpha particle with an electron. Show that the maximum energy loss and the maximum momentum transfer in one collision are small. Compute the maximum energy loss that a 10-MeV alpha particle can suffer by striking an electron at rest.

6.6. Sketch the classical derivation of the Rutherford scattering formula.

6.7. Show that Eq. (6.14) follows from Eq. (6.13) for a spherically symmetric potential.

6.8. Verify Eq. (6.16).

6.9.

(a) Show that in all experiments that can give information concerning the structure of subatomic particles the term $(\hbar/a)^2$ in Eq. (6.16) can be neglected.

(b) For what scattering angles is the correction term $(\hbar/a)^2$ important?

6.10. Rewrite Eq. (6.17) in terms of the kinetic energy of the incident particle and of the scattering angle. Verify that the resulting expression agrees with the standard Rutherford formula.

6.11. An electron of 100-MeV energy strikes a lead nucleus.

(a) Compute the maximum possible momentum transfer.

(b) Compute the recoil energy given to the lead nucleus under the conditions of part (a).

(c) Show that the electron can be treated as a massless particle for this problem.

6.12. Verify Eq. (6.28) and find the next term in the expansion.

6.13. Assume that the probability distribution is given by $(x = |\mathbf{x}|)$

$$\rho(x) = \rho_0 \qquad x \leq R$$
$$\rho(x) = 0 \qquad \text{for } x > R.$$

(a) Compute the form factor for this "uniform charge distribution."
(b) Calculate $\langle x^2 \rangle^{1/2}$.

6.14. 250-MeV electrons are scattered from ^{40}Ca.
(a) Use equations given in the text to compute numerically values of the cross section as a function of the scattering angle for the following assumptions:
    (a1) Spinless electrons, point nucleus.
    (a2) Electrons with spin, point nucleus.
    (a3) Electrons with spin, "Gaussian" nucleus [Eq. (6.31)].
(b) Find experimental values for the cross section and compare with your computations. Determine a value for b in Eq. (6.31).

6.15.
(a) What are muonic atoms?
(b) Why can muonic atoms be used to study nuclear structure?
(c) Compute the value of the $2p - 1s$ muonic transition in ^{208}Pb under the assumption that Pb is a point nucleus. Compare with the observed value of 5.8 MeV.
(d) Use the values computed and given in part (c) to estimate the nuclear radius of Pb. Compare with the actual value.

6.16. Use Eq. (6.26) to determine the normalization constant N in Eq. (6.32).

6.17. Use the values given in Eq. (6.35) to find an average value for the internucleon distance in a nucleus.

6.18. Discuss the $g - 2$ experiments for the electron and the muon.
(a) Derive Eq. (6.41) for the nonrelativistic case.
(b) Sketch the experimental arrangement for the $g - 2$ experiment for negative electrons. How were the electrons polarized? How was the polarization at the end measured?
(c) Repeat part (b) for muons.

6.19. How did Stern, Estermann, and Frisch determine the magnetic moment of the proton?

6.20.
(a) How was the magnetic moment of the neutron first determined (indirect method)?
(b) Discuss a direct method to determine the magnetic moment of the free neutron.
(c) Can storage rings for neutrons be designed? If yes, sketch a possible arrangement and describe the physical ideas.

6.21. Assume that a neutron consists part of the time of a Dirac neutron with 0 magnetic moment and part of the time of a Dirac proton (1 nuclear magneton) plus a negative pion. Assume that the negative pion and the Dirac proton form a system with an orbital angular momentum of 1. Estimate the fraction of time during which the physical neutron has to be in the proton-pion state in order to get the observed magnetic moment.

6.22. Verify Eqs. (6.50) and (6.51).

6.23. Discuss one of the methods used to determine the mean-square electric charge radius of the neutron from the scattering of slow neutrons from matter.

PART III

Symmetries and Conservation Laws

If the laws of the subatomic world were fully known, there would no longer be a need for investigating symmetries and conservation laws. The state of any part of the world could be calculated from a master equation that would contain all symmetries and conservation laws. In classical electrodynamics, for example, the Maxwell equations already contain the symmetries and the conservation laws. In subatomic physics, however, the fundamental equations are not yet established, as we shall see in Part IV. The exploration of the various symmetries and conservation laws, and of their consequences, therefore provides essential clues for the construction of the missing equations. One particular consequence of a symmetry is of the utmost importance: *Whenever a law is invariant under a certain symmetry operation there usually exists a corresponding conservation principle*. Invariance under translation in time, for instance, leads to conservation of energy; invariance under spatial rotation leads to conservation of angular momentum. This profound connection is used both ways: If a symmetry is found or suspected, the corresponding conserved quantity is searched for until it is discovered. If a conserved quantity turns up, the search is on for the corresponding symmetry principle. One word of warning is in place here: Intuitive feelings can be misleading. Often a certain symmetry principle looks attractive but turns out to be partially or completely wrong. Experiment is the only judge as to whether a symmetry principle holds.

Conserved quantities can be used to label states. A particle can be characterized by its mass or rest energy because energy is conserved. Or consider the electric charge, q. It is conserved and comes only in units of the elementary quantum e. The value of q/e can thus be used to distinguish particles of the same mass. Positive, neutral, and negative pions can be christened; pion is the family and positive the first name.

In the next three chapters we shall discuss a number of symmetries and conservation laws. Additional symmetries exist, and we shall encounter one of them later on. Some of the symmetries are perfect even under closest scrutiny, and no breakdown in the corresponding conservation law has ever been found. Rotational symmetry and conservation of angular momentum are one example of this "perfect" class. Other symmetries are "broken," and the corresponding conservation law holds only approximately. Invariance under mirroring (parity) provides one example of such a broken symmetry. At the present time it is not understood why some symmetries are broken and others are not. It is not even clear whether the question should be phrased "Why are symmetries broken?" or "Why are some symmetries perfect?" We must continue to explore symmetries and their consequences and hope that a more complete understanding will be reached at some point.[1]

1. The meaning of symmetries in physics, and more generally, in human endeavor are beautifully described in the following references:

R. P. Feynman, R. B. Leighton, and M. L. Sands, *The Feynman Lectures on Physics*, Vol. I, Addison-Wesley, Reading, Mass., 1963, Chapter 52.

H. Weyl, *Symmetry*, Princeton University Press, Princeton, N.J., 1952.

E. P. Wigner, *Symmetries and Reflections*, Indiana University Press, Bloomington, 1967.

C. N. Yang, *Elementary Particles*, Princeton University Press, Princeton, N.J., 1962.

7

Additive Conservation Laws

In this chapter we shall first discuss the connection between conserved quantities and symmetries in a general way. Such a discussion is somewhat formal, but it paves the way for an understanding of the connection between symmetries and invariances. We shall then treat some additive conservation laws, beginning with the electric charge. The electric charge is the prototype of a quantity that satisfies an additive conservation law: The charge of an assembly of particles is the algebraic sum of the charges of the individual particles. Moreover it is quantized and has only been found in multiples of the elementary quantum e. Other additive conserved and quantized observables exist, and in the present chapter we shall discuss the ones that are established beyond doubt.

7.1 Conserved Quantities and Symmetries

When Is a Physical Quantity Conserved? To answer this question, we consider a system described by a time-independent Hamiltonian H. The wave function of this system satisfies the Schrödinger equation,

$$i\hbar\frac{d\psi}{dt} = H\psi. \tag{7.1}$$

The value of an observable[1] F in the state $\psi(t)$ is given by the expectation value, $\langle F \rangle$. When is $\langle F \rangle$ independent of time? To find out, we assume that the operator F does not depend on t, and we compute $(d/dt)\langle F \rangle$:

$$\frac{d}{dt}\langle F \rangle = \frac{d}{dt}\int d^3x\, \psi^*F\psi = \int d^3x\, \frac{d\psi^*}{dt}F\psi + \int d^3x\, \psi^*F\frac{d\psi}{dt}.$$

To evaluate the last expression, the complex conjugate Schrödinger equation is needed:

$$-i\hbar\frac{d\psi^*}{dt} = (H\psi)^* = \psi^*H. \tag{7.2}$$

Here the reality of H has been used. With Eqs. (7.1) and (7.2), $(d/dt)\langle F \rangle$ becomes

$$\frac{d}{dt}\langle F \rangle = \frac{i}{\hbar}\int d^3x\, \psi^*(HF - FH)\psi. \tag{7.3}$$

The term $HF - FH$ is called the *commutator* of H and F and it is denoted by brackets:

$$HF - FH \equiv [H, F]. \tag{7.4}$$

Equation (7.3) shows that $\langle F \rangle$ is conserved (i.e., is a constant of the motion) if the commutator of H and F vanishes:

$$[H, F] = 0 \longrightarrow \frac{d}{dt}\langle F \rangle = 0. \tag{7.5}$$

If H and F commute, the eigenfunctions of H can be chosen so that they are also eigenfunctions of F,

$$\begin{aligned} H\psi &= E\psi \\ F\psi &= f\psi. \end{aligned} \tag{7.6}$$

Here, E is the energy eigenvalue and f is the eigenvalue of the operator F in state ψ.

---

1. It is a well-known fact that the concepts of *observable* and *matrix element* are at first foreign to most students. Continuous exposure and occasional rereading of a quantum mechanics text—for instance, Eisberg, Section 7.8, or, even better, Chapter 8 of Merzbacher—will remove the problem. We only remark that an observable is represented by a quantum mechanical operator F whose expectation value corresponds to a measurement. The expectation value of F in the state ψ_a is defined as

$$\langle F \rangle = \int d^3x\, \psi_a^*(\mathbf{x})F\psi_a(\mathbf{x}).$$

Since the expectation value of F can be measured, it must be real, and F therefore must be Hermitian. If two states are considered, a quantity similar to $\langle F \rangle$ can be formed by writing

$$F_{ba} = \int d^3x\, \psi_b^*(\mathbf{x})F\psi_a(\mathbf{x}).$$

F_{ba} is called the matrix element of F between states a and b. The expectation value of F in state a is the diagonal element of F_{ba} for $b = a$:

$$\langle F \rangle = F_{aa}.$$

The off-diagonal elements do not correspond directly to classical quantities. However, transitions between states a and b are related to F_{ba} (Eisberg, Section 9.2; Merzbacher, Section 5.4).

How Can Conserved Quantities Be Found? After resolving the question as to when an observable is conserved, we attack the more physical problem: *How can conserved quantities be found?* The direct approach, writing down H and inserting all observables into the commutator, is usually not feasible because H is not fully known. Fortunately, H does not have to be known explicitly; a conserved observable can be found if the invariance of H under a symmetry operation is established. To define *symmetry operation*, we introduce a transformation operator U. U changes a wave function $\psi(\mathbf{x}, t)$ into another wave function $\psi'(\mathbf{x}, t)$:

$$\psi'(\mathbf{x}, t) = U\psi(\mathbf{x}, t). \tag{7.7}$$

Such a transformation is admissible only if the normalization of the wave function is not changed:

$$\int d^3x\, \psi^*\psi = \int d^3x\, (U\psi)^*U\psi = \int d^3x\, \psi^*U^\dagger U\psi.$$

The transformation operator U consequently must be *unitary*,[2]

$$U^\dagger U = UU^\dagger = I. \tag{7.8}$$

U is a *symmetry* operator if $U\psi$ satisfies the same Schrödinger equation as ψ. From

$$i\hbar \frac{d(U\psi)}{dt} = HU\psi$$

it follows that

$$i\hbar \frac{d\psi}{dt} = U^{-1}HU\psi,$$

where U is assumed to be time-independent and where U^{-1} is the inverse operator. Comparison with Eq. (7.1) gives

$$H = U^{-1}HU = U^\dagger HU$$

or

$$HU - UH \equiv [H, U] = 0. \tag{7.9}$$

The symmetry operator U commutes with the Hamiltonian.

Comparison of Eqs. (7.5) and (7.9) shows the way to find conserved observables: If U is Hermitian, it will be an observable. If U is not Hermitian, a Hermitian operator can be found that is related to U and satisfies

---

2. *Notation and definitions:* If A is an operator, the Hermitian adjoint operator $A^\dagger$ is *defined* by

$$\int d^3x\, (A\psi)^*\phi = \int d^3x\, \psi^*A^\dagger\phi.$$

The operator A is Hermitian if $A^\dagger = A$; it is unitary if $A^\dagger = A^{-1}$ or $A^\dagger A = 1$. Unitary operators are generalizations of $e^{i\alpha}$, the complex numbers of absolute value 1 (Merzbacher, Chapter 14). *Notation:* If A is a matrix with elements a_{ik}, A^* with elements a_{ik}^* is the complex conjugate matrix. $\tilde{A}$ with elements a_{ki} is the transposed matrix. $A^\dagger$ with elements a_{ki}^* is the Hermitian conjugate (H.C.) matrix. $(AB)^\dagger = B^\dagger A^\dagger$. I is the unit matrix. The matrix F is called Hermitian if $F^\dagger = F$. The matrix U is unitary if $U^\dagger U = UU^\dagger = I$.

Eq. (7.5). Before giving an example of such a related operator, we recapitulate the essential facts about the operators F and U.

The operator F is an *observable;* it represents a physical quantity. Its expectation values must be real in order to correspond to measured values, and F consequently must be *Hermitian,*

$$F^\dagger = F. \tag{7.10}$$

The *transformation operator* U is unitary; it changes one wave function into another one, as in Eq. (7.7).

In general, transformation operators are not Hermitian and consequently do not correspond to observables. However, there exist exceptions, and to discuss these, first we note that nature contains two types of transformations, *continuous* and *noncontinuous* ones. The continuous ones connect smoothly to the unit operator; the noncontinuous ones do not. Among the latter category we find the operators that are simultaneously unitary and Hermitian. Consider, for instance, the parity operation (space inversion) which changes $\mathbf{x}$ into $-\mathbf{x}$ and represents a mirroring at the origin. Such an operation is obviously not continuous; it is impossible to mirror "just a little bit." Mirroring is either done or not done. If space inversion is performed twice, the original situation is regained; noncontinuous operators often have this property:

$$U_h^2 = 1. \tag{7.11}$$

As can be seen from Eqs. (7.8) and (7.10), U_h then is unitary *and* Hermitian and it is an observable.

A well-known example of a continuous transformation is the ordinary rotation. A rotation about a given axis can occur through any arbitrary angle, α, and α can be made as small as desired. In general, a continuous transformation can always be made so small that its operator approaches the unit operator. The operator U for a continuous transformation can be written in the form

$$U = e^{i\epsilon F} \tag{7.12}$$

where ϵ is a real parameter and where F is called the generator of U. The action of such an exponential operator on a wave function ψ is defined by

$$U\psi = e^{i\epsilon F}\psi \equiv \left[1 + i\epsilon F + \frac{(i\epsilon F)^2}{2!} + \cdots\right]\psi.$$

As a rule $e^{i\epsilon F} \neq e^{-i\epsilon F^\dagger}$ and U is not Hermitian. However, the unitarity condition, Eq. (7.8), yields (if $[F, F^\dagger] = 0$)

$$e^{-i\epsilon F^\dagger}e^{i\epsilon F} = e^{i\epsilon(F-F^\dagger)} = 1$$

or

$$F^\dagger = F. \tag{7.13}$$

The generator F of the transformation operator U is a Hermitian operator, and it is the observable connected to U if U is not Hermitian. To find F, it is usually most advantageous to consider only infinitesimally small transformations:

$$U = e^{i\epsilon F} \longrightarrow U = 1 + i\epsilon F, \qquad \epsilon F \ll 1. \tag{7.14}$$

If a system is invariant under the finite transformation, it surely is invariant under the infinitesimal transformation, and investigation of infinitesimal transformations is much less cumbersome than that of finite transformations. In particular, if U is a symmetry operator, it commutes with H, as shown by Eq. (7.9). Inserting the expansion (7.14) into Eq. (7.9) gives

$$H(1 + i\epsilon F) - (1 + i\epsilon F)H = 0$$

or

$$[H, F] = 0. \tag{7.15}$$

The generator F is a Hermitian operator that is conserved if U is conserved.

The arguments in the present section have been quite formal and abstract. The applications will show, however, that the rather dry considerations have far-reaching consequences. Continuous and noncontinuous transformations play important roles in subatomic physics. Invariance under a continuous transformation leads to an additive conservation law, and relevant examples will be discussed in the present chapter and the following chapter. Invariance under a noncontinuous transformation can lead to a multiplicative conservation law, and specific examples will be given in Chapter 9.

An Example. The treatment in the following sections and chapters is concentrated, and we therefore present first one simple example in considerable detail, in order to make the following cases easier to digest.

We consider the behavior of a particle (or system) moving in one dimension, x. Two positions of the particle, together with the corresponding wave functions, are shown in Fig. 7.1. $\psi(x)$ is the wave function of the

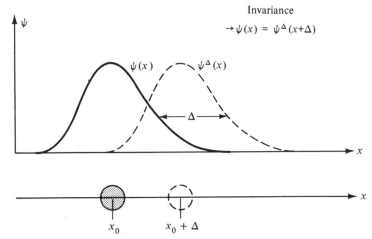

Fig. 7.1. Particle in one dimension. Two different positions and the corresponding wave functions are shown. The two positions are displaced by a distance Δ.

particle centered at position x_0 and $\psi^\Delta(x)$ is the wave function of the particle that has been displaced by the distance Δ. According to Eq. (7.7), ψ and ψ^Δ *at the same point* x are connected by a transformation operator U,

$$\psi^\Delta(x) = U(\Delta)\psi(x). \tag{7.7a}$$

So far, no invariance arguments have been used, and the wave functions ψ and ψ^Δ can have completely different shape. If the system is *invariant under translation*, ψ and ψ^Δ satisfy the same Schrödinger equation, and H and U commute. The invariance implies that the wave function does not change shape as it is displaced with the particle along x, and hence, as is apparent from Fig. 7.1,

$$\psi(x) = \psi^\Delta(x + \Delta).$$

The goal is now to find an explicit expression for the symmetry operator U and for the corresponding generator F. For infinitesimally small displacements Δ, expansion of the last equation gives

$$\psi(x) \approx \psi^\Delta(x) + \frac{d\psi^\Delta(x)}{dx}\Delta = \left(1 + \Delta\frac{d}{dx}\right)\psi^\Delta(x).$$

Multiplication from the left with $(1 - \Delta d/dx)$ and neglecting the term proportional to Δ^2 yields

$$\psi^\Delta(x) \approx \left(1 - \Delta\frac{d}{dx}\right)\psi(x).$$

Comparison with Eq. (7.7a) shows that

$$U(\Delta) \approx 1 - \Delta\frac{d}{dx}.$$

The general infinitesimal operator U is shown in Eq. (7.14); identifying the real parameter ϵ with the displacement Δ demonstrates that the generator F is proportional to the momentum operator p_x:

$$F = i\frac{d}{dx} = -\frac{i}{\hbar}p_x.$$

Since U commutes with H, so does F, as shown in Eq. (7.15). Invariance under translation along x leads to conservation of the corresponding momentum p_x.

7.2 The Electric Charge

As a first example of a conserved quantity we consider the electric charge. We are so used to the fact that electricity does not appear or disappear spontaneously that we often forget to ask: How well is electric charge conservation known? A good way to look for a possible violation of charge conservation is to search for a decay of the electron. If charge were not conserved, the decay of the electron into a neutrino and a photon,

$$e \longrightarrow \nu\gamma,$$

would be allowed by all known conservation laws. How could such a process be observed? If an electron bound in an atom decays, it will leave a hole in the shell. The hole will be filled by an electron from a higher state, and an X ray will be emitted. No such X rays have ever been seen, and the mean life of an electron is longer than $2 \times 10^{21} y$.[3] The result is generalized by saying that the total charge in any reaction is conserved; the electric charge in the initial and final state of any reaction must be the same:

$$\sum q_{\text{initial}} = \sum q_{\text{final}}. \tag{7.16}$$

The conservation law is in agreement with all observations.

Quantization of the electric charge permits us to express charge conservation in a somewhat different form. Quantization follows from Millikan's oil droplet experiment; all investigations are in agreement with the observation that the electric charge of a particle is always an integral multiple of the elementary quantum e:

$$q = Ne. \tag{7.17}$$

N is called the electric charge number, or sometimes, loosely, the electric charge. Relation (7.17) implies that the neutron charge must be exactly zero and that the charges of electron and proton must be equal in magnitude. Indeed, observation of the behavior of neutron- and neutral-atom beams in electric fields indicates that the neutron charge is less than $3 \times 10^{-20}e$ and that the electron-proton charge sum is less than $3 \times 10^{-17}e$.[4] An electric charge number N is therefore assigned to all particles. Conservation of the electric charge, Eq. (7.16), demands that N satisfy an *additive conservation law:* In any reaction

$$a + b \longrightarrow c + d + e$$

the sum of the charge numbers remains constant,

$$N_a + N_b = N_c + N_d + N_e. \tag{7.18}$$

Equation (7.16) is an example of a conservation law. We have stated in the introduction that each conservation law is related to a corresponding symmetry principle. What is the symmetry principle that gives rise to the conservation of the electric charge? To answer this question, we repeat the arguments of Section 7.1 specifically for electric charge conservation. While reading the following derivation, it is a good idea to follow the more general steps in Section 7.1, in parallel. Assume that ψ_q describes a state

3. M. K. Moe and F. Reines, *Phys. Rev.* **140B**, 992 (1965).

4. C. G. Shull, W. K. Billman, and F. A. Wedgwood, *Phys. Rev.* **153**, 1415 (1967); J. G. King, *Phys. Rev. Letters* **5**, 562 (1960).

with charge q and that it satisfies a Schrödinger equation, Eq. (7.1):

$$i\hbar\frac{d\psi_q}{dt} = H\psi_q. \tag{7.19}$$

If Q is the electric charge operator, we know from Eqs. (7.5) and (7.6) that $\langle Q \rangle$ is conserved if H and Q commute. ψ_q then can also be chosen to be an eigenfunction of Q,

$$Q\psi_q = q\psi_q, \tag{7.20}$$

and the eigenvalue q is also conserved. What symmetry guarantees that H and Q commute? The answer to this question was given by Weyl[5] who considered a transformation of the type of Eq. (7.12):

$$\psi_q' = e^{i\epsilon Q}\psi_q. \tag{7.21}$$

Here, ϵ is an arbitrary real parameter and Q the electric charge operator. The transformation is called a gauge transformation of the first kind. *Gauge invariance* means that ψ_q' satisfies the same Schrödinger equation as does ψ_q:

$$i\hbar\frac{d\psi_q'}{dt} = H\psi_q'$$

or

$$i\hbar\frac{d}{dt}(e^{i\epsilon Q}\psi_q) = He^{i\epsilon Q}\psi_q.$$

Multiplying from the left with $e^{-i\epsilon Q}$, noting that Q is a time-independent and Hermitian operator, and comparing with Eq. (7.19) give

$$e^{-i\epsilon Q}He^{i\epsilon Q} = H. \tag{7.22}$$

Since ϵ is an arbitrary parameter, it can be taken so small that $\epsilon Q \ll 1$. Expanding the exponential yields

$$(1 - i\epsilon Q)H(1 + i\epsilon Q) = H$$

or

$$[Q, H] = 0. \tag{7.23}$$

Invariance under the gauge transformation (7.21) guarantees conservation of the electric charge. The proper way to look at the connection is probably the other way around: Since the electric charge is observed to be conserved, the acceptable Hamiltonian must be invariant under the gauge transformation. The arguments given so far may appear to be rather formal, but they are very important because they connect electricity (Q) and quantum mechanics (ψ).

5. H. Weyl, *The Theory of Groups and Quantum Mechanics*, Dover, New York, 1950 pp. 100, 214.

7.3 The Baryon Number

Conservation of the electric charge alone does not guarantee stability against decay. The proton, for instance, could decay into a positron and a gamma ray without violating either charge or angular momentum conservation. What prevents such a decay? Stueckelberg first suggested that the total number of nucleons should be conserved.[6] This law can be formulated compactly by assigning a *baryon number* $A = 1$ to the proton and the neutron and $A = -1$ to the antiproton and the antineutron. (We shall discuss antiparticles in Section 7.5.) Leptons, photons, and mesons are assigned $A = 0$. (Particle physicists use B for baryon number, but we follow the convention of the nuclear physicists here.) The additive conservation law for the baryon number then reads

$$\sum A_i = \text{const.} \tag{7.24}$$

The extent to which Eq. (7.24) holds can be described by a limit on the lifetime of the nucleons. A good limit is obtained by considering the heat flow from the interior of the earth. If nucleons were to decay, heat would be released, and the heat flow could be used to determine the nucleon lifetime. If contributions from known radioactive elements are subtracted from the observed heat flow, a limit of about 10^{20} y results. A better limit is found by measuring possible decays in a large piece of matter with very large counters that are shielded from cosmic rays by being deep underground.[7] The limit then becomes about 10^{28} y. We do not have to live in fear of wasting away through the decay of nucleons—just through biological decay.

The discovery of strange particles led to a generalization of the law of nucleon conservation. Consider, for instance, the decays

$$\Lambda^0 \longrightarrow n\pi^0$$

$$\Sigma^+ \begin{cases} \longrightarrow p\pi^0 \\ \longrightarrow \Lambda e^+ \nu \end{cases}$$

$$\Sigma^- \longrightarrow n\pi^-$$

or any of the hyperon decays listed in Appendix A3. In each of these decays, the baryon number is conserved if it is generalized to read

$$A = 1 \qquad \text{for } pn\Lambda\Sigma\Xi\Omega$$

and $A = -1$ for the corresponding antiparticles. Similarly, *resonances* and *nuclei* can be characterized by their baryon number A. Since nuclei are built up from proton and neutrons, the baryon number A is identical

6. E. C. G. Stueckelberg, *Helv. Phys. Acta* **11**, 225, 299 (1938).
7. W. R. Kropp, Jr., and F. Reines, *Phys. Rev.* **137B**, 740 (1965).

to the mass number, introduced in Section 5.9. *Hypernuclei* are similar to nuclei, but one or two nucleons are replaced by a hyperon.

As in the case of the electric charge, the question of the symmetry responsible for baryon conservation arises. Again, a gauge transformation

$$\psi' = \psi e^{i\epsilon A} \tag{7.25}$$

leads formally to the conservation law, Eq. (7.24). The physical origin of this gauge transformation is a mystery.

7.4 Lepton and Muon Number

In Section 5.6 the basic characteristics of four leptons (electron, muon, and two neutrinos) were sketched, and we pointed out that four antileptons also existed. To explain the absence of some decay modes allowed by all other conservation laws, Konopinski and Mahmoud in essence introduced a lepton number, L, and lepton conservation.[8] They assigned $L = 1$ to e^-, μ^- ν_e and ν_μ, $L = -1$ to the antileptons e^+, μ^+, $\bar{\nu}_e$, and $\bar{\nu}_\mu$; and $L = 0$ to all other particles. Lepton conservation then demands

$$\sum L_i = \text{const.} \tag{7.26}$$

Leptons, like baryons, can be created or destroyed only in particle-antiparticle pairs. High-energy photons can produce pairs such as

$$\gamma \longrightarrow e^- e^+, \qquad \gamma \longrightarrow p\bar{p}, \tag{7.27}$$

but not $\gamma \longrightarrow e^- p$. (Remember that these processes can happen only in the field of a nucleus that takes up momentum; see Problem 3.22.)

Is the assignment of a lepton number meaningful and correct? If yes, how can we be certain that the chosen assignment is the right one? We first notice that a positive answer to the first question defies intuition. Altogether four neutrinos exist, electron and muon neutrino and their two antiparticles. Neutrinos have no charge and no mass; they possess only spin and momentum. How can such a simple particle appear in four versions? If, on the other hand, it turns out that neutrino and antineutrino are identical, then the assignment of a lepton number is wrong.

Evidence for lepton conservation comes from neutrino reactions and from double beta decay studies. We shall talk only about neutrino reactions here because they are in the forefront of research and will very likely lead to more exciting results in the next few years. Consider first antineutrino capture,

$$\bar{\nu}_e p \longrightarrow e^+ n. \tag{7.28}$$

This process is allowed by lepton conservation because the lepton number on both sides of the equation is -1. Antineutrino capture has been observed by Reines, Cowan, and collaborators with antineutrinos from a

8. E. J. Konopinski and H. M. Mahmoud, *Phys. Rev.* **92**, 1045 (1953).

nuclear reactor.[9] A reactor produces predominantly antineutrinos be-cause fission yields neutron-rich nuclides. (See Chapter 17.) These decay through processes involving the mode

$$n \longrightarrow pe^-\bar{\nu}_e. \qquad (7.29)$$

Since the neutron has $L = 0$, the right-hand side must also have $L = 0$, and the massless particle emitted together with the negative electron must be an antineutrino. The observation of the reaction Eq. (7.28) is in agree-ment with Eq. (7.29). However, reactions of the type $\bar{\nu}_e n \longrightarrow e^- p$ and $\nu_e p \longrightarrow e^+ n$ are forbidden by lepton conservation. Davis has searched for a reac-tion of this type,

$$\bar{\nu}_e \, {}^{37}Cl \longrightarrow e^- \, {}^{37}Ar, \qquad (7.30)$$

again using antineutrinos from reactors. Here, $L = -1$ on the left-hand side and $L = +1$ on the right-hand side, and lepton conservation would be violated if the reaction were observed to go. Davis has not seen reaction (7.30), and the limit he has been able to set on the cross section rules out the possibility that neutrino and antineutrino are identical.[10] Note, how-ever, that the reaction

$$\nu_e \, {}^{37}Cl \longrightarrow e^- \, {}^{37}Ar \qquad (7.31)$$

should occur, and Davis has used it to try to observe neutrinos emitted by the thermonuclear reactions in the sun. (See Chapter 18.)

The results from the neutrino reactions [Eqs. (7.28) and (7.30)] are corroborated by other experiments, and the fact has to be faced that neutrino and antineutrino are different. A difference has been observed in beta decay, and it is presented in Fig. 7.2: The neutrino always has its spin opposite to its direction of motion, while the antineutrino has parallel spin and momentum. In other words, the neutrino is a left-handed and the antineutrino a right-handed particle. Such a situation is compatible with lepton conservation only if the neutrinos have no mass. Massless particles move with the velocity of light, and a right-handed particle re-mains right-handed in any coordinate system. For a massive particle, a Lorentz transformation along the momentum can be performed in such a way that the momentum is reversed in the new coordinate system. The spin, however, can be pictured as a rotation of the particle around the momentum axis. The sense of this rotation is not changed by a Lorentz transformation along the momentum direction. The Lorentz transforma-tion that reverses the direction of the momentum consequently leaves the sense of rotation unchanged, and a right-handed particle changes into a left-handed one. A massive antineutrino would change into a neutrino, and the lepton number would not be conserved.

The situation shown in Fig. 7.2 gives an observable difference between

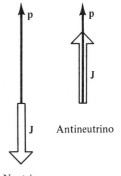

Fig. 7.2. Neutrino and anti-neutrino are always *polarized*. The neutrino has its spin always opposite to its momen-tum; the antineutrino has parallel spin and momentum.

9. F. Reines, C. L. Cowan, F. B. Harrison, A. D. McGuire, and H. W. Kruse, *Phys. Rev.* **117**, 159 (1960).
10. R. Davis, *Phys. Rev.* **97**, 766 (1955).

neutrino and antineutrino. Why have we also distinguished a muon and an electron neutrino? Both have $L = 1$. In what way are they different? To attack this question, another of the puzzles that surround neutrinos must be told. The muon decays through the mode

$$\mu \longrightarrow e\bar{v}v, \tag{7.32}$$

but the possibility

$$\mu \longrightarrow e\gamma \tag{7.33}$$

is allowed by all the conservation laws discussed so far. Over the years many groups have searched for the gamma decay of the muon, without any success, and the limit on the branching ratio is less than 2×10^{-8}. The simplest way to explain the absence of the muon gamma decay is a new conservation law, conservation of the muon number L_{μ}. $L_{\mu} = +1$ is assigned to the negative and $L_{\mu} = -1$ to the positive muon. The lepton number of the neutrinos associated with muons can then be found from the pion decays:

$$
\begin{array}{ccccc}
\pi^- \longrightarrow \mu^- \ \bar{v}_{\mu}, & & \pi^+ \longrightarrow \mu^+ v_{\mu} & \\
L_{\mu}: \quad 0 \qquad 1 \ -1 & & 0 \qquad -1 \ 1 &
\end{array} \tag{7.34}
$$

The muon neutrino has $L_{\mu} = 1$, and the muon antineutrino $L_{\mu} = -1$. All other particles are assigned $L_{\mu} = 0$. $\bar{v}_{\mu}$ is labeled an antineutrino because it is right-handed.

Conservation of the muon number accounts for the absence of the decay $\mu \rightarrow e\gamma$. However, if the introduction of the muon number does nothing else, it is not meaningful. Actually, it does lead to new predictions, as can be seen by considering the two reactions

$$
\begin{aligned}
v_{\mu}n &\longrightarrow \mu^- p \\
v_{\mu}n &\longrightarrow e^- p.
\end{aligned} \tag{7.35}
$$

If muon number is conserved, only the first one is allowed; the second one is forbidden. The reactions can be tested because the pion decay, Eq. (7.34), produces only muon neutrinos. The experimental observation is difficult because neutrinos have an extremely small cross section and the detector for the reaction equation (7.35) must be guarded against all other particles. In 1962, a Columbia group performed a successful experiment at the Brookhaven accelerator and indeed found that no electrons were produced by muon neutrinos.[11] Since this first experiment, the fact has been verified many times.

7.5 Particles and Antiparticles

The particle-antiparticle concept is one of the most fascinating ones in physics. At the same time, it leads to many questions, and confusion is

11. G. Danby, J. M. Gaillard, K. Goulianos, L. M. Lederman, N. Mistry, M. Schwartz, and J. Steinberger, *Phys. Rev. Letters* **9**, 36 (1962). See also *Adventures in Experimental Physics.*, Vol. α, 1972.

often greater after explanation than before. The present section is brief and restricted and will leave many problems unsolved. At the same time some of the aspects that are needed in later sections and chapters should become somewhat clearer.

The story begins about 1927 with Eq. (1.2):

$$E^2 = (pc)^2 + (mc^2)^2. \qquad (1.2)$$

Consider a particle with momentum **p** and mass m. What is its energy? All of us were taught early in our life to write a square root with two signs,

$$E^\pm = \pm[(pc)^2 + (mc^2)^2]^{1/2}. \qquad (7.36)$$

Two solutions appear, a positive and a negative one. What does the negative energy solution mean? In classical physics, it did not cause havoc. When the classical gods created the world, they chose the initial conditions without negative energies. Continuity then guaranteed that none would appear later. In quantum mechanics, the situation is far more serious. Consider the energy levels of a particle with mass m. Equation (7.36) states that positive and negative energy levels are possible, and these levels are shown in Fig. 7.3. The smallest possible positive energy is $E = mc^2$; the largest

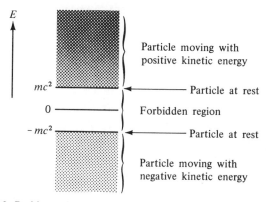

Fig. 7.3. Positive and negative energy states of a particle with mass m.

negative energy is $-mc^2$. According to Eq. (7.36), the particle can have any energy from mc^2 to $+\infty$ and from $-mc^2$ to $-\infty$. Do the negative energy states lead to observable consequences? We shall see that they do and that there is an enormous amount of experimental evidence to back up this claim. Before doing so, we mention a mathematical argument that also calls for their existence: One of the most fundamental theorems in quantum mechanics states that any observable has a complete set of eigenfunctions.[12] It can be shown in relativistic quantum mechanics that eigenfunctions do *not* form a complete set without the negative energy states.

If the negative energy states exist, what do they mean? They cannot be

12. Merzbacher, Section 8.3.

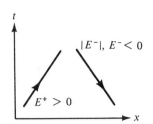

Fig. 7.4. The particle with positive energy, E^+, moves like any ordinary particle. The particle with negative energy, E^-, is represented as a particle with positive energy $|E^-|$, but moving backwards in time. Both travel to the right.

normal energy states as indicated in Fig. 7.3; otherwise, ordinary particles could make transitions to the negative energy states with emission of energy, and matter would rapidly disappear. The first workable interpretation of the negative energy states is due to Dirac,[13] who identified particles missing from the negative energy states (holes) with antiparticles. We shall not discuss his *hole theory* but proceed immediately to a more modern interpretation, first proposed by Stueckelberg and later again in more powerful form by Feynman.[14] We present this approach in a pedestrian version and first consider a particle moving along the positive x axis with positive momentum p and positive energy E^+. The trajectory of this particle is shown in an xt plot in Fig. 7.4. Its wave function is of the form

$$\psi(x, t) = e^{(i/\hbar)(px - E^+t)}. \tag{7.37}$$

The fact that it moves to the right can be seen most easily by noting that the phase of the wave function is constant if

$$px - E^+t = \text{const.}$$

or if

$$x = \frac{E^+}{p}t. \tag{7.38}$$

The point x moves to the right. (This argument can be made more rigorous by using a wave packet.) For the negative energy solution,

$$\psi(x, t) = e^{(i/\hbar)(px - E^-t)}, \qquad E^- < 0, \tag{7.39}$$

the relation (7.38) becomes

$$x = \frac{E^-}{p}t = -\frac{|E^-|}{p}t = \frac{|E^-|}{p}(-t), \tag{7.40}$$

and it can be interpreted as a particle moving backward in time but having a positive energy, $|E^-|$.

What is a particle moving backward in time? The classical equation of motion of a particle of charge $-q$ in a magnetic field becomes, with the Lorentz force [Eq. (2.13)],

$$m\frac{d^2\mathbf{x}}{dt^2} = \frac{-q}{c}\frac{d\mathbf{x}}{dt} \times \mathbf{B} = \frac{q}{c}\frac{d\mathbf{x}}{d(-t)} \times \mathbf{B}. \tag{7.41}$$

A particle with charge q moving backward in time satisfies the same equation of motion as a particle with charge $-q$ moving forward in time.[15]

The content of Eqs. (7.40) and (7.41) can be combined: Equation (7.40) suggests that a negative energy solution can be looked at as a particle moving backward in time but having a positive energy. Equation (7.41) demonstrates that a particle moving backward in time satisfies the same

13. P. A. M. Dirac, *Proc. Roy. Soc.* (*London*) **A126**, 360 (1930).

14. E. C. G. Stueckelberg, *Helv. Phys. Acta*, **14**, 588 (1941); R. P. Feynman, *Phys. Rev.* **74**, 939 (1948).

15. The argument becomes more convincing in the covariant formulation, given, for instance, in Jackson, Eq. (12.65).

equation of motion as a particle with opposite charge moving forward in time. Taken together, the two relations imply that a particle with charge q and *negative* energy behaves like a particle with charge $-q$ and *positive* energy. The particle with charge $-q$ is the antiparticle of the one with charge q. The negative energy states thus behave like antiparticles. With this interpretation the processes shown in Fig. 7.5 can be described in two

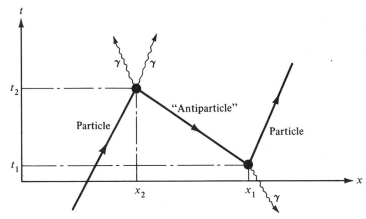

Fig. 7.5. Pair production at (x_1, t_1) and particle-antiparticle annihilation at (x_2, t_2). As noted in Chapter 3, pair production can occur only in the field of a nucleus that takes up momentum. A nucleus is implied near point (x_1, t_1).

different but equivalent ways: In the conventional language, a particle-antiparticle pair is produced at time t_1 and position x_1. The antiparticle meets another particle at time t_2 and position x_2, giving rise to two gamma quantas that propagate forward in time. In Stueckelberg-Feynman language, the particle is the primary object and it weaves through space and time, backward and forward: At time t_2, the particle emits two photons and turns back in time to reach the spot (x_1, t_1). There it is scattered by a photon and again moves forward in time.

What is the advantage of this way of looking at negative energy states? Negative energy states have disappeared from the discussion, and they are replaced by antiparticles with positive energy. The description makes it obvious that the antiparticle concept applies just as well to bosons as to fermions.

Assuming an antiparticle to be a particle moving backward in time, a number of conclusions can be drawn immediately. A particle and its anti-particle must have the same mass and the same spin because they are the same particle, just moving in a different direction in time:

$$m(\text{particle}) = m(\text{antiparticle})$$
$$J(\text{particle}) = J(\text{antiparticle}). \tag{7.42}$$

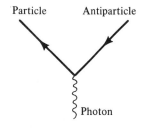

Particle   Antiparticle

Photon

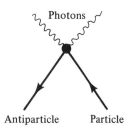

Photons

Antiparticle   Particle

Fig. 7.6. *Arrow convention* for particles and antiparticles.

However, particle and antiparticle are expected to have opposite additive quantum numbers. Consider the pair production at the time t_1 in Fig. 7.5. For times $t < t_1$, only a photon is present in the region around x_1, and its additive quantum numbers q, A, L, and L_μ are zero. If these quantum numbers are conserved, the sum of the corresponding quantum numbers for the particle-antiparticle pair must also add up to zero so that

$$N(\text{particle}) = -N(\text{antiparticle}). \tag{7.43}$$

Here, N stands for any additive quantum number whose value for the photon is zero.

A final remark about a technical point in labeling Feynman diagrams may help prevent some confusion. A pair production process is usually drawn as shown in Fig. 7.6(a). The outgoing particle has its arrow along its momentum. The antiparticle, however, is shown with the arrow reversed. This convention makes reading diagrams unambiguous, and the example in Fig. 7.6(b) should be clear.

Are the Stueckelberg-Feynman concepts of particles and antiparticles correct? Only experiment can tell, and experiment has indeed provided impressive support. Dirac predicted the antielectron in 1931, and it was found in 1933.[16] After this major success, the question arose whether an antiproton existed, but even persistent search in cosmic rays failed to turn it up. It was finally discovered in 1955 when the Bevatron in Berkeley began working.[17] Since then, antiparticles to essentially all particles have been found. A spectacular example is the observation of the antiomega.[18] This hyperon was produced in the reaction

$$dK^+ \longrightarrow \bar{\Omega}\Lambda\Lambda p\pi^+\pi^-. \tag{7.44}$$

Production and decay are shown in Figs. 7.7 and 7.8.

7.6 Hypercharge (Strangeness)

In 1947, Rochester and Butler observed the first V particles[19] (Fig. 5.21). By about 1952, many V events had been seen, and a mystery had developed: The V particles were produced copiously but decayed very slowly. The production, for instance, through Eq. (5.54), $p\pi^- \longrightarrow \Lambda^0 K^0$, occurred with a cross section of the order of mb, whereas the decays had mean lives of about 10^{-10} sec. Cross sections of the order of mb are typical of the hadronic interactions, whereas decays of the order of 10^{-10} sec are characteristic of the weak interaction: Kaons and hyperons are

16. C. D. Anderson, *Phys. Rev.* **43**, 491 (1933); *Am. J. Phys.* **29**, 825 (1961).
17. O. Chamberlain, E. Segrè, C. Wiegand, and T. Ypsilantis, *Phys. Rev.* **100**, 947 (1955).
18. A. Firestone, G. Goldhaber, D. Lissauer, B. M. Sheldon, and G. H. Trilling, *Phys. Rev. Letters* **26**, 410 (1971).
19. G. D. Rochester and C. C. Butler, *Nature* **160**, 855 (1947).

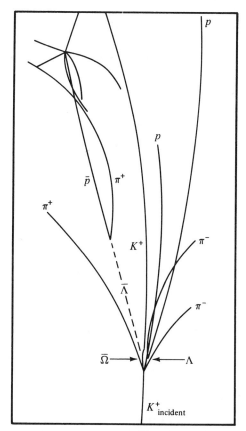

Fig. 7.7. Drawing of the reaction $dK^+ \longrightarrow \overline{\Omega}\Lambda\Lambda p\pi^+\pi^-$ and the resulting decays. [A. Firestone et al., *Phys. Rev. Letters* **26**, 410 (1971).]

produced strongly but decay weakly. Pais made the first step to the solution of the paradox by suggesting that V particles are always produced in pairs.[20] The complete solution came from Gell-Mann and from Nishijima, who both introduced a new quantum number.[21] Gell-Mann called it *strangeness* and the name stuck. We shall describe the assignment of this new additive quantum number by using well-established hadronic reactions.[22]

We begin by assigning strangeness $S = 0$ to nucleons and pions, and note that strangeness is not defined for leptons. Strangeness is assumed to

20. A. Pais, *Phys. Rev.* **86**, 663 (1952).

21. M. Gell-Mann, *Phys. Rev.* **92**, 833 (1953); T. Nakano and K. Nishijima, *Progr. Theoret. Phys.* **10**, 581 (1953).

22. It is necessary to stress that the assignment is much easier now than in 1952 or 1953. An enormous number of reactions are known now, whereas Pais, Gell-Mann, and Nishijima had to work with very few clues and had to make imaginative guesses.

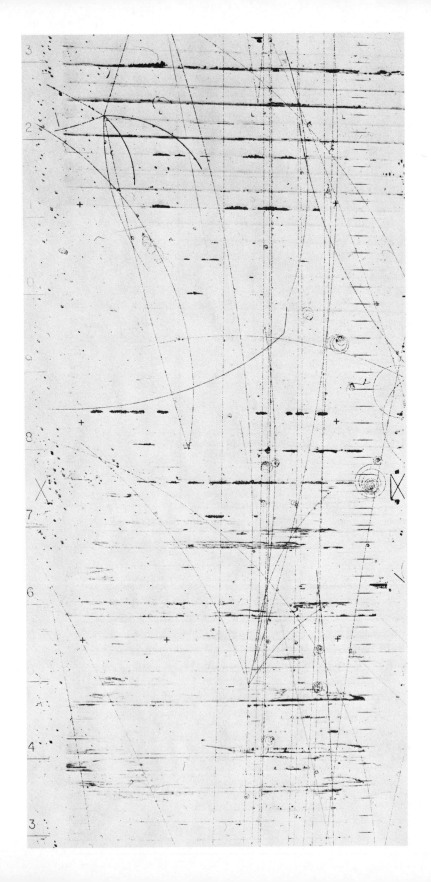

be a conserved quantity in all interactions that are not weak:

$$\sum_i S_i = \text{const. in hadronic and electromagnetic interactions.} \quad \textbf{(7.45)}$$

We have introduced here the first example of a "broken" symmetry: S is assumed to be conserved in hadronic and electromagnetic interactions but violated in weak ones. With such a quantum number, the mystery of copious production and slow decay can be explained easily. Consider the production reaction $p\pi^- \longrightarrow \Lambda^0 K^0$ and assign a strangeness $S = 1$ to K^0. The total strangeness on both sides of the reaction must be zero, since only nonstrange particles are present initially. The Λ^0 consequently must have strangeness -1. and Pais' rule is explained: In reactions involving only nonstrange particles in the initial state, strange particles must be produced in pairs. Moreover, a single strange particle cannot decay hadronically or electromagnetically to a state involving only nonstrange particles; such decays must proceed by the weak interaction, and they are therefore slow. Thus the observed long lifetime of the strange particles is also explained.

The assignment of strangeness to the various hadrons is based on reactions that are observed to proceed hadronically. *By definition*, the strangeness of the positive kaon is set equal to 1:

$$S(K^+) = 1. \quad \textbf{(7.46)}$$

The reaction

$$p\pi^- \longrightarrow nK^+K^- \quad \textbf{(7.47)}$$

is observed to proceed with a cross section characteristic of hadronic interactions, and it therefore yields

$$S(K^-) = -1. \quad \textbf{(7.48)}$$

Positive and negative kaons have opposite strangeness, and we assume, with Eq. (7.43), that they form a particle-antiparticle pair.

Next we turn to the stable baryons, listed in Table A3 in the Appendix. We first see that all have $A = 1$, and therefore they are all particles. The corresponding set of antiparticles also exists, and the strangeness quantum numbers for the antiparticles are opposite to the ones of the particles that we are about to find.

The two charged kaons are excellent tools for establishing values of S. Consider first the reaction

$$p\pi^- \longrightarrow XK. \quad \textbf{(7.49)}$$

The initial state contains only nonstrange particles, and the observation of reaction (7.49) consequently gives $S(X) = -S(K)$. The hyperon X has $S = -1$ if the kaon is positive and $S = +1$ if the kaon is negative. At

Fig. 7.8. (Left) Production of the $\overline{\Omega}$, observed in a study of K^+d interactions at a momentum of 12 GeV/c, in the 2 m SLAC (Stanford Linear Accelerator Center) bubble chamber.[18] (Courtesy Gerson Goldhaber, Lawrence Berkeley Laboratory.)

modern accelerators, separated kaon beams are available, and reactions of the type

$$pK^- \Big\langle \begin{array}{l} {}^{\nearrow} X\pi \\ {}_{\searrow} X'K^+ \end{array} \tag{7.50}$$

or the corresponding ones with positive kaons can also readily be observed. In the first of the reactions [Eq. (7.50)], $S(X) = S(K^-) = -1$ and in the second $S(X') = -2$. Reactions (7.49) and (7.50) are only two prototypes; far more involved processes occur and serve to find S.

As an example of reaction (7.49), the process

$$p\pi^- \longrightarrow \Sigma^- K^+$$

assigns $S = -1$ to the negative sigma. An example of Eq. (7.50) is

$$pK^- \longrightarrow \Sigma^+ \pi^-,$$

which gives $S(\Sigma^+) = -1$. Σ^- and Σ^+ are both baryons with $A = 1$; they have the same strangeness but opposite charge. This fact does not contradict Eq. (7.43), which demands only that antiparticles have opposite charge but does not state that a pair with opposite charges has to be a particle-antiparticle pair.

The reactions

$$pp \longrightarrow p\Sigma^0 K^+ \qquad \text{and} \qquad pK^- \longrightarrow \Lambda^0 \pi^0$$

assign strangeness -1 to Λ^0 and Σ^0. The reaction

$$pK^- \longrightarrow \Xi^- K^+$$

yields $S = -2$ for Ξ^-. Similarly, the strangeness of Ω^- is found to be -3, and the strangeness of $\bar{\Omega}^-$ follows from Eq. (7.44) as $+3$.

Now we return to the *kaons*. Reaction (5.54),

$$p\pi^- \longrightarrow \Lambda^0 K^0,$$

determines the strangeness of the K^0 as positive. The fact raises a question. We have

$$S(K^+) = 1, \qquad S(K^-) = -1$$
$$S(K^0) = 1, \qquad\qquad ?$$

Something is missing: We have two kaons with $S = 1$ and only one with $S = -1$. Gell-Mann therefore suggested that K^0 should also have an antiparticle, $\overline{K^0}$, with $S = -1$. This antiparticle was found; it can, for instance, be produced in the reaction

$$p\pi^+ \longrightarrow pK^+ \overline{K^0}.$$

The existence of the two neutral kaons, different only in their strangeness but in no other quantum number, gives rise to truly beautiful quantum mechanical interference effects; they will be discussed in Chapter 9. These effects are the subatomic analog to the inversion spectrum of ammonia.

For most discussions it has become customary to use the hypercharge Y rather than strangeness; the *hypercharge* Y is defined by

$$Y = A + S. \tag{7.51}$$

In Table 7.1 we list the values of baryon number, strangeness, and hypercharge for some hadrons. In the last column we give the average value of the charge number of the particles listed in the relevant row. This quantity will be used later.

Table 7.1 provides considerable food for thought, and a few remarkable

Table 7.1 Baryon Number A, Strangeness S, Hypercharge Y, and Average Value of the Charge Number $N_q = q/e$.

Particle		A	S	Y	$\langle N_q \rangle$
Photon	γ	0	0	0	0
Pion	$\pi^+\pi^0\pi^-$	0	0	0	0
Kaon	K^+K^0	0	1	1	$\frac{1}{2}$
Nucleon	pn	1	0	1	$\frac{1}{2}$
Lambda	Λ^0	1	-1	0	0
Sigma	$\Sigma^+\Sigma^0\Sigma^-$	1	-1	0	0
Cascade	$\Xi^-\Xi^0$	1	-2	-1	$-\frac{1}{2}$
Omega	Ω^-	1	-3	-2	-1

facts stand out. Some of these we shall be able to explain later. First we note that the number of particles in each row varies. There are three pions, two kaons, two nucleons, one lambda, and so forth. Why? We shall give an explanation in Chapter 8. Second, we remark that all antiparticles exist and have been found. In some cases the set of antiparticles is identical to the set of particles. When can this happen? Equation (7.43) states that a particle can be identical to its antiparticle only if all additive quantum numbers vanish. The only particles in Table 7.1 satisfying this condition are the photon and the neutral pion. The pion set is identical to its own antiset, and the positive pion is the antiparticle of the negative one. All other entries in Table 7.1 are different from their antiparticles. Third, we note that

$$Y = 2 < N_q > = 2 < \frac{q}{e} >, \tag{7.52}$$

and this relation will be used later.

7.7 References

A careful and interesting survey of the literature on symmetry has been given by D. Park, "Resource Letter SP-1 on Symmetry in Physics," *Am. J. Phys.* **36**, 577 (1968).

Symmetries and invariance principles are the subject of the following two books: J. J. Sakurai, *Invariance Principles and Elementary Particles*, Princeton University Press, Princeton, N.J., 1964; F. Low, *Symmetries and Elementary Particles*, Gordon & Breach, New York, 1967. Both of these are above the level of the present book (despite the introductory sentence in Low which states "These lectures are extremely elementary.") Nevertheless, for further insight into the problems discussed in the present and the two following chapters, these books are recommended.

The limits set on the various conservation laws are treated in G. Feinberg and M. Goldhaber, *Proc. Natl. Acad. Sci. U.S.* **45**, 1301 (1959).

Additional information can also be found in the various texts on particle physics, in particular in R. K. Adair and E. C. Fowler, *Strange Particles*, Wiley-Interscience, New York, 1963, and in W. R. Frazer, *Elementary Particles*, Prentice-Hall, Englewood Cliffs, N.J., 1966.

PROBLEMS

7.1. Show that the reality of the expectation value $\langle F \rangle$ demands that the operator F be Hermitian.

7.2. Discuss more carefully and in more detail than in the text
(a) Quantum mechanical operators and matrices associated with these operators. How is a matrix associated with an observable F and a transformation operator U?
(b) How is Hermiticity defined for operators and for the corresponding matrices?
(c) How is unitarity defined for operators and for matrices?

7.3. Discuss the evidence for conservation of the electric charge and the electric current in macroscopic systems (classical electrodynamics).

7.4. Devise an experiment that would measure a possible neutron charge. Use realistic values of neutron flux, neutron velocity, electric field strength, and spatial resolution of neutron counters to obtain an estimate on the limit that could be obtained.

7.5. Assume that nucleons decay with a lifetime of 10^{15} y and that all the energy of the nucleons decaying in the earth is transformed into heat. Compute the heat flow at the surface of the earth. Compare the energy produced with the energy that the earth receives from the sun during the same time.

7.6. Sketch the experimental arrangement of Reines and Kropp [*Phys. Rev.* **137B**, 740 (1965)] for measuring the lifetime of nucleons. How could the experiment be improved?

7.7. The cross section for the absorption of antineutrinos with energies as emitted by nuclear reactors is about 10^{-43} cm^2.
(a) Compute the thickness of a water absorber needed to reduce the intensity of an antineutrino beam by a factor of 2.
(b) Consider a liquid scintillator with a volume of 10^3 liters and an antineutrino beam with an intensity of 10^{13} $\bar{\nu}$/cm^2-sec. How many capture events [Eq. (7.28)] are expected per day?
(c) How can the antineutrino capture be distinguished from other reactions?

7.8. How can Eq. (7.31) be observed? [Start with Bahcall and Davis, *Phys. Rev. Letters* **26**, 662 (1971), and work your way back.]

7.9. Use wave packets to justify the interpretation of a particle with negative energy being a particle with positive energy but moving backward in time.

7.10. Use the covariant formulation of the equation of motion of a charged particle in an electromagnetic field to show that a particle with charge $-q$ moving backward in time behaves like an antiparticle of charge q moving forward in time.

7.11. Can strange particles be produced singly by reactions that involve only nonstrange particles? If yes, give a possible reaction.

7.12. Follow the production and decay of $\bar{\Omega}$ in Figs. 7.7 and 7.8 and verify that the additive quantum numbers A and q are conserved in every interaction. Where is S conserved and where not?

7.13. Discuss the reaction(s) that allows the assignment $S = -3$ to Ω^- and $S = +3$ to $\overline{\Omega^-}$.

7.14. Which of the following reactions can take place? If forbidden, state by what selection rule. If allowed, indicate through which interaction the reaction will proceed.

(a) $p\bar{p} \longrightarrow \pi^+\pi^-\pi^0\pi^+\pi^-$.

(b) $pK^- \longrightarrow \Sigma^+\pi^-\pi^+\pi^-\pi^0$.

(c) $p\pi^- \longrightarrow pK^-$.

(d) $p\pi^- \longrightarrow \Lambda^0\overline{\Sigma^0}$.

(e) $\bar{\nu}_\mu p \longrightarrow \mu^+ n$.

(f) $\bar{\nu}_\mu p \longrightarrow e^+ n$.

(g) $\nu_e p \longrightarrow e^+ \Lambda^0 K^0$.

(h) $\nu_e p \longrightarrow e^- \Sigma^+ K^+$.

8

Angular Momentum and Isospin

In this chapter we shall show that invariance under rotation in space leads to conservation of angular momentum. We shall then introduce isospin, a quantity that has many properties similar to ordinary spin, and discuss the "breaking" of isospin invariance.

8.1 Invariance Under Spatial Rotation

Invariance under spatial rotation provides an important application of the general considerations presented in Section 7.1. Consider an idealized experimental arrangement, shown in Fig. 8.1. We assume for simplicity that the equipment is in the xy plane; its orientation is described by the angle φ. We further assume that the result of the experiment is described by a wave function $\psi(\mathbf{x})$. Next, the equipment is rotated by an angle α about the z axis. This rotation is denoted by $R_z(\alpha)$, and it carries a point $\mathbf{x}$ into a point $\mathbf{x}^R$:

$$\mathbf{x}^R = R_z(\alpha)\mathbf{x}. \tag{8.1}$$

The rotation changes the wave function; the relation between the rotated and unrotated wave function at point $\mathbf{x}$ is given by Eq. (7.7) as

$$\psi^R(\mathbf{x}) = U_z(\alpha)\psi(\mathbf{x}). \tag{8.2}$$

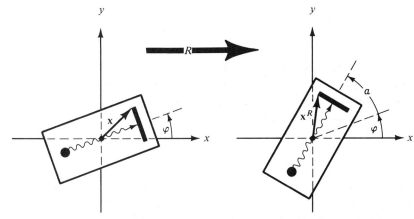

Fig. 8.1. Rotation around the z axis. The angle φ fixes the position of the original equipment axis; it does not denote a rotation. The equipment is rotated about the z axis by an angle α. Invariance under rotation means that the outcome of the experiment is not affected by the rotation.

The notation indicates that the rotation is by an angle α about the z axis. So far, no invariance properties have been used and Eqs. (8.1) and (8.2) are valid even if the system changes during rotation.

Invariance arguments can now be used to find U. If the state of the system is unaffected by rotation, the wave function at point $\mathbf{x}$ in the original system is identical to the rotated wave function at the rotated point $\mathbf{x}^R$:

$$\psi(\mathbf{x}) = \psi^R(\mathbf{x}^R). \tag{8.3}$$

Note the difference between Eqs. (8.2) and (8.3). The first connects $\psi(\mathbf{x})$ to ψ^R at the same point, and the second to ψ^R at the rotated point $\mathbf{x}^R$. U can be found if $\psi^R(\mathbf{x}^R)$ can be expressed in terms of $\psi^R(\mathbf{x})$. Because the rotation is continuous, any rotation by a finite angle can be built up from rotations by infinitely small angles. An *infinitesimal rotation* suffices to find U. If the system is rotated by an infinitesimal angle $\delta\alpha$ about the z axis, $\psi^R(\mathbf{x}^R)$ becomes

$$\psi^R(\mathbf{x}^R) = \psi^R(\mathbf{x}) + \frac{\delta\psi^R(\mathbf{x})}{\partial\varphi}\delta\alpha = \left(1 + \delta\alpha\frac{\partial}{\partial\varphi}\right)\psi^R(\mathbf{x}).$$

This relation can be inverted by multiplication with $[1 - \delta\alpha(\partial/\partial\varphi)]$. Neglecting terms in $\delta\alpha^2$ and using Eq. (8.3) then yield

$$\psi^R(\mathbf{x}) = \left(1 - \delta\alpha\frac{\partial}{\partial\varphi}\right)\psi(\mathbf{x}). \tag{8.4}$$

Comparison with Eq. (8.2) shows that the operator in front of $\psi(\mathbf{x})$ is $U_z(\delta\alpha)$. The general expression for the operator for an infinitesimal unitary transformation is given by Eq. (7.14). Identifying ϵ with $\delta\alpha$ and comparing

179

the two expressions for U yield the desired Hermitian operator F,[1]

$$F = i\frac{\partial}{\partial\varphi}. \tag{8.5}$$

If U commutes with H, so will F, according to Eq. (7.15), and we have found the desired conserved observable. We could start exploring the physical consequences of F and find the eigenfunctions and eigenvalues. This procedure is not necessary because F is an old friend. Equation (5.3) shows that

$$F = -\frac{L_z}{\hbar}. \tag{8.6}$$

Not unexpectedly, F is proportional to the z component of the orbital angular momentum. Invariance of a system under rotation around the z axis leads to conservation of F and thus also of L_z.

Two generalizations are physically reasonable, and we give them without proof: (1) If the system has a total angular moment $\mathbf{J}$ (spin plus orbital), then L_z is replaced by J_z. (2) U for a rotation by an angle δ around the arbitrary direction $\hat{\mathbf{n}}$ (where $\hat{\mathbf{n}}$ is a unit vector) is

$$U_{\mathbf{n}}(\delta) = e^{-i\,\delta\hat{\mathbf{n}}\cdot\mathbf{J}/\hbar}. \tag{8.7}$$

If the system is invariant under rotation about $\hat{\mathbf{n}}$, the Hamiltonian will commute with $U_{\mathbf{n}}$ and consequently also with $\hat{\mathbf{n}}\cdot\mathbf{J}$:

$$[H, U_{\mathbf{n}}] = 0 \longrightarrow [H, \hat{\mathbf{n}}\cdot\mathbf{J}] = 0. \tag{8.8}$$

The component of the angular momentum along $\hat{\mathbf{n}}$ is conserved. If $\hat{\mathbf{n}}$ can be taken to be any direction, all components of $\mathbf{J}$ are conserved, and $\mathbf{J}$ is a constant of the motion.

With Eq. (8.7) it is straightforward to find the commutation relations for the components of $\mathbf{J}$:

$$[J_x, J_y] = i\hbar J_z, \tag{5.6}$$
$$\text{cyclic.}$$

The steps in the derivation are outlined in Problem 8.1. The commutation relations [Eq. (5.6)] are a consequence of the unitary transformation [Eq. (8.7)], which in turn is a consequence of the invariance of H under rotation.

---

1. Some confusion can arise because formally $F^\dagger = -i\partial/\partial\varphi$ looks different from F. However, Hermiticity is not a property of an operator alone but also of the wave functions and the region of integration (Park, p. 61). For a Hermitian operator, with $F^\dagger = F$, the equation in the footnote on p. 157 reads

$$\int d^3x\,(F\psi)^*\phi = \int d^3x\,\psi^*F\phi.$$

$F = i\partial/\partial\varphi$ satisfies this relation:

$$\int d^3x\left(i\frac{\partial\psi}{\partial\varphi}\right)^*\phi = \int d^3x\left(-i\frac{\partial}{\partial\varphi}\right)\psi^*\phi = \int d^3x\,\psi^*i\frac{\partial\phi}{\partial\varphi}.$$

In the last step, a partial integration has brought the operator to the right of ψ^*. The explicit form of a Hermitian operator clearly depends on its position with respect to the wave functions.

8.2 Symmetry Breaking by Magnetic Field

A particle with spin $\mathbf{J}$ and magnetic moment $\boldsymbol{\mu}$ can be described by a Hamiltonian

$$H = H_0 + H_{\text{mag}}, \tag{8.9}$$

where H_{mag} is given in Eq. (5.20). Usually, H_0 is isotropic, and the system described by H_0 is invariant under rotations about any direction. This fact is expressed by

$$[H_0, \mathbf{J}] = 0. \tag{8.10}$$

The energy of the particle is independent of its orientation in space. If a magnetic field is switched on, the symmetry is broken, and Eq. (8.10) no longer holds:

$$[H, \mathbf{J}] = [H_0 + H_{\text{mag}}, \mathbf{J}] \neq 0. \tag{8.11}$$

[If needed, the commutator can be calculated with Eqs. (5.20) and (5.6).] The component of the angular momentum along the field, however, still remains conserved. It is customary to select the quantization axis z along the magnetic field. Equations (5.6) and (5.20) then give

$$[H_0 + H_{\text{mag}}, J_z] = 0. \tag{8.12}$$

The system is still invariant under rotations about the direction of the externally applied field, namely the z axis. However, the introduction of a preferred direction through the application of the magnetic field has broken the overall symmetry, and $\mathbf{J}$ is no longer conserved. Before the application of the field, the energy levels of the system were $(2J + 1)$-fold degenerate, as shown on the left-hand side of Fig. 5.5. The introduction of the field results in a removal of the degeneracy, and the corresponding Zeeman splitting is shown in Fig. 5.5.

8.3 Charge Independence of Hadronic Forces

In 1932, when the neutron was discovered, the nature of the forces holding nuclei together was still mysterious. By about 1936, crucial features of the nuclear force had emerged.[2] Particularly revealing was the analysis of *pp* and *np* scattering data. Of course, in these years, such scattering experiments could be performed only at very low energies, but the outcome was still surprising: After subtracting the effect of the Coulomb force in *pp* scattering, it was found that the *pp* and the *np* hadronic force were of about equal strength and had about equal range.[3] This result was

---

2. In 1936 and 1937, Bethe and collaborators surveyed the state of the art in a series of three articles, later known as the *Bethe bible*. These admirable reviews in *Rev. Modern Phys.* **8**, 82 (1936), **9**, 69 (1937), and **9**, 245 (1937), can still be read with profit.
3. G. Breit, E. U. Condon, and R. D. Present, *Phys. Rev.* **50**, 825 (1936).

corroborated by studies of the masses of ^{3}H and ^{3}He which gave approximately equal values for the *pp*, *np*, and *nn* interactions. Strong evidence for a *charge independence* of the nuclear forces was also found by Feenberg and Wigner.[4] Charge independence for nuclear forces can be formulated by stating that the forces between any two nucleons in the same state are the same, apart from electromagnetic effects. Today, the experimental evidence for charge independence is very strong, and it is known that all hadronic forces, not just the one between nucleons, are charge-independent.[5] We shall not discuss the experimental evidence for charge independence here but only point out that the concept of isospin, which will be discussed in the following sections, is a direct consequence of the charge independence of hadronic forces.

8.4 The Nucleon Isospin

Charge independence of nuclear forces leads to the introduction of a new conserved quantum number, isospin. As early as 1932, Heisenberg treated the neutron and the proton as two states of one particle, the nucleon N.[6] Without the electromagnetic interaction, the two states presumably have the same mass, but its presence makes the masses slightly different. Support for the idea that neutron and proton are two states of the same particle comes from the structure studies reported in Section 6.7. Equation (6.48) shows that the magnetic form factors, G_M of the neutron and the proton, have the same functional dependence. If it is assumed that the magnetic moment is tied to the hadronic structure, the similarity is impressive.

To describe the two states of the nucleon, an isospin space (charge space) is introduced, and the following analogy to the two spin states of a spin-$\frac{1}{2}$ particle is made:

	Spin-$\frac{1}{2}$ particle in ordinary space	Nucleon in isospin space
Orientation	Up	Up, proton
	Down	Down, neutron

The two states of an ordinary spin-$\frac{1}{2}$ particle are not treated as two particles but as two states of one particle. Similarly, the proton and the

4. E. Feenberg and E. P. Wigner, *Phys. Rev.* **51**, 95 (1937).

5. The experimental evidence for charge independence of the hadronic forces is evaluated by E. M. Henley in *Isospin in Nuclear Physics* (D. H. Wilkinson, ed.), North-Holland, Amsterdam, 1969.

6. W. Heisenberg, *Z. Physik* **77**, 1 (1932). Translated in D. M. Brink, *Nuclear Forces*, Pergamon, Elmsford, N.Y., 1965.

neutron are considered as the *up* and the *down* state of the nucleon. Formally, the situation is described by introducing a new quantity, *isospin* $\vec{I}$.[7] The nucleon with isospin $\frac{1}{2}$ has $2I + 1 = 2$ possible orientations in isospin space. The three components of the isospin vector $\vec{I}$ are denoted by I_1, I_2, and I_3. The value of I_3 distinguishes, *by definition*, between the proton and the neutron. $I_3 = +\frac{1}{2}$ is the proton and $I_3 = -\frac{1}{2}$ is the neutron.[8] The most convenient way to write the value of I and I_3 for a given state is by using a Dirac ket:

$$|I, I_3>.$$

Then proton and neutron are

$$\text{proton} |\tfrac{1}{2}, \tfrac{1}{2}>$$
$$\text{neutron} |\tfrac{1}{2}, -\tfrac{1}{2}>. \tag{8.13}$$

The charge for the particle $|I, I_3>$ is given by

$$q = e(I_3 + \tfrac{1}{2}). \tag{8.14}$$

With the values of the third component of I_3 given in Eq. (8.13), the proton has charge e, and the neutron charge 0.

8.5 Isospin Invariance

What have we gained with the introduction of isospin? So far, very little. Formally, the neutron and the proton can be described as two states of one particle. New aspects and new results appear when charge independence is introduced and when isospin is generalized to all particles.

Charge independence states that the hadronic forces do not distinguish between the proton and the neutron. As long as only the hadronic interaction is present, the isospin vector $\vec{I}$ can point in any direction. In other words, there exists rotational invariance in isospin space; the system is invariant under rotations about any direction. As in Eq. (8.10), this fact is expressed by

$$[H_h, \vec{I}] = 0. \tag{8.15}$$

With only H_h present, the $2I + 1$ states with different values of I_3 are degenerate; they have the same energy (mass). Said simply, with only the hadronic interaction present, neutron and proton would have the same mass. The electromagnetic interaction destroys the isotropy of isospin space; it breaks the symmetry, and, as in Eq. (8.11), it gives

$$[H_h + H_{em}, \vec{I}] \neq 0. \tag{8.16}$$

---

7. To distinguish spin and isospin, we write isospin vectors with an arrow.

8. In nuclear physics, isospin is often called isobaric spin; it is often denoted by T, and the neutron is taken to have $I_3 = \frac{1}{2}$ and the proton $I_3 = -\frac{1}{2}$, because there are more neutrons than protons in stable nuclei and $I_3(T_3)$ is then positive for these cases.

However, we know from Section 7.1 that the electric charge is always conserved, even in the presence of H_{em}:

$$[H_h + H_{em}, Q] = 0. \tag{8.17}$$

Q is the operator corresponding to the electric charge q; it is connected to I_3 by Eq. (8.14): $Q = e(I_3 + \frac{1}{2})$. Introducing Q into the commutator, Eq. (8.17), gives

$$[H_h + H_{em}, I_3] = 0. \tag{8.18}$$

The third component of isospin is conserved even in the presence of the electromagnetic interaction. The analogy to the magnetic field case is evident; Eq. (8.18) is the isospin equivalent of Eq. (8.12).

It was pointed out in Section 8.4 that charge independence holds not only for nucleons but for all hadrons. Before generalizing the isospin concept to all hadrons and exploring the consequences of such an assumption, a few preliminary remarks are in order concerning isospin space. We stress that $\vec{I}$ is a vector in isospin space, not in ordinary space. The direction in isospin space has nothing to do with any direction in ordinary space, and the value of the operator $\vec{I}$ or I_3 in isospin space has nothing to do with ordinary space. So far, we have related only the third component of $\vec{I}$ to an observable, the electric charge q [Eq. (8.14)]. What is the physical significance of I_1 and I_2? These two quantities cannot be connected directly to a physically measurable quantity. The reason is nature: In the laboratory, two magnetic fields can be set up. The first can point in the z directions, and the second in the x direction. The effect of such a combination on the spin of the particle can be computed, and the measurement along any direction is meaningful (within the limits of the uncertainty relations). The electromagnetic field in the isospin space, however, cannot be switched on and off. The charge is always related to one component of $\vec{I}$, and this component is traditionally taken to be I_3. Renaming the components and connecting the charge, for instance, to I_2 does not change the situation.

We now assume the general existence of an isospin space, with its third component connected to the charge of the particle by a linear relation of the form

$$q = aI_3 + b. \tag{8.19}$$

With such a relationship, conservation of the electric charge implies conservation of I_3. I_3 is therefore a good quantum number, even in the presence of the electromagnetic interaction. The unitary operator for a rotation in isospin space by an angle ω about the direction $\hat{\alpha}$ is

$$U_{\hat{\alpha}}(\omega) = e^{-i\omega\hat{\alpha}\cdot\vec{I}}. \tag{8.20}$$

Here $\vec{I}$ is the Hermitian generator associated with the unitary operator U, and we expect $\vec{I}$ to be an observable. As in the case of the angular momentum operator $\mathbf{J}$, the arguments follow the general steps outlined in Section 7.1. To study the physical properties of $\vec{I}$, we assume first that only the hadronic interaction is present. Then the electric charge is zero for all systems, and Eq. (8.19) does not determine the direction of I_3. Charge independence thus implies that a hadronic system without electromagnetic interaction is invariant under any rotation in isospin space. We know from Section 7.1, Eq. (7.9), that U then commutes with H_h:

$$[H_h, U_{\vec{\alpha}}(\omega)] = 0. \tag{8.21}$$

As in Eq. (7.15), conservation of isospin follows immediately,

$$[H_h, \vec{I}] = 0.$$

Charge independence of the hadronic forces leads to conservation of isospin.

In the case of the ordinary angular momentum, the commutation relations for **J** follow from the unitary operator (8.7) by straightforward algebraic steps. No further assumptions are involved. The same argument can be applied to $U_{\vec{\alpha}}(\omega)$, and the three components of the isospin vector must satisfy the commutation relations

$$[I_1, I_2] = iI_3$$
$$[I_2, I_3] = iI_1 \tag{8.22}$$
$$[I_3, I_1] = iI_2.$$

The eigenvalues and eigenfunctions of the isospin operators do not have to be computed because they are analogous to the corresponding quantities for ordinary spin. The steps from Eq. (5.6) to Eqs. (5.7) and (5.8) are independent of the physical interpretation of the operators. All results for ordinary angular momentum can be taken over. In particular, I^2 and I_3 obey the eigenvalue equations

$$I_{op}^2 |I, I_3\rangle = I(I + 1)|I, I_3\rangle \tag{8.23}$$

$$I_{3, op}|I, I_3\rangle = I_3 |I, I_3\rangle. \tag{8.24}$$

Here I_{op}^2 and $I_{3, op}$ on the left-hand side are operators, and I and I_3 on the right-hand side are quantum numbers. The symbol $|I, I_3\rangle$ denotes the eigenfunction ψ_{I, I_3}. (In a situation where no confusion can arise, the subscripts "op" will be omitted.) The allowed values of I are the same as for J, Eq. (5.9), and they are

$$I = 0, \tfrac{1}{2}, 1, \tfrac{3}{2}, 2, \ldots \tag{8.25}$$

For each value of I, I_3 can assume the $2I + 1$ values from $-I$ to I.

In the following sections, the results expressed by Eqs. (8.22)–(8.25) will be applied to nuclei and to particles. It will turn out that isospin is essential in understanding and classifying subatomic particles.

● We have noted above that the components I_1 and I_2 are not directly connected to observables. However, the linear combinations

$$I_\pm = I_1 \pm iI_2 \tag{8.26}$$

have a physical meaning. Applied to a state $|I, I_3\rangle$, I_+ raises and I_- lowers the value of

I_3 by one unit:

$$I_{\pm}\,|\,I, I_3\rangle = [(I \mp I_3)(I \pm I_3 + 1)]^{1/2}\,|\,I, I_3 \pm 1\rangle. \tag{8.27}$$

Equation (8.27) can be derived with the help of Eqs. (8.22)–(8.24).[9] ●

8.6 Isospin of Particles

The isospin concept was first applied to nuclei, but it is easier to see its salient features in connection with particles. As stated in the previous section, isospin is presumably a good quantum number as long as only the hadronic interaction is present. The electromagnetic interaction destroys the isotropy of isospin space, just as a magnetic field destroys the isotropy of ordinary space. Isospin and its manifestations should consequently appear most clearly in situations where the electromagnetic interaction is small. For nuclei, the total electric charge number Z can be as high as 100, whereas for particles it is usually 0 or 1. Isospin should therefore be a better and more easily recognized quantum number in particle physics.

If isospin is an observable that is realized in nature, then Eqs. (8.15) and (8.23)–(8.25) predict the following characteristics: The quantum number I can take on the values $0, \frac{1}{2}, 1, \frac{3}{2}, \ldots$. For a given particle, I is an immutable property. In the absence of the electromagnetic interaction, a particle with isospin I is $(2I + 1)$-fold degenerate, and the $2I + 1$ *subparticles* all have the same mass. Since H_h and $\vec{I}$ commute, all subparticles have the same hadronic properties and are differentiated only by the value of I_3. The electromagnetic interaction partially or completely lifts the degeneracy, as shown in Fig. 8.2, and it thus given rise to the isospin analog of the Zeeman effect. The $2I + 1$ subparticles belonging to a given state with isospin I are said to form an *isospin multiplet*. The electric charge of each member is related to I_3 by Eq. (8.19). Quantum numbers that are conserved by the electromagnetic interaction are unaffected by the switching on of H_{em}. Since most quantum numbers have this property, the members of an isospin multiplet have very nearly identical properties; they have, for instance, the same spin, baryon number, hypercharge, and intrinsic parity. (Intrinsic parity will be discussed in Section 9.2.) The different members of an isospin multiplet are in essence the same particle appearing with different orientations in isospin space, just as the various Zeeman levels are states of the same particle with different orientations of its spin with respect to the applied magnetic field. The determination of the quantum number I for a given state is straightforward if all subparticles belonging to the multiplet can be found: Their number is $2I + 1$ and thus yields I. Sometimes counting is not possible, and it is then necessary to resort to other approaches, such as the use of selection rules.

The arguments given so far can be applied most easily to the pion. The

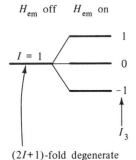

H_{em} off H_{em} on

$I = 1$

1
0
-1

I_3

$(2I+1)$-fold degenerate

Fig. 8.2. A particle with isospin I is $(2I + 1)$-fold degenerate in the absence of the electromagnetic interaction. H_{em} lifts the degeneracy, and the resulting subparticles are labeled by I_3.

9. Merzbacher, Section 16.2; Messiah, Section XIII.I.

possible values of the isospin of the pion can be found by looking at Fig. 5.19: If virtual pions are exchanged between nucleons, the basic Yukawa reaction

$$N \longrightarrow N' + \pi$$

should conserve isospin. Nucleons have isospin $\frac{1}{2}$; isospins add vectorially like angular momenta, and the pion consequently must have isospin 0 or 1. If I were 0, only one pion would exist. The assignment $I = 1$, on the other hand, implies the existence of three pions.[10] Indeed, three and only three hadrons with mass of about 140 MeV are known, and the three form an *isovector* with the assignment

$$I_3 = \begin{cases} +1 & \pi^+, \quad m = 139.576 \text{ MeV}/c^2, \\ 0 & \pi^0, \quad m = 134.972 \text{ MeV}/c^2, \\ -1 & \pi^-, \quad m = 139.576 \text{ MeV}/c^2. \end{cases}$$

The charge is connected to I_3 by the relation

$$q = eI_3, \tag{8.28}$$

which is a special case of Eq. (8.19). The pion shows particularly clearly that the properties in ordinary and in isospin space are not related because it is vector in isospin space but a scalar (spin 0) in ordinary space.

In the ordinary Zeeman effect, it is easy to demonstrate that the various sublevels are members of one Zeeman multiplet: If the applied magnetic field is reduced to zero, they coalesce into one degenerate level. This method cannot be applied to an isospin multiplet because the electromagnetic interaction cannot be switched off. It is necessary to resort to calculations to show that the observed splitting can be blamed solely on H_{em}. Comparison of the pion and the nucleon shows that the problem is not straightforward: The proton is lighter than the neutron, whereas the charged pions are heavier than the neutral one. Nevertheless, the computations performed up to the present time make it very likely that the mass splitting can be accounted for by the electromagnetic interaction.[11]

After having spent considerable time on the isospin of the pion, the other hadrons can be discussed more concisely.

The *kaon* appears in two particle and two antiparticle states. The assignment $I = \frac{1}{2}$ is in agreement with all known facts.

The assignment of I to *hyperons* is also straightforward. It is assumed that hyperons with approximately equal masses form isospin multiplets. The lambda occurs alone, and it is a singlet. The sigma shows three charge states, and it is an isovector. The cascade particle is a doublet, and the omega is a singlet.

The hadrons encountered so far can all be characterized by a set of

10. N. Kemmer, *Proc. Cambridge Phil. Soc.* **34**, 354 (1938).

11. R. P. Feynman and G. Speisman, *Phys. Rev.* **94**, 500 (1954); F. E. Low, *Comments Nucl. Particle Phys.* **2**, 111 (1968). A. Zee, *Phys. Rept.* **3C**, 127 (1973). For a modern approach see eg. K. H. Georgi and T. Goldman, *Phys. Rev. Letters* **30**, 514 (1973).

additive quantum numbers, A, q, Y, and I_3. For pions, charge and I_3 are connected by Eq. (8.28). Gell-Mann and Nishijima showed how this relation can be generalized to apply also to strange particles. They assumed charge and I_3 to be connected by a linear relation as in Eq. (8.19). The constant a in Eq. (8.19) is determined from Eq. (8.28) as e. To find the constant b, we note that I_3 ranges from $-I$ to $+I$. The average charge of a multiplet is therefore equal to b:

$$\langle q \rangle = b.$$

The average charge of a multiplet has already been determined in Eq. (7.52):

$$\langle q \rangle = \tfrac{1}{2} e Y. \tag{8.29}$$

Only particles with zero hypercharge have the center of charge of the multiplet at $q = 0$; for all others, it is displaced. Consequently the generalization of Eqs. (8.14) and (8.28) is

$$q = e(I_3 + \tfrac{1}{2} Y) = e(I_3 + \tfrac{1}{2} A + \tfrac{1}{2} S). \tag{8.30}$$

This equation is called the Gell-Mann-Nishijima relation. If q is considered to be an operator, it can be said that the electric charge operator is composed of an isoscalar ($\tfrac{1}{2} e Y$) and the third component of an isovector ($e I_3$).

The Gell-Mann-Nishijima relation can be visualized in a Y versus q/e diagram, shown in Fig. 8.3. A few isospin multiplets are plotted. The multiplets with $Y \neq 0$ are *displaced:* Their center of charge is not at zero but, as expressed by Eq. (8.29), at $\tfrac{1}{2} e Y$.

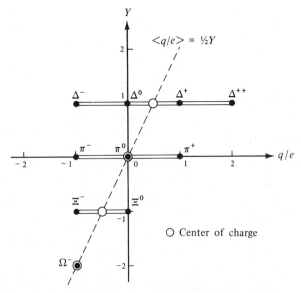

Fig. 8.3. Isospin multiplets with $Y \neq 0$ are displaced: Their center of charge (average charge) is at $\tfrac{1}{2} e Y$. A few representative multiplets are shown, but many more exist.

The considerations in the present section have shown that isospin is a useful quantum number in particle physics. The value of I for a given particle determines the number of subparticles belonging to this particular isospin multiplet. The third component, I_3, is conserved in all interactions, whereas $\vec{I}$ is conserved only by the hadronic force. In the following section we shall demonstrate that isospin is also a valuable concept in nuclear physics.

8.7 Isospin in Nuclei[12]

A nucleus with A nucleons, Z protons, and N neutrons, has a total charge Ze. The total charge can be written as a sum over all A nucleons with the help of Eq. (8.14):

$$Ze = \sum_{i=1}^{A} q_i = e(I_3 + \tfrac{1}{2}A), \tag{8.31}$$

where the third component of the total isospin is obtained by summing over all nucleons,

$$I_3 = \sum_{i=1}^{A} I_{3,i}. \tag{8.32}$$

The isospin $\vec{I}$ behaves algebraically like the ordinary spin $\mathbf{J}$, and the total isospin of the nucleus A is the sum over the isospins from all nucleons:

$$\vec{I} = \sum_{i=1}^{A} \vec{I}_i. \tag{8.33}$$

Do these equations mean something? All states of a given nuclide are characterized by the same values of A and Z. What are the values of I and I_3? According to Eq. (8.31), all states of a nuclide have the same value of I_3, namely

$$I_3 = Z - \tfrac{1}{2}A = \tfrac{1}{2}(Z - N). \tag{8.34}$$

The assignment of the total isospin quantum number I is not so simple. There are A isospin vectors with $I = \tfrac{1}{2}$, and, since they add vectorially, they can add up to many different values of I. The maximum value of I is $\tfrac{1}{2}A$, and it occurs if the contributions from all nucleons are parallel. The minimum value is $|I_3|$, because a vector cannot be smaller than one of its components. I therefore satisfies

$$\tfrac{1}{2}|Z - N| \leq I \leq \tfrac{1}{2}A. \tag{8.35}$$

Can a value of I be assigned to a given nuclear level, and can it be determined experimentally? To answer these questions, we return to a world where all but the hadronic interactions are switched off, and we consider a nucleus formed from A nucleons. I is a good quantum number in a purely hadronic world, and each state of the nucleus can be characterized by a

---

12. E. P. Wigner, *Phys. Rev.* **51**, 106, 947 (1937).

value of I. Equation (8.35) shows that I is integer if A is even and half-integer if A is odd. The state is $(2I + 1)$-fold degenerate. If the electromagnetic interaction is switched on, the degeneracy is broken, as indicated in Fig. 8.4. Each of the substates is characterized by a unique value of I_3

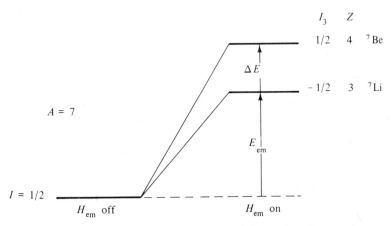

Fig. 8.4. Isospin doublet. Without the electromagnetic interaction, the two substates are degenerate. With H_{em} switched on, the degeneracy is lifted, and each sublevel appears in a different isobar. The levels in the real nuclides are said to form an isospin multiplet.

and, as shown by Eq. (8.31), appears in a different isobar. As long as the electromagnetic interaction is reasonably small $[(Ze^2/\hbar c) \ll 1]$ it is expected that real nuclear states will behave as described and consequently can be labeled by I. It turns out that I can even be assigned to states in heavy nuclei where this condition is not fulfilled. Such states are called *isobaric analog states*; they were discovered in 1961.[13] Figure 8.4 is the nuclear analog to Fig. 8.2. Both are the isospin analogs of the Zeeman effect shown in Fig. 5.5. In the magnetic (spin) case, the levels are labeled by J and J_z, and in the isospin case by I and I_3. In the magnetic case, the splitting is caused by the magnetic field, and in the isospin case by the Coulomb interaction.

The way to find the value of I is similar to the one used for particles: If all members of an isospin multiplet can be found, their number can be counted; it is $2I + 1$, and I is determined. As pointed out in Section 8.6, all members of an isospin multiplet are expected to have the same quantum numbers, apart from I_3 and q. Properties other than discrete quantum numbers can be affected by the electromagnetic force but should still be approximately alike. The search is started in a given isobar, and levels with similar properties are looked for in neighboring isobars. In contrast to

13. J. D. Anderson and C. Wong, *Phys. Rev. Letters* **7**, 250 (1961). Isobaric analog states are discussed in Section 15.6.

particle physics, where the effect of the electromagnetic interaction is difficult to compute, the positions of the levels can be predicted with confidence: The electromagnetic force produces two effects, a repulsion between the protons in the nucleus and a mass difference between neutron and proton. The Coulomb repulsion can be calculated, and the mass difference is taken from experiment. The energy difference between members of an isospin multiplet in isobars $(A, Z + 1)$ and (A, Z) is

$$\Delta E = E(A, Z + 1) - E(A, Z) \approx \Delta E_{\text{Coul}} - (m_n - m_p)c^2, \quad (8.36)$$

with $(m_n - m_p)c^2 = 1.293$ MeV. The simplest estimate of the Coulomb energy is obtained by assuming that the charge Ze is distributed uniformly through a sphere of radius R. The classical electrostatic energy is then given by

$$E_{\text{Coul}} = \frac{3}{5} \frac{(Ze)^2}{R}, \quad (8.37)$$

and it gives rise to the shift shown in Fig. 8.4. The energy difference between isobars with charges $Z + 1$ and Z becomes approximately

$$\Delta E_{\text{Coul}} \approx \frac{6}{5} \frac{e^2}{R} Z \quad (8.38)$$

if both nuclides have equal radii. (They should have equal radii since their hadronic structures are alike.) Values for R can be taken from Eq. (6.38), and the Coulomb energy difference can then be calculated.

The values of nuclear spins vary all the way from 0 to more than 10. Does a similar richness exist in the values of isospin? It does, many iso-

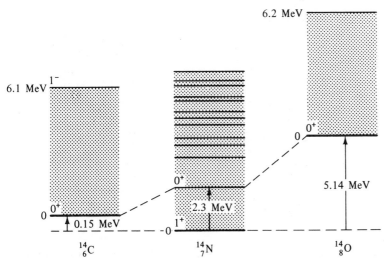

Fig. 8.5. $A = 14$ isobars. The labels denote spin and parity, for instance, 0^+. The ground state of ^{14}N is an isospin singlet; the first excited state is a member of an isospin triplet.

spin values occur, and we shall discuss a few in order to show the importance of the isospin concept. All examples will show one regularity: The isospin of the nuclear ground state always assumes the smallest value allowed by Eq. (8.35), $I_{min} = |Z - N|/2$.

Isospin *singlets*, $I = 0$, can appear only in nuclides with $N = Z$, as is evident from Eq. (8.35). Such nuclides are called self-conjugate. The ground states of ^{2}H, ^{4}He, ^{6}Li, ^{8}Be, ^{12}C, ^{14}N, and ^{16}O have $I = 0$. ^{14}N is a good example, and the lowest levels of the $A = 14$ isobars are shown in Fig. 8.5. Since A is even, only integer isospin values are allowed. If the ^{14}N ground state had a value of $I \neq 0$, similar levels would have to appear in ^{14}C and ^{14}O, with $I_3 = \pm 1$. These levels should have the same spin and parity as the ^{14}N ground state, namely 1^+. Equation (8.36) permits a calculation of the approximate position: The level in ^{14}O should be about 2.4 MeV higher, and the level in ^{14}C should be about 1.8 MeV lower than

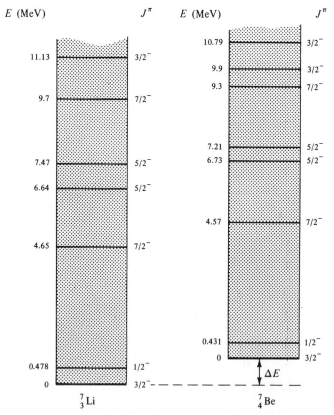

Fig. 8.6. Level structure in the two isobars ^{7}Li and ^{7}Be. These two nuclides contain the same number of nucleons; apart from electromagnetic effect, their level schemes should be identical. J^π denotes spin and parity of a level. Parity will be discussed in Chapter 9.

the ^{14}N ground state. No such states exist. On the oxygen side, the first level appears at 5.1 MeV and it has spin 0 and positive parity. On the ^{14}C side, the first level is higher and not lower, and it also has spin 0. All evidence indicates that the ^{14}N ground state has isospin 0.

Isospin *doublets* occur in mirror nuclides for which $Z = (A \pm 1)/2$. A few examples are shown in Fig. 8.6. The ground state and the first five excited states have isospin $\frac{1}{2}$. Equation (8.36) predicts an energy shift of 0.6 MeV, which is in reasonable agreement with the observed shift of 0.86 MeV.

An example of an isospin *triplet* is shown in Fig. 8.5. The ground states of ^{14}C and ^{14}O form an $I = 1$ triplet with the first excited state of ^{14}N. All three states have spin 0 and positive parity. The energies agree reasonably well with the prediction of Eq. (8.36). *Quartets* and *quintets* have also been found,[14] and the existence of isospin multiplets in isobars is well established.

8.8 References

General references to invariance properties are given in Section 7.7. In addition to these, the following books and articles are recommended.

Rotations in ordinary space and the ensuing quantum mechanics of angular momentum are important in all parts of subatomic physics. We have only scratched the surface. For further details, the texts by Messiah and Merzbacher are useful. The subject is treated in more detail in D. M. Brink and G. R. Satchler, *Angular Momentum*, Oxford University Press, London, 1968.

The early ideas concerning isospin are lucidly described in E. Feenberg and E. P. Wigner, *Rept. Progr. Phys.* **8**, 274 (1941), and in W. E. Burcham, *Progr. Nucl Phys.* **4**, 171 (1955). A more recent review is D. Robson, *Ann. Rev. Nucl. Sci.* **16**, 119 (1966). The book *Isospin in Nuclear Physics* (D. H. Wilkinson, ed.), North-Holland, Amsterdam, 1969, provides an up-to-date review of the entire field. Even though many contributions in this volume are far above the level of the present course, the book can be consulted if questions arise.

Equation (8.37) for the Coulomb energy is good enough for estimates. For detailed arguments, it must be improved. A thorough discussion of Coulomb energies is given in a review by J. A. Nolan, Jr., and J. P. Schiffer, *Ann. Rev. Nucl. Sci.* **19**, 471 (1969).

14. J. Cerny, *Ann. Rev. Nucl. Sci.* **18**, 27 (1968).

PROBLEMS

8.1. Derive the commutation relation between J_x and J_y:

(a) Equation (8.2) gives the relation between a wave function before and after rotation, $\psi^R = U\psi$. Matrix elements of an operator F can be taken between the original and the rotated states. It is, however, also possible to consider rotation of the operator F and leave the states unchanged. Justify that the relation between the rotated and the original operator is given by

$$F^R = U^\dagger F U.$$

(b) Assume $\mathbf{J} \equiv (J_x, J_y, J_z)$ to be a vector. Consider an infinitesimal rotation of $\mathbf{J}$ by the angle ϵ about the y axis. Express $\mathbf{J}^R \equiv (J_x^R, J_y^R, J_z^R)$ in terms of $\mathbf{J}$ and ϵ.

(c) Assume $\mathbf{J}$ to be the generator of the rotation U, Eq. (8.7). Use infinitesimal rotations to derive the commutation relation between J_x and J_y by setting $F = J_x$ in part (a) and using the result of part (b).

8.2. Consider the operator $U = \exp(-i\mathbf{a} \cdot \mathbf{p}/\hbar)$, where $\mathbf{a}$ is a displacement in real space and $\mathbf{p}$ is a momentum vector.

(a) What operation is described by U?

(b) Assume that H is invariant under translation in space. Find the conserved quantity corresponding to this symmetry operation and discuss its eigenfunctions and eigenvalues.

8.3. Discuss some evidence for charge independence in the pion-nucleon interaction.

8.4. Verify the steps in footnote 1.

8.5. Calculate the commutator (8.11).

8.6. Justify that the isospin of the deuteron is zero

(a) By using experimental information.

(b) By considering the generalized Pauli principle stating that the total wave function, assumed to be a product of space, spin, and isospin parts, must be antisymmetric under the exchange of the two nucleons.

8.7. The reaction

$$dd \longrightarrow \alpha \pi^0$$

has not been observed. The isospin of the deuteron and the alpha particle are known to be zero. What does the absence of the reaction tell us?

8.8. Verify Eq. (8.37), Eq. (8.38).

8.9. Study the energy levels of the $A = 12$ isobars.

(a) Sketch the energy level diagrams.

(b) Justify that the ground state and the first few excited states of ^{12}C have isospin zero.

(c) Find the first $I = 1$ state in ^{12}C and justify that it forms an isospin triplet with the ground states of ^{12}B and ^{12}N.

8.10. Consider the reactions

$$d\,^{16}O \longrightarrow \alpha\,^{14}N$$
$$d\,^{12}C \longrightarrow p\,^{13}C.$$

Assume isospin invariance. What are the values of I of the states in ^{14}N and ^{13}C that can be reached by these reactions? (^{16}O, ^{12}C, α, and d denote ground states; ^{14}N and ^{13}C can be excited.)

8.11. Consider the beta decay of ^{14}O. Normally, a beta decay will have a lifetime that is approximately proportional to E^{-5}, where E is the maximum energy of the beta particles. Use isospin invariance to explain the observed branching ratio.

8.12. Compare ΔE_{Coul} for $A = 10, 80,$ and 200. Why is it more difficult (or impossible) to find all the members of an isospin multiplet in heavy nuclei than in light nuclei?

8.13. Consider the reactions

$$\gamma A \longrightarrow nA'$$
$$dA \longrightarrow pA'$$
$$dA \longrightarrow \alpha A'$$
$$^3He\,A'' \longrightarrow {}^3H\,A'.$$

If A is a self-conjugate ($N = Z$) nuclide, what are the isospin states in A' that can be reached by these reactions? The photon "carries" isospin 0 and 1. If A'' has isospin 0, or $\frac{1}{2}$, or $\frac{3}{2}$, what are the possible values of the isospin states in A'?

8.14.

(a) Prove the commutation relations

$$[I_\pm, I^2] = 0, \quad [I_3, I_\pm] = \pm I_\pm, \quad [I_+, I_-] = 2I_3.$$

(b) Use these commutation relations and Eq. (8.24) to prove Eq. (8.27).

9

P, *C*, and *T*

In the previous chapter we have discussed two continuous symmetry operations: rotations in ordinary space and in isospin space. These rotations can be made as small as desired and consequently can be studied by employing infinitesimal transformations. Invariance under these rotations leads to conservation of spin and isospin, respectively. In this chapter we shall discuss examples of discontinuous transformations, which can lead to operators of the type already given in Eq. (7.11), namely

$$U_h^2 = 1.$$

Such operators are Hermitian *and* unitary. Invariance under U_h leads to a multiplicative conservation law in which the product of quantum numbers is an invariant.

9.1 The Parity Operation

Parity invariance, loosely stated, means invariance under an interchange left $\rightleftharpoons$ right, or symmetry of mirror image and object. For many years, physicists were convinced that all natural laws should be invariant under such mirror reflections. Clearly this belief has little to do with everyday observations because our world is not left/right-invariant. Keys, screws, and DNA have a handedness. Vitamin C comes in two forms, L-ascorbic acid and D-ascorbic acid, and we are told that only one is useful against the cold.[1] Why, then, the belief in invariance under space reflection? The history of the parity operation shows how a concept is found, how a concept is understood, how a concept becomes a dogma, and how finally the dogma falls: In 1924, Laporte discovered that atoms have two different classes of levels; he established selection rules for transitions between the two classes, but he could not explain their existence. Wigner then showed that the two classes follow from invariance of the wave function under space reflection.[2] This symmetry was so appealing that it was elevated to a dogma. The observed left/right asymmetries in nature were all blamed on initial conditions. It came, therefore, as a rude shock when Lee and Yang, in 1956, showed that no evidence for parity conservation in the weak interaction existed[3] and when parity nonconservation was subsequently found by Wu and collaborators in beta decay.[4] The fall of parity, however, was only partial. Parity is conserved in hadronic and electromagnetic processes.

The *parity operation* (space inversion), P, changes the sign of any true (polar) vector:

$$\mathbf{x} \xrightarrow{P} -\mathbf{x}, \qquad \mathbf{p} \xrightarrow{P} -\mathbf{p}. \tag{9.1}$$

Axial vectors, however, remain unchanged under P. An example is the orbital angular momentum, $\mathbf{L} = \mathbf{r} \times \mathbf{p}$. Under P, both $\mathbf{r}$ and $\mathbf{p}$ change sign, and $\mathbf{L}$ consequently remains unchanged. A general angular momentum vector, $\mathbf{J}$, behaves the same way:

$$\mathbf{J} \xrightarrow{P} \mathbf{J}. \tag{9.2}$$

This behavior follows from the observation that P commutes with an infinitesimal rotation and hence also with $\mathbf{J}$. Moreover, the transformation (9.2) leaves the commutation relations for angular momentum, Eq. (5.6), invariant. The effect of the parity operation on momentum and on angular momentum is shown in Fig. 9.1.

The parity operator is a special case of the transformation operator U discussed in Section 7.1; P changes a wave function into another wave function:

$$P\psi(\mathbf{x}) = \psi(-\mathbf{x}). \tag{9.3}$$

If P is applied a second time to Eq. (9.3), the original state is regained,[5]

$$P^2\psi(\mathbf{x}) = P\psi(-\mathbf{x}) = \psi(\mathbf{x}), \tag{9.4}$$

1. L. Pauling, *Vitamin C and the Common Cold*, W. H. Freeman, San Francisco, 1970.
2. E. P. Wigner, *Z. Physik* **43**, 624 (1927).
3. T. D. Lee and C. N. Yang, *Phys. Rev.* **104**, 254 (1956).
4. C. S. Wu, E. Ambler, R. W. Hayward, D. D. Hoppes, and R. P. Hudson, *Phys. Rev.* **105**, 1413 (1957).
5. For relativistic wave functions, Eq. (9.4) must be generalized.

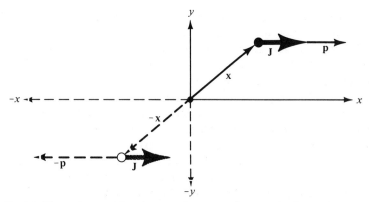

Fig. 9.1. The parity operation changes $\mathbf{x}$ into $-\mathbf{x}$, $\mathbf{p}$ into $-\mathbf{p}$, but leaves the angular momentum $\mathbf{J}$ unchanged. For clarity, only two dimensions are shown.

and P consequently satisfies the operator equation

$$P^2 = I. \tag{9.5}$$

P is an example of the operator (7.11), which was denoted by U_h and which is Hermitian and unitary at the same time. Equation (9.5) shows that the eigenvalues of P are $+1$ and -1.

Up to this point, no invariance arguments have been introduced. The discussion was restricted to the parity operation, and it dealt only with what happens under P. The wave functions $\psi(\mathbf{x})$ and $\psi(-\mathbf{x})$ can be wildly different. The situation becomes orderly when invariance under parity is introduced. Assume a system to be described by a Hamiltonian H that commutes with P:

$$[H, P] = 0. \tag{9.6}$$

In this case, the wave function $\psi(\mathbf{x})$ can be chosen to be an eigenfunction of the parity operator, as can be seen as follows. $\psi(\mathbf{x})$ is an eigenfunction of H,

$$H\psi(\mathbf{x}) = E\psi(\mathbf{x}).$$

Operating with P and using Eq. (9.6) give

$$HP\psi(\mathbf{x}) = PH\psi(\mathbf{x}) = PE\psi(\mathbf{x}),$$

or

$$H\psi'(\mathbf{x}) = E\psi'(\mathbf{x}),$$

where

$$\psi'(\mathbf{x}) \equiv P\psi(\mathbf{x}).$$

The wave functions $\psi(\mathbf{x})$ and $P\psi(\mathbf{x})$ satisfy the same Schrödinger equation with the same energy eigenvalue E, and two possibilities now exist. The state with energy E can be degenerate so that two different physical states described by the wave functions, $\psi(\mathbf{x})$ and $\psi'(\mathbf{x}) \equiv P\psi(\mathbf{x})$, have the same energy. If the state is *not* degenerate, then $\psi(\mathbf{x})$ and $P\psi(\mathbf{x})$ must describe the same physical situation, and they must be proportional to each other:

$$P\psi(\mathbf{x}) = \pi\psi(\mathbf{x}). \tag{9.7}$$

This relation has the form of an eigenvalue equation, and the eigenvalue π is called the parity of the wave function $\psi(\mathbf{x})$. The argument following Eq. (9.5) implies that the eigenvalue must be $+1$ or -1:

$$\pi = \pm 1. \tag{9.8}$$

The corresponding wave functions are said to have even ($+$) or odd ($-$) parity. Since P commutes with H, according to Eq. (9.6), parity is conserved, and π is the observable eigenvalue associated with the Hermitian operator P.

A particularly useful example of a parity eigenfunction is $Y_l^m(\theta, \varphi)$, the eigenfunction of the orbital angular momentum operator. In Eq. (5.4), we wrote this eigenfunction as $\psi_{l,m}$ and defined it as the eigenfunction of the operators L^2 and L_z. The function Y_l^m is called a spherical harmonic. Properties of the Y_l^m and their explicit form up to $l = 3$ are given in Table A8 in the Appendix. In polar coordinates, the parity operation $\mathbf{x} \rightarrow -\mathbf{x}$ is given by

$$r \longrightarrow r$$
$$\theta \longrightarrow \pi - \theta \tag{9.9}$$
$$\varphi \longrightarrow \pi + \varphi,$$

and under such a transformation, Y_l^m changes sign if l is odd and remains unchanged if l is even:

$$PY_l^m = (-1)^l Y_l^m. \tag{9.10}$$

Conservation of parity leads to a *multiplicative* conservation law, as can be seen by considering a reaction

$$a + b \longrightarrow c + d.$$

Symbolically, the initial state can be described as

$$|\text{initial}\rangle = |a\rangle |b\rangle |\text{relative motion}\rangle,$$

where $|a\rangle$ and $|b\rangle$ describe the internal state of the two subatomic particles and $|\text{relative motion}\rangle$ is the part of the wave function characteristic of the relative motion of a and b. Space inversion affects each factor so that

$$P|\text{initial}\rangle = P|a\rangle P|b\rangle P|\text{relative motion}\rangle. \tag{9.11}$$

Equation (9.9) shows that the radial part of the relative-motion wave function is unaffected by P and the orbital part gives the contribution $(-1)^l$, where l is the relative orbital angular momentum of the two particles a and b. The expressions $P|a\rangle$ and $P|b\rangle$ refer to the internal wave functions of the two particles. Since the structures of these particles are predominantly determined by hadronic and electromagnetic forces, we can assign intrinsic parities to particles so that, for instance,

$$P|a\rangle = \pi_a |a\rangle.$$

Equation (9.11) then becomes

$$\pi_{\text{initial}} = \pi_a \pi_b (-1)^l. \tag{9.12}$$

A similar equation holds for the final state, and parity conservation in the reaction demands that

$$\pi_a \pi_b (-1)^l = \pi_c \pi_d (-1)^{l'}, \tag{9.13}$$

where l' is the relative orbital angular momentum of the particles c and d in the final state. Equation (9.13) implies that parity is a conserved multiplicative quantum number.

Why does a gauge transformation lead to an additive quantum number while P leads to a multiplicative one? P is a Hermitian operator in itself, while in a gauge transformation, the Hermitian operator appears in the exponent. A product of exponentials leads to a sum of exponents and hence to an additive law.

9.2 The Intrinsic Parities of Subatomic Particles

Can intrinsic parities be assigned to subatomic particles? We shall show that such assignments are feasible, but we shall also encounter a fine example of an unsuspected trap.

As in all cases where a sign is involved, the starting point must be defined. In electricity, the charge on cat fur is defined to be positive, whence the proton acquires a positive charge. The intrinsic parity of the proton is also defined to be positive,

$$\pi(\text{proton}) = +. \tag{9.14}$$

The determination of the parity of other particles is based on relations of the type of Eq. (9.13). As an example, we consider the capture of negative pions by deuterium.[6] Low-energy negative pions impinge on a deuterium target, and the reaction products are observed. Of the three reactions,

$$d\pi^- \longrightarrow nn \tag{9.15}$$

$$d\pi^- \longrightarrow nn\gamma \tag{9.16}$$

$$d\pi^- \longrightarrow nn\pi^0 \tag{9.17}$$

only the first two are observed; the third one is absent. Parity conservation for the first reaction leads to the relation

$$\pi_d \pi_{\pi^-} (-1)^l = \pi_n \pi_n (-1)^{l'} = (-1)^{l'}.$$

First consider spin and parity of the initial state. The deuteron is the bound state of a proton and a neutron. The nucleon spins are parallel and add up to a deuteron spin 1. The relative orbital angular momentum of the two nucleons is predominantly zero. (We shall discuss the deuteron in more detail in Chapter 12.) Consequently the deuteron parity is $\pi_d = \pi_p \pi_n$. The

6. W. K. H. Panofsky, R. L. Aamodt, and J. Hadley, *Phys. Rev.* **81**, 565 (1951).

negative pion slows down in the target and is finally captured around a
deuteron, forming a pionic atom. With emission of photons, the pion
rapidly falls to an orbit with zero orbital angular momentum from where
reactions (9.15) and (9.16) occur. Consequently the orbital angular mo-
mentum l is zero, and the parity of the initial state is given by $\pi_{\pi^-}\pi_p\pi_n$. The
angular momentum l' in the final state can also be obtained easily: The
total wave function in the final state must be antisymmetric (two identical
fermions). If the spins of the two neutrons are antiparallel, the spin state is
antisymmetric, and the space state must be symmetric; consequently l' must
be even, and the possible total angular momenta are $0, 2, \ldots$. The total
augular momentum in the initial state is 1; angular conservation therefore
rules out the antisymmetric spin state. For the symmetric spin state,
where the two spins are parallel, the angular momentum l' must be odd,
$l' = 1, 3, \ldots$. Only in the state $l' = 1$ can the total angular momentum be 1,
and the final state therefore is $^3P_1$. With $l' = 1$ the parity relation becomes

$$\pi_p\pi_n\pi_{\pi^-} = -1. \tag{9.18}$$

Two solutions exist, and with the standardization (9.14) they are

$$\pi_p = \pi_n = 1, \qquad \pi_{\pi^-} = -1, \tag{9.19}$$

and

$$\pi_p = \pi_{\pi^-} = 1, \qquad \pi_n = -1. \tag{9.19a}$$

The two solutions are equivalent, experimentally. It turns out that no experiment can be devised that gets around the ambiguity and measures the
relative parity between proton and neutron. The choice is made on theoretical grounds: Proton and neutron form an isodoublet. According to Eq.
(8.15), the members of an isospin multiplet should have the same hadronic
properties, and it is assumed that they do have the same intrinsic parity.
By setting

$$\pi(\text{neutron}) = + \tag{9.20}$$

the parity of the pion becomes negative; the pion is a *pseudoscalar* particle.
The absence of the reaction (9.17) indicates that the neutral pion is also a
pseudoscalar.

● Why can the relative parity of the proton and the neutron, or of the positive and the
neutral pion, not be measured? The reason is connected with the existence of additive
conservation laws. Consider the parity equations for the proton and the neutron,

$$P|p\rangle = |p\rangle$$
$$P|n\rangle = |n\rangle.$$

A modified parity operator, P', is introduced through the definition

$$P' = Pe^{i\pi Q}, \tag{9.21}$$

where Q is the electric charge operator. Physically, the new operator P' is indistinguishable from P. It performs the same function (for instance, changes $\mathbf{x}$ into $-\mathbf{x}$), and,
according to Eq. (7.22), it commutes with H. P and P' therefore are equally good parity
operators. Applied to $|p\rangle$ and $|n\rangle$, P' gives

$$P'|p\rangle = Pe^{i\pi Q}|p\rangle = -P|p\rangle = -|p\rangle, \qquad P'|n\rangle = |n\rangle.$$

The modified parity operator assigns negative intrinsic parity to the proton and leaves the neutron parity unchanged. Since P and P' are equally good parity operators and we have no reason to prefer one over the other, we conclude that the relative parity between systems of different electric charge is not a measurable concept. There, then, is no way to determine experimentally which of the two solutions given in Eq. (9.19) is correct; the assignment of equal parities to the proton and neutron cannot be verified by a measurement, but it rests on firm theoretical grounds.

Instead of the modification (9.21), parity operators of the form

$$P'' = Pe^{i\pi A} \qquad \text{or} \qquad P''' = Pe^{i\pi Y}$$

can be introduced, where A is the baryonic number operator and Y the hypercharge operator. The arguments proceed as above, and it becomes clear that the relative parity is observable only for systems that have equal additive quantum numbers Q, A, and Y. ●

We have just shown that the relative parity of two system is measurable only if the two systems have equal quantum numbers Q, A, and Y. This restriction limits the usefulness of the parity concept, but not as much as could be suspected. It is only necessary to fix the intrinsic parities of three hadrons; the parities of all other hadrons can be found by building composite systems of the *standard particles* and measuring the relative parities of all other states with respect to these. The parities of the proton and the neutron have already been set positive; it is customary to add the lambda as the third standard particle so that

$$\pi(\text{proton}) = \pi(\text{neutron}) = \pi(\text{lambda}) = +. \qquad \textbf{(9.22)}$$

With these definitions, the parities of all other hadrons, including all nuclear states, can be measured experimentally, at least in principle. (Leptons have been omitted, and the reason will become clear in Section 9.3.)

A first example of the determination of the parity of a particle has already been given above where it was shown that the reaction (9.15) leads to the assignment of negative parity to the pion. As a second example, consider the following reactions:

$$dd \longrightarrow p \; {}^3\text{H} \qquad \textbf{(9.23)}$$

$$dd \longrightarrow n \; {}^3\text{He} \qquad \textbf{(9.24)}$$

$$d \; {}^3\text{H} \longrightarrow n \; {}^4\text{He}. \qquad \textbf{(9.25)}$$

Spin and parity of the deuteron, d, have already been discussed above where it was found that the assignment is 1^+. The spins of ${}^3\text{H}$, ${}^3\text{He}$, and ${}^4\text{He}$ can be measured with standard techniques; studies of reactions (9.23)–(9.25) yield values of l and l' and the assignments J^π become $\frac{1}{2}^+$ for ${}^3\text{H}$ and ${}^3\text{He}$ and 0^+ for ${}^4\text{He}$.

In principle, parities of other states can be investigated with similar reactions. One more example is shown in Fig. 9.2. Assume that the assignment 0^+ for ${}^{228}\text{Th}$ is known and that the spins of the various states in ${}^{224}\text{Ra}$ have also been determined. As stated above, the alpha particle has spin 0 and positive parity. If it is emitted with orbital angular momentum L, it carries a parity $(-1)^L$. Since the initial state of the decay has spin 0,

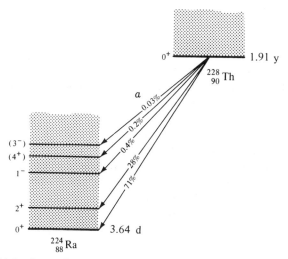

Fig. 9.2. Alpha decay of ²²⁸Th. The intensities of the various alpha branches are given in %. Spin and parity assignments that are not fully established are given in parentheses.

an alpha emitted with angular momentum L can only reach states with spin $J = L$. The parities of these states then must be $(-1)^L = (-1)^J$, or $0^+, 1^-, 2^+, 3^-, 4^+, \ldots$. Such states indeed are seen to be populated by the alpha decay in Fig. 9.2.

The examples given so far are simple. In the actual assignment of parities to particles and excited nuclear states, more complex methods are often necessary, but the basic ideas remain the same. The various methods used in nuclear and in particle physics are described in the references listed in Section 5.12.

9.3 Conservation and Breakdown of Parity

In the previous section we have discussed the experimental determination of the intrinsic parities of some subatomic particles. Implied in all arguments was conservation of parity in the processes used to find π. How good is the evidence for parity conservation in the various interactions? To answer this question in a quantitative way, a measure for the degree of parity conservation must be introduced. If $|\alpha\rangle$ is a nondegenerate state of a system with, for instance, even parity, it is written as

$$|\alpha\rangle = |\text{even}\rangle.$$

If parity is not conserved, $|\alpha\rangle$ can be written as a superposition of an even and an odd part,

$$|\alpha\rangle = c\,|\text{even}\rangle + d\,|\text{odd}\rangle$$
$$|c|^2 + |d|^2 = 1. \tag{9.26}$$

A state of this form, with $c \neq 0$ and $d \neq 0$, is no longer an eigenstate of the parity operator P because

$$P|\alpha\rangle = c|\text{even}\rangle - d|\text{odd}\rangle \neq \pi|\alpha\rangle.$$

$\mathfrak{F} = d/c$ is a measure for the degree of parity nonconservation ($d \leq c$). Parity violation is maximal if the state contains equal amplitudes of $|\text{even}\rangle$ and $|\text{odd}\rangle$, or if $|\mathfrak{F}| = 1$.

Sensitive tests for parity conservation in the *hadronic* and the *electromagnetic* interaction are based on selection rules for alpha decay. In Fig. 9.2 it was shown how the occurrence of an alpha decay can be used to determine the parity of a state to which a transition occurs. The approach can be inverted: Since an alpha particle with orbital angular momentum L carries a parity $(-1)^L$, decays such as $1^+ \xrightarrow{\alpha} 0^+$ or $2^- \xrightarrow{\alpha} 0^+$ are parity-forbidden. They can occur only if one or both of the states involved contain an admixture of the opposite parity. Figure 9.3 shows the levels

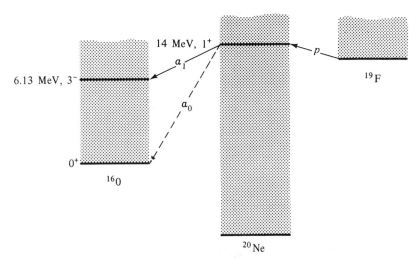

Fig. 9.3. Alpha decays from a 1⁺ level in ²⁰Ne. Only levels of interest are shown.

used in the first experiment.[7] A 1^+ level in ²⁰Ne at an excitation energy of about 14 MeV can be reached by bombarding ¹⁹F with protons, and it decays by alpha emission to the 3^- state at 6.13 MeV in ¹⁶O. This transition is parity-allowed, because vector addition of angular momenta permits emission of an alpha particle with $L = 3$ in a transition $1^+ \xrightarrow{\alpha} 3^-$. However, the transition to the ground state can only go with $L = 1$; the corresponding parity is negative, and the decay $1^+ \xrightarrow{\alpha} 0^+$ is parity-forbidden. The search for such a parity-forbidden branch consequently constitutes a search for $|\mathfrak{F}|^2$. In the experiment just discussed, a limit on $|\mathfrak{F}|^2$ of $\approx 4 \times 10^{-8}$ was found. Later experiments with different decays have

7. N. Tanner, *Phys. Rev.* **107**, 1203 (1957).

given a better value,[8]

$$|\mathcal{F}|^2 \lesssim 3 \times 10^{-13}. \tag{9.27}$$

Such a small number provides very good evidence for parity conservation in the hadronic interaction. At the same time, it shows that parity is also conserved in the electromagnetic interaction. If parity were violated electromagnetically, the nuclear wave functions would also be of the form of Eq. (9.26), and parity-forbidden alpha decays would become possible. Since the electromagnetic force is weaker than the hadronic one by about a factor of 100, the limit on the corresponding violation is less stringent than Eq. (9.27), but still very low.

The limit (9.27) is a recent one; before 1957, the limits were much less convincing. However, since parity conservation had already achieved the position of a dogma, very few physicists were willing to spend their time improving a number that was considered to be safe anyway. The astonishment was therefore great when it was found early in 1957 that parity was not conserved in the *weak interaction*.[9] The puzzle that motivated the crucial thinking developed before 1956. By 1956, it had become clear that two strange particles with remarkable properties existed. They were called the tau and the theta, and they appeared to be identical in every respect (mass, production cross section, spin, charge) except in their decay. One decayed to a state of negative parity, and the other to a state of positive parity. The dilemma thus was as follows: Either two practically identical particles with opposite parities existed or parity conservation had to be given up. Lee and Yang studied the problem in depth[3] and found, much to their surprise, an overlooked fact: Evidence for parity conservation existed, but only for the hadronic and the electromagnetic interaction, and not for the weak one. And the decays of the tau and the theta were so slow that they were known to be weak; Lee and Yang suggested experiments to test parity conservation specifically in the weak interaction. The first experiment was performed by Wu and collaborators, and it brilliantly showed the correctness of Lee and Yang's conjecture.[4]

The concept underlying the Wu et al. experiment is explained in Fig. 9.4. ^{60}Co nuclei are polarized so that their spins **J** point along the positive z axis. When the nuclei decay through

$$^{60}\text{Co} \longrightarrow {}^{60}\text{Ni} + e^- + \bar{\nu},$$

the intensity of the emitted electrons is measured in the two directions 1 and 2. The electron momenta are denoted by $\mathbf{p}_1$ and $\mathbf{p}_2$, and the corre-

---

8. The tests are discussed in E. M. Henley, *Ann. Rev. Nucl. Sci.* **19**, 367 (1969).

9. The discovery of parity nonconservation in the weak interaction came as a great shock to most physicists. The background and the story is described in a number of books and reviews. We recommend C. N. Yang, *Elementary Particles*, Princeton University Press, Princeton, N.J., 1962. A letter from Pauli to Weisskopf (German but with English translation) is reprinted in W. Pauli, *Collected Scientific Papers*, Vol. 1 (R. Kronig and V. F. Weisskopf, eds.), Wiley-Interscience, New York, 1964, p. xii. The letter shows how much the fall of parity affected physicists.

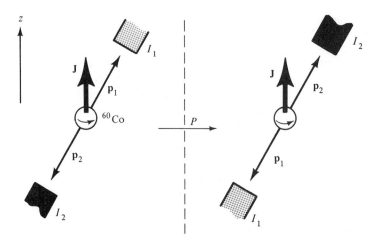

Fig. 9.4. Concept of the Wu et al. experiment. A polarized nucleus emits electrons with momenta $\mathbf{p}_1$ and $\mathbf{p}_2$. The original situation is shown at the left, and the parity-transformed one at the right. Invariance under parity means that the two situations cannot be distinguished.

sponding intensities by I_1 and I_2. Under the parity transformation, the spins remain unchanged, but the momenta $\mathbf{p}_1$ and $\mathbf{p}_2$, and the intensities I_1 and I_2, are interchanged. Invariance under the parity operation means that the original and the parity-transformed situations cannot be distinguished. Figure 9.4 shows that the two situations give identical intensities if $I_1 = I_2$. Parity conservation demands that the intensity of electrons emitted parallel to $\mathbf{J}$ is the same as for electrons emitted antiparallel to $\mathbf{J}$.

In a more formal way, the essential aspect of the experiment is the observation of the expectation value of the operator

$$\mathcal{P} = \mathbf{J} \cdot \mathbf{p}, \tag{9.28}$$

where $\mathbf{J}$ is the spin of the nucleus and $\mathbf{p}$ is the momentum of the emitted electron. $\mathcal{P}$ is a pseudoscalar; under the parity operation it transforms as

$$\mathbf{J} \cdot \mathbf{p} \xrightarrow{\;P\;} -\mathbf{J} \cdot \mathbf{p}. \tag{9.29}$$

Invariance under the parity operation means that the transition rates in the two situations, $\mathbf{J} \cdot \mathbf{p}$ and $-\mathbf{J} \cdot \mathbf{p}$, are identical. Equation (9.29) instructs the experimental physicist how to test parity invariance: Measure the transition rate for a fixed orientation of $\mathbf{J}$ and $\mathbf{p}$ and compare the result to the transition rate for the state $-\mathbf{J} \cdot \mathbf{p}$. The state $-\mathbf{J} \cdot \mathbf{p}$ can be reached by inverting $\mathbf{J}$ or $\mathbf{p}$. The experiment of Wu and collaborators consisted of comparing the transition rates for $\mathbf{J} \cdot \mathbf{p}$ and $-\mathbf{J} \cdot \mathbf{p}$ by inverting $\mathbf{J}$ through inverting the polarization of the ^{60}Co nuclei.

In a radioactive source at room temperature, the nuclear spins are randomly oriented. It is necessary to polarize the nuclei so that all spins $\mathbf{J}$ point in the same direction. The transition rate for electron emission parallel and antiparallel to $\mathbf{J}$ can then be compared. To describe the experimental approach, we use a hypothetical decay, shown in Fig. 9.5(a). A nuclide with spin 1 and g factor $g > 0$ decays by emission of an electron and an antineutrino to a state with spin 0. To polarize the nuclei, the sample is placed in a strong magnetic field **B** and cooled to a very low temperature T. The magnetic sublevels of the initial state split as in Fig. 5.5;

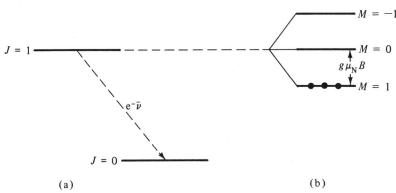

(a)   (b)

Fig. 9.5. (a) Beta decay from a state with spin 1 to a state with spin 0. (b) At very low temperatures in a high magnetic field, only the lowest Zeeman level is populated, and the nucleus (with $g > 0$) is fully polarized and points in the direction of **B**.

the energy of a state with magnetic quantum number M is given by Eq. (5.21) as $E(M) = E_0 - g\mu_N BM$. The ratio of populations, $N(M')/N(M)$, of two states, M' and M, is determined by the Boltzmann factor,

$$\frac{N(M')}{N(M)} = e^{-\{E(M')-E(M)\}/kT}, \tag{9.30}$$

or, with Eq. (5.21),

$$\frac{N(M')}{N(M)} = e^{(M'-M)g\mu_N B/kT}. \tag{9.31}$$

If the condition

$$kT \ll g\mu_N B \tag{9.32}$$

is satisfied, only the lowest Zeeman level is populated, the nucleus is fully polarized, and its spin points in the direction of the magnetic field [Fig. 9.5(b)]. The change $\mathbf{J} \cdot \mathbf{p} \rightarrow -\mathbf{J} \cdot \mathbf{p}$ is obtained by reversing the direction of the external field, **B**. The experimental arrangement requires mastery of many techniques. The radioactive nuclei are introduced into a cerium-magnesium-nitrate crystal and cooled to a temperature of 0.01 K by adiabatic demagnetization. The magnetic field required to satisfy Eq. (9.32) is very high. To obtain such a high field, paramagnetic atoms are chosen, and the field at the nucleus is then predominantly produced by its own electronic shell. The radioactive source must be thin so that the electrons can escape and be counted in a detector placed in the cryogenic system

[Fig. 9.6(a)]. Data are reproduced in Fig. 9.6(b). The result is striking. The expectation value of $\mathcal{P} = \mathbf{J} \cdot \mathbf{p}$ does not vanish, and parity is not conserved in beta decay. Many additional experiments have borne out the remarkable result that parity is violated in weak interactions. We can now return to an earlier figure and understand it better. In Fig. 7.2, neutrino and anti-neutrino are shown to be fully polarized. Full polarization means that neutrino and antineutrino have a nonvanishing value of $\mathbf{J} \cdot \mathbf{p}$ and therefore are a permanent expression of parity nonconservation in the weak interaction.

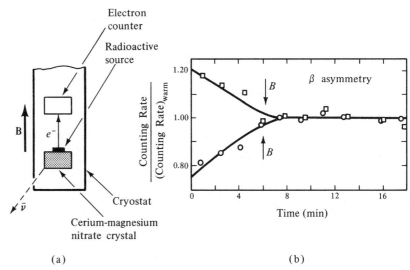

(a)   (b)

Fig. 9.6. (a) Arrangement to measure beta emission from polarized nuclei. (b) Result of the earliest experiment showing parity nonconservation [C. S. Wu, E. Ambler, R. W. Hayward, D. D. Hoppes, and R. P. Hudson, *Phys. Rev.* **105**, 1413 (1957).] A normalized counting rate in the beta detector is shown for two directions of the external magnetic field. After adiabatic demagnetization, the source warms up, the polarization decreases, and the effect disappears.

● It is customary to describe the polarization of a spin-$\frac{1}{2}$ particle not by $\mathbf{J} \cdot \mathbf{p}$ but by the helicity operator

$$\mathcal{H} = 2\frac{\mathbf{J} \cdot \hat{\mathbf{p}}}{\hbar}, \tag{9.33}$$

where $\hat{\mathbf{p}}$ is a unit vector in the direction of the momentum. The expectation value of $\mathcal{H}$ for a particle that has its spin along its momentum is $+1$; $\langle |\mathcal{H}| \rangle = -1$ characterizes a particle with spin opposite to $\hat{\mathbf{p}}$. Particles with nonvanishing helicity can be produced in many experiments; common to all these is the existence of a preferred direction, for instance, given by a magnetic field. If no preferred direction exists, a nonvanishing value of $\langle |\mathbf{J} \cdot \hat{\mathbf{p}}| \rangle$ and hence also of $\langle |\mathcal{H}| \rangle$ is a sign of parity nonconservation. An example is the helicity of leptons emitted from isotropic weak sources, such as beta or muon decay. The helicity of the charged leptons in such weak decays has been measured.[10] The

10. H. Frauenfelder and R. M. Steffen, in *Alpha-, Beta- and Gamma-Ray Spectroscopy*, Vol. 2, (K. Siegbahn, ed.), North-Holland, Amsterdam, 1965.

result,

$$\langle \mathcal{H}(e^-) \rangle = -\frac{v}{c}, \qquad \langle \mathcal{H}(e^+) \rangle = +\frac{v}{c}, \tag{9.34}$$

where v is the lepton velocity, confirms parity nonconservation in the weak interaction. ●

9.4 Charge Conjugation

In Section 7.5, the concept of antiparticles was introduced. This concept gives rise to long and mainly philosophical discussions centered around questions such as "Is there really a sea of negative energy states?" or "Can a particle really move backward in time?" The important features, however, are not connected with such vague aspects but concern the undeniable fact that antiparticles exist. In the present section, the particle-antiparticle connection will be put into a more formal frame than in Section 7.5. Many of the ideas are similar to the ones already introduced in connection with parity in Section 9.1 so that the discussion can be brief.

We describe a particle by the ket $|N\rangle$, where N stands for the additive quantum numbers A, q, S, L, and L_μ. The operation of charge conjugation, C, is then defined by

$$C|N\rangle = |-N\rangle. \tag{9.35}$$

Charge conjugation reverses the sign of the additive quantum numbers but leaves momentum and spin unchanged. C is sometimes also called particle-antiparticle conjugation to express the fact that not only the electric charge but also baryon number, strangeness, lepton number, and muon number change sign. The situation is depicted in Fig. 9.7. If C is applied a second

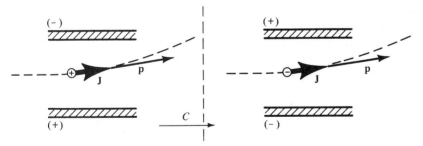

Fig. 9.7. Charged particle traversing an electric field. Charge conjugation, acting on the whole system, reverses the additive quantum numbers of a particle but leaves space-time properties $(\mathbf{p}, \mathbf{J})$ unchanged. The charges of the external field are also reversed so that the trajectories of particle and antiparticle are the same.

time, the original charges are regained so that

$$C^2 = 1. \tag{9.36}$$

C, like P, is a discontinuous operator of the type of Eq. (7.11), and it is unitary and Hermitian.

Equation (9.36) indicates that the eigenvalues of the charge conjugation operator are $+1$ and -1. However, as we shall see now, there is a considerable difference between P and C because C does *not* always have eigenstates. To explore this new feature, we write tentatively

$$C|N\rangle \stackrel{?}{=} \eta_c |N\rangle \qquad (9.37)$$

and ask when such a relation is meaningful. As an example, the state $|N\rangle$ is taken to be an eigenstate of the charge operator, Q. For a particle with charge q, described by $|q\rangle$, the eigenvalue equation

$$Q|q\rangle = q|q\rangle \qquad (9.38)$$

holds. But by Eq. (9.35), C applied to $|q\rangle$ gives

$$C|q\rangle = |-q\rangle.$$

The commutator of the two operators Q and C, when operating on $|q\rangle$, can now be obtained in a straightforward way:

$$CQ|q\rangle = qC|q\rangle = q|-q\rangle$$
$$QC|q\rangle = Q|-q\rangle = -q|-q\rangle$$

or

$$(CQ - QC)|q\rangle = 2q|-q\rangle = 2CQ|q\rangle. \qquad (9.39)$$

The operators C and Q do not commute; this result can be expressed as an operator equation,

$$[C, Q] = 2CQ. \qquad (9.40)$$

Since the two operators C and Q do not commute, it is, in general, not possible to find states that are simultaneous eigenstates. A charged particle cannot satisfy an eigenvalue equation of the form of Eq. (9.37) since nature has chosen particles to be eigenstates of Q. The argument just given applies also to the baryonic number A and to the hypercharge Y. Particles appear in nature as eigenstates of A and of Y, and A and Y also do not commute with C. These is one loophole, however. *Fully neutral* particles, that is, particles with $q = A = Y = 0$, can be in an eigenstate of C, because the states $|q = A = Y = 0\rangle$ and $C|q = A = Y = 0\rangle$ have the same additive quantum numbers, namely 0. For such systems, Eq. (9.37) applies:

$$C|N = 0\rangle = \eta_c |N = 0\rangle, \qquad \eta_c = \pm 1, \qquad (9.41)$$

and η_c is called the *charge parity* (or charge conjugation quantum number). It satisfies a multiplicative conservation law.

What is the charge parity of the fully neutral particles, the photon, the neutral pion, and η^0? A satisfactory answer requires quantum field theory, but the correct values can be obtained with some hand waving. The photon is described by its vector potential $\mathbf{A}$. The potential is produced by charges and currents and consequently changes sign under C:

$$\mathbf{A} \xrightarrow{\ c\ } -\mathbf{A}. \qquad (9.42)$$

An example of this sign change has already been shown in Fig. 9.7. Equation (9.42) suggests the assignment

$$\eta_c(\gamma) = -1. \tag{9.43}$$

π^0 and η^0 decay electromagnetically into two photons,

$$\pi^0 \longrightarrow 2\gamma \quad \text{and} \quad \eta^0 \longrightarrow 2\gamma,$$

and therefore must have positive C parity if C is conserved in the decay:

$$\eta_c(\pi^0) = 1, \quad \eta_c(\eta^0) = 1. \tag{9.44}$$

If C parity were applicable only to the photon, π^0 and η^0, it would not be very useful. However, there exist many particle-antiparticle systems that are fully neutral. Examples are positronium $(e^+e^-), \pi^+\pi^-, p\bar{p}, n\bar{n}$. The C parity of these systems depends on angular momentum and spin, and it is a useful quantity for discussing the possible decay modes.

Use of charge parity for discussion of a decay requires η_c to be a good quantum number. It is conserved if C commutes with the Hamiltonian H. It is easy to see that C is not conserved in the weak interaction,

$$[H_{\text{weak}}, C] \neq 0. \tag{9.45}$$

In Fig. 7.2 it is shown that neutrino and antineutrino have opposite polarization (helicity). If charge conjugation were conserved in the weak interaction, the two particles would have to have the same helicity.

To test conservation of C in the electromagnetic interaction, charge-parity-forbidden decays are looked for. Consider the decays

$$\pi^0 \longrightarrow 3\gamma \quad \text{and} \quad \eta^0 \longrightarrow 3\gamma.$$

π^0 and η^0 have positive charge parity; the three photons in the final states have negative charge parity, and the decay is forbidden. The decays have not been found, but the actual limit set on a possible charge-parity violation is not very good yet.

C conservation in the hadronic interaction has been tested in reactions of the type

$$p\bar{p} \longrightarrow \pi^+\pi^-\pi^0. \tag{9.46}$$

C acting on the reaction gives

$$\bar{p}p \longrightarrow \pi^-\pi^+\pi^0. \tag{9.46a}$$

If the hadronic interaction Hamiltonian commutes with C, the two reactions should give the same result. The initial state, $p\bar{p}$, is the same in both. C invariance then requires that the positive and the negative pions have the same energy spectrum. Comparison of the two distributions, and of the corresponding distributions in other similar reactions, shows no difference; the result can be stated as,[11]

$$\left| \frac{C - \text{nonconserving amplitude}}{C - \text{conserving amplitude}} \right| \lesssim 0.01. \tag{9.47}$$

11. C. Baltay, N. Barash, P. Franzini, N. Gelfand, L. Kirsch, G. Lütjens, J. C. Severiens, J. Steinberger, D. Tycko, and D. Zanello, *Phys. Rev. Letters* **15**, 591 (1965).

The present evidence indicates that charge conjugation and the hadronic Hamiltonian commute.

9.5 Time Reversal

In the two previous sections, the discrete transformations P and C were introduced. Both operations are unitary and Hermitian and give rise to multiplicative quantum numbers. In the present section, a third discrete transformation is introduced, time reversal, T. It will turn out that T is not unitary, and a complication is thus introduced; no conserved quantity such as parity or charge parity is associated with it. Nevertheless, time-reversal *invariance* is a very useful symmetry in subatomic physics.

Formally, the time-reversal operation is defined by

$$t \xrightarrow{\ T\ } -t, \qquad \mathbf{x} \xrightarrow{\ T\ } \mathbf{x}. \tag{9.48}$$

Since classically $\mathbf{p} = d\mathbf{x}/dt$, momentum and angular momentum change sign under T:

$$\mathbf{p} \xrightarrow{\ T\ } -\mathbf{p}$$
$$\mathbf{J} \xrightarrow{\ T\ } -\mathbf{J}. \tag{9.49}$$

In classical mechanics and electrodynamics, the basic equations are invariant under T: Newton's law of motion and Maxwell's equations are second-order differential equations in t and are therefore unaffected by the replacement of t by $-t$.

The essential aspects of time-reversal invariance appear already in the treatment of a nonrelativistic spinless particle, described by the Schrödinger equation,

$$i\hbar \frac{d\psi(t)}{dt} = H\psi(t). \tag{9.50}$$

This equation is formally similar to the diffusion equation which is *not* invariant under $t \rightarrow -t$. The feature that distinguishes T from P and C turns up when the connection between ψ and $T\psi$ is explored. According to the arguments given in Section 7.1, T is a symmetry operator and satisfies

$$[H, T] = 0 \tag{9.51}$$

if $T\psi(t)$ and $\psi(t)$ obey the same Schrödinger equation. The Schrödinger equation for $T\psi(t)$ is

$$i\hbar \frac{dT\psi(t)}{dt} = HT\psi(t). \tag{9.52}$$

The simplest attempt to satisfy this equation,

$$T\psi(t) = \psi(-t), \tag{9.53}$$

is incorrect: Inserting Eq. (9.53) into Eq. (9.52) and writing $-t = t'$ give

$$-i\hbar\frac{d\psi(t')}{dt'} = H\psi(t'). \tag{9.54}$$

This equation is not the same as Eq. (9.50). The fact that Eq. (9.54) is written in terms of t' rather than t is immaterial because t is only a parameter. What counts is *form invariance:* $\psi(t)$ and $T\psi(t)$ must satisfy equations that have the same form.

The correct time-reversal transformation was found by Wigner, who set[12]

$$T\psi(t) = \psi^*(-t). \tag{9.55}$$

Inserting $\psi^*(-t)$ into Eq. (9.52) and taking the complex conjugate of the entire equation produce a relation that has the same form as the original Schrödinger equation if H is real.

The simplest application of the time-reversal transformation (9.55) is to a free particle with momentum $\mathbf{p}$, described by the wave function

$$\psi(\mathbf{x}, t) = e^{i(\mathbf{p}\cdot\mathbf{x} - Et)/\hbar}.$$

The time-reversed wave function is

$$T\psi(\mathbf{x}, t) = \psi^*(\mathbf{x}, -t) = e^{-i(\mathbf{p}\cdot\mathbf{x} + Et)/\hbar} = e^{i(-\mathbf{p}\cdot\mathbf{x} - Et)/\hbar}. \tag{9.56}$$

The time-reversed wave function describes a particle with momentum $-\mathbf{p}$, in accord with Eq. (9.49). It is not necessary to consider the function $T\psi(\mathbf{x}, t)$ to describe a particle going backward in time. The more physical interpretation of T is *motion reversal:* T reverses momentum and angular momentum,

$$T|\mathbf{p}, \mathbf{J}\rangle = |-\mathbf{p}, -\mathbf{J}\rangle. \tag{9.57}$$

When we played the game with P and C, at this point we asked for conserved eigenvalues. The answers were parity π and charge parity η_c. Does T have observable and conserved eigenvalues? Such eigenvalues would be solutions of the equation

$$T\psi(t) = \eta_T\psi(t).$$

Equation (9.55) shows, however, that T changes ψ into its complex conjugate, and the eigenvalue equation makes no sense. This fact is connected with the *antiunitarity* of T. P and C are unitary operators; unitary operators are linear and satisfy the relation

$$U(c_1\psi_1 + c_2\psi_2) = c_1 U\psi_1 + c_2 U\psi_2. \tag{9.58}$$

Antiunitary operators, however, obey the relation

$$T(c_1\psi_1 + c_2\psi_2) = c_1^* T\psi_1 + c_2^* T\psi_2. \tag{9.59}$$

The time-reversal transformation is antiunitary. Why are P and C unitary but not T? In Sections 9.1 and 9.4 we justified the choice of P and C as

---

12. E. Wigner, *Nachr. Akad. Wiss. Goettingen, Math. Physik. Kl. IIa,* **31**, 546 (1932).

unitary operators by saying that they must leave the norm $\mathfrak{N}$ invariant, where $\mathfrak{N}$ is

$$\mathfrak{N} = \int d^3x \, \psi^*(\mathbf{x})\psi(\mathbf{x}).$$

An antiunitary operator also leaves $\mathfrak{N}$ invariant, as can be seen by inserting Eq. (9.55) into $\mathfrak{N}$. The choice between the two possibilities is dictated by the physical nature of the transformation. For P and C, the transformed wave functions satisfy the original equations if the transformation is unitary. For T, form invariance demands that it be antiunitary.

We have just seen that T does not have observable eigenvalues; states can therefore not be labeled with such eigenvalues, and invariance under T cannot be tested by searching for *time-parity-forbidden decays*. Fortunately there are other approaches. Time-reversal invariance predicts, for instance, equality of transition probabilities for a reaction and its inverse (principle of detailed balance). A great deal of effort has gone into testing time-reversal invariance in the various interactions, and the result can be summarized by stating that no violation has been found in the hadronic, the electromagnetic, and the ordinary weak interaction.[8] Two remarks must be added, however. The first concerns the accuracy of these tests. Time-reversal tests are very difficult to perform, and the conservation is only established to about 1 part in 10^2–10^3. The second remark concerns some experiments performed with neutral kaons. We have mentioned before that the neutral kaon system has remarkable properties. These properties make it a sensitive test to time-reversal (actually CP) invariance. Because of the importance of this test, we shall describe the neutral kaon system in the following sections and discuss the time-reversal violation in Section 9.8.

9.6 The Two-State Problem

As an introduction to the discussion of neutral kaons, we consider two identical unconnected potential wells L and R shown in Fig. 9.8(a). The energies of the stationary states $|L\rangle$ and $|R\rangle$ are given by the Schrödinger equations,

$$H_0|L\rangle = E_0|L\rangle, \qquad H_0|R\rangle = E_0|R\rangle.$$

Since H_0 does not connect the two wells, we write

$$\langle L|H_0|R\rangle = \langle R|H_0|L\rangle = 0.$$

For simplicity it is assumed that only the states $|L\rangle$ and $|R\rangle$ play a role. All other states are assumed to have so much higher energies that they can be neglected. If we switch on a perturbing interaction, H_{int}, that lowers the barrier between the wells and induced transitions $L \rightleftharpoons R$, the stationary states of the system are determined by

$$H|\psi\rangle \equiv (H_0 + H_{int})|\psi\rangle = E|\psi\rangle. \tag{9.60}$$

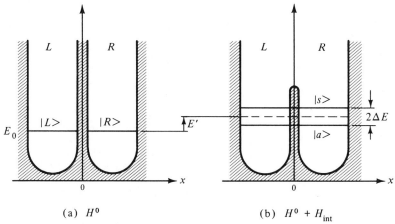

(a) H^0 (b) $H^0 + H_{\text{int}}$

Fig. 9.8. Eigenvalues and eigenfunctions of a particle in two identical potential wells, without and with transmission through the barrier.

The problem consists of finding the eigenvalues and eigenfunctions of the total Hamiltonian $H \equiv H_0 + H_{\text{int}}$. Since the two unperturbed states $|L\rangle$ and $|R\rangle$ are degenerate, the solution requires use of the correct linear combinations of the unperturbed eigenfunctions.[13] These combinations can be found by symmetry considerations. Since the potentials are placed symmetrically about the origin, the Hamiltonian is invariant under reflections through the origin, and H and the parity operator P commute,

$$[H, P] = [H_0 + H_{\text{int}}, P] = 0. \tag{9.61}$$

With the choice of coordinates shown in Fig. 9.8, the parity operator gives

$$P|L\rangle = |R\rangle, \qquad P|R\rangle = |L\rangle. \tag{9.62}$$

The simultaneous eigenfunctions of H_0 and P are easy to find; they are the symmetric and antisymmetric combinations of the unperturbed states $|L\rangle$ and $|R\rangle$:

$$\begin{aligned}|s\rangle &= \sqrt{\tfrac{1}{2}}\{|L\rangle + |R\rangle\} \\ |a\rangle &= \sqrt{\tfrac{1}{2}}\{|L\rangle - |R\rangle\}.\end{aligned} \tag{9.63}$$

These combinations indeed are eigenstates of P,

$$\begin{aligned}P|s\rangle &= +|s\rangle \\ P|a\rangle &= -|a\rangle.\end{aligned} \tag{9.64}$$

Eqs. (9.61) and (9.64) together prove that H does not connect $|a\rangle$ and $|s\rangle$:

$$\langle a|H|s\rangle = \langle a|HP|s\rangle = \langle a|PH|s\rangle = -\langle a|H|s\rangle,$$

13. Merzbacher, Section 17.5; Park, Section 8.4; Eisberg, Section 9.4.

or

$$\langle a | H | s \rangle = 0. \qquad (9.65)$$

Ordinary perturbation theory can consequently be applied to the states $|a\rangle$ and $|s\rangle$. The energy shift caused by the perturbation, H_{int}, is given by the expectation value of H_{int}, or

$$\langle s | H_{\text{int}} | s \rangle = E' + \Delta E$$
$$\langle a | H_{\text{int}} | a \rangle = E' - \Delta E, \qquad (9.66)$$

where

$$\langle L | H_{\text{int}} | L \rangle = \langle R | H_{\text{int}} | R \rangle = E'$$
$$\langle L | H_{\text{int}} | R \rangle = \langle R | H_{\text{int}} | L \rangle = \Delta E. \qquad (9.67)$$

The interaction shifts the center of the energy levels by E' and splits the degenerate levels by an amount $2\Delta E$, as indicated in Fig. 9.8(b). The splitting shows up in the hydrogen molecule ion and particularly clearly in the inversion spectrum of ammonia.[14]

What happens to a particle that is dropped into *one* potential well, say L, at time $t = 0$? Equation (9.63) gives its state at $t = 0$ as

$$|\psi(0)\rangle = |L\rangle = \sqrt{\tfrac{1}{2}}\{|s\rangle + |a\rangle\}; \qquad (9.68)$$

the state does not have definite parity and is not an eigenstate of H. To investigate the behavior of the particle at later times, we use the time-dependent Schrödinger equation

$$i\hbar \frac{d}{dt} |\psi(t)\rangle = (H_0 + H_{\text{int}}) |\psi(t)\rangle \qquad (9.69)$$

and the expansion

$$|\psi(t)\rangle = \alpha(t) |L\rangle + \beta(t) |R\rangle$$
$$|\alpha(t)|^2 + |\beta(t)|^2 = 1. \qquad (9.70)$$

Inserting the expansion (9.70) into the Schrödinger equation (9.69) and multiplying in turn from the left by $\langle L|$ and $\langle R|$, yield a system of two coupled differential equations for $\alpha(t)$ and $\beta(t)$:

$$i\hbar\dot{\alpha}(t) = (E_0 + E')\alpha(t) + \Delta E \beta(t)$$
$$i\hbar\dot{\beta}(t) = \Delta E \alpha(t) + (E_0 + E')\beta(t). \qquad (9.71)$$

The solution of these equations with initial conditions $\alpha(0) = 1$ and $\beta(0) = 0$ gives

$$|\psi(t)\rangle = e^{-i(E_0 + E')t/\hbar}\left\{\cos\left(\frac{\Delta E t}{\hbar}\right)|L\rangle - i \sin\left(\frac{\Delta E t}{\hbar}\right)|R\rangle\right\}. \qquad (9.72)$$

The probability of finding the particle, dropped into well L at $t = 0$, in well R at a time t is given by the absolute square of the expansion coeffi-

14. Two-state systems and the ammonia MASER are beautifully treated in R. P. Feynman, R. B. Leighton, and M. Sands, *The Feynman Lectures on Physics*, Vol. III, Addison-Wesley, Reading, Mass., 1965, Chapters 8–11.

cient of $|R\rangle$, or

$$\mathrm{prob}(L) = \sin^2\left(\frac{\Delta E t}{\hbar}\right).\tag{9.73}$$

The particle hence oscillates between the two wells with a circular frequency

$$\omega = \frac{\Delta E}{\hbar} = \langle L|H_{\mathrm{int}}|R\rangle\frac{1}{\hbar}.\tag{9.74}$$

9.7 The Neutral Kaons

Hypercharge is the only quantum number that distinguishes the neutral kaon from its antiparticle: $Y(K^0) = 1$, $Y(\overline{K^0}) = -1$. Since the hadronic and the electromagnetic interactions conserve hypercharge, K^0 and $\overline{K^0}$ appear as two distinctly different particles in all experiments involving these two forces. However, the weak interaction does not conserve hypercharge, and virtual weak transitions between the two particles can occur. Both particles decay, for instance, into two pions, $K^0 \rightarrow 2\pi$ and $\overline{K^0} \rightarrow 2\pi$. They are therefore connected by virtual second-order weak transitions,

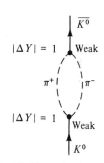

Fig. 9.9. Example of a virtual second-order weak transition $K^0 \longrightarrow \overline{K^0}$.

$$K^0 \rightleftharpoons 2\pi \rightleftharpoons \overline{K^0},\tag{9.75}$$

shown in Fig. 9.9. The existence of these virtual transitions leads to remarkable effects, as first pointed out by Gell-Mann and Pais.[15] The effects are easy to understand if the analogy to the two-well problem is recognized: In the absence of the weak interaction, $|K^0\rangle$ and $|\overline{K^0}\rangle$ are two unconnected degenerate states just like $|L\rangle$ and $|R\rangle$ before switching on H_{int}. The weak interaction, H_w, then plays the same role as H_{int} and connects the two states $|K^0\rangle$ and $|\overline{K^0}\rangle$. With minor changes, the equations and results of the previous section can be applied to the neutral kaon system by setting

$$H_0 = H_h + H_{em} \equiv H_s, \qquad H_{\mathrm{int}} = H_w.\tag{9.76}$$

To find the transformation that corresponds to Eq. (9.62), we note that charge conjugation changes K^0 into $\overline{K^0}$ and vice versa,

$$C|K^0\rangle = |\overline{K^0}\rangle, \qquad C|\overline{K^0}\rangle = |K^0\rangle.\tag{9.77}$$

Gell-Mann and Pais used these relations in their original work in place of Eq. (9.62) in order to find the proper linear combinations of the unperturbed eigenstates $|K^0\rangle$ and $|\overline{K^0}\rangle$. When the breakdown of parity was discovered it became clear that C does not commute with the total Hamiltonian, and this fact is expressed in Eq. (9.45). The combined parity, CP, is a better choice, as can be seen as follows. C applied to a neutrino with negative

15. M. Gell-Mann and A. Pais, *Phys. Rev.* **97**, 1387 (1955).

helicity changes it into an antineutrino with negative helicity, in disagreement with experiment. CP, however, changes a negative helicity neutrino into an antineutrino with positive helicity, in agreement with observation. To find the effect of CP on states $|K^0\rangle$ and $|\overline{K^0}\rangle$, we note that the intrinsic parity of the kaons is negative,

$$P|K^0\rangle = -|K^0\rangle, \qquad P|\overline{K^0}\rangle = -|\overline{K^0}\rangle, \tag{9.78}$$

so that the effect of the combined parity is given by

$$CP|K^0\rangle = -|\overline{K^0}\rangle, \qquad CP|\overline{K^0}\rangle = -|K^0\rangle. \tag{9.79}$$

If the total Hamiltonian conserves CP,

$$[H, CP] = [H_s + H_w, CP] = 0, \tag{9.80}$$

then the eigenstates of H can be chosen to also be eigenstates of CP. (We shall return to the question of CP conservation in Section 9.8.) Just as in Eq. (9.63), we write these eigenstates as[16]

$$\begin{aligned} |K_1^0\rangle &= \sqrt{\tfrac{1}{2}}\{|K^0\rangle - |\overline{K^0}\rangle\} \\ |K_2^0\rangle &= \sqrt{\tfrac{1}{2}}\{|K^0\rangle + |\overline{K^0}\rangle\}, \end{aligned} \tag{9.81}$$

with

$$CP|K_1^0\rangle = +|K_1^0\rangle, \qquad CP|K_2^0\rangle = -|K_2^0\rangle. \tag{9.82}$$

K_1^0 has a combined parity η_{CP} of $+1$, and K_2^0 one of -1.

The analogy with the two-well problem in Section 9.6 is obvious: States $|K^0\rangle$ and $|\overline{K^0}\rangle$, just as states $|L\rangle$ and $|R\rangle$, are eigenstates of the unperturbed Hamiltonian. States $|K_1^0\rangle$ and $|K_2^0\rangle$, just as $|s\rangle$ and $|a\rangle$, are simultaneous eigenstates of the total Hamiltonian and of the relevant symmetry operator. The results of Section 9.6 can be applied to the neutral kaons and remarkable predictions ensue:

1. K^0 is the antiparticle of $\overline{K^0}$. The two should therefore have the same mass and the same lifetime. K_1^0, however, is not the antiparticle of K_2^0, and the two particles can have very different properties.

2. The thought experiment of "dropping the particle at $t = 0$ into one well," discussed in Section 9.6, can be realized with kaons. Kaons are produced by hadronic interactions, for instance, by $\pi^- p \longrightarrow K^0 \Lambda^0$. Such a production in a state of well-defined hypercharge corresponds to dropping the particle into one well. Equations (9.72) and (9.73) predict that the particle will tunnel into the other well. The other well corresponds to the opposite hypercharge: A neutral kaon, produced in a state of $Y = 1$, should partially transform to a state with $Y = -1$ after a certain time.

3. States $|s\rangle$ and $|a\rangle$ have slightly different energies, as is shown by Eq. (9.66) and Fig. 9.8. The corresponding kaon states, $|K_1^0\rangle$ and $|K_2^0\rangle$, should therefore have slightly different rest energies.

---

16. The freedom allowed by the arbitrary phases in the definitions of C and P has led to different ways of writing the linear combinations (9.81). The observable consequences are unchanged by the phase choice.

In the following we shall describe the verification of two of these three predictions.

1. *K_1^0 and K_2^0 decay differently.* Energetically, kaons can decay into two or three pions. Since the kaon spin is zero, the total angular momentum of the pions in the final state must also be zero. Consider first the two-pion system, $\pi^+\pi^-$. In the c.m. of the two pions, the parity operation exchanges π^+ and π^-. Charge conjugation exchanges π^- and π^+ again so that the combined operation *CP* leads back to the original state. The same argument holds for two neutral pions so that

$$CP\,|\pi\pi\rangle = +|\pi\pi\rangle \qquad \text{in all states with } J = 0. \qquad (9.83)$$

Two pions with total angular momentum zero have a combined parity $\eta_{CP} = +1$. If the total Hamiltonian conserves *CP*, as assumed by Eq. (9.80), *CP* must be conserved in the decays of the neutral kaons. K_1^0, with $\eta_{CP} = 1$, then can decay into two pions. K_2^0, with $\eta_{CP} = -1$, *cannot* decay into two pions; it must decay into at least three:

$$K_2^0 \nrightarrow 2\pi \qquad \text{if } CP \text{ conserved.} \qquad (9.84)$$

The decay energy available for the two-pion mode is about 220 MeV, and for the three-pion mode about 90 MeV. The phase space available for decay into three pions is therefore considerably smaller than for that into two pions (Chapter 10), and the mean life τ_1 of K_1^0 is expected to be much smaller than the mean life τ_2 of K_2^0.

The decay of K^0 (or of $\overline{K^0}$) is more complicated. Consider, for instance, K^0 produced by a reaction such as $\pi^- p \rightarrow K^0 \Lambda^0$. At $t = 0$, the state has hypercharge $Y = 1$; with Eq. (9.81) the initial state is

$$|t = 0\rangle \equiv |K^0\rangle = \sqrt{\tfrac{1}{2}}\{|K_1^0\rangle + |K_2^0\rangle\}. \qquad (9.85)$$

If the particle is allowed to decay freely, it will do so through the weak interactions. We have observed above that K_1^0 and K_2^0 are expected to decay with different lifetimes τ_1 and τ_2. K^0 will therefore *not* decay with a single lifetime. Gell-Mann and Pais expressed their prediction in these words[15]: "To sum up, our picture of the K^0 implies that it is a particle mixture exhibiting two distinct lifetimes, that each lifetime is associated with a different set of decay modes, and that *not more than half of all K^0's* can undergo the familiar decay into two pions." They also stated "Since we should properly reserve the word 'particle' for an object with a unique lifetime, it is the K_1^0 and the K_2^0 quanta that are the true 'particles.' The K^0 and the $\overline{K^0}$ must, strictly speaking, be considered 'particle mixtures.' "

The unequivocal predictions of Gell-Mann and Pais concerning the decay properties of K^0 posed a challenge to the experimental physicists: Does K^0 possess a long-lived component that decays into three pions? At the time of Gell-Mann and Pais' paper, neutral kaons were known to decay with a lifetime of about 10^{-10} sec. A longer-lived component was found by a Columbia-Brookhaven group using a cloud chamber.[17] The experimental arrangement is sketched in Fig. 9.10. A 90-cm cloud chamber was exposed to the neutral beam emitted from a copper target hit by 3-GeV protons. Charged particles were eliminated by a sweeping magnet. The 6-m flight path from target to chamber corresponded to about 100 mean lives for the known decay component; the K_1^0 component hence was absent in the chamber. The observation of many V events that could not be fitted kinematically by two-pion decays established the existence of a long-lived three-pion decay of K_2^0 and constituted a clear verification of the brilliant proposal by Gell-Mann and Pais. Later experiments substantiated this conclusion, and the mean lives of the two components were found to be $\tau(K_2^0) = 0.52 \times 10^{-7}$ sec and $\tau(K_1^0) = 0.86 \times 10^{-10}$ sec.

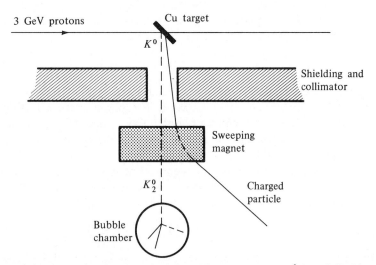

Fig. 9.10. Observation of the long-lived neutral kaon component, K_2^0, by a Columbia-Brookhaven group in a cloud chamber. [K. Lande et al., *Phys. Rev.* **103**, 1901 (1956); **105**, 1925 (1957).] The charged particles are swept out of the beam by a magnet; the neutral particles in the beam are observed after a flight of about 3×10^{-8} sec. The observed V events cannot be explained by two-particle decays.

2. *Hypercharge oscillations.*[18] Equation (9.72) predicts that a particle that was dropped into one well at time $t = 0$ will continuously oscillate between the two wells, with a circular frequency given by Eq. (9.74). If neutral kaons were stable, they would do the same. However, they decay, and the oscillations are damped. Consider a situation where at time $t = 0$ a K^0 was produced, as described by Eq. (9.85). After a time that is

17. K. Lande, E. T. Booth, J. Impeduglia, L. M. Lederman, and W. Chinowsky, *Phys. Rev.* **103**, 1901 (1956); **105**, 1925 (1957).

18. A. Pais and O. Piccioni, *Phys. Rev.* **100**, 1487 (1955).

long compared to $\tau(K_1^0)$, all $K_1^0 s$ will have decayed, and only $K_2^0 s$ are left, as shown in Fig. 9.10. Equation (9.81) expresses K_2^0 in terms of the eigenstates of hypercharge as

$$|K_2^0\rangle = \sqrt{\tfrac{1}{2}}\{|K^0\rangle + |\overline{K^0}\rangle\}.$$

The kaon beam will consist of equal parts K^0 and $\overline{K^0}$. A kaon beam that has been produced in a pure $Y = 1$ state has changed to one containing equal parts $Y = 1$ and $Y = -1$. Experimentally, the appearance of the $\overline{K^0}$ component can be verified through the observation of hadronic interactions such as $\overline{K^0}p \rightarrow \pi^+\Lambda^0$. Since nucleons have $Y = 1$ and the Λ^0 has $Y = 0$, a state $\pi^+\Lambda^0$ can be produced only by $\overline{K^0}$, not by K^0. The features of the observation of the $\overline{K^0}$ component are shown in Fig. 9.11.

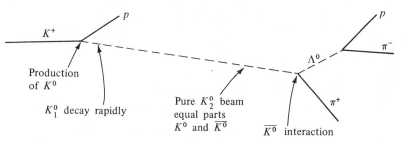

Fig. 9.11. Observation of the $\overline{K^0}$ component of an initially pure K^0 beam.

9.8 The Fall of *CP* Invariance

Kaons are a wonderful source of surprising effects. In Section 9.3 we described how the observation of two different decay modes of the charged kaons led to the fall of parity invariance. In the previous section, we showed that the coherence properties of the neutral kaons give rise to two different decay mean lives and to hypercharge oscillations. The coherence properties were predicted theoretically, and the subsequent experimental verification was exciting but not unexpected. The breakdown of parity was unexpected, but it was taken in stride and was quickly incorporated into the theoretical framework. In this section we shall treat the next major surprise, the fall of *CP* invariance. Its impact is best described by the accompanying cartoon, and it has not yet been possible to explain it in a satisfactory way.

Three features that were discussed in the previous section underlie the experiments demonstrating *CP* violation:

1. A neutral kaon beam far away from the point of production is in a pure $|K_2^0\rangle$ state.

J. Fabergé, *CERN Courier*, **6**, No. 10, 193 (October 1966). (Courtesy of Madame Fabergé.)

2. State $|K_2^0\rangle$ is an eigenstate of the total Hamiltonian. In vacuum, no transitions from $|K_2^0\rangle$ to $|K_1^0\rangle$ can occur.[19] For the two wells, the absence of such transitions is expressed by Eq. (9.65). The corresponding relation for kaons follows from Eqs. (9.80) and (9.81) as

$$\langle K_1^0 | H | K_2^0 \rangle = 0. \tag{9.86}$$

3. As stated by Eq. (9.84), K_2^0 cannot decay into two pions if *CP* is conserved.

In 1964, a Princeton group performed an experiment to set a lower limit on the two-pion decay of K_2^0.[20] Another experiment was simultaneously done by an Illinois group.[21] Both gave the astounding result that decays into two pions do occur; the branching ratio was found to be ap-

19. If a K_2^0 beam passes through an absorber, a small fraction of K_1^0 is regenerated. Regeneration occurs because K_2^0 consists of equal parts K^0 and $\overline{K^0}$ and $\overline{K^0}$ interacts differently with the absorber than K^0. After passage, the beam no longer contains equal parts K^0 and $\overline{K^0}$, and this imbalance corresponds to the presence of some K_1^0. In a vacuum, where no hadronic interactions can occur, there is no difference between K^0 and $\overline{K^0}$, and the statements leading to Eq. (9.86) hold.

20. J. H. Christenson, J. W. Cronin, V. L. Fitch, and R. Turlay, *Phys. Rev. Letters* **13**, 138 (1964).

21. A. Abashian, R. J. Abrams, D. W. Carpenter, G. P. Fisher, B. M. K. Nefkens, and J. H. Smith, *Phys. Rev. Letters* **13**, 243 (1964).

proximately

$$\frac{\text{Int}(K_L^0 \longrightarrow \pi^+\pi^-)}{\text{Int}(K_L^0 \longrightarrow \text{all charged modes})} \approx 2 \times 10^{-3}. \qquad \textbf{(9.87)}$$

We have switched notation here and denote the long-lived neutral kaon with K_L^0 and the short-lived one with K_S^0. The reason for the switch is Eq. (9.82), which *defines* K_1^0 and K_2^0 to be eigenstates of *CP*. Equation (9.87) indicates, however, that the long-lived kaon is *not* an eigenstate of *CP*. It is customary to retain the notation K_1^0 and K_2^0 for the eigenstates of *CP* and to denote the real particles with K_S^0 and K_L^0.

The news of violation of *CP* traveled through the world of physics with nearly the speed of light, just as, seven years earlier, had the news of parity breakdown. It was greeted with even more scepticism. To describe the reason for the disbelief, we digress to describe the celebrated *TCP theorem*. The *TCP* theorem is easy to understand but difficult to prove. In a somewhat sloppy way, it can be stated as follows: The product of the three operations *T*, *C*, and *P* commutes with practically every conceivable Hamiltonian, or

$$[TCP, H] = 0. \qquad \textbf{(9.88)}$$

In other words, our world and a time-reversed parity-reflected antiworld must behave identically. The order of the three operators *T*, *C*, and *P* is irrelevant.[22] The operation *TCP* is thus very different from the individual operations *T*, *C*, and *P*. It is easy to construct a Lorentz-invariant Hamiltonian that violates, for instance, *P* and *C*, and we shall discuss one in Chapter 11. However, it is nearly impossible to construct a Lorentz-invariant Hamiltonian that violates *TCP*. (These statements are somewhat oversimplified, but the essential features are correct.)

The *TCP* theorem was something of a sleeper. In preliminary form, it was discovered independently by Schwinger and by Lüders.[23] Pauli then generalized the theorem.[24] Up to 1956, however, it was considered to be rather esoteric. Dogma held that the three operations *T*, *C*, and *P* were separately conserved, and the *TCP* theorem was assumed to give little experimentally usable information. When violation of parity became a possibility, the *TCP* theorem suddenly acquired more meaning[25]: Equation (9.88) states that if *P* is violated, some other operation must also be violated. Indeed, we have mentioned in Section 9.4 that *C* is also not conserved in the weak interaction.

22. Since the order of the operations *T*, *C*, and *P* does not matter, there exist 3! possibilities of naming the theorem. Lüders and Zumino checked that their choice agreed with the name of a well-known gasoline additive. We use their choice. [G. Lüders, *Physikalische Blätter* **22**, 421 (1966).]

23. J. Schwinger, *Phys. Rev.* **82**, 914 (1951); **91**, 713 (1953); G. Lüders, *Kgl. Danske Videnskab Selskab, Mat.fys. Medd.* **28**, No. 5 (1954).

24. W. Pauli, in *Niels Bohr and the Development of Physics*, (W. Pauli, ed.) McGraw-Hill, New York, 1955.

25. T. D. Lee, R. Oehme, and C. N. Yang, *Phys. Rev.* **106**, 340 (1957).

After this digression, we return to the situation in 1964. The observed *CP* violation in the decay of the neutral kaons together with the *TCP* theorem leads nearly inescapably to one of two conclusions: Either *T* is not conserved or the *TCP* theorem is wrong. Theorists had in the meantime found even stronger proofs for it[26] and were rather reluctant to give it up. On the other hand, time reversal is also a cherished symmetry. Certainly the easiest way out would have been capitulation of the experimentalists with an admission that the experiments were wrong. Additional data, however, strengthened the earliest conclusions. Detailed analysis of all the information from the decays of the neutral kaons at least provides some further insight. The analysis implies that the *TCP* theorem holds but that not only *CP* but also *T* invariance is violated.[27] While it thus appears virtually certain that time-reversal invariance is violated in the decay of the neutral kaons, the cause of this violation remains obscure. Despite heroic efforts, no evidence for either *CP* or *T* violation has been found in any other system. It is also not clear which interaction is responsible for the violation. We shall again briefly return to this problem in Chapter 11.

9.9 References

General references concerning invariance properties are given in Section 7.7. A survey of the literature on parity nonconservation is given in L. M. Lederman, "Resource Letter Neu-1 History of the Neutrino," *Am. J. Phys.* **38**, 129 (1970).

Discontinuous transformations and unitary and antiunitary operators are treated in detail in Messiah, Vol. II, Chapter XV. The present status of the experimetal and theoretical aspects concerning the *T* and *P* invariances are discussed in several reviews: E. M. Henley, "Parity and Time-Reversal Invariance in Nuclear Physics," *Ann. Rev. Nucl. Sci.* **19**, 367 (1969), W. D. Hamilton, "Parity Violation in Electromagnetic and Strong Interaction Processes," *Progr. Nucl. Phys.* **10**, 1 (1969), E. Fischbach and D. Tadić, "Parity—Violating Nuclear Interactions and Models of the Weak Hamiltonian," *Phys. Reports* **6C**, 124 (1973) and M. Gari, "Parity Non-Conservation in Nuclei," *Phys. Reports* **6C**, 318 (1973). While these reviews are written at a higher level, much useful information can be extracted even at the level of the present book.

26. Proofs of the *TCP* theorem require relativistic field theory and are never easy. For the reader who wants to convince himself of this fact, we list here a few references, approximately in order of increasing difficulty: J. J. Sakurai, *Invariance Principles and Elementary Particles*, Princeton University Press, Princeton, N.J., 1964; G. Lüders, *Ann. Phys.* (New York) **2**, 1 (1957); R. F. Streater and A. S. Wightman, *PCT, Spin, and Statistics, and All That*, Benjamin, Reading, Mass., 1964.

27. R. C. Casella, *Phys. Rev. Letters* **21**, 1128 (1968); **22**, 554 (1969); K. R. Schubert, B. Wolff, J. C. Chollet, J. M. Gaillard, M. R. Jane, T. J. Ratcliffe, and J.-P. Repellin, *Phys. Letters* **31B**, 662 (1970); G. V. Dass, *Fortschritte Phys.* **20**, 77 (1972).

Neutral kaons and *CP* violation are treated in P. K. Kabir, *The CP Puzzle*, Academic Press, New York, 1968. A more popular account of *CP* and *T* violations is given in R. G. Sachs, *Science* **176**, 587 (1972).

PROBLEMS

9.1.
(a) Show that an infinitesimal rotation, R, and space inversion (parity), P, commute by showing in a sketch that PR and RP transform an arbitrary vector $\mathbf{x}$ into the same vector $\mathbf{x}'$.
(b) Use part (a) to show that P and $\mathbf{J}$ commute, where $\mathbf{J}$ is the generator of the infinitesimal rotation R.

9.2. Show that the commutation relations for angular momentum remain invariant under the parity operation.

9.3. Use the Schrödinger equation with a Hamiltonian $H = (p^2/2m) + V(\mathbf{x})$. Show that $\psi(-\mathbf{x})$ satisfies the Schrödinger equation if $\psi(\mathbf{x})$ does, provided that $V(\mathbf{x}) = V(-\mathbf{x})$.

9.4. Show that the eigenfunctions ψ_{lm} given in Problem 5.3 are eigenfunctions of P. Compute the eigenvalues and compare the result with Eq. (9.10).

9.5. Use a gauge transformation of the form of Eq. (7.25), with a properly chosen value of ϵ, to show that the relative parity of the proton and the positive pion is not a measurable quantity.

9.6. Would it be possible to assign meaningful intrinsic parities to all hadrons if in Eq. (9.22) instead of the parity of the lambda the parity of
(a) π^0 or
(b) K^+
had been chosen? Justify your answers.

9.7. Discuss the reaction

$$np \longrightarrow d\gamma$$

and use information in the literature to determine the intrinsic parity of the deuteron.

9.8. Find information on the reactions

$$dd \longrightarrow \text{p } {}^3\text{H}$$
$$dd \longrightarrow n \, {}^3\text{He}$$

and discuss the parities of the ^{3}H and ^{3}He.

9.9. Discuss the determination of the parity of a hyperon (not the lambda).

9.10. How would you determine the parity of the kaon? Compare your proposal with actual experiments.

9.11. The operator for the emission of electric dipole gamma radiation is of the form $q\mathbf{x}$, where q is a charge. The matrix element for a transition $i \longrightarrow f$ is of the form

$$F_{fi} = \int d^3x \psi_f^*(\mathbf{x}) q\mathbf{x} \, \psi_i(\mathbf{x}).$$

Use this expression to find the parity selection rule for electric dipole radiation.

9.12. Discuss the arguments and facts that assign spin 0 and positive parity to the alpha particle (ground state of ^{4}He).

9.13. Electrons and positrons emitted in weak interactions can be characterized by their momenta and their spins.
(a) Show that a nonvanishing value of the expectation value $\langle \mathbf{J} \cdot \mathbf{p} \rangle$ implies parity nonconservation.
(b) Discuss an experiment that can be used to measure the helicity of electrons.

9.14. Assume a nucleus with a g factor of $g = 1$ to be in a magnetic field of 1 MG. Compute the temperature at which at least 99% of the nuclei are polarized.

9.15. Use the information given in Figs. 7.1 and 9.6 to answer the following question. Are electron and antineutrino emitted predominantly in the same direction or in opposite directions?

9.16. Discuss the evidence for parity nonconservation in the decay $\pi^+ \longrightarrow \mu^+ \nu$:
(a) What polarization of the muon is expected?
(b) How can the muon polarization be observed?

9.17. Electrons emitted in nuclear beta decay are found to have negative helicity, whereas positrons show positive helicity. What can be deduced from this observation?

9.18. Consider a system consisting of a positive and a negative pion, with orbital angular momentum l in their c.m.
(a) Determine the C parity of this $(\pi^+\pi^-)$ system.
(b) If $l = 1$, can the system decay into two photons? Justify your answer.

9.19. Show that Maxwell's equations are invariant under time reversal.

9.20. Assume

$$\psi = \begin{pmatrix} \psi_1 \\ \psi_2 \end{pmatrix}$$

to be a two-component Pauli spinor, satisfying the Pauli equation. Find the wave function $T\psi$ that satisfies the Pauli equation.

9.21. Discuss one test of time-reversal invariance in the hadronic and one in the electromagnetic interaction.

9.22. Show that the helicity $\mathbf{J} \cdot \hat{\mathbf{p}}$ is invariant under the time-reversal operation.

9.23. A very small violation of parity invariance has been observed in nuclear decays ($\mathcal{F} \approx 10^{-7}$). How can this violation be explained without giving up parity conservation in the hadronic interaction?

9.24. Sketch the application of the two-well model to ammonia. How big is the total splitting $2\Delta E$ between states $|a\rangle$ and $|s\rangle$? Which state lies higher? Are transitions between states $|a\rangle$ and $|s\rangle$ observed? If yes, where are these transitions important?

9.25.
(a) Find the general solution of Eqs. (9.71).
(b) Verify that Eq. (9.72) is the special solution of Eq. (9.71) with the initial conditions $\alpha(0) = 1$ and $\beta(0) = 0$.

9.26. Neutron and antineutron are neutral antiparticles, just as K_0 and $\overline{K}_0$ are. Why is it not meaningful to introduce linear combinations N_1 and N_2, similar to K_1^0 and K_2^0?

9.27. Assume that K^0 is produced at $t = 0$.
(a) Justify that the wave function of K^0 at rest at time t can be written as

$$|t\rangle = \sqrt{\frac{1}{2}} \left\{ |K_1^0\rangle \exp\left(\frac{-im_1 c^2 t}{\hbar} - \frac{t}{2\tau_1}\right) \right.$$
$$\left. + |K_2^0\rangle \exp\left(\frac{-im_2 c^2 t}{\hbar} - \frac{t}{2\tau_2}\right) \right\},$$

where m_i and τ_i are mass and lifetime of K_i.
(b) Express $|t\rangle$ as a function of $|K^0\rangle$ and $|\overline{K^0}\rangle$.
(c) Compute the probability of finding $\overline{K^0}$ at time t as a function of $\Delta m = m_1 - m_2$.
(d) Sketch the probability for

$$\Delta m = 0, \qquad \Delta m = \frac{\hbar}{c^2 \tau_1}, \qquad \Delta m = \frac{2\hbar}{c^2 \tau_1}.$$

9.28. K_1^0 and K_2^0 have slightly different rest masses.
(a) Estimate the magnitude of the mass difference by assuming that the splitting is due to a second-order weak effect and that the weak interaction is about a factor of 10^7 weaker than the hadronic one.
(b) Describe how the magnitude of the mass difference can be determined.
(c) Compare the actually observed value with your estimate.

9.29.
(a) Assume that K^0 and $\overline{K^0}$ beams, of equal energy, pass through a slab of matter. Will the beams be attenuated equally? If not, why not?
(b) A pure K_2^0 beam passes through a slab of matter. Will the emerging beam still be a pure K_2^0 beam? Explain your answer.
(c) How can it be experimentally decided if the K_2^0 beam is still pure after passage through the slab?

9.30. Describe the experimental arrangements that were used to detect the two-pion decay of the long-lived neutral kaon.

9.31. Assume that you are in contact with physicists on another galaxy. The contact is restricted to exchange of information. Can you find out if the other physicists are built from matter or antimatter? Discuss the following three possibilities:
(a) C, P, and T are conserved in all interactions.
(b) C and P are violated in the weak interaction.
(c) C, P, and CP are violated, as discussed in Section 9.8.

9.32. Show that TCP invariance guarantees that a particle and its antiparticle have equal mass.

PART IV

Interactions

In the previous nine chapters, we have used the concept of *interaction* without discussing it in detail. In the present part, we shall rectify this omission, and we shall outline the important aspects of the three interactions that rule subatomic physics, the hadronic, the electromagnetic, and the weak.

It is useful in the treatment of interactions to distinguish between bosons and fermions. Bosons can be created and destroyed singly. Lepton and baryon conservation guarantee that fermions are always emitted or absorbed in pairs. The simplest interaction is thus one in which a boson is emitted or absorbed. Two examples are shown in Fig. IV.1. The interactions occur at the vertices where three particle lines are joined. The fermion does not disappear, but the boson either is created or destroyed. In both cases, the strength of the interaction can be characterized by a coupling constant. This coupling constant is written next to the vertex. A boson can also transform into another boson, as shown in Fig. IV.2. There a photon disappears, and a vector meson, for instance, a rho, takes its place. Again the coupling constant is indicated near the vertex.

The force between two particles is usually assumed to be mediated by

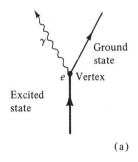

(a)

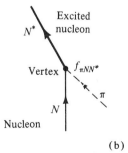

(b)

Fig. IV.1. Emission and absorption of a boson by a fermion. The coupling constants are denoted by e and $f_{\pi NN^*}$.

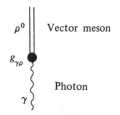

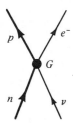

Fig. IV.2. Transformation of one boson into another.

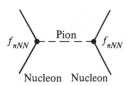

Fig. IV.3. The force between two nucleons is mediated by the exchange of mesons, for instance, pions, as shown here.

Fig. IV.4. Example of a four-particle interaction, involving four fermions.

particles, as discussed in Section 5.8. The exchange of a pion between two nucleons, shown in Fig. 5.19, is again represented in Fig. IV.3. At the present time it is not clear if all forces are transmitted by virtual particles or if other types of interactions exist. If the exchanged particle becomes extremely heavy, the range of the force becomes correspondingly small, and the process will look like a four-particle interaction, shown in Fig. IV.4. Of course it is possible that some forces are indeed correctly described by a four-particle interaction. The examples given here indicate the nature and properties of the forces involved only in a crude way. In the following three chapters, we shall study interactions in more detail.

10

The Electromagnetic
Interaction

Wake a theoretical physicist from his dreams and ask him about an interaction. Chances are that he will say "jay-dot-A" or "minimal electromagnetic interaction." These two terms belong to the language of subatomic physics, and we shall discuss them by starting from classical physics. Actually, the electromagnetic interaction is important in subatomic physics for two reasons. First, it enters whenever a charged particle is used as a probe. Second, it is the only interaction whose form can be studied in classical physics, and it provides a model after which other interactions can be patterned.

Without at least some approximate computations, interactions cannot be understood. In the simplest form, such computations are based on quantum mechanical perturbation theory and, in particular, on the expression for the transition rate from an initial state α to a final state β:

$$w_{\beta\alpha} = \frac{2\pi}{\hbar} |\langle \beta | H_{\text{int}} | \alpha \rangle|^2 \rho(E). \tag{10.1}$$

Fermi called this expression *the golden rule*, because of its usefulness and importance. In Section 10.1 we shall derive this relation; in Section 10.2, we shall discuss the density-of-states factor $\rho(E)$. Readers who are familiar with these topics can omit these two sections.

10.1 The Golden Rule

Consider a system that is described by a time-independent Hamiltonian H_0; its Schrödinger equation is

$$ih\frac{\partial\varphi}{\partial t} = H_0\varphi. \tag{10.2}$$

The stationary states of this system are found by inserting the ansatz,

$$\varphi = u_n(\mathbf{x})e^{-iE_nt/\hbar}, \tag{10.3}$$

into Eq. (10.2). The result is the time-independent Schrödinger equation

$$H_0 u_n = E_n u_n. \tag{10.4}$$

For the further discussion it is assumed that this equation has been solved, that the eigenvalues E_n and the eigenfunctions u_n are known, and that the eigenfunctions form a complete orthonormal set, with

$$\int d^3x\, u_N^*(\mathbf{x})u_n(\mathbf{x}) = \delta_{Nn}. \tag{10.5}$$

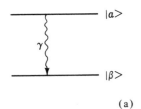

If the system is produced in one of the eigenstates u_n, it will remain in that state forever and no transitions to other states will occur.

We next consider a system that is similar to the one just discussed, but its Hamiltonian, H, differs from H_0 by a small term, the interaction Hamiltonian, H_{int},

$$H = H_0 + H_{int}.$$

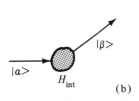

Fig. 10.1. The interaction Hamiltonian H_{int} is responsible for transitions from the unperturbed eigenstate $|\alpha\rangle$ to the unperturbed eigenstate $|\beta\rangle$.

The state of this system can, in zeroth approximation, still be characterized by the energies E_n and the eigenfunctions u_n. It is still possible to form the system in a state described by one of the eigenfunctions u_n, and we shall call a particular such initial state $|\alpha\rangle$. However, such a state will in general no longer be stationary; the perturbing Hamiltonian H_{int} will cause transitions to other states, for instance, $|\beta\rangle$. In the following we shall derive an expression for the transition rate $|\alpha\rangle \rightarrow |\beta\rangle$. Two examples of such transitions are shown in Fig. 10.1. In Fig. 10.1(a), the interaction is responsible for the decay of the state via the emission of a photon. In Fig. 10.1(b), an incident particle in state $|\alpha\rangle$ is scattered into the state $|\beta\rangle$.

To compute the rate for a transition, we use the Schrödinger equation,

$$ih\frac{\partial\psi}{\partial t} = (H_0 + H_{int})\psi. \tag{10.6}$$

To solve this equation, ψ is expanded in terms of the complete set of unperturbed eigenfunctions, Eq. (10.3):

$$\psi = \sum_n a_n(t)u_n e^{-iE_nt/\hbar}. \tag{10.7}$$

The coefficients $a_n(t)$ generally depend on time and $|a_n(t)|^2$ is the probability of finding the system at time t in state n with energy E_n. Inserting

ψ into the Schrödinger equation gives ($\dot{a} \equiv da_n/dt$)

$$i\hbar \sum_n \dot{a}_n u_n e^{-iE_n t/\hbar} + \sum_n E_n a_n u_n e^{-iE_n t/\hbar} = \sum_n a_n (H_0 + H_{\text{int}}) u_n e^{-iE_n t/\hbar}.$$

Because of Eq. (10.4), the second term of the left-hand side and the first term on the right-hand side cancel. Multiplying by u_N^* from the left, integrating over all space, and using the orthonormality relation produce the result

$$i\hbar \dot{a}_N = \sum_n \langle N | H_{\text{int}} | n \rangle a_n e^{i(E_N - E_n)t/\hbar}. \tag{10.8}$$

Here, a convenient abbreviation for the matrix element of H_{int} has been introduced:

$$\langle N | H_{\text{int}} | n \rangle = \int d^3 x \, u_N^*(\mathbf{x}) H_{\text{int}} u_n(\mathbf{x}). \tag{10.9}$$

The set of relations (10.8) for all N is equivalent to the Schrödinger equation (10.6) and no approximation is involved.

A useful approximate solution of Eq. (10.8) is obtained if it is assumed that the interacting system is initially in one particular state of the unperturbed system and if the perturbation H_{int} is weak. In Fig. 10.1, the initial state is $|\alpha\rangle$; it can, for instance, be a well-defined excited level. In terms of the expansion (10.7), the situation is described by

$$a_\alpha(t) = 1, \quad \text{all other } a_n(t) = 0, \quad \text{for } t < t_0. \tag{10.10}$$

Only one of the expansion coefficients is different from zero; all others vanish. The assumption that the perturbation is weak means that, during the time of observation, so few transitions have occurred that the initial state is not appreciably depleted, and other states are not appreciably populated. In lowest order it is then possible to set

$$a_\alpha(t) \approx 1, \quad a_n(t) \ll 1, \quad n \neq \alpha, \quad \text{all } t. \tag{10.11}$$

Equation (10.8) then simplifies to

$$\dot{a}_N = (i\hbar)^{-1} \langle N | H_{\text{int}} | \alpha \rangle e^{i(E_N - E_\alpha)t/\hbar}.$$

If H_{int} is switched on at the time $t = t_0$ and is time-independent thereafter, integration, for $N \neq \alpha$, gives

$$a_N(T) = (i\hbar)^{-1} \langle N | H_{\text{int}} | \alpha \rangle \int_0^T dt \, e^{i(E_N - E_\alpha)t/\hbar}$$

or

$$a_N(T) = \frac{\langle N | H_{\text{int}} | \alpha \rangle}{E_N - E_\alpha} [1 - e^{i(E_N - E_\alpha)T/\hbar}]. \tag{10.12}$$

The probability of finding the system in the particular state N after time T is given by the absolute square of $a_N(T)$, or

$$P_{N\alpha}(T) = |a_N(T)|^2 = 4 |\langle N | H_{\text{int}} | \alpha \rangle|^2 \frac{\sin^2 [(E_N - E_\alpha)T/2\hbar]}{(E_N - E_\alpha)^2}. \tag{10.13}$$

If the energy E_N is different from E_α, then the factor $(E_N - E_\alpha)^{-2}$ depresses the transition probability so much that transitions to the corresponding states can be neglected for large times T. However, there may be a group of states with energies $E_N \approx E_\alpha$, such as shown in Fig. 10.2(a), for which the matrix element $\langle N | H_{\text{int}} | \alpha \rangle$ is almost independent of N. This case occurs, for instance, if the states N lie in the continuum. To express the fact

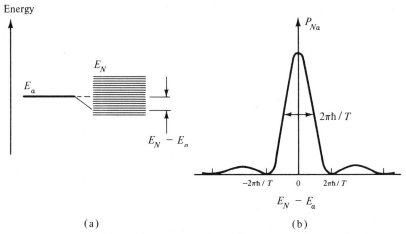

Fig. 10.2. (a) Transitions occur mainly to states with energies E_N that are close to the initial energy E_α. (b) Transition probability as a function of the energy difference $E_N - E_\alpha$.

that the matrix element is assumed to be independent of N, it is written as $\langle \beta | H_{\text{int}} | \alpha \rangle$. The transition probability is then determined by the factor $\sin^2 [(E_N - E_\alpha)T/2\hbar](E_N - E_\alpha)^{-2}$, and it is shown in Fig. 10.2(b). The transition probability is appreciable only within the energy region

$$E_\alpha - \Delta E \quad \text{to} \quad E_\alpha + \Delta E, \qquad \Delta E = \frac{2\pi\hbar}{T}. \qquad (10.14)$$

As time increases, the spread becomes smaller: Within the limits given by the uncertainty relation, energy conservation is a consequence of the calculation and does not have to be added as a separate assumption.

Equation (10.13) gives the transition probability from one initial state to one final state. The total transition probability to all states E_N within the interval (10.14) is the sum over all individual transitions,

$$P = \sum_N P_{N\alpha} = 4 |\langle \beta | H_{\text{int}} | \alpha \rangle|^2 \sum_N \frac{\sin^2 [(E_N - E_\alpha)T/2\hbar]}{(E_N - E_\alpha)^2}, \qquad (10.15)$$

where it has been assumed that the matrix element is independent of N. This assumption is good as long as $\Delta E/E_\alpha$ is small compared to 1. With Eq. (10.14), the condition becomes

$$T \gg \frac{2\pi\hbar}{E_\alpha} \approx \frac{4 \times 10^{-21} \text{Mev–sec}}{E_\alpha \text{ (in Mev)}}, \qquad (10.16)$$

where T is the time of observation. In most experiments, this condition is satisfied.

Now we return to the original problem, shown, for instance, in Fig. 10.1(a). Here, the energy in the initial state is well defined, but in the final state, the emitted photon is free and can have an arbitrary energy (Fig. 10.3). The discrete energy levels E_N of Fig. 10.2(a) consequently are replaced by a continuum. This fact is expressed by writing the energy as

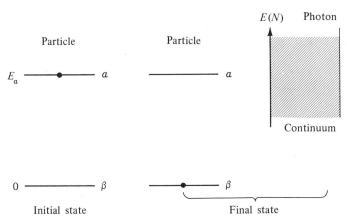

Fig. 10.3. In the initial state the subatomic particle is in the excited state α, and no photon is present. In the final state, the subatomic system is in state β, and a photon with energy $E(N)$ has been emitted. The energy of the photon "is in the continuum."

$E(N)$. N now labels the energy levels of the photon in the continuum, and it is a continuous variable. The total transition probability follows from Eq. (10.15) if the sum is replaced by an integral, $\sum_N \longrightarrow \int dN$:

$$P(T) = 4|\langle\beta|H_{\text{int}}|\alpha\rangle|^2 \int \frac{\sin^2[(E(N) - E_\alpha)T/2\hbar]}{(E(N) - E_\alpha)^2}\, dN. \qquad \textbf{(10.17)}$$

The integral extends over the states to which the transitions can occur. Since the integral converges very rapidly, the limits can be extended to $\pm\infty$. With

$$x = \frac{(E(N) - E_\alpha)T}{2\hbar},$$

$$dN = \frac{dN}{dE}\, dE = \frac{2\hbar}{T}\frac{dN}{dE}\, dx,$$

the transition probability becomes

$$P(T) = 4|\langle\beta|H_{\text{int}}|\alpha\rangle|^2 \frac{dN}{dE}\frac{T}{2\hbar} \int_{-\infty}^{+\infty} dx\, \frac{\sin^2 x}{x^2}.$$

The integral has the value π, so that the transition probability finally becomes

$$P(T) = \frac{2\pi T}{\hbar} |\langle \beta | H_{\text{int}} | \alpha \rangle|^2 \frac{dN}{dE}. \tag{10.18}$$

The notation $\langle \beta | H_{\text{int}} | \alpha \rangle$ indicates that the transition occurs from states $|\alpha\rangle$ to states $|\beta\rangle$. Since H_{int} is assumed to be time-independent, the transition probability is proportional to the time T. The transition *rate* is the transition probability per unit time, and it is

$$w_{\beta\alpha} = \dot{P}(T) = \frac{2\pi}{\hbar} |\langle \beta | H_{\text{int}} | \alpha \rangle|^2 \frac{dN}{dE}. \tag{10.19}$$

We have thus derived the golden rule. (Actually Fermi called it the *golden rule No. 2*.) It is extremely useful in all discussions of transition processes and we shall refer to it frequently. The factor

$$\frac{dN}{dE} \equiv \rho(E) \tag{10.20}$$

is called the *density-of-states factor;* it gives the number of available states per unit energy, and it will be discussed in Section 10-2.

● In some applications it happens that the matrix element $\langle \beta | H_{\text{int}} | \alpha \rangle$, connecting states of equal energy, vanishes. The approximation that leads to Eq. (10.18) can then be taken one step further. Fermi called this result the *golden rule No. 1*, and it can be stated simply: Replace the matrix element $\langle \beta | H_{\text{int}} | \alpha \rangle$ in Eq. (10.19) by

$$\langle \beta | H_{\text{int}} | \alpha \rangle \longrightarrow - \sum_n \frac{\langle \beta | H_{\text{int}} | n \rangle \langle n | H_{\text{int}} | \alpha \rangle}{E_n - E_\alpha}. \tag{10.21}$$

The one-step transition $|\alpha\rangle \longrightarrow |\beta\rangle$ from the initial to the final state is replaced by a sum over two-step transitions. These proceed from the initial state $|\alpha\rangle$ to all accessible *intermediate* states $|n\rangle$ and from there to the final state $|\beta\rangle$. ●

10.2 Phase Space

In the present section, we shall derive an expression for the density-of-states factor $\rho(E) \equiv dN/dE$. We consider first a one-dimensional problem, where a particle moves along the x direction with momentum p_x. Position and momentum of the particles are described simultaneously in an x-p_x plot (phase space). The representation is different in classical and in quantum mechanics. In classical mechanics, position and momentum can be measured simultaneously to arbitrary accuracy, and the state of a particle can be represented by a point [Fig. 10.4(a)]. Quantum mechanics, however, limits the description in phase space. The uncertainty relation

$$\Delta x \, \Delta p_x \geq \hbar$$

states that position and momentum cannot be simultaneously measured to unlimited accuracy. The product of uncertainties must be bigger than $\hbar$, and a particle consequently must be represented by a cell rather than a point in phase space. The shape of the cell depends on the measurements

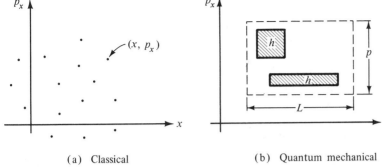

(a) Classical   (b) Quantum mechanical

Fig. 10.4. Classical and quantum mechanical one-dimensional phase space. In the classical case, the state of a particle can be described by a point. In the quantum case, a state must be described by a cell of *volume* $h = 2\pi\hbar$.

that have been made, but the volume turns out to always be equal to $h = 2\pi\hbar$. In Fig. 10.4(b), a volume Lp is shown. The maximum number of cells that can be crammed into this volume is given by the total volume divided by the cell volume,

$$N = \frac{Lp}{2\pi\hbar}. \tag{10.22}$$

N is the *number of states* in the volume Lp.[1]

Equation (10.22) can be verified by considering a one-dimensional infinitely high square well, as shown in Fig. 10.5. The Schrödinger equation in the well is

$$\frac{d^2\psi}{dx^2} + \frac{2m}{\hbar^2}E\psi = 0, \tag{10.23}$$

and the wave function must vanish at the walls,

$$\psi(0) = \psi(L) = 0.$$

The ansatz $\psi(x) = A \sin kx$ satisfies the boundary condition at $x = 0$. Inserted into the Schrödinger equation (10.23) it gives

$$E = \frac{\hbar^2 k^2}{2m}.$$

The boundary condition $\psi(L) = 0$ yields $\sin kL = 0$ or $kL = N\pi$, where N is an integer. The energy eigenvalues thus become

$$E = \frac{\pi^2\hbar^2}{2mL^2}N^2, \qquad N = 1, 2, \ldots. \tag{10.24}$$

The number of levels up to an energy E is given by N. For each energy, there are two possible values of the momentum, $p = \pm(2mE)^{1/2}$, where

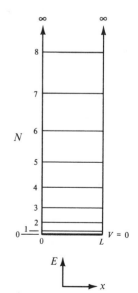

Fig. 10.5. Energy levels in an infinite square well.

---

1. Note that N is the number of states, not particles. One state can accommodate one fermion but an arbitrary number of bosons.

235

the sign refers to the direction of motion along the x axis. Equation (10.24) gives

$$N = \frac{Lp}{2\pi\hbar},$$

in agreement with Eq. (10.22).

Equation (10.22) is valid for a particle with one degree of freedom. For a particle in three dimensions, the volume of a cell is given by $h^3 = (2\pi\hbar)^3$, and the number of states in a volume $\int d^3x \, d^3p$ in the six-dimensional phase space is

$$N_1 = \frac{1}{(2\pi\hbar)^3} \int d^3x \, d^3p. \tag{10.25}$$

The subscript 1 indicates that N_1 is the number of states for one particle. If the particle is confined to a spatial volume V, integration over d^3x gives

$$N_1 = \frac{V}{(2\pi\hbar)^3} \int d^3p. \tag{10.26}$$

The density-of-states factor, Eq. (10.20), can now be computed easily:

$$\rho_1 = \frac{dN_1}{dE} = \frac{V}{(2\pi\hbar)^3} \frac{d}{dE} \int d^3p = \frac{V}{(2\pi\hbar)^3} \frac{d}{dE} \int p^2 \, dp \, d\Omega. \tag{10.27}$$

Here, $d\Omega$ is the solid-angle element. With $E^2 = (pc)^2 + (mc^2)^2$, d/dE becomes

$$\frac{d}{dE} = \frac{E}{pc^2} \frac{d}{dp}$$

and consequently (with $(d/dp) \int dp \rightarrow 1$)

$$\rho_1 = \frac{V}{(2\pi\hbar)^3} \frac{pE}{c^2} \int d\Omega. \tag{10.28}$$

For transitions to all final states, regardless of the direction of the momentum $\mathbf{p}$, the density-of-states factor for one particle is

$$\rho_1 = \frac{VpE}{2\pi^2 c^2 \hbar^3}. \tag{10.29}$$

Next we consider the density of states for *two particles*, 1 and 2. If the total momentum of the two particles is fixed, the momentum of one determines the momentum of the other and the extra degrees of freedom are not really there. The total number of states in momentum space is the same as for one particle, namely N_1, as in Eq. (10.26). However, the density-of-states factor, ρ_2, is different from Eq. (10.28) because E is now the total energy of the *two* particles:

$$\rho_2 = \frac{V}{(2\pi\hbar)^3} \frac{d}{dE} \int d^3p_1 = \frac{V}{(2\pi\hbar)^3} \frac{d}{dE} \int p_1^2 \, dp_1 \, d\Omega_1, \tag{10.30}$$

where

$$dE = dE_1 + dE_2 = \frac{p_1 c^2}{E_1} dp_1 + \frac{p_2 c^2}{E_2} dp_2.$$

The evaluation is easiest in the c.m. where $\mathbf{p}_1 + \mathbf{p}_2 = 0$, or

$$p_1^2 = p_2^2 \longrightarrow p_1\,dp_1 = p_2\,dp_2,$$

and

$$dE = p_1\,dp_1\frac{(E_1 + E_2)}{E_1 E_2}c^2.$$

The density-of-states factor is then given by

$$\rho_2 = \frac{V}{(2\pi\hbar)^3 c^2}\frac{E_1 E_2}{(E_1 + E_2)p_1}\frac{d}{dp_1}\int p_1^2\,dp_1\,d\Omega_1$$

or

$$\rho_2 = \frac{V}{(2\pi\hbar)^3 c^2}\frac{E_1 E_2 p_1}{(E_1 + E_2)}\int d\Omega_1. \tag{10.31}$$

The extension of Eq. (10.30) to three or more particles is straightforward. Consider three particles; in their c.m. the momenta are constrained by

$$\mathbf{p}_1 + \mathbf{p}_2 + \mathbf{p}_3 = 0. \tag{10.32}$$

The momenta of two particles can vary independently, but the third one is determined. The number of states therefore is

$$N_3 = \frac{V^2}{(2\pi\hbar)^6}\int d^3p_1\int d^3p_2, \tag{10.33}$$

and the density-of-states factor becomes

$$\rho_3 = \frac{V^2}{(2\pi\hbar)^6}\frac{d}{dE}\int d^3p_1\int d^3p_2. \tag{10.34}$$

For n particles, the generalization of Eq. (10.34) is:

$$\rho_n = \frac{V^{n-1}}{(2\pi\hbar)^{3(n-1)}}\frac{d}{dE}\int d^3p_1\cdots\int d^3p_{n-1}. \tag{10.35}$$

We shall encounter an application of Eq. (10.34) in Chapter 11, and we shall discuss futher evaluation there.

10.3 The Classical Electromagnetic Interaction

The energy (Hamiltonian) of a free nonrelativistic particle with mass m and momentum $\mathbf{p}_{\text{free}}$ is given by

$$H_{\text{free}} = \frac{\mathbf{p}_{\text{free}}^2}{2m}. \tag{10.36}$$

How does the Hamiltonian change if the particle is subject to an electric field $\boldsymbol{\varepsilon}$ and a magnetic field $\mathbf{B}$? The resulting modification can best be expressed in terms of potentials rather than the fields $\boldsymbol{\varepsilon}$ and $\mathbf{B}$. A scalar potential A_0 and a vector potential $\mathbf{A}$ are introduced and the fields are related to the potentials through the relations[2]

$$\mathbf{B} = \mathbf{\nabla} \times \mathbf{A} \tag{10.37}$$

$$\mathbf{\mathcal{E}} = -\mathbf{\nabla} A_0 - \frac{1}{c}\frac{\partial \mathbf{A}}{\partial t}. \tag{10.38}$$

The Hamiltonian of a point particle with charge q in the presence of the external fields is obtained from the free Hamiltonian by a procedure introduced by Larmor.[3] Energy and momentum of the free particle are replaced by

$$H_{\text{free}} \longrightarrow H_{\text{free}} - qA_0, \qquad \mathbf{p}_{\text{free}} \longrightarrow \mathbf{p} - \frac{q}{c}\mathbf{A}. \tag{10.39}$$

The resulting interaction is called *minimal electromagnetic interaction*. The term was coined by Gell-Mann to express the fact that only the charge q is introduced as a fundamental quantity. All currents are produced by the motion of particles. In particular, the current of a point particle is given by $q\mathbf{v}$. All higher moments (dipole moment, quadrupole moment, etc.) are assumed to be due to the particle's *structure;* they are not introduced as fundamental constants.

With the substitution (10.39), the Hamiltonian (10.36) changes to

$$H = \frac{1}{2m}\left(\mathbf{p} - \frac{q}{c}\mathbf{A}\right)^2 + qA_0 \tag{10.40}$$

or

$$H = H_{\text{free}} + H_{\text{int}} + \frac{q^2 A^2}{2mc^2}, \tag{10.41}$$

where H_{free} is given by Eq. (10.36) and H_{int} is

$$H_{\text{int}}(\mathbf{x}) = -\frac{q}{mc}\mathbf{p}\cdot\mathbf{A} + qA_0. \tag{10.42}$$

For all practical field strengths, the last term in Eq. (10.41) is so small that it can be neglected. If no external charges are present, the scalar potential vanishes, and the interaction energy becomes

$$H_{\text{int}}(\mathbf{x}) = -\frac{q}{mc}\mathbf{p}\cdot\mathbf{A} = -\frac{q}{c}\mathbf{v}\cdot\mathbf{A}. \tag{10.43}$$

$H_{\text{int}}(\mathbf{x})$ in Eq. (10.42) is the interaction energy of the nonrelativistic point particle at the position $\mathbf{x}$ with the fields characterized by the potentials $\mathbf{A}$ and A_0. For many applications, this form is already sufficient. In particular, it allows a description of the emission and absorption of photons. For some other applications, for instance, the electromagnetic interaction between two particles, the equations must be rewritten by expressing the potentials in terms of the currents and charges producing

2. Jackson, Section 6.4.

3. J. Larmor, *Aether and Matter*, Cambridge University Press, Cambridge 1900. See also Messiah, Sections 20.4 and 20.5; Jackson, Section 12.5; and Park, Section 7.6. Note that q can be positive or negative, whereas e is always positive.

them. Rather than deriving the general expression, we shall treat specific examples that are useful later.

The simplest situation arises if the electromagnetic field is produced by a point charge, q', at rest at $\mathbf{x}'$. The potential is then given by

$$A_0(\mathbf{x}) = \frac{q'}{|\mathbf{x} - \mathbf{x}'|}, \tag{10.44}$$

and the interaction is the ordinary *Coulomb energy*, already encountered in Eq. (6.15). If the charge q' is distributed over a volume, for instance, the volume of a nucleus, the scalar potential is given by[4]

$$A_0(\mathbf{x}) = q' \int d^3x' \frac{\rho'(\mathbf{x}')}{|\mathbf{x} - \mathbf{x}'|}, \tag{10.45}$$

and the interaction is of the form found in Eq. (6.23). The charge contained in the volume d^3x' at point $\mathbf{x}'$ is given by $q'\rho'(\mathbf{x}')\,d^3x'$, and the probability density $\rho'(\mathbf{x}')$ is normalized by Eq. (6.26).

The interaction of a point particle with a vector potential is given by Eq. (10.43). For a particle with an extended structure described by the charge distribution $q\rho(\mathbf{x})$, the factor $q\mathbf{p}/m = q\mathbf{v}$ in Eq. (10.43) must be replaced by

$$q \int d^3x\, \rho(\mathbf{x})\mathbf{v}(\mathbf{x}).$$

It is straightforward to see that

$$q\rho(\mathbf{x})\mathbf{v}(\mathbf{x}) = q\mathbf{j}(\mathbf{x}), \tag{10.46}$$

where $q\mathbf{j}(\mathbf{x})$ is the *charge current density*, namely the charge flowing through unit area per unit time. With Eq. (10.46), the interaction with an external potential $A(\mathbf{x})$ becomes

$$H_{\text{int}} = -\frac{q}{c} \int d^3x\, \mathbf{j}(\mathbf{x}) \cdot \mathbf{A}(\mathbf{x}). \tag{10.47}$$

Here the famous jay-dot-A has turned up. Equation (10.47) is one of the fundamental equations on which many calculations are based.

The vector potential $\mathbf{A}(\mathbf{x})$ produced by a current density $q'\mathbf{j}'(\mathbf{x}')$ is given by[5]

$$\mathbf{A}(\mathbf{x}) = \frac{q'}{c} \int d^3x'\, \mathbf{j}'(\mathbf{x}') \frac{1}{|\mathbf{x} - \mathbf{x}'|}. \tag{10.48}$$

Inserting this expression into Eq. (10.47) yields

$$H_{\text{int}} = -\frac{qq'}{c^2} \int d^3x\, d^3x'\, \mathbf{j}(\mathbf{x}) \cdot \mathbf{j}'(\mathbf{x}') \frac{1}{|\mathbf{x} - \mathbf{x}'|}. \tag{10.49}$$

---

4. Jackson, Eq. (1.17). The equations here differ from Jackson by a factor q or q' because our $\rho(\mathbf{x})$ is the probability density and not a charge density. Similarly, our $\mathbf{j}(\mathbf{x})$, introduced below, is a probability current density and not a charge current density.
5. Jackson, Eq. (5.32).

Such a *current-current interaction* was first written down by Ampère, and it will be a helpful guide in elucidating the weak interaction.

One additional classical relation is a useful guide in subatomic physics, namely the *continuity equation*. Maxwell's equations show that the density ρ and the current density $\mathbf{j}$ satisfy

$$\frac{\partial \rho}{\partial t} + \mathbf{\nabla} \cdot \mathbf{j} = 0. \tag{10.50}$$

A connection between the continuity equation and the conservation of the electric charge is established by integrating Eq. (10.50) over a volume V:

$$\int_V d^3x \, \frac{\partial \rho(\mathbf{x})}{\partial t} = - \int_V d^3x \, \mathbf{\nabla} \cdot \mathbf{j} = - \int_S d\mathbf{S} \cdot \mathbf{j}.$$

Here, S is the surface bounding the volume V. If the surface is far away from the system under consideration, the current through it will vanish. Interchanging integration and differentiation on the left-hand side and multiplication by the constant q give

$$\frac{\partial}{\partial t} \int_V d^3x \, q \rho(\mathbf{x}) = \frac{\partial}{\partial t} Q_{\text{total}} = 0. \tag{10.51}$$

The continuity equation implies conservation of the total electric charge.

10.4 Photon Emission[6]

The relations in the previous section are classical and consequently cannot be applied to the elementary processes in quantum mechanics. The task facing us then is a twofold one. First, the interaction energy must be translated into quantum mechanics where it becomes an operator, the interaction Hamiltonian. Second, once H_{int} is found, the transition rate or the cross section for a particular process must be computed so that it can be compared with experiment. We cannot proceed very far with the solution of these tasks without hand waving. A major part of the problem lies with the photon. It always moves with the velocity of light, and a nonrelativistic description of the photon makes no sense. In addition, in most of the processes of interest, the particles involved have energies large compared to their rest energies, and they also must be treated relativistically. A proper discussion of quantum electrodynamics consequently is far above our level. We shall only treat one process here in some detail, namely the emission of a photon by a quantum mechanical system. Many of the ideas

6. The problems inherent in any treatment of radiation theory make it difficult to write a really easy introduction. Probably the easiest-to-read first article is the beautiful review by E. Fermi, *Rev. Modern Phys.* **4**, 87 (1932). A modern readable introduction is R. P. Feynman, *Quantum Electrodynamics*, Benjamin, Reading, Mass., 1962. The present section is somewhat more difficult than the others, and parts of it can be omitted without losing information that is essential for later chapters.

that are important in quantum electrodynamics will show up in this simple problem.

The elementary radiation process, the emission or absorption of a quantum, is shown in Fig. 10.6. Two types of questions can be asked about such a process, kinematical and dynamical ones. The *kinematical* ones are of the type "What is the energy and momentum of the photon if it is emitted at a certain angle?" They can be answered by using energy and momentum conservation. The *dynamical* ones concern, for instance, the probability of decay or the polarization of the emitted radiation; they can be answered only if the form of the interaction is known. In the present section we shall solve the simplest dynamical problem, the computation of the lifetime of an electromagnetic decay, by using the golden rule, Eq. (10.1). The first step is the choice of the proper interaction Hamiltonian, H_{int}. An appealing candidate is Eq. (10.43) in Section 10.3.[7] For an electron, with charge $q = -e$, $e > 0$, the interaction Hamiltonian, now denoted as H_{em}, is

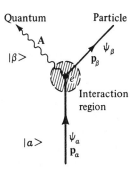

Fig. 10.6. Emission of a photon by an atomic or subatomic system in a transition $|\alpha\rangle \longrightarrow |\beta\rangle$.

$$H_{em} = e\frac{\mathbf{p}}{mc} \cdot \mathbf{A}. \qquad (10.52)$$

The three factors in this expression can be associated with the elements of the diagram in Fig. 10.6: The vector potential $\mathbf{A}$ describes the emitted photon, $(\mathbf{p}/mc)$ characterizes the particle, and the constant e gives the strength of the interaction.

The classical quantity H_{em} becomes an operator by translating $\mathbf{p}$ and $\mathbf{A}$ into quantum mechanics. The momentum $\mathbf{p}$ is straightforward; it becomes the momentum operator

$$\mathbf{p} \longrightarrow -i\hbar\mathbf{\nabla}. \qquad (10.53)$$

This substitution is well known from nonrelativistic quantum mechanics. The corresponding substitution for $\mathbf{A}$ depends on the process under consideration. Two kinds of emission events occur from the state $|\alpha\rangle$. The first takes place in the presence of an external electromagnetic field, produced, for instance, by photons incident on the system. $\mathbf{A}$ is the field due to these photons, and it gives rise to *stimulated* or *induced* emission of photons. Stimulated photon emission is the basic physical process involved in lasers. Here we are interested in the second kind of emission, called *spontaneous*. The state $|\alpha\rangle$ can decay even in the absence of an external electromagnetic field. The expression for $\mathbf{A}$ for spontaneous emission cannot be obtained from nonrelativistic quantum mechanics, because photons are always relativistic. We circumvent quantum electrodynamics by

7. Many students claim that the best way to solve physics problems in undergraduate courses is the following: List the physical quantities that appear in the problem. Find the equation in the text that contains the same symbols. Insert. Hand in. We are apt to laugh at such a naive approach but do the same when confronted with a new phenomenon. We see what observables nature has given us and then form the combination that has the properties expected from invariance laws.

postulating that **A** is the wave function of the created photon.[8] The form of **A** can be found by considering the vector potential of a classical electromagnetic plane wave,

$$\mathbf{A} = a_0 \hat{\boldsymbol{\epsilon}} \cos(\mathbf{k} \cdot \mathbf{x} - \omega t). \tag{10.54}$$

Here, $\hat{\boldsymbol{\epsilon}}$ is the polarization vector and a_0 the amplitude. If this wave is contained in a volume V, the average energy is given by

$$W = \frac{V}{4\pi} \overline{|\boldsymbol{\varepsilon}|^2},$$

or, with Eq. (10.38),

$$W = \frac{V\omega^2 a_0^2}{4\pi c^2} \overline{\sin^2(\mathbf{k} \cdot \mathbf{x} - \omega t)} = \frac{V\omega^2 a_0^2}{8\pi c^2}. \tag{10.55}$$

If **A** is to describe one photon in the volume V, W must be equal to the energy $E_\gamma = \hbar\omega$ of this photon. This condition fixes the constant a_0 as

$$a_0 = \left[\frac{8\pi\hbar c^2}{\omega V} \right]^{1/2}. \tag{10.56}$$

With $E_\gamma = \hbar\omega$ and $\mathbf{p}_\gamma = \hbar\mathbf{k}$, the wave function of the photon, Eq. (10.54), is determined. **A** is real because classically it is connected to the observable, and therefore real, fields $\boldsymbol{\varepsilon}$ and **B** by Eqs. (10.37) and (10.38). For the application to emission and absorption it will turn out to be convenient to rewrite Eq. (10.54) into the form

$$\mathbf{A}(\text{one photon}) = \left[\frac{2\pi\hbar^2 c^2}{E_\gamma V} \right]^{1/2} \hat{\boldsymbol{\epsilon}} \{ e^{i(\mathbf{p}_\gamma \cdot \mathbf{x} - E_\gamma t)/\hbar} + e^{-i(\mathbf{p}_\gamma \cdot \mathbf{x} - E_\gamma t)/\hbar} \}. \tag{10.57}$$

Here, **A** is no longer a classical vector potential, but it is postulated to be the wave function of the emitted photon. **A** is a vector, as is appropriate for photons which are spin-1 particles (Section 5.5). The next step is the construction of the matrix element of H_{em},

$$\langle \beta | H_{em} | \alpha \rangle \equiv \int d^3x \psi_\beta^* H_{em} \psi_\alpha$$

$$= \frac{e}{mc} \int d^3\psi_\beta^* \mathbf{p}\psi_\alpha \cdot \mathbf{A} = -i\frac{e\hbar}{mc} \int d^3x \psi_\beta^* \nabla\psi_\alpha \cdot \mathbf{A}. \tag{10.58}$$

To evaluate $\langle \beta | H_{em} | \alpha \rangle$, we make approximations. The first is the *electric dipole approximation*. The momentum part of the exponent in **A** can be expanded,

$$e^{\pm i\mathbf{p}_\gamma \cdot \mathbf{x}/\hbar} = 1 \pm i\frac{\mathbf{p} \cdot \mathbf{x}}{\hbar} + \cdots. \tag{10.59}$$

The exponential can be replaced by unity if $\mathbf{p}_\gamma \cdot \mathbf{x} \ll \hbar$. To obtain an approximate idea of what this condition implies, we assume that x has roughly the size of the system that emits the photon, and we denote this

---

8. This step can be justified by using quantum electrodynamics. Here we have no choice but to postulate it without further explanation. See Merzbacher, Chapter 22; Messiah, Section 21.27.

dimension by R. The condition imposed on the gamma-ray energy then is

$$E_\gamma = p_\gamma c \ll \frac{\hbar c}{R} \simeq \frac{197 \text{ MeV-fm}}{R \text{ (in fm)}}. \tag{10.60}$$

The second approximation applies to the decaying system. We assume it to be spinless and so heavy that it is at rest before and after the emission of the photon. The wave functions ψ_α and ψ_β can then be written as

$$\begin{aligned} \psi_\alpha(\mathbf{x}, t) &= \Phi_\alpha(\mathbf{x})e^{-iE_\alpha t/\hbar} \\ \psi_\beta(\mathbf{x}, t) &= \Phi_\beta(\mathbf{x})e^{-iE_\beta t/\hbar}, \end{aligned} \tag{10.61}$$

where $\Phi_\alpha(\mathbf{x})$ and $\Phi_\beta(\mathbf{x})$ describe the spatial extension of the system before and after the photon emission (Chapter 6). E_α and E_β are the rest energies of the initial and final states. Energy conservation demands that

$$E_\alpha = E_\beta + E_\gamma. \tag{10.62}$$

With Eqs. (10.57), (10.59), and (10.61), the matrix element, Eq. (10.58), becomes

$$\begin{aligned} \langle\beta|H_{em}|\alpha\rangle = \frac{-i\hbar^2 e}{m}&\left[\frac{2\pi}{E_\gamma V}\right]^{1/2}(e^{i(E_\beta-E_\gamma-E_\alpha)t/\hbar} \\ &+ e^{i(E_\beta+E_\gamma-E_\alpha)t/\hbar})\hat{\boldsymbol{\epsilon}} \cdot \int d^3x\, \Phi_\beta^* \nabla\Phi_\alpha. \end{aligned} \tag{10.63}$$

The two exponential factors that appear in the matrix element behave very differently. With Eq. (10.62), the first one becomes $\exp[-2iE_\gamma t/\hbar]$. Perturbation theory in the form derived in Section 10.1 is valid only if, according to Eq. (10.16), the time t is large compared to $2\pi\hbar/E_\gamma$. For such times, the exponential factor is a very rapidly oscillating function of time. Any observation involves an averaging over times satisfying Eq. (10.16), and the rapid oscillation wipes out any contribution to the matrix element from the first term. The second exponential factor is unity because of energy conservation, Eq. (10.62), and the emission matrix element becomes

$$\langle\beta|H_{em}|\alpha\rangle = -i\frac{\hbar^2 e}{m}\left[\frac{2\pi}{E_\gamma V}\right]^{1/2}\hat{\boldsymbol{\epsilon}} \cdot \int d^3x\, \Phi_\beta^* \nabla\Phi_\alpha. \tag{10.64}$$

If a photon is absorbed rather than emitted in the transition $|\alpha\rangle \rightarrow |\beta\rangle$, Eq. (10.62) reads $E_\alpha + E_\gamma = E_\beta$. The first exponential in Eq. (10.63) is then unity, and the second one does not contribute. The transition rate for spontaneous emission is now obtained with the golden rule, Eq. (10.19), which we write as

$$dw_{\beta\alpha} = \frac{2\pi}{\hbar}|\langle\beta|H_{em}|\alpha\rangle|^2\rho(E_\gamma). \tag{10.65}$$

With $p_\gamma = E_\gamma/c$, the density-of-states factor $\rho(E_\gamma)$ is given by Eq. (10.28) as

$$\rho(E_\gamma) = \frac{E_\gamma^2 V\, d\Omega}{(2\pi\hbar c)^3}. \tag{10.66}$$

Here $dw_{\beta\alpha}$ is the probability per unit time that the photon is emitted with momentum $\mathbf{p}_\gamma$ into the solid angle $d\Omega$. With the matrix element Eq. (10.64), the transition rate becomes

$$dw_{\beta\alpha} = \frac{e^2 E_\gamma}{2\pi m^2 c^3} |\hat{\boldsymbol{\epsilon}} \cdot \int d^3x \; \Phi_\beta^* \boldsymbol{\nabla} \Phi_\alpha|^2 \; d\Omega. \qquad (10.67)$$

If the wave functions Φ_α and Φ_β are known, the transition rate can be computed. However, the integral containing the wave functions can be changed into a form that expresses the salient facts more clearly. Assume that the Hamiltonian H_0 describing the decaying system, but not the electromagnetic interaction, is

$$H_0 = \frac{p^2}{2m} + V(\mathbf{x}),$$

where $V(\mathbf{x})$ does not depend on the momentum and hence commutes with $\mathbf{x}$. H_0 satisfies the eigenvalue equations

$$H_0 \Phi_\alpha = E_\alpha \Phi_\alpha, \qquad H_0 \Phi_\beta = E_\beta \Phi_\beta. \qquad (10.68)$$

With the commutation relation,

$$x p_x - p_x x = i\hbar, \qquad (10.69)$$

and the corresponding relations for the y and z components, the comutator of $\mathbf{x}$ and H_0 becomes

$$\mathbf{x} H_0 - H_0 \mathbf{x} = \frac{i\hbar}{m} \mathbf{p} = \frac{\hbar^2}{m} \boldsymbol{\nabla}. \qquad (10.70)$$

With this expression, the gradient operator in Eq. (10.67) can be replaced, and, with Eq. (10.68), the integral becomes

$$\int d^3x \; \Phi_\beta^* \boldsymbol{\nabla} \Phi_\alpha = \frac{m}{\hbar^2} \int d^3x \; \Phi_\beta^* (\mathbf{x} H_0 - H_0 \mathbf{x}) \Phi_\alpha = \frac{m}{\hbar^2} (E_\alpha - E_\beta) \int d^3x \; \Phi_\beta^* \mathbf{x} \Phi_\alpha$$

$$= \frac{m}{\hbar^2} E_\gamma \int d^3x \; \Phi_\beta^* \mathbf{x} \Phi_\alpha.$$

The integral is the matrix element of the vector $\mathbf{x}$, and it is written as

$$\int d^3x \; \Phi_\beta^* \mathbf{x} \Phi_\alpha \equiv \langle \beta \,|\, \mathbf{x} \,|\, \alpha \rangle. \qquad (10.71)$$

The transition rate into the solid angle $d\Omega$ is thus

$$dw_{\beta\alpha} = \frac{e^2}{2\pi \hbar^4 c^3} E_\gamma^3 |\hat{\boldsymbol{\epsilon}} \cdot \langle \beta \,|\, \mathbf{x} \,|\, \alpha \rangle|^2 \; d\Omega. \qquad (10.72)$$

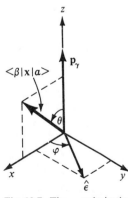

Fig. 10.7. The polarization vector $\hat{\boldsymbol{\epsilon}}$ of a photon emitted along the z axis lies in the xy plane. The vector $\langle \beta | \mathbf{x} | \alpha \rangle$, describing the decaying system, is taken to lie in the xz plane.

For a moment, we can place e^2 into the matrix element, which then becomes $\langle \beta | e\mathbf{x} | \alpha \rangle$. Since $e\mathbf{x}$ is the electric dipole moment, the radiation described by Eq. (10.72) is called electric dipole radiation, as mentioned above. The vector $\langle \beta | \mathbf{x} | \alpha \rangle$ characterizes the decaying system; the energy E_γ and the polarization vector $\hat{\boldsymbol{\epsilon}}$ describe the emitted photon. For a free photon, the unit vector $\hat{\boldsymbol{\epsilon}}$ is perpendicular to the photon momentum $\mathbf{p}_\gamma$ (Section 5.5). The vectors $\langle \beta | \mathbf{x} | \alpha \rangle$, $\mathbf{p}_\gamma$, and $\hat{\boldsymbol{\epsilon}}$ are shown in Fig. 10.7. Without loss of generality the coordinate system can be so chosen that $\mathbf{p}_\gamma$ points into the z direction and $\langle \beta | \mathbf{x} | \alpha \rangle$ lies in the xz plane; the polarization vector $\hat{\mathbf{e}}$ must be in the xy plane. With the angles θ and φ as defined in Fig. 10.7, the components of $\langle \beta | \mathbf{x} | \alpha \rangle$ and $\hat{\mathbf{e}}$ are $\langle \beta | \mathbf{x} | \alpha \rangle = |\langle \beta | \mathbf{x} | \alpha \rangle| (\sin \theta, 0, \cos \theta)$, $\hat{\mathbf{e}} = (\cos \varphi, \sin \varphi, 0)$. Performing the scalar product in Eq. (10.72) then gives

$$dw_{\beta\alpha} = \frac{e^2}{2\pi \hbar^4 c^3} E_\gamma^3 |\langle \beta | \mathbf{x} | \alpha \rangle|^2 \sin^2 \theta \cos^2 \varphi \, d\Omega. \tag{10.73}$$

If the polarization of the emitted photon is not observed, $dw_{\beta\alpha}$ must be integrated over the angle φ and summed over the two polarization states. The sum introduces a factor 2; with $d\Omega = \sin \theta \, d\theta \, d\varphi$ and $\int_0^{2\pi} d\varphi \cos^2 \varphi = \pi$, the transition rate for an unpolarized photon becomes

$$dw_{\beta\alpha} = \frac{e^2}{\hbar^4 c^3} E_\gamma^3 |\langle \beta | \mathbf{x} | \alpha \rangle|^2 \sin^3 \theta \, d\theta. \tag{10.74}$$

The total transition rate $w_{\beta\alpha}$ is obtained by integration over $d\theta$,

$$w_{\beta\alpha} = \int_0^\pi dw_{\beta\alpha} = \frac{4}{3} \frac{e^2}{\hbar^4 c^3} E_\gamma^3 |\langle \beta | \mathbf{x} | \alpha \rangle|^2. \tag{10.75}$$

The lifetime (mean life) is the reciprocal of $w_{\beta\alpha}$.

The physical content of the expression (10.75) for the total transition rate becomes more transparent if appropriate units are introduced. If the decaying system or particle has a mass m, then the characteristic length associated with it is the Compton wavelength, $\lambdabar_c = \hbar/mc$, and $E_0 = mc^2$ is the characteristic energy. The time that it takes light to move the distance $\lambdabar_c$ is given by $t_0 = \hbar/mc^2$, and the inverse of this time, $w_0 = 1/t_0 = mc^2/\hbar$, is the characteristic transition rate. With $\lambdabar_c$, $E_0 = mc^2$, and w_0, the transition rate is rewritten as

$$\frac{w_{\beta\alpha}}{w_0} = \frac{4}{3} \left(\frac{e^2}{\hbar c} \right) \left(\frac{E_\gamma}{mc^2} \right)^3 \frac{|\langle \beta | \mathbf{x} | \alpha \rangle|^2}{\lambdabar_c^2}. \tag{10.76}$$

The transition rate, expressed in terms of the "natural" rate w_0, becomes a product of three dimensionless factors, each of which has a clear physical interpretation. The last term, $|\langle \beta | \mathbf{x} | \alpha \rangle|^2/\lambdabar_c^2$, contains the information about the structure of the decaying system. If the wave functions Φ_α and Φ_β are known, the electric dipole matrix element $\langle \beta | \mathbf{x} | \alpha \rangle$ can be computed. Even without calculation, however, some properties can be deduced. For instance, the states $|\alpha\rangle$ and $|\beta\rangle$ must have opposite parities; otherwise $\langle \beta | \mathbf{x} | \alpha \rangle$ vanishes, and no electric dipole radiation can be emitted.

The term $(E_\gamma/mc^2)^3$ gives the dependence of the electric dipole radiation on the energy of the emitted photon. Equation (10.66) shows that two of the three powers of E_γ are contributed by the density-of-states factor: With increasing photon energy, the accessible volume in phase space becomes larger, and the decay consequently becomes faster. The third factor E_γ is introduced by the matrix element $\langle\beta|\mathbf{V}|\alpha\rangle$, and it is said to be of dynamical origin.

The factor

$$\frac{e^2}{\hbar c} \equiv \alpha \approx \frac{1}{137} \tag{10.77}$$

characterizes the strength of the interaction between the charged particle and the photon, and it is usually called the fine structure constant. A number of remarks concerning α are in order here. The first one concerns the fact that α, formed from three natural constants, is a dimensionless number. Since α is a pure number, it must have the same value everywhere, even on Trantor or Terminus.[9] Moreover, its value should be calculable in a truly fundamental theory. At the present time, no such theory exists that is generally accepted and understood. The second remark concerns the magnitude of α. Fortunately, α is small compared to 1, and this fact makes the application of perturbation theory successful. The expression (10.76) for the transition rate has been computed with the first-order expression, Eq. (10.1), and the result is proportional to α. The second-order term, Eq. (10.21), involves H_{em} twice, and its contribution will therefore be of order α^2 and considerably smaller than the first-order term. An example of this rapid convergence has already been presented in the discussion of the g factor of the electron, Eq. (6.40). As the third remark we note that the electric charge e plays two different roles. In Section 7.2, the charge appeared as an additive quantum number; in the present section, the strength of the electromagnetic interaction was shown to be proportional to e^2; e is therefore called a *coupling constant*. The dual role of the electric charge, as additive quantum number and as coupling constant, raises a question: In Chapter 7, other additive quantum numbers were introduced, namely the baryonic, the hypercharge, and the leptonic ones. It is not yet clear if these properties are also connected to interaction constants and, if not, why not.

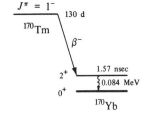

10.5 Multipole Radiation

In the previous section, a simple example of the action of the electromagnetic interaction, namely the emission of electric dipole radiation, has been computed in some detail. In the present section, the decay of actual subatomic systems will be discussed, and it will turn out that the previous considerations must be generalized. Two subatomic electromagnetic decays are shown in Fig. 10.8. In the nuclear example, the nuclide ^{170}Tm decays with a half-life of $130d$ to an excited state of ^{170}Yb, which then decays to its ground state with emission of a gamma ray of 0.084 MeV.

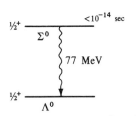

Fig. 10.8. Two examples of subatomic gamma decays. Note that the energy scales differ by about a factor 100.

9. I. Asimov, *Foundation*, Avon Books, New York, 1951.

The second example is the decay of the neutral sigma; in the transition $\Sigma^0 \xrightarrow{\gamma} \Lambda^0$, a gamma ray of 77-MeV energy is emitted.

The lifetime of the sigma has not yet been measured; only a limit of $\tau < 1.0 \times 10^{-14}$ sec has been set. The half-life of the 84-keV state in ^{170}Yb, on the other hand, has been determined as 1.57 nsec. (It is customary to quote mean lives in particle physics and half-lives in nuclear physics.) The basic idea underlying the half-life measurement is shown in Fig. 10.9.[10]

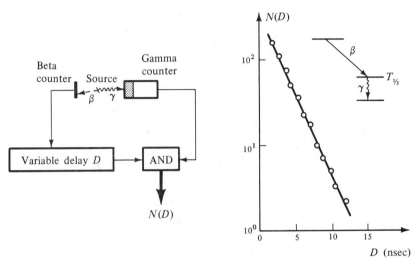

Fig. 10.9. Determination of the half-life of a short-lived nuclear state, decaying by gamma emission. The block diagram is shown at the left; a typical curve of coincidence counting rate $N(D)$, taken as a function of the delay time D, is given at the right.

The radioactive source, in the example ^{170}Tm, is placed between two counters. The beta counter detects the beta ray that populates the 2^+ state in ^{170}Yb. After some delay, the excited state decays with the emission of a 0.084-MeV photon. This photon has a certain probability of being delayed by a time D, and the coincidence rate between the delayed beta pulse and the gamma pulse is detected with an AND circuit (Section 4.7). The coincidence count rate $N(D)$ is recorded on a semilogarithmic plot versus D, and the slope of the resulting curve gives the desired half-life. The corresponding ideas have already been discussed in Section 5.7, and the plot shown in Fig. 10.9 is a specific example of an exponential decay as sketched in Fig. 5.15.

The method shown here, in which the decay curve is measured point by point, is only one possible approach. Many other techniques for in-

10. The measurement of short mean lives is discussed by R. E. Bell, in *Alpha-, Beta- and Gamma-Ray Spectroscopy*, Vol. 2 (K. Siegbahn, ed.), North-Holland, 1965, Amsterdam, and by A. Z. Schwarzschild and E. K. Warburton, *Ann. Rev. Nucl. Sci.* **18**, 265 (1968).

vestigating decay lifetimes have been evolved[10] and at present the half-lives of more than 1500 states are known.

After this brief excursion into the experimental aspects of electromagnetic transitions of subatomic particles, we return to theory and ask: Can the decays shown as examples in Fig. 10.8 be explained by the treatment given in Section 10.4? It can be seen immediately that the transition $\Sigma^0 \longrightarrow \Lambda^0$ cannot be caused by electric dipole transitions: The matrix element that appears in the electric dipole transition rate, Eq. (10.75), has the form

$$\langle \beta | \mathbf{x} | \alpha \rangle \equiv \langle \Lambda^0 | \mathbf{x} | \Sigma^0 \rangle \equiv \int d^3x \, \psi_\Lambda^* \mathbf{x} \psi_\Sigma.$$

The wave functions ψ_Λ and ψ_Σ have the same parity, and their product is even under the parity operation. The vector $\mathbf{x}$, however, is odd under parity, and the integrand is therefore also odd; the integral consequently must vanish. Similarly, it can be shown that dipole radiation cannot explain the $2^+ \longrightarrow 0^+$ transition in ^{170}Yb. The treatment given in the previous section must therefore be generalized if it is to explain all electromagnetic radiation emitted by subatomic systems.

The approximation that leads to electric dipole radiation is introduced by keeping only the first term in the expansion (10.59). Removal of this restriction is straightforward but lengthy, and we shall quote only the final result:[11] The emitted radiation can be characterized by its parity, π, and by its angular momentum quantum number, j. For any given value of j, the photon can carry away even or odd parity. It is customary to call one of these two an electric and the other a magnetic transition. Parity and angular momentum are related by

$$\text{electric radiation:} \quad \pi = (-1)^j$$
$$\text{magnetic radiation:} \quad \pi = -(-1)^j. \tag{10.78}$$

As an example, the electric dipole radiation carries an angular momentum $j = 1$ and, according to Eq. (10.78), a negative parity; it is written as $E1$. More generally, an electric (magnetic) radiation with quantum number j is written as $Ej(Mj)$. [We remind the reader that the quantum number j is defined by Eq. (5.4): If $\mathbf{J}$ is the photon angular momentum operator, $j(j + 1)\hbar^2$ is the eigenvalue of J^2.]

The values of j and π of the photons emitted in a transition $\alpha \longrightarrow \beta$ are limited by the conservation of angular momentum and parity

$$\mathbf{J}_\alpha = \mathbf{J}_\beta + \mathbf{J}$$
$$\pi_\alpha = \pi_\beta \pi. \tag{10.79}$$

11. Introductions to the theory of multipole radiation can be found in the following references: G. Baym, *Lectures on Quantum Mechanics*, Benjamin, Reading, Mass., 1959, pp. 281, 376; Jackson, Chapter 16; Blatt and Weisskopf, Chapter 12 and Appendix; S. A. Moszkowski, in *Alpha-, Beta- and Gamma-Ray Spectroscopy*, Vol. 2, (K. Siegbahn, ed.), North-Holland, Amsterdam, 1965, Chapter 15.

A few examples of possible values of j and π are given in Fig. 10.10. Note that initial and final spins are vectors. The various values of the angular momentum of the emitted radiation are obtained by vector addition, as also shown in Fig. 10.10.

The selection rules equation (10.79) state which transitions are allowed

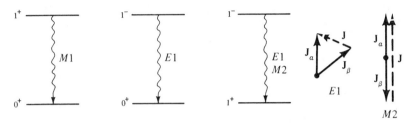

Fig. 10.10. A few examples of the possible values of angular momentum and parity emitted in a given transition. The vector diagrams for the transition $1^- \longrightarrow 1^+$ are shown at the right.

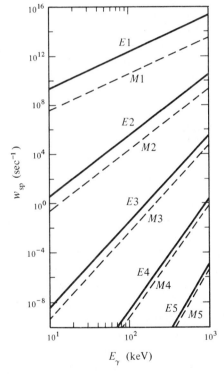

Fig. 10.11. Single proton transition rate (in sec^{-1}) as a function of the gamma-ray energy (in keV) for various multipolarities. (After S. A. Moszkowski, in *Alpha-, Beta- and Gamma-Ray Spectroscopy*, Vol. 2 (K. Siegbahn, ed.), North-Holland, Amsterdam, 1965, Chapter 15, p. 882.)

in a given decay, but they do not give information about the rate with which they occur. To find the rate, dynamical computations must be performed. In the previous section, the transition rate for $E1$ radiation was found, and Eq. (10.75) expresses this rate in terms of the matrix element $\langle \beta | \mathbf{x} | \alpha \rangle$. Expressions similar to Eq. (10.75) can be found for all multipole orders Ej and Mj. The real problem then begins: The relevant matrix elements must be evaluated, and this step requires a knowledge of the wave functions ψ_α and ψ_β. Finding the correct wave functions for a particular subatomic system is usually a long and tedious process, and only in a few cases has it come to a satisfactory conclusion. For an estimate of the transition rate, a crude model is therefore a necessity; it will provide at least an approximate value with which observed half-lives can be compared. For nuclei, the single-particle model is often used to get estimates for the half-lives of the various multipole orders. In the single-particle model it is assumed that the transition of one nucleon gives rise to the radiation. (We shall treat the single-particle model in Chapter 15.) Using a simple form for the single-nucleon wave function, the transition rates can be computed;[11] a result is shown in Fig. 10.11. The curves in Fig. 10.11 are calculated for a single proton in a nucleus with $A = 100$. Under these assumptions it is seen that the lowest multipole allowed by parity and by angular momentum selection rules dominates. Care must be taken in using the single-particle transition rates; in actual nuclei, deviations of one or even more orders of magnitude occur.

10.6 Electromagnetic Scattering of Leptons

Electromagnetic processes that involve only leptons and photons have been encountered a few times. Photoeffect, Compton scattering, pair production, and Bremsstrahlung were mentioned in Sections 3.3 and 3.4. The g factor of the leptons, discussed in Section 6.6, also involves only the electromagnetic interaction of leptons. In the present section we shall outline some of the aspects of the electromagnetic interaction of leptons without performing computations. The process to be discussed is the scattering of electrons. The diagrams for the scattering of electrons by electrons (Møller scattering) or electrons by positrons (Bhabha scattering) are shown in Fig. 10.12. The two electrons in Møller scattering are indistinguishable, and the graphs shown in Fig. 10.12(a) and (b) must both be taken into account. Since it is impossible to tell which process has taken place, the *amplitudes* for the two diagrams in Fig. 10.12(a) and (b) must be added, not the intensities. The particles in Bhabha scattering can be distinguished by their charge. Nevertheless, two graphs appear, and it is impossible to tell through which one scattering has occurred. Again the amplitudes for the two processes must be added. The contribution from

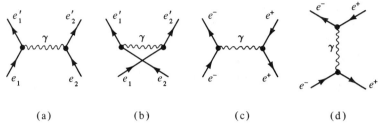

(a)   (b)   (c)   (d)

Fig. 10.12. Diagrams for the scattering $e^-e^- \longrightarrow e^-e^-$ and $e^+e^- \longrightarrow e^+e^-$.

Fig 10.12(c) is called the photon-exchange term, and the one from Fig. 10.12(d) the annihilation term.

The annihilation term, Fig. 10.12(d), deserves closer attention. It appears because the additive quantum numbers of an electron-positron pair are the same as those of the photon, namely $A = q = S = L = L_\mu = 0$. Once the virtual photon has been "formed" it no longer remembers where it came from, and it can give rise to a number of processes:

$$e^-e^+ \longrightarrow e^-e^+$$
$$2\gamma$$
$$\mu^+\mu^-$$
$$\pi^+\pi^-, \qquad \pi^+\pi^-\pi^0$$
$$K^+K^-, \qquad \bar{p}p, \qquad \bar{n}n$$
$$\vdots$$

Only the first three involve the electromagnetic interaction exclusively, and only the first one is shown in Fig. 10.12(d).

The computation of the cross section for Møller and Bhabha scattering is straightforward but requires knowledge of quantum electrodynamics and of Dirac theory. The cross sections depend on the total energy of the two electrons and on the scattering angle θ. If E is the energy of one of the two leptons in the c.m., then the cross section for Møller scattering for large energies $(E \gg m_e c^2)$ is of the form

$$\frac{d\sigma}{d\Omega} = \frac{\alpha^2}{E^2}(\hbar c)^2 f(\theta). \qquad (10.80)$$

Here α is the fine structure constant and $f(\theta)$ is a function of θ that is given explicitly in various texts on quantum electrodynamics. We note that $\alpha = e^2/\hbar c$ occurs *squared* in Eq. (10.80), in agreement with the fact that *two* vertices appear in all graphs in Fig. 10.12.

The form of Eq. (10.80) follows unambiguously from dimensional arguments: At very high energies, the electron mass can no longer play a role, and the only quantities that can enter the cross section are the coupling constant, in the dimensionless form α, and the energy, E. From these two quantities and the natural constants $\hbar$ and c, the only combination with the dimension of a cross section (area) is as given in Eq. (10.80). Only the dimensionless function $f(\theta)$ is dependent on the theory.

Experimentally, Møller and Bhabha scattering can be studied in two different ways. The straightforward approach is to employ a beam of electrons or positrons and observe the scattering from the electrons in a metal foil, as indicated in Fig. 10.13. One difficulty of this approach turns up

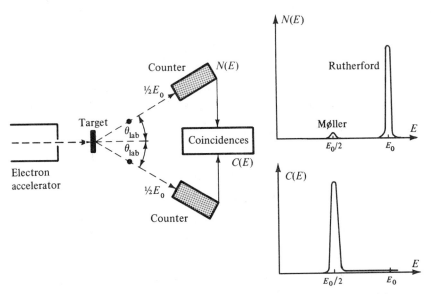

Fig. 10.13. Observation of Møller and Bhabha scattering by observing collisions with electrons in matter. $N(E)$ denotes the number of electrons with energy E observed in one counter. $C(E)$ denotes the number of coincidences in which both electrons have the energy E.

when the cross sections of Møller and Rutherford scattering are compared. For a material with atomic number Z, the ratio of cross sections is approximately $1/Z^2$. For most reasonable target materials, Rutherford scattering will be much more frequent than Møller scattering. How can the two processes be separated? For simplicity we assume the incoming energy, E_0, to be much larger than the binding energy of the electrons in the atom. The electrons in the target are thus essentially free. In symmetric scattering, shown in Fig. 10.13, both outgoing electrons make the same angle θ_{lab} with the beam axis, have energies $E_0/2$, and are simultaneous. If two counters are set at the proper angles, accepting only electrons with energies $E_0/2$, and if the signals are required to be simultaneous, Møller and Bhabha scattering can be separated cleanly from Rutherford scattering. A second disadvantage of the approach just outlined is not so easily overcome: The energy available in the c.m. to explore the structure of the electromagnetic interaction is small because of the small electron rest mass. We have studied this problem in Section 2.6; in Eq. (2.32) we found the total energy available in the c.m.,

$$W \approx [2E_0 m_e c^2]^{1/2}. \tag{10.81}$$

With $E_0 = 10$ GeV, the total energy available in the c.m. becomes

$$W \approx 100 \text{ MeV}.$$

Even at 10-GeV incident energy there is not enough c.m. energy to even produce a muon pair. The path around this difficulty has already been shown in Section 2.7; it is the use of colliding beams.

Colliding beams using electrons, or electrons and positrons, are in operation, and experiments studying Møller and Bhabha scattering have been performed.[12] The basic arrangement, in simplified form, is shown in Fig. 10.14. The two beams, for instance, electrons and positrons, collide

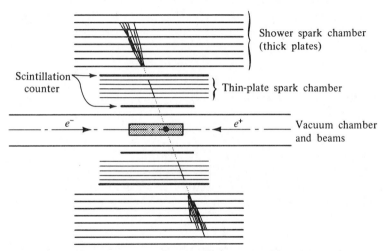

Fig. 10.14. Investigation of Bhabha scattering with colliding beams.

in an interaction volume. The volume is defined by scintillation counters that allow the spark chambers to be triggered only if two particles emerge simultaneously into the opposite counter arrangements. The particles are observed in spark chamber arrangements. Particles are identified by their range and by the presence or absence of showers. The total energy available is $2E_0$.

Measurements of Møller and Bhabha scattering involve angular distributions and absolute cross sections. All experiments agree with the predictions of quantum electrodynamics.[13] One interesting point occurs

12. For a survey of colliding beam experiments, see B. Touschek, ed., *Physics with Intersecting Storage Rings*, Academic Press, New York, 1971.

13. S. J. Brodsky and S. D. Drell, *Ann. Rev. Nucl. Sci.* **20**, 147 (1970). R. Madaras et al., *Phys. Rev. Letters* **30**, 507 (1973).

with Bhabha scattering. The virtual photons in the photon-exchange and in the annihilation diagram [Fig. 10.12(c) and (d)] have very different properties. Both photons are virtual and do not satisfy the relation $E = pc$. Consider both reactions in the c.m. In the exchange diagram, the incoming and the outgoing electrons have the same energies but opposite momenta. Consequently, energy and momentum of the virtual photon are given by

$$E_\gamma = E_e - E'_e = 0$$
$$\mathbf{p}_\gamma = \mathbf{p}_e - \mathbf{p}'_e = +2\mathbf{p}_e. \tag{10.82}$$

If we *define* a "mass" for the virtual photon through the relation $E^2 = (pc)^2 + (mc^2)^2$, we find[14]

$$(mc^2)^2 = -(2p_e c)^2 < 0. \tag{10.83}$$

The virtual photon carries only momentum—no energy. The square of its mass is negative. Such a photon is called *spacelike*. In the annihilation diagram, the situation is reversed,

$$E_\gamma = E_{e^-} + E_{e^+} = 2E$$
$$\mathbf{p}_\gamma = \mathbf{p}_{e^-} + \mathbf{p}_{e^+} = 0. \tag{10.84}$$

The virtual photon carries only energy—no momentum. The square of its mass is given by

$$(mc^2)^2 = (2E)^2 > 0; \tag{10.85}$$

it is positive and the photon is called *timelike*. In electron-positron scattering, both spacelike and timelike photons enter. The agreement of experiment with theory indicates that these concepts are correct, even if they sound strange at first.

10.7 The Photon-Hadron Interaction: Vector Mesons

The changing of Bodies into Light, and Light into Bodies, is very conformable to the Course of Nature, which seems delighted with Transmutations.

Newton, *Opticks*

The previous section and Section 6.6 have dealt with quantum electrodynamics and the interaction of photons and leptons. Before turning to the electromagnetic interaction involving hadrons, we shall review one of the central assumptions of quantum electrodynamics, namely the form of the interaction Hamiltonian. As pointed out in Section 10.3, the Hamiltonian is obtained from the principle of minimal electromagnetic interaction, Eq. (10.39). The principle introduces only the electric charge as a

---

14. Students who know four-vectors will realize that the "mass" defined here is related to the four-momentum transfer, q, by $m^2 = (q/c)^2$. It is equal to the actual particle mass only for free particles.

fundamental constant, and currents are assumed to be due to the motion of charges. Leptons are pictured as point particles and the probability current density of a lepton with velocity **v** is given by Eq. (10.46),

$$\mathbf{j}_{em}(\text{lepton}) = \rho \mathbf{v}. \tag{10.86}$$

The interaction Hamiltonian, Eq. (10.42), can be written as

$$H_{em}(\text{lepton}) = \frac{q}{c} \int d^3x \, (c\rho A_0 - \mathbf{j}_{em} \cdot \mathbf{A}). \tag{10.87}$$

The current is *conserved;* it satisfies the continuity equation (10.50).

● Interaction Hamiltonian and continuity equation can be expressed succinctly with four-vectors.[15] The quantity $A \equiv A_\mu = (A_0, \mathbf{A})$ is called a four-vector if it transforms under Lorentz transformations like $(ct, \mathbf{x})$. The scalar product of two four-vectors A_μ and B_μ is defined by

$$A \cdot B \equiv A_\mu B_\mu = A_0 B_0 - \mathbf{A} \cdot \mathbf{B}. \tag{10.88}$$

The scalar product of any two four-vectors is a Lorentz scalar or invariant; it remains constant under any Lorentz transformation. The four-vectors that occur most often are

time-space $\qquad\qquad x_\mu = (ct, \mathbf{x})$

four-momentum $\qquad\quad p_\mu = \left(\dfrac{E}{c}, \mathbf{p}\right)$

four-current $\qquad\qquad j_\mu = (c\rho, \mathbf{j})$ $\qquad$ (10.89)

four-potential $\qquad\qquad A_\mu = (A_0, \mathbf{A})$

four-gradient (note the sign) $\quad \nabla_\mu = \left(\dfrac{1}{c}\dfrac{\partial}{\partial t}, -\nabla\right).$

Using four-vectors, the minimal electromagnetic interaction, Eq. (10.39), is written as

$$(p_\mu)_{\text{free}} \longrightarrow \left(p_\mu - \frac{q}{c}A_\mu\right); \tag{10.90}$$

the interaction Hamiltonian becomes a Lorentz invariant,

$$H_{em}(\text{leptons}) = \frac{q}{c}\int d^3x \, j_\mu A_\mu, \tag{10.91}$$

and the continuity equation takes on the Lorentz invariant form,

$$\nabla_\mu j_\mu = 0. \tag{10.92}$$

In the following, we shall continue to use ordinary three-vectors. However, the reader who knows four-vectors should remember the underlying simpler form that involves four-currents and four-potentials. ●

We already know that the electromagnetic current of *hadrons* is not as simple as the one of leptons. The g factor and the elastic form factor of nucleons, both discussed in Section 6.7, indicate that the interaction of nucleons with the electromagnetic field is not directly given by the minimal electromagnetic interaction. Consequently, we write the total electromagnetic current density of a system as

$$e\mathbf{j}_{em} = e\mathbf{j}_{em}(\text{leptons}) + e\mathbf{j}_{em}(\text{hadrons}) \tag{10.93}$$

15. Feynman Lectures, I, Chapter 17; II, Chapter 25.

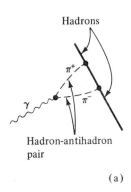

Hadrons

Hadron-antihadron
pair

(a)

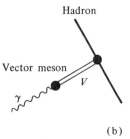

Hadron

Vector meson

(b)

Fig. 10.15. Interaction of a photon with a hadron. (a) The photon can produce a hadron-antihadron pair. (b) The photon can produce a vector meson which then interacts with the hadron.

and ask: What experiments will tell us about the hadronic contribution? Since it is assumed that the electromagnetic interaction is mediated by photons, the question can be rephrased: What experiments give information about the interaction of photons with hadrons? How does the photon interact with hadrons?

The interaction of the photon with a hadron does not occur through the electric charge alone, as is evidenced by the electromagnetic decay of the *neutral* pion into two photons. One possible way in which a photon can interact with a hadron current is indicated in Fig. 10.15. In Fig. 10.15(a) the photon produces a hadron-antihadron pair, and the partners of the pair interact hadronically with the hadron current. As early as 1960, Sakurai suggested that the two hadrons of the pair should be strongly coupled and form a vector meson, as shown in Fig. 10.15(b).[16] The photon thus would transform into a vector meson, as already anticipated in Fig. IV. 2. Sakurai made his suggestion long before the vector mesons were discovered experimentally. At the present time, theoretical suggestions can be useful guides for planning experiments, but only the result of experiments can provide the clues as to the nature of the interaction between the photon and hadrons.

Three types of experiments that can provide information about the photon-hadron interaction are illustrated by the Feynman diagrams shown in Fig. 10.16. Two of these involve virtual photons; the third one is performed with real ones. In all three cases the object of interest is the photon-hadron vertex. In the present section, we discuss timelike photons in electron-positron scattering; in the next section, real and spacelike photons will be treated.

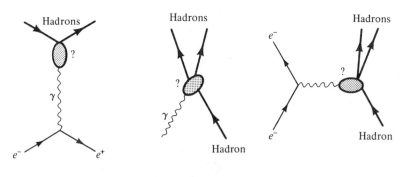

(a) Time-like photon   (b) Real photon   (c) Space-like photon

"m_γ^2" > 0 $m_\gamma = 0$ "m_γ^2" < 0

Fig. 10.16. Diagrams of three experimental possibilities to study the interaction of photons with hadrons. Details are discussed in the text.

16. J. J. Sakurai, *Ann. Phys.* (New York) **11**, 1 (1960); J. J. Sakurai, *Currents and Mesons*, University of Chicago Press, Chicago, 1969.

The experimental arrangement to study the production of hadrons in electron-positron collisions is similar to the one shown in Fig. 10.14. Only one essential difference exists: The recordings of the spark chamber events are scanned for hadrons rather than leptons.[12,17] The virtual photon produced in electron-positron collisions is timelike, as follows from Eqs. (10.84) and (10.85); in the $e^- - e^+$ c.m., it has energy but no momentum. The system of hadrons produced by timelike photons must possess quantum numbers that are determined by those of the photon. Since the electromagnetic interaction conserves hypercharge, parity, and charge conjugation, only final states with hypercharge 0, negative parity, and negative charge parity can be produced. In addition, angular momentum conservation requires the final state to have angular momentum unity. Are there such final states that are produced copiously? The experiments indicate that hadrons satisfying all conditions are indeed produced. Consider first Fig. 10.17. It shows the number of pion pairs observed at a given total energy of the colliding electrons, normalized by division by the number of electrons observed at the same energy. A pronounced peak appears at about 770 MeV, with a width of about 100 MeV. The reader with a good memory will say "Aha" and will turn back to Fig. 5.12 where a similar peak is shown at the same energy and with the same width. This peak was

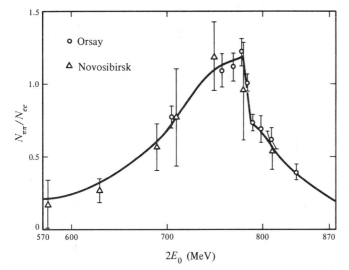

Fig. 10.17. The process $e^+e^- \longrightarrow \pi^+\pi^-$. The number of pions observed at a given energy $2E_0$ is normalized by division through the number of electron pairs observed at the same energy. $2E_0$ is the energy of the two colliding particles. [After D. Benaksas et al., *Phys. Letters* **39B**, 289 (1972).] The Novosibirsk data referred to is from V. L. Auslander et al., *Soviet J. Nucl. Phys.* **9**, 69 (1969).

17. V. L. Auslander, et al., *Phys. Letters* **25B**, 433 (1967); *Soviet J. Nucl. Phys.* **9**, 144 (1969). J. E. Augustin et al., *Phys. Letters* **28B**, 508 (1969). D. Benaksas, et al., *Phys. Letters* **39B**, 289 (1972); V. E. Balakin et al., *Phys. Letters* **34B**, 328 (1971).

identified with the rho meson. Why does the rho turn up here? Before answering this question, two more experiments will be discussed to provide additional information. In Fig. 10.18, the cross section for the process $e^+e^- \longrightarrow K^+K^-$ is shown as a function of the total energy $2E_0$. Again a resonance peak appears but this time with a peak energy of about 1020

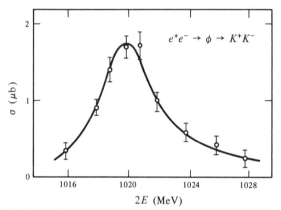

Fig. 10.18. Cross section for the process $e^+e^- \longrightarrow K^+K^-$. [From V. A. Sidorov (NOVOSIBIRSK), *Proceedings of the 4th International Symposium on Electron and Photon Interactions*, (D. W. Braben, ed.), Daresbury Nuclear Phys. Lab, 1969.]

MeV and a width of about 4 MeV. Table A4 in the Appendix shows that the ϕ^0 meson has these two properties. Observation of the reaction $e^+e^- \longrightarrow \pi^+\pi^-\pi^0$ yields a peak at about 780 MeV with a width of about 10 MeV. These values point to the ω^0. The virtual photon in the reaction $e^+e^- \longrightarrow$ hadrons produces resonances at the positions of the ρ^0, the ω^0, and the ϕ^0. To see what these three mesons have in common, we list their properties in Table 10.1.

The three mesons in Table 10.1 satisfy the conditions set out above: They have spin $J = 1$, negative parity, negative charge parity, and hypercharge 0. Since a vector has negative parity and the same number of independent components as a spin-1 particle, the mesons are called *vector mesons*. The rho has isospin 1 and is an isovector, whereas the two others

Table 10.1 VECTOR MESONS. π is the parity and η_c the charge parity of the vector mesons.

Meson	I	J	π	η_c	Y	Rest energy (MeV)	Width (MeV)	Dominant decay mode
ρ^0	1	1	-1	-1	0	770	146	$\pi\pi$
ω^0	0	1	-1	-1	0	784	10	$\pi^+\pi^-\pi^0$
ϕ^0	0	1	-1	-1	0	1019	4	$K\bar{K}$

are isoscalars. As pointed out in Section 8.6, after Eq. (8.30), the electric
charge operator is composed of an isoscalar and the third component of
an isovector. The photon, as carrier of the electromagnetic force, should
have the same transformation properties, and it matches the vector
mesons in their isospin properties. The diagrams for the production of
the three vector mesons listed in Table 10.1 are given in Fig. 10.19.

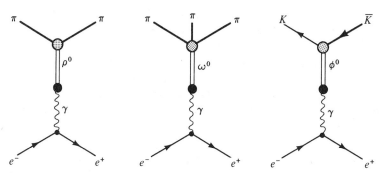

Fig. 10.19. The transformation of a virtual photon into a vector meson gives rise to
the resonances and their decays observed in colliding beam experiments.

10.8 The Photon-Hadron Interaction:
Real and Spacelike Photons

*Are there not other original Properties of the Rays of Light, besides those
already described?*

Newton, *Opticks*

The interaction of *real* photons with hadrons at low and moderate
photon energies (say below 20 MeV) has formed a considerable part of
nuclear physics for at least 40 years. One example, multipole radiation,
was sketched in Section 10.5. Another celebrated case is the photodisin-
tegration of the deuteron,,

$$\gamma d \longrightarrow pn,$$

which was discovered in 1934 by Chadwick and Goldhaber[18] and used
by them for a measurement of the neutron mass. A third example is the exploration of the excited states of nuclei with incident gamma rays. The
cross section for gamma-ray absorption shows the existence of individual
excited states and the occurence of the giant dipole resonance.[19] The
basic features of the resulting cross section have already been shown in
Fig. 5.27. Such studies produce a great deal of information concerning
nuclear structure, but they teach us little new about the nature of the
photon-hadron interaction: The photon interacts with the electric charges

18. J. Chadwick and M. Goldhaber, *Proc. Roy. Soc.* (London) **A151**, 479 (1935).
19. See, for instance, F. W. K. Firk, *Ann. Rev. Nucl. Sci.* **20**, 39 (1970).

and currents in the nucleus. The distributions of the charges and currents are determined by the hadronic force. If they are assumed to be given, then the interaction with the probing photon can be described by the Hamiltonian (10.87). Below, say, 100-MeV incident photon energy, this behavior can be understood: The (reduced) photon wave-length is of the order 2 fm or longer: short enough to probe some details of the nuclear charge and current distributions but not short enough to probe details of the photon-nucleon interaction.[20]

The interaction of photons with very high energies ($E \geq$ a few GeV) with hadrons presents a different picture and new aspects emerge: *The photon shows hadronlike properties.*[21] The roots of these properties can be understood with concepts that have been introduced earlier. In Section 3.3, the production of real electron-positron pairs by real photons was mentioned. In the previous section, it was found that timelike photons can produce hadrons, as indicated in Fig. 10.19. To describe the high-energy behavior of real photons, we now consider such processes in more detail. As already stated in Section 3.3 (Problem 3.22), a photon cannot produce a real pair of massive particles in free space. A nucleus must be present to take up momentum in order to satisfy energy and momentum conservation. However, the uncertainty principle permits violation of energy conservation by an amount ΔE during times smaller than $\hbar/\Delta E$. A photon can therefore produce a *virtual* pair or a *virtual* particle with the same quantum numbers as the photon and with total energy ΔE, but such a state can exist only for a time less than $\hbar/\Delta E$. Consider as a simple example the virtual decay of a photon of energy E_γ into a hadron h, with mass m_h. Momentum conservation demands that photon and hadron have the same momentum $p \equiv p_\gamma = E_\gamma/c$. The energy of a free hadron with mass m_h and momentum p is

$$E_h = [(pc)^2 + m_h^2 c^4]^{1/2} = [E_\gamma^2 + m_h^2 c^4]^{1/2},$$

and the energy difference between the photon and the virtual hadron becomes

$$\Delta E = E_h - E_\gamma = [E_\gamma^2 + m_h^2 c^4]^{1/2} - E_\gamma. \tag{10.94}$$

The limiting cases for photon energies small and large compared to $m_h c^2$ are

$$\Delta E = m_h c^2, \qquad E_\gamma \ll m_h c^2, \tag{10.95a}$$

$$\Delta E = \frac{m_h^2 c^4}{2E_\gamma}, \qquad E_\gamma \gg m_h c^2. \tag{10.95b}$$

20. ● It has been shown by various calculations that the scattering of photons in the limit of zero photon energy is given entirely by the *static* particle properties, mass, charge, and higher moments. The hadron structure dynamics does not enter, and the limit agrees with the classical result. W. Thirring, *Phil. Mag.* **41**, 1193 (1950); F. E. Low, *Phys. Rev.* **96**, 1428 (1954); M. Gell-Mann and M. L. Goldberger, *Phys. Rev.* **96**, 1433 (1954). ●

21. L. Stodolsky, *Phys. Rev. Letters* **18**, 135 (1967); S. J. Brodsky and J. Pumplin, *Phys. Rev.* **182**, 1794 (1969); V. N. Gribov, *Soviet Phys. JETP* **30**, 709 (1970).

The times during which the hadron can "virtually exist" are

$$T = \frac{\hbar}{m_h c^2}, \qquad E_\gamma \ll m_h c^2, \tag{10.96a}$$

$$T = \frac{2\hbar E_\gamma}{m_h^2 c^4}, \qquad E_\gamma \gg m_h c^2. \tag{10.96b}$$

The hadron can travel at most with the velocity of light, and the distance traversed during its virtual existence is limited by

$$L \lesssim \frac{\hbar}{m_h c} = \lambdabar_h, \qquad E_\gamma \ll m_h c^2, \tag{10.97a}$$

$$L \lesssim \frac{2\hbar E_\gamma}{m_h^2 c^3} = 2\lambdabar_h \frac{E_\gamma}{m_h c^2}, \qquad E_\gamma \gg m_h c^2, \tag{10.97b}$$

where $\lambdabar_h$ is the reduced Compton wavelength of the hadron. The quantum numbers of the photon do not allow a decay into one pion; the lowest possible hadron state consists of two pions, and $\lambdabar_h$ is consequently limited by

$$\lambdabar_h \lesssim \frac{\hbar}{2 m_\pi c} \approx 0.7 \text{ fm}. \tag{10.98}$$

The lowest-mass particle with $J^\pi = 1^-$ is the rho meson, for which $\lambdabar_h \approx$ 0.3 fm. Equation (10.97a) then shows that the pathlength of virtual hadrons associated with low-energy photons is much smaller than nuclear and even smaller than nucleon dimensions. Equation (10.97b) indicates, however, that the pathlength can become much larger than nuclear diameters at photon energies exceeding a few GeV.

The argument given so far reveals how far a virtual hadron accompanying the photon can propagate, but it does not predict how often a hadronic fluctuation arises. To describe the second property, we write for the normalized state function, $|\gamma\rangle$, of the real photon:

$$|\gamma\rangle = c_0 |\gamma_0\rangle + c_h |h\rangle. \tag{10.99}$$

Here $c_0 |\gamma_0\rangle$ is the purely electromagnetic part of the photon (*bare photon*) and $c_h |h\rangle$ is the hadronic part (*hadron cloud*). The absolute square $c_h^* c_h$ gives the probability of finding the photon in a hadronic state; as we shall see later, it is proportional to α. We shall return to a more detailed discussion of $|h\rangle$ below but note here that we expect, for instance, by analogy to the production of real lepton pairs (Fig. 3.7), that the ratio c_h / c_0 increases with increasing energy. Even a small contribution will become experimentally observable because the hadronic force is so much stronger than the electromagnetic one. To summarize, we picture low-energy and high-energy photons in Fig. 10.20.

The question as to whether the photon indeed is accompanied by a hadron cloud must be answered by experiments. We shall discuss two examples that demonstrate the existence of a hadronic component. The first one is the scattering of photons from nucleons. The total cross sec-

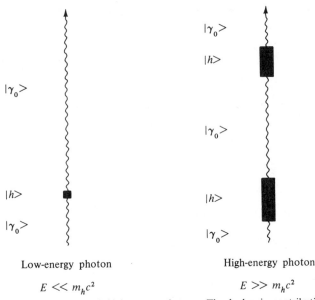

Low-energy photon

High-energy photon

$$E \ll m_h c^2 \qquad\qquad E \gg m_h c^2$$

Fig. 10.20. Low-energy and high-energy photons. The hadronic contribution for low-energy photons is insignificant. The high-energy photon is accompanied by a hadron cloud that leads to observable effects.

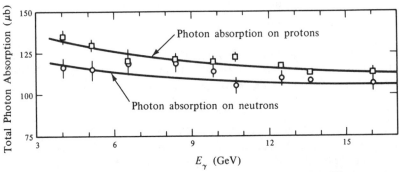

Fig. 10.21. Total absorption cross section for photons on nucleons. Very different cross sections are expected if the photon interacts with the electric charge. If the absorption occurs via vector mesons (hadrons), the absorption should be essentially the same for neutron and proton targets. [After D. O. Caldwell et al., *Phys. Rev. Letters* **25**, 613 (1970).]

tions for scattering of photons with energies up to 16 GeV from protons and neutrons have been measured, and the result is shown in Fig. 10.21.[22] As the energy increases above a few GeV, the two cross sections begin to coalesce. If the photons were to interact solely with the electric charge,

22. D. O. Caldwell, et al., *Phys. Rev. Letters* **25**, 609, 613 (1970); *Phys. Rev.* **D7**, 1362 (1973).

proton and neutron should have different total cross sections because their electromagnetic properties are different, as indicated by the behavior of their form factors G_E and G_M, Eqs. (6.46) and (6.48): The electric form factor of the neutron vanishes, indicating that the neutron is not only overall neutral but that it contains very little electric charge at all. The magnetic form factor of the neutron is smaller than that of the proton in the ratio $|\mu_n/\mu_p| \approx 0.7$. If the photon were to interact only with the electric charges and currents, scattering from the neutron would be much smaller than from the proton. The situation is different for the hadronic component, $c_h|h\rangle$. Proton and neutron form an isospin doublet. According to Eq. (8.15), the hadronic Hamiltonian commutes with $\vec{I}$ and the hadronic structure is independent of the orientation in isospin space. Proton and neutron consequently have the same hadronic structure. The forces between hadrons are charge-independent and do not depend on the orientation of the nucleon isospin vector. Indeed, it is known experimentally that hadron-proton and hadron-neutron cross section are about equal at very high energies[23]. The component $c_h|h\rangle$ therefore should produce equal scattering from protons and neutrons. Indeed, as Fig. 10.21 shows, at energies where $E_\gamma \gg m_h c^2$, the cross sections $\sigma(\gamma, p)$ and $\sigma(\gamma, n)$ approach each other and indicate that the term $c_h|h\rangle$ becomes dominant.

The behavior of the total cross section for photons on nuclei as a function of the scatterer baryon number, A, provides a second striking demonstration of the hadronic traits of high-energy photons. Below an energy of a few GeV, the total cross section is proportional to A,

$$\sigma_{\text{tot}}(\gamma) \propto A, \qquad E < \text{GeV}, \tag{10.100}$$

while at about 15 GeV, it increases less rapidly with mass number A[24,25]:

$$\sigma_{\text{tot}}(\gamma) \propto A^{0.9}, \qquad E \approx 15\,\text{GeV}. \tag{10.101}$$

The deviation from exponent 1 is at least six times outside the error limits. To show that this experimental result provides more evidence for the existence of a hadronic contribution to the photon, we shall discuss the behavior of the two components, $|\gamma_0\rangle$ and $|h\rangle$, separately. Consider first the bare photon, $|\gamma_0\rangle$. The mean free path of photons of about 15-GeV energy in nuclear matter (an infinitely large nucleus) is about 600 fm. This number follows from the values of the photon-nucleon cross section of Fig. 10.21, $\sigma \approx 10^{-2}\,\text{fm}^2$, and the nuclear density given in Eq (6.36), $\rho_n \approx 0.17$ nucleon/fm^3. Since the nuclear diameter of even the heaviest nucleus is less than 20 fm, bare photons "illuminate" nuclei uniformly, and the contribution of the term $c_0|\gamma_0\rangle$ to the cross section is proportional to A (Fig. 10.22). The hadronic term, $c_h|h\rangle$, produces two contributions to

23. J. V. Allaby et al., *Phys. Letters* **30B**, 500 (1969).

24. E. M. Henley, *Comments Nucl. Particle Phys.* **4**, 107 (1970); F. V. Murphy and D. E. Yount, *Sci. Amer.* **224**, 94 (July 1971).

25. D. O. Caldwell, V. B. Elings, W. P. Hesse, G. E. Jahn, R. J. Morrison, F. V. Murphy, and D. E. Yount, *Phys. Rev. Letters* **23**, 1256 (1969).

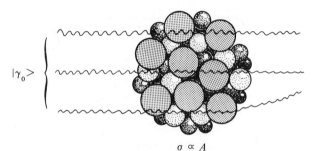

$$\sigma \propto A$$

Fig. 10.22. The bare photon, without hadronic interaction, has a mean free path in nuclear matter of about 600 fm, and it illuminates nuclei uniformly. The corresponding cross section is proportional to the mass number A.

the total cross section. As will be shown in Chapter 12, the cross section for hadrons is of the order of 3 fm², and the mean free path is about 2 fm. If the photon transforms to the hadron state *inside* the nucleus, the hadron will interact near the position of production (Fig. 10.23). Since the production can occur anywhere, the contribution to the total cross section is proportional to A, just like that of bare photons. On the other hand, virtual hadrons created *before* striking the nucleus interact with nucleons in the nuclear surface layer because of their short mean free path. The corresponding contribution to the total cross section consequently is proportional to the nuclear area, or to $A^{2/3}$. At a given photon energy, the total cross section is the sum of the three contributions, and it should be of the form

$$\sigma(\gamma A) = aA + bA^{2/3}. \qquad (10.102)$$

As stated above, the second term is due to photons that transform into hadrons before striking the nucleus; Fig. 10.23 makes it clear that such hadrons have a chance to interact if they are produced within the distance

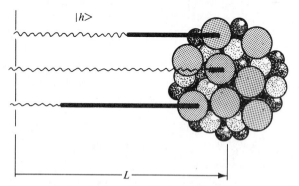

Fig. 10.23. Photons transforming into virtual hadrons interact in two ways with nuclei: If the photon transforms inside the nucleus, the contribution to the cross section is proportional to A. If the transformation occurs before the nucleus, the hadrons interact in the surface layer, and the cross section is proportional to $A^{2/3}$. The figure shows only interacting photons. Many more photons will pass through the nucleus without transformation into hadrons.

L. According to Eq. (10.97b), at high photon energies, L is large compared to nuclear diameters and proportional to E_γ. Other things being equal, the coefficient b should thus be proportional to E_γ, and the surface term should become dominant at energies large compared to $m_h c^2$. The unanticipated behavior of the cross section as expressed by Eqs. (10.100) and (10.101) therefore can be understood in terms of virtual hadrons.

The expression for the hadron cloud of the photon, $c_h|h\rangle$, can be written in an informative form by using perturbation theory. We assume the states of the various hadrons and of the photon, in the absence of the electromagnetic interaction, to be given by the Schrödinger equations

$$H_h|\gamma_0\rangle = 0$$
$$H_h|n\rangle = E_n\,|n\rangle. \tag{10.103}$$

H_h is the hadronic Hamiltonian, $|\gamma_0\rangle$ the state function of the bare photon, and $|n\rangle$ represents a hadronic state. If the electromagnetic interaction is switched on, hadronic states are superimposed onto the bare photon state:

$$|\gamma\rangle = c_0|\gamma_0\rangle + \sum_n c_n|n\rangle$$
$$|c_0|^2 + \sum_n |c_n|^2 = 1. \tag{10.104}$$

Since H_{em} is much weaker than H_h, the expansion coefficients c_n are small and $c_0 \approx 1$. The state of the physical photon is a solution of the complete Schrödinger equation,

$$(H_h + H_{em})|\gamma\rangle = E_\gamma|\gamma\rangle. \tag{10.105}$$

Inserting the expansion (10.104) into Eq. (10.105) gives, with Eq. (10.103) and with $\langle n|\gamma_0\rangle = 0$, $c_n \ll 1$,

$$c_n = \frac{\langle n|H_{em}|\gamma_0\rangle}{E_\gamma - E_n}. \tag{10.106}$$

The energy difference between the photon energy E_γ and the hadron energy E_n is given by Eq. (10.94); for large photon energies, the expansion coefficient becomes, with Eq. (10.95b),

$$c_n = \langle n|H_{em}|\gamma_0\rangle \frac{2E_\gamma}{m_h^2 c^4}. \tag{10.107}$$

The square of the matrix element is of order $\alpha \approx \frac{1}{137}$; if it is constant, the contribution from the hadronic state $|n\rangle$ to the photon state should be proportional to the photon energy. At values of E_γ that are small compared to $m_h c^2$, the photon behaves like an ordinary light quantum.

To compute actual values of c_n and thus find the hadron cloud, the wave functions of the states $|n\rangle$ and H_{em} must be known. At present it is believed that H_{em} is given by the minimal electromagnetic interaction and that all difficulties in computing the matrix elements stem from the absence of a detailed understanding of the structure of the hadron states $|n\rangle$.

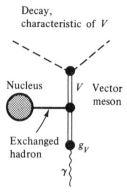

Decay,
characteristic of V

Nucleus V Vector
                      meson

Exchanged g_V
hadron

γ

Fig. 10.24. Forward scattering of a vector meson. The high-energy incident photon transforms into a vector meson, which then interacts with the nucleus via exchange of a hadron.

● Since, as we have just stated, there is no general theory that permits a complete computation of $|h\rangle$, calculations are carried out with simplified models. No one model describes all experiments at the present time, but the *vector dominance model* (VDM) is reasonably successful in correlating many aspects. This model was introduced by Sakurai,[16] and it is based on the assumption that the only hadronic states of importance in the sum in Eq. (10.104) are the lightest vector mesons ρ, ω, and ϕ. Only three matrix elements, of the form $\langle V|H_{em}|\gamma_0\rangle$, thus appear, and approximate values of these can be obtained from the experiments on vector meson production in colliding beam experiments.

As an example of the application of VDM, we shall again discuss briefly the A dependence of the total photon-nucleus cross section.[24] Consider first the process shown in Fig. 10.24. An incident photon transforms into a vector meson V with a probability amplitude g_V, and the interaction with the nucleus occurs through the vector mesons V. The total cross section, $\sigma_{tot}(V)$, for the interaction of V with the nucleus can be obtained from the optical theorem, Eq. (6.63):

$$\sigma_{tot}(V) = \frac{4\pi}{k}\, \text{Im}\, f_V(0°). \qquad (10.108)$$

Here, $f_V(0°)$ is the elastic scattering amplitude at $0°$ (forward scattering). At this point a new problem arises: Scattering can occur through transformation of a photon into ρ, ω, or ϕ. Can we add the three cross sections, or must we add the amplitudes? The answer is well known from optics and quantum mechanics: The intensities (cross sections) can be added if it can be decided through which process (channel) a given event has occurred. In the present case, such a decision is possible: The forward-scattered vector meson can be identified through its decay products. It then follows from Eq. (10.108) that no interference between the three cross sections $\sigma(\rho)$, $\sigma(\omega)$, and $\sigma(\phi)$ occurs. If the probability amplitude of the photon transforming into the vector meson ρ is given by g_ρ, the cross section for interaction through production of a rho is $|g_\rho|^2\sigma_{tot}(\rho)$. The total cross section for the gamma rays is the sum over the three contributions,

$$\sigma_{tot}(\gamma) = \sum_V |g_V|^2\sigma_{tot}(V). \qquad (10.109)$$

Since the vector mesons have a mean free path in nuclear matter of about 1 fm[26] they interact only with the nucleons on the surface, and the total cross section for vector mesons is proportional to the nuclear area:

$$\sigma_{tot}(V) \propto A^{2/3}. \qquad (10.110)$$

Equation (10.109) states that the total photon cross section should be proportional to that of vector mesons, or to $A^{2/3}$. We have therefore regained the surface term of Eq. (10.102) with a specific model.

Finally, we turn to the third case shown in Fig. 10.16, the exchange of spacelike photons in lepton-hadron scattering. One special case, elastic electron-nucleon scattering, has already been discussed in Section 6.7. There we pointed out that the nucleon structure gives rise to a deviation of the scattering cross section from that expected for point particles and that the deviation is expressed in terms of form factors. One example, the proton magnetic form factor, is shown in Fig. 6.13. The blob appearing in Fig. 10.16(c) is nothing but an expression for the same fact, and the form factor describes the features of the interaction of the spacelike photon with the nucleon. It is interesting to realize that the first suggestion for the existence of an isoscalar vector meson was made by Nambu in 1957 in order to explain the nucleon form factor[27]; an isovector-vector

26. J. G. Asbury, U. Becker, W. K. Bertram, P. Joos, M. Rohde, A. S. S. Smith, C. L. Jordan, and S. C. C. Ting, *Phys. Rev. Letters* **19**, 865 (1967).

27. Y. Nambu, *Phys. Rev.* **106**, 1366 (1957).

meson was postulated also.[28] Since, then, vector mesons were "invented" to describe the nucleon form factors and since nucleon form factors involve spacelike photons, one would guess that the VDM model would describe spacelike photons particularly well. However, this expectation is not fulfilled. The simple VDM model does not describe the elastic form factors adequately. Moreover, electron scattering can also give rise to inelastic events, where the proton in the final state is accompanied by other hadrons produced during the interaction. Of particular interest is the *deep inelastic electron scattering*, where the momentum transfer to the nucleon is large and where the nucleon is highly excited.[29] (Section 6.8.) Attempts to explain the features of these processes using vector dominance have failed, and a different interaction mechanism must be taking place. How the photon interacts is not yet understood. ●

The results of the last two sections can be summarized by saying that the photon definitely has hadronlike traits in its high-energy interaction with hadrons but that no detailed explanation of these traits exists at present.

10.9 Summary and Open Problems

> *Are not all Hypotheses erroneous which have hitherto been invented for explaining the Phaenomena of Light, by new Modifications of the Rays?*
>
> Newton, *Opticks*

Quantum electrodynamics, the description of the *interaction of photons and leptons*, is an extremely successful theory. Brodsky and Drell state the situation as follows[13]: "QED has been applied with complete and fantastic success over a range of 24 decades from the subnuclear realm of 10^{-14} cm out to a limit of 5.5×10^{10} cm (about 80 earth radii) for the wave length of the photon." Quantum electrodynamics gives precise answers to questions involving the interaction between leptons and photons. Problems remain, but they are much deeper and appear to lie outside the framework of QED: Why is charge quantized? What determines the charge and the mass of the electron? Why do only two charged leptons exist? Is there a fundamental reason for this limitation or do heavier leptons exist in nature? Questions of this type will probably have to await deeper insight into the nature of all interactions, not just the electromagnetic one.[30]

The interaction of *photons with hadrons* is no longer determined by quantum electrodynamics. Some experiments with high-energy photons can be explained by assuming that photons transform into vector mesons

28. W. R. Frazer and J. Fulco, *Phys. Rev. Letters* **2**, 365 (1959).

29. H. W. Kendall and W. K. H. Panofsky, *Sci. Amer.* **224**, 61 (June 1971).

30. The Solvay Congress of 1961 was devoted to the quantum theory of fields (*The Quantum Theory of Fields*, Wiley-Interscience, New York, 1961). While some of the reports and discussions are outdated, and some well above the level of the present book, most of the book makes fascinating reading and gives an insight into the thinking of the men who created much of the existing theory.

and that these couple to the hadrons. Other experiments, particularly the deep inelastic scattering of electrons, cannot be described by vector dominance. When is the vector dominance model applicable, and what are its limitations? What interaction explains the deep inelastic electron scattering? Is there a unifying principle that allows the description of the photon-hadron interaction at all energies and in the space- and timelike regions? We stand here at the beginning of a field, and further surprises can be expected.

Finally we come to another unsolved aspect of the electromagnetic interaction, the possible existence of *magnetic monopoles*. Classical electrodynamics is based on the observation that electric, but no magnetic, charges exist. The magnetic field is always produced by magnetic dipoles, never by magnetic charges (*monopoles*). This fact is expressed through the Maxwell equation

$$\mathbf{V} \cdot \mathbf{B} = 0. \tag{10.111}$$

Since this relation states an experimental result, the question as to its validity must be asked. As early as 1931, Dirac proposed a theory with magnetic monopoles.[31] In this theory, Eq. (10.111) is replaced by

$$\mathbf{V} \cdot \mathbf{B} = 4\pi \rho_m, \tag{10.112}$$

where ρ_m is the magnetic charge density. In an extension of his work, Dirac showed that the rules of quantum mechanics lead to a quantization of the electric charge e and the magnetic charge g[32]:

$$eg = \tfrac{1}{2}n\hbar c, \tag{10.113}$$

where n is an integer. Schwinger has confirmed this relation[33] (but with a factor of 2 instead of $\tfrac{1}{2}$) and has drawn some speculative conclusions from it. Two important features follow from Eq. (10.113): (1) The existence of a magnetic monopole would explain the quantization of the electric charge. (2) Squaring Eq. (10.113) gives approximately

$$\frac{g^2}{\hbar c} \approx \frac{\hbar c}{e^2} \approx 137. \tag{10.114}$$

The dimensionless constant describing the interaction between two magnetic monopoles is enormous. This interaction could therefore be responsible for the hadronic force. Despite considerable efforts, no magnetic monopoles have been seen.[34] The gargantuan strength of the monopole-monopole interaction could explain this failure: Monopoles probably are very heavy and could not have been produced with present accelerators.

31. P. A. M. Dirac, *Proc. Roy. Soc.* (London) **A133**, 60 (1931).

32. P. A. M. Dirac, Phys. Rev. **74**, 817 (1948).

33. J. Schwinger, *Science* **165**, 757 (1969).

34. H. H. Kolm, F. Villa, and A. Odian, *Phys. Rev.* **D4**, 1285 (1971); L. W. Alvarez, P. H. Eberhard, R. Ross, and R. D. Watt, *Science* **167**, 701 (1970); P. H. Eberhard, R. R. Ross, L. W. Alvarez, and R. D. Watt, *Phys. Rev.* **D4**, 3260 (1971). See also A. O. Barut, *Phys. Letters* **38B**, 97 (1972).

Experiments with higher energies will hopefully give more insight into this question.

10.10 References

A lucid introduction to classical electrodynamics is given in the Feynman lectures, Vol. II. A more complete and sophisticated treatment can be found in Jackson.

No very easy introduction to quantum electrodynamics exists. However, as already mentioned in Footnote 6, the article by Fermi [*Rev. Modern Phys.* **4**, 87 (1932)] and the book by Feynman (*Quantum Electrodynamics*) can be read even by undergraduate students if they do not give up easily. Brief introductions to the quantization of the electromagnetic field can also be found, for instance, in the quantum mechanics texts by Merzbacher, Messiah, or E. G. Harris, *A Pedestrian Approach to Quantum Field Theory*, Wiley-Interscience, New York, 1972.

On a more sophisticated level, a number of excellent books exist, but they are not easy to read. Nevertheless, we list here three such texts for the reader who is anxious to understand quantum electrodynamics well: (1) W. Heitler, *The Quantum Theory of Radiation*, Oxford University Press, London, 1954. This book has gone through three editions, and many physicists have learned most of their radiation theory from it. It is somewhat old-fashioned, but the physical points are brought out clearly. (2) J. D. Bjorken and S. D. Drell, *Relativistic Quantum Mechanics*, McGraw-Hill, New York, 1964. This book is much more modern than Heitler's, and it provides a thorough exposition of the physical ideas and the calculational techniques of relativistic quantum mechanics. (3) J. J. Sakurai, *Advanced Quantum Mechanics*, Addison-Wesley, Reading, Mass., 1967. This book is an excellent companion to Bjorken and Drell. It illuminates many of the same problems from a different point of view.

The classic papers on QED are collected and introduced in J. Schwinger, ed., *Quantum Electrodynamics*, Dover, New York, 1958. The present evidence for the validity of quantum electrodynamics is discussed in S. J. Brodsky and S. D. Drell, *Ann. Rev. Nucl. Sci.* **20**, 147 (1970); R. Gatto, in *High Energy Physics*, Vol. 5 (E. H. S. Burhop, ed.), Academic Press, New York, 1972; and B. E. Lautrup, A. Peterman, and E. de Rafael, *Phys. Rept.* **3C**, 196 (1972). The literature concerning tests of quantum electrodynamics is reviewed in M. M. Sternheim, "Resource Letter TQE-1," *Am. J. Phys.* **40**, 1363 (1972).

The specific features of the interaction of high-energy photons with hadrons have not yet found a place in textbooks. The reader who desires to learn more about this rapidly advancing field must turn to original papers and to the proceedings of the various international conferences and summer schools, for instance, J. Cumming and H. Osborn, eds., *Hadronic*

Interactions of Photons and Electrons, Academic Press, New York, 1971. See also D. R. Yennie, *1972 Cargèse Summer Institute on Electromagnetic Interactions of Elementary Systems* (to be published), and K. Gottfried, in *1971 International Symposium on Electron and Photon Interactions at High Energies*, Cornell Univ. 1972. The physics of colliding beams is treated in B. Touscheck, ed., *Physics with Intersecting Storage Rings* (Italian Phys. Soc. Course 46), Academic Press, New York, 1971.

A detailed account of the interaction of photons in the MeV energy range with nuclei is contained in J. M. Eisenberg and W. Greiner, *Nuclear Theory*, Vol. 2: *Excitation Mechanisms of the Nucleus.*, North-Holland, Amsterdam, 1970. A large number of articles and books review the nuclear photoeffect; we mention only one of the most recent where additional references are given: F. W. K. Firk, *Ann. Rev. Nucl. Sci.* **20**, 39 (1970).

PROBLEMS

10.1. Draw the transition probability factor $P_{N\alpha}(T)/4|\langle N|H_{int}|\alpha\rangle|^2$ of Eq. (10.13) for the following times T:
(a) $T = 10^{-7}$ sec.
(b) $T = 10^{-22}$ sec.

10.2. Derive the golden rule No. 1, Eq. (10.21), by developing the approximation involved in Eq. (10.19) to second order.

10.3. Consider the nonrelativistic scattering of a particle with momentum $\mathbf{p} = m\mathbf{v}$ by a fixed potential $H_{int} \equiv V(\mathbf{x})$ [Fig. 10.1(b)]. Assume that the incident and the scattered particles can be described by plane waves (Born approximation). L^3 is the quantization volume.
(a) Use the golden rule to show that the transition rate into the solid angle $d\Omega$ is given by

$$dw = \frac{v}{L^3}\left|\frac{m}{2\pi\hbar^2}\int d^3x\, e^{i(\mathbf{p}_\alpha-\mathbf{p}_\beta)\cdot\mathbf{x}/\hbar}H_{int}\right|^2 d\Omega.$$

(b) Show that the connection between cross section $d\sigma$ and transition rate is given by

$$w_{\beta\alpha} = F\, d\sigma,$$

where F is the incident flux [Eq. (6.2)].
(c) Verify the Born approximation expression, Eq. (6.13), for the scattering amplitude $f(\mathbf{q})$.

10.4. Verify Eq. (10.26) by computing the number of states in a three-dimensional box of volume L^3.

10.5. Derive the Lorentz force by starting from the Hamiltonian (10.40).

10.6. Show that the term $q^2A^2/2mc^2$ in Eq. (10.41) can be neglected in realistic situations.

10.7. Veryify that $q\rho(\mathbf{x})\mathbf{v}(\mathbf{x})$ in Eq. (10.46) is the charge that traverses unit area per unit time.

10.8. Show that the continuity equation, Eq. (10.50), is a consequence of Maxwell's equations.

10.9. Justify that the total energy in a plane electromagnetic wave in a volume V is given by

$$W = V\frac{\overline{|\boldsymbol{\mathcal{E}}|^2}}{4\pi},$$

where $\boldsymbol{\mathcal{E}}$ is the electric field vector.

10.10. Equation (10.67) describes the transition rate for the spontaneous *emission* of dipole radiation in the transition $\alpha \longrightarrow \beta$.
(a) Compute the corresponding expression for the *absorption* of a photon by dipole radiation inducing the transition $\beta \longrightarrow \alpha$.
(b) Compare the transition rates for emission

and absorption. Compare the ratio with the ratio expected from time-reversal invariance.

10.11. Prove Eq. (10.69) and (10.70).

10.12. Sketch the radiation pattern predicted by Eq. (10.73) and (10.74) for dipole radiation, assuming that the vector $\langle \beta | \mathbf{x} | \alpha \rangle$ points along the z direction. Compare to the radiation pattern for classical dipole radiation.

10.13. Use Eq. (10.75) to make a crude estimate for the mean life of an electric dipole transition
(a) In an atom, $E_\gamma = 10$ eV.
(b) In a nucleus, $E_\gamma = 1$ MeV.
Find relevant transitions in nuclei and atoms and compare your result with the actual values.

10.14. Discuss an accurate method for determining the fine structure constant.

10.15. Why do nuclei and particles not have permanent electric dipole moments? Why can some molecules have permanent electric dipole moments?

10.16. Why does the transition $\Sigma^0 \longrightarrow \Lambda^0$ occur through an electromagnetic and not a hadronic decay?

10.17. What kind of multipole transition is involved in the decay $\Sigma^0 \longrightarrow \Lambda^0$? Use an extrapolation of Fig. 10.11 to estimate the mean life. Compare to the presently known limit.

10.18. Discuss time-to-amplitude converters (TACs).
(a) Describe the function of a TAC.
(b) How can a TAC be used to measure lifetimes?
(c) Sketch the block diagram of a TAC.

10.19. Show that a $2^+ \xrightarrow{\gamma} 0^+$ transition, as, for example, shown in Fig. 10.8, cannot occur through dipole radiation.

10.20. Verify that the selection rules of Eq. (10.78) and the conservation laws of Eq. (10.79) together lead to the multipole assignments shown in Fig. 10.10.

10.21. The transition from an excited to a nuclear ground state can usually proceed by two competing processes, photon emission and emission of *conversion electrons*.
(a) Discuss the process of internal conversion.

(b) Assume that a particular decay has a half-life of 1 sec and a conversion coefficient of 10. What is the nuclear half-life for bare nuclei, i.e., nuclei stripped of all their electrons?
(c) The nuclide ^{111}Cd has a first excited state at 247-keV excitation energy. If the electron spectrum of this nuclide is observed, lines appear. Sketch the position of the conversion electron lines produced by the 247-keV transition.

10.22. Consider Møller scattering as shown in Fig. 10.13 (symmetric case).
(a) Assume that the incident electron has a kinetic energy of 1 MeV. Compute the angle θ_{lab}.
(b) Repeat the problem for an incident electron energy of 1 GeV.
(c) Compute the ratio of cross sections for parts (a) and (b) assuming that the angular function $f(\theta)$ in Eq. (10.80) has the same value for both cases.

10.23. Consider Møller scattering. Assume that the electrons in the target foil are completely polarized along the direction of the incident electrons. Use the Pauli principle to get an idea how longitudinally polarized incident electrons will scatter if their spin is (a) parallel and (b) antiparallel to the target spins. Consider only the symmetric scattering shown in Fig. 10.13.

10.24. To study the high-energy behavior of photons, monoenergetic beams are required. An ingenious way of producing such photons involves an intense laser pulse that collides head-on with a well-focused electron beam. The photons that are scattered by 180° acquire considerable energy. Compute the energy of the photons from a ruby laser that are scattered by 180° from an electron beam of energy
(a) 1 MeV.
(b) 1 GeV.
(c) 20 GeV.

10.25. Estimate the ratio of probabilities for the emission of a rho to that of a gamma ray from a high-energy nucleon that passes close to another one.

10.26. Magnetic monopoles (magnetic charges) would have remarkable properties:

(a) How would a magnetic monopole interact with matter?

(b) How would the track of a monopole look in a bubble chamber?

(c) How could a monopole be detected?

(d) Compute the energy of a monopole accelerated in a field of 20 kG.

10.27. Estimate the mass of a magnetic monopole by using the following, very speculative,

approach: The *classical electron radius* r_e is given by

$$r_e = \frac{e^2}{m_e c^2}.$$

Assume that a magnetic monopole has a similar radius, with e replaced by g and m_e by the monopole mass.

10.28. Prove Eq. (10.106).

11

The Weak Interaction

The history of the weak interaction is a series of mystery stories. In each story, a puzzle appears, at first only in a vague form and then more and more clearly. Clues to the solution are present but are overlooked or discarded, usually for the wrong reason. Finally, the hero comes up with the right explanation and everything is clear until the next corpse is unearthed. In the treatment of the electromagnetic interaction, the well-understood classical theory provided an example which, properly translated and reformulated, guided the development of quantum electrodynamics. No such classical analog is present in the weak interaction, and the correct features have to be taken from experiment and from analogies to the electromagnetic interaction. We shall describe some of the puzzles and their solutions. In doing so we are hampered by the self-imposed constraint of not using the Dirac theory. We shall therefore not be able to write the interaction properly but shall have to be satisfied by explaining the crucial concepts.

11.1 The Continuous Beta Spectrum

The continuous β-spectrum would then be understandable under the assumption that during β-decay a light neutral particle is emitted with every electron such that the sum of energies of neutrino and electron are constant.

W. Pauli

Radioactivity was discovered in 1896 by Becquerel, and it became clear within a few years that the decaying nuclei emitted three types of radiation, called α, β, and γ rays. The outstanding puzzle was connected with the beta rays. Very careful measurements over more than 20 years indicated that the beta particles were electrons and that they were not emitted with discrete energies but as a continuum. An example of such a beta spectrum is shown in Fig. 11.1. We have discussed nuclear energy levels in Chapter 5, and we

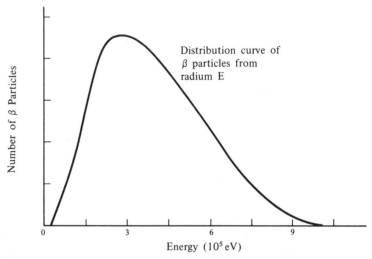

Fig. 11.1. Example of a beta spectrum. [This figure is taken from one of the classic papers: C. D. Ellis and W. A. Wooster, *Proc. Roy. Soc.* (London) **A117**, 109 (1927).] Present experimental techniques yield more accurate energy spectra, but all essential aspects are already contained in the curve reprinted here.

have found quantized states. The existence of quantized levels was well known in 1920, and the first puzzle posed by the continuous beta spectrum thus was: Why is the spectrum of electrons continuous and not discrete? A second equally serious puzzle arose a few years later when it was realized that no electrons are present inside nuclei. Where, then, do the electrons come from?

The first puzzle was solved by Pauli, who suggested the existence of a new, very light, uncharged, and penetrating particle, the *neutrino*.[1] Today, with so many particles known, proposing a new particle scarcely raises eyebrows. In 1930, however, it was a revolutionary step. Only two particles were known, the electron and the proton. Destroying the simplicity of

1. Pauli first suggested the neutrino in a letter addressed to some of his friends who were attending a physics meeting in Tübingen. He declared that he was unable to be present at the gathering because he wanted to attend the famous annual ball of the Swiss Federal Institute of Technology. The letter should be read by every physicist; it is reprinted in R. Kronig and V. F. Weisskopf, eds., *Collected Scientific Papers by Wolfgang Pauli*, Vol. II, Wiley-Interscience, New York 1964, p. 1316.

the subatomic world by addition of a third citizen was considered to be heresy, and very few people took the idea seriously. One of the ones who did was Fermi; he used Pauli's neutrino hypothesis to solve the second puzzle. Fermi assumed with Pauli that a neutrino is emitted together with the beta particle in every beta decay. Consequently, the simplest nuclear beta decay, the one of the neutron, is written as

$$n \longrightarrow pe^-\bar{\nu}.$$

Since the neutrino is chargeless, it is not observed in a spectrometer. Electron and neutrino share the decay energy, and the observed electrons sometimes have very little of it and sometimes nearly the maximum energy. The spectrum shown in Fig. 11.1 is thus qualitatively explained. To avoid the problems posed by electrons inside nuclei, Fermi postulated that the electron and the neutrino were *created* in the decay, just as a photon is created when an atom or a nucleus decays from an excited to the ground state or two photons are created in the decay of the neutral pion.

Fermi did not simply speculate how beta decay could occur; he performed the computations to find the expressions for the electron spectrum and the decay probability. His original treatment[2] is above our level, and it has to be watered down here. In the present section, we shall only show that even a crude approach reproduces the shape of the beta spectrum. Since the interaction responsible for beta decay is weak, perturbation theory can be used, and the transition rate is given by the golden rule, Eq. (10.1),

$$dw_{\beta\alpha} = \frac{2\pi}{h} |\langle \beta | H_w | \alpha \rangle|^2 \rho(E).$$

Here H_w is the Hamiltonian responsible for beta decay, and we have written $dw_{\beta\alpha}$ rather than $w_{\beta\alpha}$ in order to indicate that we are interested in the transition rate for transitions with electron energies between E_e and $E_e + dE_e$. We first consider the density-of-states factor $\rho(E)$. Three particles are present in the final state, and $\rho(E)$ is given by Eq. (10.34) as

$$\rho(E) = \frac{V^2}{(2\pi h)^6} \frac{d}{dE_{\max}} \int p_e^2 \, dp_e \, d\Omega_e p_\nu^2 \, dp_\nu \, d\Omega_{\bar\nu}. \tag{11.1}$$

V is the quantization volume. Since the final results are independent of this volume, it is set equal to 1. The differentiation $d/dE_{\max}$ requires a word of explanation. $E_{\max}$ is constant, and it thus appears at first sight that $d/dE_{\max}$ should vanish. However, it has the meaning of a variation; $(d/dE_{\max}) \int \cdots$ indicates how the integral changes under a variation of the maximum energy.

To evaluate $\rho(E)$, we must first decide what we are interested in. Figure 11.1 shows the electron spectrum, that is, the number of electrons emitted with an energy between E_e and $E_e + dE_e$. To calculate the corresponding

2. E. Fermi, *Z. Physik* **88**, 161 (1934); translated in *The Development of Weak Interaction Theory* (P. K. Kabir, ed.), Gordon & Breach, New York, 1963.

transition rate, E_e and consequently also p_e are kept constant. The d/dE_{max} in Eq. (11.1) then does not affect the terms relating to the electron, and Eq. (11.1) becomes

$$\rho(E) = \frac{d\Omega_e \, d\Omega_{\bar{v}}}{(2\pi\hbar)^6} \, p_e^2 \, dp_e p_{\bar{v}}^2 \, \frac{dp_{\bar{v}}}{dE_{max}}. \tag{11.2}$$

The next step is simplified by the fact that the nucleon in the final state is much heavier than either lepton and therefore receives very little recoil energy. To a good approximation, electron and neutrino share the total energy:

$$E_e + E_{\bar{v}} = E_{max}. \tag{11.3}$$

For a massless neutrino, $E_{\bar{v}} = p_{\bar{v}}c$, and for constant E_e,

$$\frac{dp_{\bar{v}}}{dE_{max}} = \frac{1}{c} \frac{dE_{\bar{v}}}{dE_{max}} = \frac{1}{c},$$

so that

$$\rho(E) = \frac{d\Omega_e d\Omega_{\bar{v}}}{(2\pi\hbar)^6 c} \, p_e^2 p_{\bar{v}}^2 \, dp_e. \tag{11.4}$$

As written, $\rho(E)$ is the density-of-states factor for a transition in which the electron has a momentum between p_e and $p_e + dp_e$ and is emitted into the solid angle $d\Omega_e$. With Eq. (11.3), $p_{\bar{v}}^2$ is replaced by $(E_{max} - E_e)^2/c^2$. Moreover, if the matrix element $\langle \beta | H_w | \alpha \rangle$ is averaged over the angle between the electron and the neutrino, $dw_{\beta\alpha}$ can be integrated over $d\Omega_e d\Omega_{\bar{v}}$ and with Eq. (11.4) the result is

$$dw_{\beta\alpha} = \frac{1}{2\pi^3 c^3 \hbar^7} \, |\langle pe^-\bar{v} | H_w | n \rangle|^2 p_e^2 (E_{max} - E_e)^2 \, dp_e. \tag{11.5}$$

This expression gives the transition rate for the decay of a neutron into a proton, an electron, and an antineutrino, with the electron having a momentum between p_e and $p_e + dp_e$. Does the expression agree with experiment? Since at this point we know nothing about the matrix element, the simplest approach is to assume that it is independent of the electron momentum and to see how the other factors in Eq. (11.5) fit the observed beta spectra. In principle, then, a function

$$p_e^2 (E_{max} - E_e)^2 \, dp_e$$

could be fitted to the experimental data. There exists an easier way: Equation (11.5) is rewritten into the form

$$\left[\frac{dw_{\beta\alpha}}{p_e^2 \, dp_e} \right]^{1/2} = \text{const.} \, \left(|\langle pe^-\bar{v} | H_w | n \rangle|^2 \right)^{1/2} (E_{max} - E_e). \tag{11.6}$$

If the expression on the left-hand side is determined experimentally and plotted versus the electron energy E_e, a straight line results if the matrix element is momentum-independent. Such a plot is called a Fermi or Kurie plot. Figure 11.2 shows the Kurie plot for the neutron decay. It is indeed a straight line over most of the energy range. The deviation at the low-energy end was caused by experimental difficulties in this early experiment: The electron counter had a window 5 mg/cm² thick, and it absorbed low-energy electrons. [See Fig. 3.8 and Eq. (3.7).] The number of electrons shown in Fig. 11.2 is not corrected for this loss.

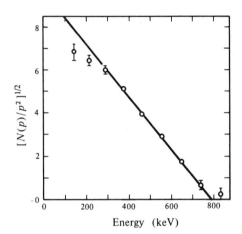

Fig. 11.2.   Kurie plot for the neutron decay. [From J. M. Robson, *Phys. Rev.* **83**, 349 (1951).]

The technique just described can also be applied to beta decays other than that of the neutron, but a problem not yet mentioned must be faced first. If a nucleus with charge Ze decays by beta emission, the charged lepton experiences the Coulomb force once it has left the nucleus. This force will decelerate negative and accelerate positive electrons. The spectrum will be distorted: There will be more positrons of high energy and more electrons of low energy than predicted by Eq. (11.5). Fortunately, the effect of the Coulomb potential on the emitted electrons can be computed accurately. The *Coulomb correction* introduces an additional factor in Eq. (11.5), and for a decay $N \rightarrow N'ev$ it becomes

$$dw_{\beta\alpha} = \frac{1}{2\pi^3 c^3 \hbar^7} \overline{|\langle N'ev|H_w|N\rangle|^2} F(\mp, Z, E_e) p_e^2 (E_{max} - E_e)^2 \, dp_e. \qquad \textbf{(11.7)}$$

$F(\mp, Z, E)$ is called the Fermi function; the sign indicates whether it applies to electrons or positrons, Ze is the nuclear charge, and E_e is the electrons' energy. Extensive tables of the Fermi function have been computed and published.[3]

The Fermi function also corrects Kurie plots for the Coulomb distortion, and the momentum dependence of the matrix element can be tested in many decays. It turns out that the matrix element is essentially momentum-independent in all cases of interest, for decay energies up to a few

3. H. Behrens and J. Jänecke, *Numerical Tables for Beta Decay and Electron Capture*, *Landolt-Börnstein*, New Series, Vol. I/4, Springer, Berlin, 1969.

Mev. The shape of the electron spectrum in beta decay is dominated by phase–space considerations and not by properties of the matrix element. Consequently, the shape cannot tell us much about the structure of the weak interaction.

11.2 Beta Decay Lifetimes

While the shape of the beta spectrum is not too useful, information about the magnitude of the matrix element can be obtained from the lifetimes of beta emitters. Since the matrix element has been shown to be momentum-independent, the total transition rate $w_{\beta\alpha}$ and the mean life τ can be obtained from Eq. (11.7) by integration over the momentum:

$$w = \frac{1}{\tau} = \frac{1}{2\pi^3 c^3 \hbar^7} \overline{|\langle N'ev|H_w|N\rangle|^2} \int_0^{p_{max}} dp_e F(\mp, Z, E_e) p_e^2 (E_{max} - E_e)^2.$$

$$\textbf{(11.8)}$$

The integral can be computed once F is known. In particular, for very large energies, where $E_{max} \approx cp_{max}$, and for small Z, where $F \approx 1$, it becomes

$$\int_0^{p_{max}} dp_e p_e^2 (E_{max} - E_e)^2 \simeq \frac{1}{30c^3} E_{max}^5. \qquad \textbf{(11.9)}$$

While this relation is sometimes useful for estimates, accurate values of the integral are needed for a meaningful treatment of the data. Fortunately the integral has been tabulated;[3] to read the tables, we have to know that the following abbreviation is used:

$$\int_0^{p_{max}} dp_e F(\mp, Z, E_e) p_e^2 (E_{max} - E_e)^2 = m_e^5 c^7 f(E_{max}). \qquad \textbf{(11.10)}$$

The factor $m_e^5 c^7$ has been inserted in order to make f dimensionless. With Eqs. (11.10) and (11.8), the matrix element becomes

$$\overline{|\langle N' ev|H_w|N\rangle|^2} = \frac{2\pi^3}{f\tau} \frac{\hbar^7}{m_e^5 c^4}. \qquad \textbf{(11.11)}$$

If τ is measured and f computed[3] then the square of the matrix element can be obtained from Eq. (11.11) It is unfortunately customary to use $ft_{1/2}$ and not $f\tau$ in tabulations. $ft_{1/2}$ is called the *comparative half-life*. The name stems from the fact that all beta-decaying states would have the same value of $ft_{1/2}$ if all matrix elements were equal. Nature provides an enormous range of values of $ft_{1/2}$, from about 10^3 to 10^{23} sec. If such a variation were caused by the fact that the weak interaction, H_w, were not universal but would change from decay to decay, an understanding of the weak processes would be hopeless. It is assumed that H_w is the same for all decays and that the nuclear wave functions that enter the calculation of $\langle N'ev|H_w|N\rangle$ are responsible for the variations. In general it is then to be expected that the most fundamental decays have the "best" wave functions and give rise to the largest matrix elements. These decays consequently should have the smallest $ft_{1/2}$ values. A few important cases are listed in Table 11.1.

Table 11.1 COMPARATIVE HALF-LIVES OF A FEW BETA DECAYS

Decay	Spin-parity sequence	$t_{1/2}$	E_{max} (MeV)	$ft_{1/2}$ (sec)
$n \longrightarrow p$	$\frac{1}{2}^+ \longrightarrow \frac{1}{2}^+$	10.6 min	0.782	1100
$^6He \longrightarrow {}^6Li$	$0^+ \longrightarrow 1^+$	0.813 sec	3.50	810
$^{14}O \longrightarrow {}^{14}N$	$0^+ \longrightarrow 0^+$	71.4 sec	1.812	3100

With $ft_{1/2} = (\ln 2)f\tau$ [Eq. (5.33)] and with the numerical values of the constants, Eq. (11.11) becomes

$$\overline{|\langle N' \, e\nu | H_w | N \rangle|^2} = \frac{43 \times 10^{-6} \text{ MeV}^2\text{-fm}^6\text{-sec}}{ft_{1/2} \text{ (in sec)}}. \qquad (11.12)$$

Now consider the decay of the neutron. With the value of $ft_{1/2}$ given in Table 11.1, the magnitude of the matrix element of H_w becomes

$$\overline{|\langle pe\bar{\nu}| H_w \,|n\rangle|} \approx 2 \times 10^{-4} \text{ MeV-fm}^3. \qquad (11.13)$$

The matrix element (11.13) gives an energy times a volume. The volume of the proton follows from Eq. (6.51) as approximately 2 fm^3. The weak energy, distributed over the volume of the proton, is of the order

$$H_w \approx 10^{-4} \text{ MeV}. \qquad (11.14)$$

This number demonstrates the weakness of the weak interaction: Presumably the mass of the proton, about 1 GeV, is given by the hadronic interaction. The weak interaction is consequently about a factor of 10^7 smaller.

11.3 The Current-Current Interaction

Two facts have become clear in the previous two sections: The dominant feature of the beta spectrum is given by the phase-space factor, and the beta decay interaction is so weak that perturbation theory can be used. However, we have learned very little about the Hamiltonian responsible for beta decay. Is it nevertheless possible to make a stab at the construction of a *weak Hamiltonian*? We have said above that the first successful theory of beta decay was formulated by Fermi[2] and that even less was known about beta decay in 1933 than we have described so far. It is therefore only proper to show how Fermi's genius led to a profound understanding of the weak interaction. We shall follow Fermi's reasoning but use more modern language.

Fermi assumed that electron and neutrino were created during the process of beta decay. This act of creation is similar to the process of photon emission. By 1933 the quantum theory of radiation was well understood, and Fermi patterned his theory after it. The result was incredibly successful and withstood all assaults for nearly 25 years. When parity fell in 1957, Fermi's theory finally required modification. The most successful extension was put forward by Feynman and Gell-Mann and, in somewhat different form, by Marshak and Sudarshan.[4] Amazingly enough, the modified theory, which is the basis of the currently accepted formulation, is very close to the original Fermi version. It can be said that the weak interaction tries as hard as possible to look like its stronger cousin, the electromagnetic one.

Figure 11.3(a) shows the diagram for the decay of the neutron. Such

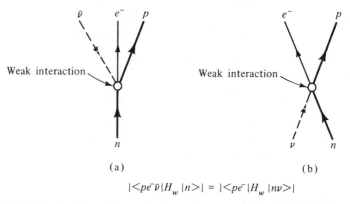

$$|<pe^-\bar{\nu}|H_w|n>| = |<pe^-|H_w|n\nu>|$$

Fig. 11.3. Neutron decay and neutrino absorption. It is assumed that the absolute values of the matrix elements for the two processes are the same.

a decay is not the most convenient one for writing down an interaction, because one particle comes in and three particles leave. It is easier to see the analogy to the electromagnetic force in a case where two particles are destroyed and two are created. We learned in Section 7.5 that antiparticles can be looked at as particles going backward in time. It is therefore reasonable to assume that one of the outgoing antiparticles, say the antineutrino, can be replaced by an incoming particle, in this case a neutrino. The process then appears as in Fig. 11.3(b). It is assumed that the matrix elements for the two processes in Fig. 11.3(a) and (b) have the same magnitude. [The transition rates are different because of unequal phase space factors $\rho(E)$.]

In the next step, the electromagnetic and the weak interaction are

4. R. P. Feynman and M. Gell-Mann, *Phys. Rev.* **109**, 193 (1958); E. C. G. Sudarshan and R. E. Marshak, *Phys. Rev.* **109**, 1860 (1958).

compared (Fig. 11.4). The electromagnetic interaction has the by-now familiar form where the force is transmitted by a virtual photon. The weak interaction has been changed from Fig. 11.3(b), and a hypothetical particle, the *intermediate boson* or *W* (for *weak*), has been inserted. The assumption of such a force-carrying particle makes the analogy to electromagnetism more direct. Most of the following arguments can be made without assuming the existence of the *W*, but they become more transparent and easier to remember with it.

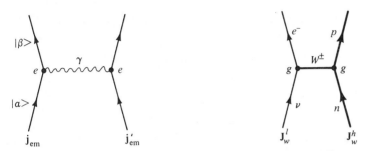

Fig. 11.4. Comparison of the electromagnetic and the weak interactions. The superscripts *l* and *h* indicate the weak currents of leptons and hadrons, respectively.

Consider first the electromagnetic case where two currents, each produced by a particle of charge *e*, interact via a virtual photon. The interaction energy is given by Eq. (10.49):

$$H_{em} = -\frac{e^2}{c^2} \int d^3x \, d^3x' \, \mathbf{j}(\mathbf{x}) \cdot \mathbf{j}'(\mathbf{x}') \frac{1}{|\mathbf{x} - \mathbf{x}'|}. \tag{11.15}$$

Charge conservation demands that the charge in each branch be conserved since the photon is neutral. The long range of the force, given by $|\mathbf{x} - \mathbf{x}'|^{-1}$, is caused by the vanishing mass of the photon.

The weak interaction, as shown in the graph in Fig. 11.4, is assumed to arise from a *weak current-current interaction*, and the form of H_w is patterned after H_{em}. Lepton conservation in the weak case corresponds to charge conservation in the electromagnetic interaction, and each weak current retains its lepton number. Consequently, the lepton number of the *W* must be zero. [Had we, in going from Fig. 11.3(a) to (b), replaced the outgoing proton by an incoming antiproton, the currents would not satisfy such a conservation law.] The weak currents shown in Fig. 11.4 change the value of the electric charge by one unit at the vertex; the neutrino, for instance, changes into an electron. Such a change of the electric charge occurs in all weak processes observed so far. Since the electric charge must be conserved in the overall reaction, the *W* must always be charged. This observation is expressed by saying that *weak neutral currents* are absent.

The modifications that distinguish H_w from H_{em} are the following: (1) The coupling constant *e* is replaced by the weak coupling constant *g*. (2) The electromagnetic current $\mathbf{j}_{em}$ is replaced by the weak currents $\mathbf{J}_w^l$ and

$\mathbf{J}_w^h$. The superscripts l and h indicate whether $\mathbf{J}_w$ is the weak current of leptons or hadrons. (3) The W is not massless and neutral but massive and electrically charged. Since the W has not yet been observed, its mass is unknown. The searches that have been performed indicate, however, that it must be larger than a few GeV/c^2. Equation (5.51) then indicates that the range, R_W, of the weak force must be less than

$$R_W = \frac{\hbar}{m_W c} \lesssim 0.1 \text{ fm.} \tag{11.16}$$

In analogy to the electromagnetic interaction, Eq. (11.15), the weak Hamiltonian can now be written as

$$H_w = -\frac{g^2}{c^2} \int d^3x\, d^3x'\, \mathbf{J}_w^l(\mathbf{x}) \cdot \mathbf{J}_w^h(\mathbf{x}') f(r), \tag{11.17}$$

where $r = |\mathbf{x} - \mathbf{x}'|$ and where $f(r)$ gives the dependence of the weak force on distance. The exact behavior of $f(r)$ is unknown, but if the W exists, then the range of $f(r)$ will be given by Eq. (11.16). Since the range is so small, the exact form of $f(r)$ is unimportant at energies presently available. It is customary to describe such short-range forces by a Yukawa shape,

$$f(r) = \frac{e^{-r/R_W}}{r}. \tag{11.18}$$

We shall return to this form in Chapter 12. Here it is sufficient to note that it is a function that is appreciably different from zero only for distances of the order of, or less than, R_W. If we further assume that the weak currents vary very little over distances of the order of R_W, then $\mathbf{J}_w^h(\mathbf{x}') \approx \mathbf{J}_w^h(\mathbf{x})$, Eq. (11.18) can be inserted into Eq. (11.17), and the integral over d^3x' can be performed. The result is

$$H_w = -4\pi \frac{g^2 R_W^2}{c^2} \int d^3x\, \mathbf{J}_w^l(\mathbf{x}) \cdot \mathbf{J}_w^h(\mathbf{x}). \tag{11.19}$$

Since the mass of W and hence R_W are not known, Eq. (11.19) is rewritten as

$$H_w = -\frac{G}{\sqrt{2}\, c^2} \int d^3x\, \mathbf{J}_w^l(\mathbf{x}) \cdot \mathbf{J}_w^h(\mathbf{x}), \tag{11.20}$$

with

$$G = \sqrt{2}\, 4\pi g^2 R_w^2 = \sqrt{2}\, 4\pi \left(\frac{\hbar}{m_W c}\right)^2 g^2. \tag{11.21}$$

The factor $1/\sqrt{2}\, c^2$ in Eq. (11.20) is introduced by convention. G is a new weak coupling constant that no longer has the same dimension as the electric charge e.

As Eq. (11.20) stands, it is not yet correct for the following reason: H_w is an operator that must be Hermitian. If the currents $\mathbf{J}_w^l$ and $\mathbf{J}_w^h$ were Hermitian, H_w would be Hermitian. In the electromagnetic interaction, Hermiticity of $\mathbf{j}_{em}$ is guaranteed because the electromagnetic current can be observed; the photon is neutral. No such guarantee exists in the weak interaction, and, in fact, as already indicated, the weak current is *not* Hermitian. H_w must therefore be made Hermitian. There are two ways of achieving this goal. One is to add the Hermitian conjugate expression to Eq. (11.20). The second one is again patterned in analogy to the electromagnetic case. In Eq. (10.93) the electromagnetic current was written as the sum of two contributions, one from leptons and the other from hadrons. Similarly, it is assumed that the total weak current is the sum of two contributions, one from leptons and the other from hadrons,

$$\mathbf{J}_w = \mathbf{J}_w^l + \mathbf{J}_w^h. \tag{11.22}$$

The weak Hamiltonian is then Hermitian if Eq. (11.20) is generalized to

$$H_w = -\frac{G}{\sqrt{2}\,c^2} \int d^3x\, \mathbf{J}_w(\mathbf{x}) \cdot \mathbf{J}_w^\dagger(\mathbf{x}). \tag{11.23}$$

This form is not yet complete. Our starting point, the electromagnetic interaction in the form of Eq. (11.15), describes only the energy due to two currents but leaves out the Coulomb interaction. The Coulomb energy between two charges described by electric charge densities $e\rho(\mathbf{x})$ and $e\rho'(\mathbf{x})'$ is given by

$$H_c = e^2 \int d^3x\, d^3x' \frac{\rho(\mathbf{x})\rho'(\mathbf{x}')}{|\mathbf{x} - \mathbf{x}'|}.$$

If *weak charges* $g\rho_w$ exist, then the arguments leading to Eq. (11.23) can be repeated, and the complete weak Hamiltonian becomes

$$H_w = \frac{G}{\sqrt{2}\,c^2} \int d^3x[c^2\rho_w(\mathbf{x})\rho_w^\dagger(\mathbf{x}) - \mathbf{J}_w(\mathbf{x})\cdot\mathbf{J}_w^\dagger(\mathbf{x})]. \tag{11.24}$$

It is possible to treat weak interactions by using H_w in this form. However, a more general notation makes arguments simpler and more transparent. The probability density and the probability current together form a four-vector, as already indicated in Eq. (10.89):

$$J_w = (c\rho_w, \mathbf{J}_w).$$

For the rest of this chapter we denote four-vectors with ordinary letters. The scalar product of two four-vectors is defined by Eq. (10.88); the product $J_w \cdot J_w^\dagger$ is

$$J_w \cdot J_w^\dagger = c^2 \rho_w \rho_w^\dagger - \mathbf{J}_w \cdot \mathbf{J}_w^\dagger,$$

and the weak Hamiltonian becomes

$$H_w = \frac{G}{\sqrt{2}\,c^2} \int d^3x\, J_w(\mathbf{x}) \cdot J_w^\dagger(\mathbf{x}). \tag{11.25}$$

This equation makes it obvious that H_w is a Lorentz invariant.

The reader who is not comfortable with four-vectors can consider the product to be an ordinary scalar product and little harm is done.

The concepts of a weak current and a weak charge require some reassuring remarks. We are used to electric charges and currents: They can be observed and measured, and they form part of our everyday surroundings. Weak currents and weak charges, on the other hand, have no classical analog. The only way to become familiar with them is to assume their existence and explore the consequences. If all experiments agree with the predictions based on a weak current-current interaction as given in Eq. (11.25), confidence in the existence of weak charges and currents is justified. If experiments disprove the Hamiltonian H_w, a different approach will have to be found. In the following sections, we shall inquire into three questions related to H_w: (1) What phenomena are described by H_w? (2) What is the form of the weak current J_w? (3) What is the value of the coupling constant G?

11.4 A Survey of Weak Processes

The discussion so far has been restricted to beta decay, the oldest and best known example of a weak interaction. If it were the only manifestation of the weak force, interest would be limited. However, a surprising variety of weak processes is already known, and they have been a rich source of unexpected new phenomena. We mention only one, the breakdown of parity conservation. Experiments with very-high-energy accelerators, such as NAL, promise to add to this store of surprises. In the present section, we shall categorize the weak processes, list a few examples, and state why they all are called weak.

A *classification* of weak processes can be based on the separation of the weak current into a leptonic and a hadronic part, as in Eq. (11.22). Inserting Eq. (11.22) in the form $J_w = J_w^l + J_w^h$ into the weak Hamiltonian (11.25) produces four scalar products; one involves only leptons and one only hadrons, and two couple lepton and hadron currents. The classification is performed according to these terms:

$$\text{leptonic processes:} \quad J_w^l \cdot J_w^{l\dagger}$$

$$\text{semileptonic processes:} \quad J_w^l \cdot J_w^{h\dagger} + J_w^h \cdot J_w^{l\dagger} \qquad \textbf{(11.26)}$$

$$\text{hadronic processes:} \quad J_w^h \cdot J_w^{h\dagger}.$$

Weak processes of each of these three classes are known. In Chapter 10, in the treatment of the electromagnetic interaction, we have learned that life is easy as long as only leptons are present. The story repeats itself in the weak interaction: Leptonic processes can be calculated, and theory and experiment agree. Semileptonic processes produce many difficulties, and the weak processes involving only hadrons cannot yet be explained in detail. We shall now list processes in each of the three classes.

Leptonic Processes. The only purely leptonic decay that has been studied so far is that of the muon,

$$\mu \longrightarrow e\bar{\nu}\nu. \qquad \textbf{(11.27)}$$

Muon decay will be discussed in the following section, where it will be seen that the maximum energy of the emitted electrons is about 53 MeV, the lifetime is 2.2 μsec, and parity is not conserved.

The scattering of neutrinos with charged leptons also involves only leptons: The processes

$$\nu_e e^- \longrightarrow \nu_e e^-$$
$$\nu_\mu e^- \longrightarrow \nu_e \mu^-$$

(11.28)

are without electromagnetic or hadronic complications, and they, and the corresponding ones involving antineutrinos, are ideal for studying the weak interaction at high energies. However, presently available neutrino fluxes and detector efficiencies are too small so that neutrino-lepton scattering has not yet been seen.

Semileptonic Processes. In semileptonic processes, one current is leptonic and the other one hadronic. A typical example is shown in Fig. 11.4, and the prototype is the neutron decay. Three semileptonic decays are listed in Table 11.2. Many additional ones are given in Table A.3 in the Appendix.

Table 11.2 Decay Properties of Three Semileptonic Decays. $t_{1/2}$ denotes the partial half-life.

Decay	Spin-parity sequence	$t_{1/2}$ (sec)	E_{max} (MeV)	$ft_{1/2}$ (sec)
$\pi^\pm \longrightarrow \pi^0 e\nu$	$0^- \longrightarrow 0^-$	1.8	4.1	2×10^3
$n \longrightarrow pe\bar{\nu}$	$\frac{1}{2}^+ \longrightarrow \frac{1}{2}^+$	640	0.78	1.2×10^3
$\Sigma^- \longrightarrow \Lambda^0 e^- \bar{\nu}$	$\frac{1}{2}^+ \longrightarrow \frac{1}{2}^+$	1.7×10^{-6}	79	5×10^3

In connection with Table 11.2, a question comes to mind: Can decays give sufficient information to study the weak interaction completely? The maximum energy listed in Table 11.2 is 72 MeV, but the electromagnetic interaction taught us that energies of the order of many GeV are necessary to explore some of the properties. Do weak decays with such energies exist? None are listed in the tables, and the reason is obvious: If a state has such a high excitation energy, it can decay hadronically, and the weak interaction never gets a chance. The high-energy behavior of the weak interaction can be studied only with neutrino reactions. Semileptonic neutrino reactions have been observed, as, for instance,

$$\nu_\mu n \longrightarrow \mu^- p$$
$$\bar{\nu}_\mu p \longrightarrow \mu^+ n$$

(11.29)

In the semileptonic processes listed so far, the weak decay has never involved a change of strangeness. True, the decay $\Sigma^+ \longrightarrow \Lambda^0 e^+ \nu$ listed in

Table 11.2 involves strange particles, but the hadrons in the initial and final states have the same strangeness. We have, however, mentioned in Section 7.6 that strangeness or hypercharge is not necessarily conserved in the weak interaction. Indeed, strangeness-changing weak decays exist, and three are listed in Table 11.3.

Table 11.3 Hypercharge Changing Semileptonic Decays.
$t_{1/2}$ is the partial half-life.

Decay	Spin-parity sequence (of hadron)	$t_{1/2}$ (sec)	$E_{max}(e)$ (MeV)	$ft_{1/2}$ (sec)
$K^+ \longrightarrow \pi^0 e^+ \nu_e$	$0^- \longrightarrow 0^-$	1.8×10^{-7}	230	1×10^5
$\Lambda^0 \longrightarrow p e^- \bar{\nu}_e$	$\frac{1}{2}^+ \longrightarrow \frac{1}{2}^+$	2×10^{-7}	160	2×10^4
$\Sigma^- \longrightarrow n e^- \bar{\nu}_e$	$\frac{1}{2}^+ \longrightarrow \frac{1}{2}^+$	0.95×10^{-7}	230	7×10^4

Hadronic Processes. Examples of weak decays in which only hadrons are involved are

$$K^+ \longrightarrow \pi^+ \pi^0$$
$$\longrightarrow \pi^+ \pi^+ \pi^-$$
$$\longrightarrow \pi^+ \pi^0 \pi^0 \qquad (11.30)$$

and

$$\Lambda^0 \longrightarrow p \pi^-$$
$$\longrightarrow n \pi^0. \qquad (11.31)$$

Other weak decays involving only hadrons are given in the tables in the Appendix. All of these obey the hypercharge selection rule

$$|\Delta Y| = 1. \qquad (11.32)$$

The absence of observed $\Delta Y = 0$ transitions is easily explained: Transitions without change of hypercharge can proceed by hadronic or electromagnetic decays, and the weak branch is hidden.

Why are all the processes listed in the present section called weak, regardless of whether they involve leptons, hadrons, or both? The justification comes from the fact that the strength of the interaction responsible for the various processes appears to be the same. Additional support comes from considerations of selection rules and from the observation that all processes that are weak according to the strength classification also show violations of parity and charge conjugation invariance.

The *strength* of the interaction responsible for a decay expresses itself in the lifetime, other things being equal. The decays in Table 11.2 are of the type $A \longrightarrow B e \nu$. While the decay energies vary by about a factor of 100 and the density-of-states factors by a factor of 10^{10}, the ft values are approximately the same. It is therefore likely that the three very different decays in Table 11.2 are caused by the same force. A discrepancy appears when the *ft* values in Table 11.2 and 11.3 are compared. While the decays appear to be similar, the *ft* values for hypercharge-changing decays are between one and two orders of magnitude larger than the corresponding ones for hypercharge-conserving decays. We shall return to this discrepancy in Section 11.8 and show that it has an explanation within the framework of the weak current-current interaction.

Parity violation has already been treated in Section 9.3; the electromagnetic and the hadronic force conserve parity, but a violation appears in the weak one. The example discussed in Section 9.3 was a semileptonic decay. The original evidence for parity nonconservation came from the decay of the charged kaons into two and three pions; these weak decays involve hadrons. In the next section we shall show that the purely leptonic decay of the muon also does not conserve parity. These examples indicate that the various processes all violate parity conservation. This fact alone would not justify classing them all into one category. However, it indicates a similarity in the *form* of the interaction that causes these decays, and it supports the conclusion already reached from a consideration of the lifetimes.

Conservation of strangeness or hypercharge in the hadronic and the electromagnetic interaction was postulated in Eq. (7.45). The examples of weak decays discussed in Section 7.6 and in the present section indicate that many cases are known where the hypercharge changes by one unit; no case has been found where a change of two units occurs. The *selection rule* for hypercharge,

$$\Delta Y = 0 \qquad \text{in hadronic and electromagnetic interaction}$$

$$\Delta Y = 0, \pm 1 \qquad \text{in the weak interaction,} \qquad \textbf{(11.33)}$$

thus establishes another characteristic feature of the weak interaction.

11.5 The Muon Decay

In the previous section we have surveyed weak processes, and we have partially answered the first question posed at the end of Section 11.3, namely what phenomena are described by H_w. The form of the weak current and the value of the weak coupling constant remain to be studied. We can expect that the fundamental features of the weak interaction will be easiest to explore in purely leptonic processes because no serious interference from the hadronic force is present there. Up to now only one such phenomenon, namely muon decay, has been observed. In this section, the salient features of the muon decay will be described.

Muons do not interact strongly, and it is consequently not possible to produce them directly and copiously through a reaction. However, the decay of charged pions is a convenient source of muons. Assume, for instance, that positive pions are produced at an accelerator. The pions are selected in a pion channel and slowed down in an absorber (Fig. 11.5). They usually come to rest before decaying through the mode

$$\pi^+ \longrightarrow \mu^+ \nu_\mu. \qquad (11.34)$$

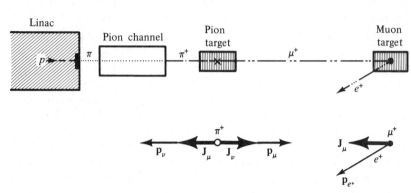

Fig. 11.5. A positive pion is selected in the pion channel and comes to rest in the pion target. The pion decay results in a fully polarized muon. The muon escapes from the pion target and comes to rest in the muon target. Its spin points in the direction from which it came. The decay electron is then observed.

Conservation laws determine much of what is now happening: Conservation of the lepton and muon numbers requires the neutral particle to be a muon neutrino. Momentum conservation demands that the muon and the muon neutrino have equal and opposite momenta in the c.m. of the decaying pion. The muon neutrino has its spin opposite to its momentum, as shown in Fig. 7.2. Since the pion has spin 0, angular momentum conservation consequently insists that the positive muon must be fully polarized, with its spin pointing opposite to its momentum. The muons escape from the pion target; some are stopped in the muon target, and their decay positron can be detected. With proper choice of the muon target, the decaying muon is still polarized, and its spin **J** points into the direction from which it came.

The processes just described and shown in Fig. 11.5 permit a number of measurements that all give information concerning the weak interaction. We shall discuss three aspects here, parity nonconservation, the lifetime of the muon, and the spectrum of the decay electron.

Parity Nonconservation. As Fig. 11.5 is drawn, it shows the breakdown of parity in two different places. The muon is expected to be polarized because the neutrino emitted together with it is polarized. A longitudinally polarized muon violates parity conservation, as was explained in Section 9.3. A measurement of the polarization of the muon thus demonstrates that parity is not conserved in the weak decay of the pion. Such a polarization has been detected.[5] The second place where parity non-conservation shows up is in the decay of the muon. As sketched in Fig. 11.5, the muon spin points into a well-defined direction, and the probability of positron emission can now be determined with respect to this direction. This experiment is analogous to the one discussed in Section 9.3 and shown in Fig. 9.6. Indeed, as in the Wu-Ambler experiment, it was found that the positron is preferentially emitted parallel to the spin of the incoming muon, indicating that parity is also violated in the muon decay.[6]

Muon Lifetime. The experimental arrangement for determining the muon lifetime has already been described in Chapter 4. In Fig. 4.15, the logic elements are shown, and it is easy to see how they fit into the setup of Fig. 11.5. Observation of the number of electrons detected in counter D as a function of the delay time between counters B and D gives a curve of the form shown in Fig. 5.15, and the slope of the curve determines the muon lifetime. The present best value is listed in Table A2 in the Appendix. For most estimates it is sufficient to remember that the muon mean life is 2.2 μsec.

Electron Spectrum. To investigate the electron spectrum, the number of electrons is measured as a function of momentum. To determine the momentum, the electron path in a magnetic field is observed, for instance, by using wire spark chambers.[7] The result of one such experiment is

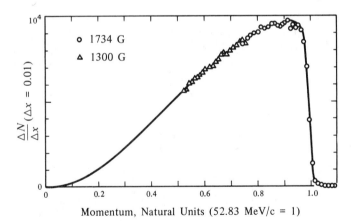

Fig. 11.6. Electron spectrum from unpolarized muons. [B. A. Sherwood, *Phys. Rev.* **156**, 1475 (1967).] The momentum is measured in units of the maximum electron momentum.

5. G. Backenstoss, B. D. Hyams, G. Knop, P. C. Marin, and U. Stierlin, *Phys. Rev. Letters* **6**, 415 (1961); M. Bardon, P. Franzini, and J. Lee, *Phys. Rev. Letters* **7**, 23 (1961).

6. R. L. Garwin, L. M. Lederman, and M. Weinrich, *Phys. Rev.* **105**, 1415 (1957); J. L. Friedman and V. L. Telegdi, *Phys. Rev.* **105**, 1681 (1957).

7. M. Bardon, P. Norton, J. Peoples, A. M. Sachs, and J. Lee-Franzini, *Phys. Rev. Letters* **14**, 449 (1965); B. A. Sherwood, *Phys. Rev.* **156**, 1475 (1967).

shown in Fig. 11.6. Some similarity to the electron spectrum in beta decay, Fig. 11.1, exists but the drop-off at high electron momenta is much steeper. The electron spectrum is no longer determined by the phase-space factor alone. It is therefore to be expected that a detailed comparison with theory will provide some information on the form of the weak Hamiltonian.

11.6 The Weak Current of Leptons

In the previous section, some of the salient features of the muon decay have been discussed. These data and some additional information will now be used to construct the weak Hamiltonian, Eq. (11.25), in more detail. In particular, we shall have to find the form of the weak current, J_w^l, as far as we can with our limited tools. The first fact to be used is the uncanny similarity between electron and muon, a fact often stated by the words *muon-electron universality*. This universality is expressed by writing the total weak current of leptons as the sum of an electron and a muon current,

$$J_w^l = J_w^e + J_w^\mu \tag{11.35}$$

and assuming that both behave alike. Inserting the sum into, for the leptonic part of the weak Hamiltonian H_w, Eq. (11.25), gives

$$H_w = \frac{G}{\sqrt{2}\,c^2} \int d^3x \{ J_w^e \cdot J_w^{e\dagger} + J_w^e \cdot J_w^{\mu\dagger} + J_w^\mu \cdot J_w^{e\dagger} + J_w^\mu \cdot J_w^{\mu\dagger} \}. \tag{11.36}$$

For the explicit construction of the weak current J_w^e, we use the analogy to electromagnetism. In Chapter 10, we systematically went from the classical Hamiltonian, Eq. (10.47),

$$H_{em} = \frac{e}{c} \int d^3x\, \mathbf{j} \cdot \mathbf{A}$$

to the matrix element, Eq. (10.58),

$$\langle \beta | H_{em} | \alpha \rangle = -i\frac{e\hbar}{mc} \int d^3x\, \psi_\beta^* \nabla \psi_\alpha \cdot \mathbf{A}.$$

Comparison of these two expressions shows that the substitution

$$\mathbf{j}_{em} = -i\frac{\hbar}{m} \psi_\beta \nabla \psi_\alpha = \psi_\beta^* \left(\frac{\mathbf{p}_{op}}{m} \right) \psi_\alpha = \psi_\beta^* \mathbf{v}_{op} \psi_\alpha \tag{11.37}$$

provides the transition from the classical Hamiltonian to the quantum mechanical matrix element. The analogous substitution for the probability density is

$$\rho_{em} = \psi_\beta^* \psi_\alpha. \tag{11.38}$$

Equations (11.37) and (11.38) are valid for nonrelativistic electrons. To allow for generalizations, we introduce two operators, V_0 and $\mathbf{V}$, and write

$$\rho_{em} = \psi_\beta^* V_0 \psi_\alpha, \qquad \mathbf{j}_{em} = c\psi_\beta^* \mathbf{V} \psi_\alpha$$

The velocity of light, c, has been inserted in order to make $\mathbf{V}$ dimensionless. Charge density and current density combine to form a four-vector,

$$j_{em} = (c\rho, \mathbf{j}),$$

or, with the operators V_0 and $\mathbf{V}$,

$$j_{em} = c\psi_\beta^* V \psi_\alpha. \tag{11.39}$$

The notation $V \equiv (V_0, \mathbf{V})$ is a reminder that the "sandwich" $\psi^* V \psi$ transforms like a four-vector. With Eqs. (11.37) and (11.38), the explicit form of V for a nonrelativistic electron is

$$V \equiv (V_0, \mathbf{V}), \qquad V_0 = 1, \qquad \mathbf{V} = \frac{\mathbf{p}}{mc}. \tag{11.40}$$

The weak current density J_w^e is now written in analogy to the electromagnetic one:

$$J_w^e = c\psi_e^* V \psi_{\nu_e}. \tag{11.41}$$

The essential difference between the electromagnetic and the weak current densities consists in the choice of the wave functions: ψ_α and ψ_β in Eq. (11.39) describe the initial and the final state of the same electron, whereas ψ_ν and ψ_e in Eq. (11.41) refer to the neutrino in the initial state and the electron in the final state, respectively. The difference can best be seen in Fig. 11.4, where it should be noted that the weak current does not conserve the electric charge.

It will turn out soon that the weak current is more complicated than the electromagnetic one and that the form Eq. (11.41) provides only half of the correct answer. Before coming to this aspect we shall use the simplicity of Eq. (11.41) to learn more about the physics that is hidden in it. To do so, we consider the Hermitian conjugate of the current J_w^e. The operator V is Hermitian; noting that for a one-component wave function $\psi^\dagger = \psi^*$ then gives

$$J_w^{e\dagger} = c(\psi_e^* V \psi_{\nu_e})^\dagger = c\psi_{\nu_e}^* V \psi_e. \tag{11.42}$$

Comparison with J_w^e and with Fig. 11.4 shows that $J_w^{e\dagger}$ describes the destruction of an electron and the creation of an electron neutrino. The product $J_w^e \cdot J_w^{e\dagger}$ in H_w^e is thus responsible for the scattering of electron neutrinos by electrons, $\nu_e e^- \rightarrow \nu_e e^-$, a process that has already been listed in Eq. (11.28). The two currents and the scattering process are displayed in Fig. 11.7.

The operator $J_w^e \cdot J_w^{e\dagger}$ can do more than just induce neutrino scattering. Since incoming particles and outgoing antiparticles are assumed to be equivalent, it is also responsible for the reaction

$$e^+ \bar{\nu}_e \longrightarrow e^+ \bar{\nu}_e, \tag{11.43}$$

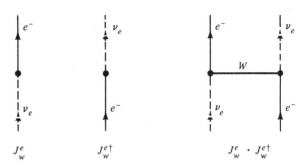

Fig. 11.7. Interpretation of the currents J_w^e and $J_w^{e\dagger}$ and of the product $J_w^e J_w^{e\dagger}$.

namely the scattering of antineutrinos from positrons. The other terms in the Hamiltonian (11.36) similarly give rise to weak processes involving only leptons. One term that is responsible for muon decay is easily seen to be

$$J_w^e \cdot J_w^{\mu\dagger} = c^2 \psi_e^* V \psi_{\nu_e} \cdot \psi_{\nu_\mu}^* V \psi_\mu. \tag{11.44}$$

In the previous section, the muon decay has been discussed and predictions based on the scalar product, Eq. (11.44), must now be compared to the experimental facts. It quickly becomes obvious that the form of Eq. (11.44) is not general enough to describe the effects of parity nonconservation observed in the muon decay. According to the golden rule, the transition rate is proportional to the square of the matrix element, or

$$w_\mu \propto |\int d^3 x \, \psi_e^* V \psi_{\nu_e} \cdot \psi_{\nu_\mu}^* V \psi_\mu|^2.$$

The operator $V = (V_0, \mathbf{V})$ behaves under the parity operation as

$$V_0 \xrightarrow{P} V_0 \qquad \mathbf{V} \xrightarrow{P} -\mathbf{V}. \tag{11.45}$$

The fact that the vector part changes sign follows from Eq. (9.1). V_0, on the other hand, is a probability density, and it remains unchanged under the parity operation. The vector product $V \cdot V = V_0 V_0 - \mathbf{V} \cdot \mathbf{V}$ remains unchanged under P; if w_μ^P denotes the transition rate after the parity operation, it is equal to w_μ:

$$w_\mu^P = w_\mu.$$

This result disagrees with the electron asymmetry observed in muon decay. How can the expression for the weak current be generalized in such a way that the analogy to the electromagnetic current is not completely destroyed but that parity nonconservation is included? A hint to the answer comes from comparing linear and angular momentum. Under ordinary rotations, both behave in the same way. We have not demonstrated this fact explicitly, but the proof is straightforward if the arguments given in Section 8.2 are used. Under the parity operation, the polar vector $\mathbf{p}$ and the axial vector $\mathbf{J}$ reveal their difference: $\mathbf{p}$ changes sign, whereas $\mathbf{J}$ does not. These properties remain true for general operators V and A: V and A behave identically under ordinary rotations but differently under space inversion. The properties of a general axial four-vector A under P are given by

$$A_0 \xrightarrow{P} -A_0 \qquad \mathbf{A} \xrightarrow{P} \mathbf{A}. \qquad (11.46)$$

The behavior of the *axial probability density* cannot be visualized as easily as the one for the ordinary probability density: The electric charge provides an example for the properties of V_0, but no classical example for an axial charge exists.[8] The suggested generalization of the weak current, Eq. (11.41), is

$$J_w^e = c\psi_e^*(V + A)\psi_{v_e}. \qquad (11.47)$$

With Eqs. (10.1) and (11.25) the transition rate for the muon decay then becomes

$$w_\mu = \frac{\pi G^2}{\hbar} \left| \int d^3x \, \psi_e^*(V + A)\psi_{v_e} \cdot \psi_{v_\mu}^*(V + A)\psi_\mu \right|^2 \rho(E),$$

or

$$w_\mu = \frac{\pi G^2}{\hbar} |M_{\text{even}} + M_{\text{odd}}|^2 \rho(E), \qquad (11.48)$$

with

$$M_{\text{even}} = \int d^3x \, [\psi_e^* V \psi_{v_e} \cdot \psi_{v_\mu}^* V \psi_\mu + \psi_e^* A \psi_{v_e} \cdot \psi_{v_\mu}^* A \psi_\mu]$$

$$M_{\text{odd}} = \int d^3x \, [\psi_e^* V \psi_{v_e} \cdot \psi_{v_\mu}^* A \psi_\mu + \psi_e^* A \psi_{v_e} \cdot \psi_{v_\mu}^* V \psi_\mu].$$

Under the parity operation, M_{even} remains unchanged, M_{odd} changes sign, and the transition rate becomes

$$w_\mu^P = \frac{\pi G^2}{\hbar} |M_{\text{even}} - M_{\text{odd}}|^2 \rho(E). \qquad (11.49)$$

Comparison of Eqs. (11.48) and (11.49) shows that

$$w_\mu^P \neq w_\mu.$$

The presence of both a vector *and* an axial vector operator in the weak current permits description of the observed violation of parity invariance. The violation becomes maximal if V and A have equal magnitudes.

The detailed computation of a transition rate or cross section can be performed only if the explicit form of the operators V and A is known. This form depends on the type of particles that carry the weak current. For *nonrelativistic electrons*, the operators V_0 and $\mathbf{V}$ are given in Eq. (11.40). The axial vector current is usually not treated in introductory quantum mechanics, and we are forced to establish its form by using invariance arguments. An electron is described by its energy, its momentum $\mathbf{p}$, and

8. If magnetic monopoles exist, they provide an example for an axial charge. The magnetic charge density ρ_m, introduced in Eq. (10.112), changes sign under the parity operation. This fact can be proved by considering the energy of a magnetic monopole in a magnetic field and assuming invariance of the corresponding Hamiltonian under P.

its spin **J**. It is customary to use instead of the spin **J** the dimensionless Pauli spin operator $\boldsymbol{\sigma}$; it is connected to **J** by

$$\boldsymbol{\sigma} = \frac{2\mathbf{J}}{\hbar}. \tag{11.50}$$

The only axial vector available is **J**, or $\boldsymbol{\sigma}$. The operator **A** must therefore be proportional to $\boldsymbol{\sigma}$. The axial charge operator, A_0, changes sign under the parity operation, as indicated by Eq. (11.46); since $\boldsymbol{\sigma} \cdot \mathbf{p}$ has this property, we set

$$A = (A_0, \mathbf{A}), \qquad A_0 = \frac{\boldsymbol{\sigma} \cdot \mathbf{p}}{mc}, \qquad \mathbf{A} = \boldsymbol{\sigma}. \tag{11.51}$$

The factor $1/mc$ in A_0 is chosen to make the operator dimensionless.

The nonrelativistic operators, as given in Eqs. (11.40) and (11.51), cannot be used for the evaluation of the muon decay because there all particles in the final state must be treated relativistically. The generalization of the operators V and A to relativistic leptons is well known.[9] Calculations with the relativistic operators are, however, beyond our means here, and we therefore give the transition rate for the muon decay without proof. The rate $dw_\mu(E_e)$ for the emission of an electron with energy between E_e and $E_e + dE_e$ becomes, for $E_e \gg m_e c^2$,

$$dw_\mu(E_e) = G^2 \frac{m_\mu^2 c^4}{4\pi^3 \hbar^7} E_e^2 \left[1 - \frac{4}{3} \frac{E_e}{m_\mu c^2} \right] dE_e. \tag{11.52}$$

This expression, after replacing the electron energy by the electron momentum, agrees very well with the spectrum shown in Fig. 11.6.

11.7 The Weak Coupling Constant *G*

The electromagnetic coupling constant e can be determined by observing the force on a charged particle in a known field, by measuring the Rutherford or Mott cross section [Eqs. (6.17) or (6.19)] from a point scatterer, or by determining the lifetime of a decay with well-known matrix element $\langle f | \mathbf{x} | i \rangle$ [Eq. (10.75)]. What is the best way of determining the weak coupling constant G? Again there are a number of possibilities, but the total lifetime of the muon is a good choice. The reason is twofold: The muon decay involves no hadrons so that complications due to the hadronic interaction do not have to be considered, and the muon lifetime has been measured very accurately.

The total transition rate for the muon decay is obtained by integrating Eq. (11.52),

$$w_\mu = \int_0^{E_{\max}} dw_\mu(E_e) = G^2 \frac{m_\mu^2 c^4}{4\pi^3 \hbar^7} \int_0^{E_{\max}} dE_e E_e^2 \left[1 - \frac{4}{3} \frac{E_e}{m_\mu c^2} \right] = \frac{G^2 m_\mu^5 c^4}{192\pi^3 \hbar^7}. \tag{11.53}$$

9. Merzbacher, Section 23.3; Messiah, Section 20.10; W. A. Blanpied, *Modern Physics*, Holt, Rinehart and Winston, New York, 1971, Sections 15.4 and 25.5.

The muon mean life is the inverse of w_μ, or

$$\tau_\mu = \frac{192\pi^3\hbar^7}{G^2 m_\mu^5 c^4}.$$

With the observed muon lifetime the coupling constant becomes[10]

$$G = (1.435 \pm 0.001) \times 10^{-49} \text{ erg-cm}^3$$
$$= (0.896 \pm 0.001) \times 10^{-4} \text{ MeV-fm}^3. \tag{11.54}$$

In the electromagnetic case we have expressed the strength of the interaction by making e^2 dimensionless as in Eq. (10.77):

$$\alpha = \frac{e^2}{\hbar c} \approx \frac{1}{137}.$$

Comparison of Eqs. (11.15) and (11.17) makes it clear that the weak analog to the electric charge is g, not G. Like e^2, g^2 is made dimensionless by division by $\hbar c$. The connection with G, as given in Eq. (11.21), then permits us to write $g^2/\hbar c$ in terms of G and the mass m,

$$\frac{g^2}{\hbar c} = \frac{1}{\sqrt{2}} \frac{1}{4\pi} \frac{1}{\hbar c} \left(\frac{m_W c}{\hbar}\right)^2 G.$$

With the numerical value of G, the dimensionless weak coupling is described by

$$\frac{g^2}{\hbar c} \approx 0.65 \times 10^{-12} (m_W c^2)^2, \tag{11.55}$$

where the rest energy $m_W c^2$ is expressed in MeV. At present it is not known if W exists. To still get an idea of the value of the dimensionless coupling constant, m_W is replaced by the nucleon mass, and the dimensionless weak coupling constant becomes

$$\frac{g^2}{\hbar c} \approx 0.57 \times 10^{-6}.$$

● If the expressions for g^2 given here are compared to those in the literature, it must be remembered that we use unrationalized units, whereas most papers employ rationalized units. Rationalized and unrationalized charges are connected by

$$e^2(\text{rat.}) = 4\pi e^2(\text{unrat.}), \qquad g^2(\text{rat.}) = 4\pi g^2(\text{unrat.}). \qquad ●$$

11.8 Nonstrange and Strange Weak Currents

The universal Fermi interaction in the current-current form, Eq. (11.36), with $V - A$ currents is successful in describing all observed leptonic weak interactions. While the computations involve more tricks than we have available here, no new physical principle enters, and we can claim that at energies below a few GeV the leptonic weak interaction is as well understood as the electromagnetic one. The situation is less clear in the

10. M. Roos and A. Sirlin, *Nucl. Phys.* **B29**, 296 (1971).

treatment of the semileptonic and the hadronic weak processes, where troubles are introduced by hadronic effects. A first problem arises when weak decays without and with a change in hypercharge are compared. Decays between similar particle states, but without and with change of hypercharge, are listed in Tables 11.2 and 11.3. The salient features of the six decays are collected in Table 11.4. In the last column of this table, the ratio of the comparative half-lives of the hypercharge-conserving to the hypercharge-changing decays of the same type are shown. The decays that are compared are in all cases similar. Neutron and lambda, for instance, have similar hadronic and electromagnetic properties, and the decays $n \rightarrow pe\nu$ and $\Lambda \rightarrow pe\nu$ should have similar ft values. However, the ft values of the $|\Delta Y| = 1$ decays in Table 11.4 are at least one order of magnitude larger than the corresponding $\Delta Y = 0$ ft values. In other words, the rates for the $|\Delta Y| = 1$ transitions are systematically about a factor of 20 slower than the rates of the corresponding $\Delta Y = 0$ ones. Such a difference could in principle be understood if the matrix elements did depend strongly on the decay energy because decays with $|\Delta Y| = 1$ have larger energies than the corresponding $\Delta Y = 0$ decays. No evidence for such an energy dependence exists, however.

Table 11.4 COMPARISON OF THE $ft_{1/2}$ VALUES FOR SIMILAR DECAYS WITHOUT AND WITH CHANGE OF HYPERCHARGE

Type	*Hypercharge-conserving,* $\Delta Y = 0$	*Hypercharge-changing,* $\|\Delta Y\| = 1$	$\dfrac{ft(\|\Delta Y\| = 1)}{ft(\Delta Y = 0)}$
$0^- \rightarrow 0^-$	$\pi^{\pm} \rightarrow \pi^0 e\nu$	$K^+ \rightarrow \pi^0 e^+\nu$	50
$\frac{1}{2}^+ \rightarrow \frac{1}{2}^+$	$n \rightarrow pe^-\bar{\nu}$	$\Lambda \rightarrow pe^-\bar{\nu}$	17
$\frac{1}{2}^+ \rightarrow \frac{1}{2}^+$	$\Sigma^- \rightarrow \Lambda^0 e^-\bar{\nu}$	$\Sigma^- \rightarrow ne^-\bar{\nu}$	12

A modification of the weak current that explains the observed facts was proposed by Cabibbo.[11] We describe some features of his theory by using an analogy to an electric current, I, flowing into two resistors, as shown in Fig. 11.8(a). If the resistor R_1 is infinite, the entire electric current will flow through R_0. If R_1 is finite, current will flow in this branch; if the total current I is kept constant, the current I_1 will be robbed from the other branch. Kirchhoff's law states that $I = I_0 + I_1$. Now consider the weak current of hadrons and assume first that the weak interaction conserves hypercharge. The total weak current of hadrons then flows into the hypercharge-conserving branch. However, if hypercharge conservation is violated, decays with $|\Delta Y| = 1$ become possible. The new decays rob some strength from the decays with $\Delta Y = 0$ if the total weak current, J_w^h, is assumed to be unchanged. To express this weak analog to Kirchhoff's

11. N. Cabibbo, *Phys. Rev. Letters* **10**, 531 (1963).

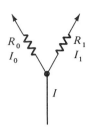

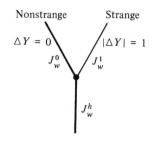

(a) Electromagnetic current

(b) Weak current of hadrons

Fig. 11.8. Kirchhoff's and Cabibbo's rules: Branching of the electromagnetic and the weak current. (a) Electromagnetic current. (b) Weak current of hadrons.

law, the weak current of hadrons is written as

$$J_w^h = a J_w^0 + b J_w^1.$$

Here J_w^0 and J_w^1 are the currents corresponding to $\Delta Y = 0$ and $|\Delta Y| = 1$ transitions, respectively. J_w^0 and J_w^1 are normalized so that the strengths of the corresponding transitions are given by the coefficients a and b. It is assumed that the total weak probability current J_w^h is not changed, and this condition restricts the coefficients a and b to

$$|a|^2 + |b|^2 = 1. \qquad (11.56)$$

Following Cabibbo, it has become customary to write $a = \cos\theta$, $b = \sin\theta$. The normalization condition, Eq. (11.56), then is automatically satisfied, and the weak current of hadrons becomes

$$J_w^h = \cos\theta \, J_w^0 + \sin\theta \, J_w^1. \qquad (11.57)$$

To find an approximate value for the *Cabibbo angle* θ, we note that the transition rate is proportional to $|\langle\beta|H_w|\alpha\rangle|^2$. The rate for $\Delta Y = 0$ transitions consequently is proportional to $G^2 \cos^2\theta$ and that for $|\Delta Y| = 1$ to $G^2 \sin^2\theta$. The ratio of ft values listed in Table 11.3 gives $\cot^2\theta$:

$$\cot^2\theta \approx \frac{ft(|\Delta Y| = 1)}{ft(\Delta Y = 0)}.$$

The decays of the neutron and of the lambda are best known: From $\cot^2\theta = 17$ follows $\theta = 0.24$. A more sophisticated evaluation gives[12] $\theta = 0.188 \pm 0.006$. Past experience with error limits calls for caution, however, and we shall use

$$\theta = 0.20 \pm 0.02. \qquad (11.58)$$

In the present section we have introduced the Cabibbo form of the weak current of hadrons in a rather arbitrary way. The main support comes

12. E. Fischbach, M. M. Nieto, H. Primakoff, C. K. Scott, and J. Smith, *Phys. Rev. Letters* **27**, 1403 (1971).

from the fact that computations based on Eq. (11.57) give very good agreement with experiment.

11.9 Weak Currents in Nuclear Physics

The exploration of the structure of the weak current of hadrons is a problem that has occupied many physicists for a very long time. Even at the present time it is not fully known, but a number of features have emerged. In the present section, some of the information that can be extracted from nuclear beta decay is treated. Actually, the amount of experimental material available in beta decay alone is staggering, but we present only one example, the decay $^{14}O \xrightarrow{\beta^+} {}^{14}N$. The interest in this particular decay will be made clear shortly; we shall already be able to learn important facts from this one transition. Figure 8.5 displays the $A = 14$ isobars ^{14}C, ^{14}N, and ^{14}O. The ground states of ^{14}C and ^{14}O and the first excited state of ^{14}N form an isospin triplet. The positron decay of interest leads from the ground state of ^{14}O to the first excited state of ^{14}N. The maximum positron energy is 1.81 MeV, the half-life of ^{14}O is 73 sec, and the ft value is 3100 sec (Table 11.1). Two reasons make this decay useful: (1) The transition occurs between members of an isospin multiplet. Apart from electromagnetic corrections, the wave functions of the initial and final state of the decay consequently describe the same hadronic state and thus are identical in their spin and space properties. Matrix elements involving them can be computed accurately. Such transitions are called *superallowed*. (2) Initial and final states have spin-parity $J^\pi = 0^+$. Parity and angular momentum selection rules then severely restrict the matrix elements, as we shall see soon.

The goal is to verify as far as possible Eq. (11.57) for the weak current of hadrons. In nuclear beta decay, hypercharge is always conserved so that

$$J_w^h \text{ (nuclear physics)} = \cos\theta \, J_w^0. \tag{11.59}$$

Denoting the wave functions of the initial and final nuclear states by $\psi_{0^+\alpha}$ and $\psi_{0^+\beta}$ and writing the weak current J_w^0 in the same form as J_w^e, Eq. (11.47), $\mathbf{J}_w^h$ becomes

$$J_w^h(0^+ \longrightarrow 0^+) = c \cos\theta \, \psi_{0^+\beta}^* (V + A) \psi_{0^+\alpha}.$$

With Eqs. (11.25) and (11.47), the matrix element of H_w then becomes

$$\langle \beta | H_w | \alpha \rangle = \frac{1}{\sqrt{2}} G \cos\theta \int d^3x \, \psi_{e^+}^*(V+A)\psi_{\bar\nu_e} \cdot \psi_{0^+\beta}^*(V+A)\psi_{0^+\alpha}.$$

The positron and the neutrino are leptons, and they do not interact hadronically with the nucleus. After emission, they can therefore be described by plane waves, like free particles:

$$\psi_{e^+} = u_e e^{i\mathbf{p}_e \cdot \mathbf{x}/\hbar}, \qquad \psi_{\bar\nu} = u_{\bar\nu} e^{i\mathbf{p}_\nu \cdot \mathbf{x}/\hbar}. \tag{11.60}$$

Here the spin wave functions u_e and $u_{\bar{\nu}}$ are no longer functions of $\mathbf{x}$. (The plane wave for the electron is slightly distorted by the Coulomb field of the nucleus. This distortion results in a small correction that has been discussed in Section 11.2 and is given by the function F introduced there.) The energies of the leptons are less than a few MeV, the reduced wavelengths $\lambdabar = \hbar/p$ are long compared to the nuclear radius, and the lepton wave functions can be replaced by their values at the origin, u_e and $u_{\bar{\nu}}$. The matrix element then becomes

$$\langle \beta | H_w | \alpha \rangle = \frac{1}{\sqrt{2}} G \cos \theta u_e^*(V + A)u_{\bar{\nu}} \cdot \int d^3x \psi_{0^+\beta}^*(V + A)\psi_{0^+\alpha}.$$

$$(11.61)$$

Parity and angular momentum conservation simplify this expression considerably. Consider parity first.[13] Under P, the nuclear wave functions $\psi_{0^+\alpha}$ and $\psi_{0^+\beta}$ remain unchanged. According to Eqs. (11.45) and (11.46), $\mathbf{V}$ and A_0 change sign. Consequently, the corresponding integrands are odd under P and the integrals vanish. To see that the term involving $\mathbf{A}$ also vanishes, we note that the wave functions are scalars under rotation, whereas $\mathbf{A}$ behaves like a vector. The average of a vector over a spherical surface vanishes: Scalars transform like Y_0, vectors like Y_1, and the integral $\int d^3x Y_0^* Y_1 Y_0$ vanishes. The only term left under the integral is V_0, and the matrix element takes on the form

$$\langle \beta | H_w | \alpha \rangle = \frac{1}{\sqrt{2}} G \cos \theta u_e^*(V_0 + A_0)u_{\bar{\nu}}\langle 1 \rangle,   (11.62)$$

where $\langle 1 \rangle$ is the symbol used in nuclear physics for the integral

$$\langle 1 \rangle = \int d^3x \psi_{0^+\beta}^* V_0 \psi_{0^+\alpha}.   (11.63)$$

The recoil energy imparted to the decaying nucleus is very small so that the nuclear matrix element $\langle 1 \rangle$ can be computed nonrelativistically; the result is

$$\langle 1 \rangle = \sqrt{2}.   (11.64)$$

● To verify Eq. (11.64), we use the nonrelativistic operator $V_0 = 1$ from Eq. (11.40) so that

$$\langle 1 \rangle = \int d^3x \psi_{0^+\beta}^* \psi_{0^+\alpha}.$$

A new problem arises here: The wave functions ψ_β and ψ_α belong to different isobars and hence are orthogonal. As written, the integral vanishes. The solution to the problem is simple if the isospin formalism is introduced. The states in ^{14}O and ^{14}N belong to the same $I = 1$ isospin multiplet, with I_3 values of 1 and 0, respectively. They have the same

---

13. At first sight, the parity argument seems inappropriate, because the weak interaction does not conserve parity. However, the parity of the initial and final nuclear states is given by the hadronic interaction, and V and A have well-defined transformation properties under P. The argument is therefore correct.

spatial wave function so that the total wave functions can be written

$$^{14}\text{O}: \quad \psi_\alpha = \psi_0(\mathbf{x})\Phi_{1,1}$$

$$^{14}\text{N}: \quad \psi_\beta = \psi_0(\mathbf{x})\Phi_{1,0}.$$

Here, $\Phi_{1,1}$ and $\Phi_{1,0}$ denote the normalized isospin functions. The weak current changes ^{14}O into ^{14}N; it lowers the I_3 value by one unit. This lowering is expressed by the operator I_-, given in Eq. (8.26). In the isospin formalism the complete matrix element $\langle 1 \rangle$ thus becomes

$$\langle 1 \rangle = \int d^3x \psi_0^*(x)\psi_0(x)\Phi_{1,0}^* I_- \Phi_{1,1}.$$

The isospin part is evaluated with Eq. (8.27):

$$\Phi_{1,0}^* I_- \Phi_{1,1} = \sqrt{2}\,\Phi_{1,0}^* \Phi_{1,0} = \sqrt{2}.$$

The spatial wave function is normalized to 1 so that the final result, $\langle 1 \rangle = \sqrt{2}$, verifies Eq. (11.64). ●

With Eq. (11.64), the square of the matrix element of H_w becomes

$$|\langle \beta | H_w | \alpha \rangle^2 = G^2 \cos^2 \theta \, |u_e^*(V_0 + A_0)u_{\bar{\nu}}|^2.$$

The magnitude of the lepton matrix element can be obtained by assuming spinless nonrelativistic electrons and by first considering only the vector term, proportional to V_0. Equation (11.40) then gives

$$u_e^* V_0 u_{\bar{\nu}} = u_e^* u_{\bar{\nu}} \quad \text{and} \quad |u_e^* V_0 u_{\bar{\nu}}|^2 = u_e^* u_e u_{\bar{\nu}}^* u_{\bar{\nu}}.$$

If the leptons are normalized to one particle per unit volume, Eq. (11.60) gives $u_e^* u_e = u_{\bar{\nu}}^* u_{\bar{\nu}} = 1$. The matrix element of A_0 vanishes nonrelativistically, as is evident from Eq. (11.51) with $p/m \rightarrow 0$. For highly relativistic electrons, $p/mc \rightarrow pc/E \rightarrow 1$, and the matrix element of A_0 approaches that of V_0. There is no interference between A_0 and V_0 in this case so that the square of the lepton matrix element becomes

$$|u_e^*(V_0 + A_0)u_{\bar{\nu}}|^2 = 2. \tag{11.65}$$

The square of the matrix element for a weak $0^+ \rightarrow 0^+$ transition thus is

$$|\langle \beta | H_w | \alpha \rangle|^2 = 2G^2 \cos^2 \theta. \tag{11.66}$$

With Eq. (11.11) and $ft_{1/2} = f\tau \ln 2$, the final result becomes

$$G^2 \cos^2 \theta = \pi^3 \ln 2 \frac{\hbar^7}{m_e^5 c^4} \frac{1}{ft_{1/2}}. \tag{11.67}$$

The ft value of ^{14}O is given in Table 11.1. A number of other $0^+ \rightarrow 0^+$ superallowed transitions have been investigated carefully. Taking into account some small corrections, the value of $G \cos \theta$ becomes[12,14]

$$G_V \cos \theta_V = (1.410 \pm 0.002) \times 10^{-49} \text{ erg-cm}^3. \tag{11.68}$$

The subscripts V on G and θ indicate that the constants have been determined from a decay involving only the vector interaction in the nuclear

14. R. J. Blin-Stoyle and J. M. Freeman, *Nucl. Phys.* **A150**, 369 (1970); S. A. Fayans, *Phys. Letters* **37B**, 155 (1971); C. J. Christensen, A. Nielsen, A. Bahnsen, W. K. Brown, and B. M. Rustad, *Phys. Rev.* **D5**, 1628 (1972).

(hadronic) matrix element. Investigations of decays to which the axial vector interaction contributes, for instance that of the neutron, yield a value for the corresponding coupling constant G_A. The ratio $|G_A/G_V|$ is found to have the value[14]

$$\left|\frac{G_A}{G_V}\right| = 1.24 \pm 0.01. \tag{11.69}$$

In many mystery stories, the essential clues are hidden in aspects that appear, at first sight, completely normal, and the obviously guilty party often turns out to be innocent. We now have G, $\cos\theta$, $G_V \cos\theta_V$, and $|G_A/G_V|$, given in Eqs. (11.54), (11.58), (11.68), and (11.69). Within the given limits of error, the following relations hold:

$$G_V = G$$
$$\theta_V = \theta \tag{11.70}$$
$$G_A \neq G.$$

What do these relations tell us about the weak interaction? At first sight it appears that the equal coupling constants for the vector current (G_V) and for the purely leptonic current (G) simply express the *universality* of the weak interaction and that $G_A \neq G$ requires an explanation. However, the situation is not so straightforward. A proton, for instance, is not just a simple Dirac proton, but it is clothed by a meson cloud (Fig. 6.10). Why should the meson cloud have the same weak interaction as the bare Dirac particle? There is no a priori reason why G_V and G should be identical. The result $G_A \neq G$ appears to be more in agreement with intuitive arguments, and the primary puzzle is the explanation of $G_V = G$. The solution to the puzzle is the *conserved vector current hypothesis* (CVC). It was first proposed in a tentative way by Gershtein and Zeldovich[15] and put into a powerful form by Feynman and Gell-Mann.[4] To explain CVC, consider first the electromagnetic case. In Section 7.2, it was pointed out that the electromagnetic charge is conserved. The positron and the proton have the same electric charge, despite the fact that the proton is surrounded by a meson cloud, whereas the positron is not. In other words, the coupling constant e which characterizes the interaction with the electromagnetic field is the same for particles of the same charge regardless of their other interaction properties. The virtual mesons do not change the value of the coupling constant e. The classical expression for this fact is current conservation, Eq. (10.50). A specific example is given by Kirchhoff's law, shown in Fig. 11.8(a). The CVC hypothesis postulates that the *weak vector* current is also conserved:

$$\frac{1}{c}\frac{\partial V_0}{\partial t} + \mathbf{\nabla}\cdot\mathbf{V} = 0. \tag{11.71}$$

15. S. S. Gershtein and Y. B. Zeldovich, *Zh. Eksperim. i Teor. Fiz.* **29**, 698 (1955); (Tr.) *Soviet Phys. JETP* **2**, 576 (1957).

The equality of the coupling constants G_V and G then follows: Whenever a hadron virtually decomposes into another set of hadrons (for instance, a proton into a neutron and a negative pion), the weak current is conserved. The weak interactions of the bare hadron and of the hadron plus cloud are the same. The equality of G_V and G is not the only evidence for CVC; many additional experiments support Eq. (11.71).[16]

The hypothesis of the conservation of the *vector* current is based on the analogy to the electromagnetic current, which is also a vector current. No electromagnetic *axial* vector current exists, and it is thus not possible to refer to a well-known theory for guidance. Indeed, $G_A \neq G$ shows that the axial vector current is not conserved. The fact, however, that G_A does not differ from G by more than 25% shows that the axial current is almost conserved. The detailed description of this fact is called the PCAC hypothesis or the *partially conserved axial vector current hypothesis*.

Finally, we note that the agreement between G_V and G implies that the hadronic hypercharge-conserving current is characterized by a coupling constant $G \cos \theta$ and not simply G. Without the factor $\cos \theta$, theory and experiment would disagree by a few percent, well outside the limits of error. The Cabibbo angle is not only needed to explain the slowness of the rate of hypercharge-changing decays but also to get agreement between the decay rates for the muon and the superallowed nuclear transitions.

11.10 The Weak Current of Hadrons at High Energies

High energies are crucial for the exploration of two aspects of the weak interaction: (1) Nucleons and nuclei carry weak charges and weak currents. To investigate their distributions (weak form factors), weakly interacting probes with wavelengths smaller than the dimensions of interest are required. The problem is similar to the study of the electromagnetic structure of subatomic particles discussed in Chapter 6. If the weak form factors have about the same behavior as the electromagnetic ones, then the discussion in Chapter 6 shows that weak probes with energies of the order of a few GeV are needed. (2) The range of the weak interaction is not yet known. If, for the sake of argument, it is assumed that the weak interaction is mediated by a W, then the range is given by $\hbar/m_w c$. The energy needed to study the form of the interaction thus depends on m_w, and it can be extremely high. The first aspect, the investigation of the weak structure of hadrons, is within the energy range available now. The second aspect, the study of the range of the weak interaction, may well be out of reach for many years.

In the electromagnetic case, structure investigations use charged leptons (electrons or muons) and photons. How can corresponding high-

16. C. S. Wu, *Rev. Modern Phys.* **36**, 618 (1964).

energy studies of weak effects be performed? The electron possesses a weak interaction. Can it be used as a probe? The answer is, unfortunately, no. Electromagnetic effects would swamp any weak effect, at least at energies that are presently available. Can weak decays be used? Again, the answer is no. The energies of weak decays never extend into the GeV region. The neutrino is the obvious choice. It is in many ways a beautiful probe, but neutrino cross sections are very small. Large fluxes and large counters are needed to obtain results. In the following, we shall discuss some theoretical aspects of neutrino scattering, sketch an experiment, and present some results.

Consider an antineutrino beam incident on a hydrogen target. The antineutrinos can undergo capture as described in Eq. (11.29):

$$\bar{\nu}_e p \longrightarrow e^+ n, \qquad \bar{\nu}_\mu p \longrightarrow \mu^+ n,$$

or, written more generally,

$$\bar{\nu} p \longrightarrow l^+ n, \tag{11.72}$$

where l^+ is a positive lepton. The transition rate for this semileptonic process is given by the golden rule,

$$dw = \frac{2\pi}{\hbar} |\langle nl^+ | H_w | p\bar{\nu} \rangle|^2 \rho(E).$$

The transition rate gives the number of particles scattered per unit time by one scattering center. Equation (6.5) then shows that cross section and transition rate are connected by

$$d\sigma = \frac{dw}{F}. \tag{11.73}$$

Antineutrinos move with the velocity of light; with the normalization of one particle per unit volume, the flux F is equal to the velocity of light, $F = c$. Consequently, the cross section becomes

$$d\sigma = \frac{2\pi}{\hbar c} |\langle nl^+ | H_w | p\bar{\nu} \rangle|^2 \rho(E). \tag{11.74}$$

The density-of-states factor for two particles in the final state, in their c.m., is given by Eq. (10.31). With $V = 1$, $\rho(E)$ is given by

$$\rho(E) = \frac{E_n E_l p_l}{(2\pi\hbar)^3 c^2 (E_n + E_l)} d\Omega_l,$$

where $d\Omega_l$ is the solid-angle element into which the lepton is scattered. The differential cross section for antineutrino capture in the c.m. becomes

$$d\sigma_{\text{cm}}(\bar{\nu}p \longrightarrow ln) = \frac{1}{4\pi^2\hbar^4 c^3} \frac{E_n E_l p_l}{E_n + E_l} |\langle nl | H_w | p\bar{\nu} \rangle|^2 d\Omega_l. \tag{11.75}$$

We shall first apply this expression to the *capture of low-energy electron antineutrinos*. This application will not give the promised information about the weak structure of hadrons, but it will show that Eq. (11.75) predicts the correct cross section. We have pointed out earlier that the magnitude of the matrix element $\langle ne^+ | H_w | p\bar{\nu} \rangle$ is presumed to be the same as that for the neutron decay, $\langle pe^-\bar{\nu} | H_w | n \rangle$. The neutron decay matrix element is connected to the neutron $f\tau$ value by Eq. (11.11). Integrating Eq. (11.75) over $d\Omega_l$, inserting Eq. (11.11) into Eq. (11.75), and noting that for low electron energies $E_n \approx m_n c^2$, $E_e \ll m_n c^2$, give

$$\sigma(\bar{\nu}_e p \longrightarrow e^+ n) = \frac{2\pi^2 \hbar^3}{m_e^5 c^7} \frac{p_e E_e}{(f\tau)_{\text{neutron}}}. \tag{11.76}$$

With the numerical values of the constants and the observed $f\tau$ (Table 11.1) and using convenient energy and momentum units, the cross section is

$$\sigma \,(\text{cm}^2) = 2.3 \times 10^{-44} \frac{p_e}{m_e c} \frac{E_e}{m_e c^2}.$$

At the antineutrino energies occurring at a reactor, the recoil energy of the neutron in the reaction $\bar{\nu} p \longrightarrow e^+ n$ can be neglected, and the total energy of the positron is connected to the antineutrino energy by $E_{e^+} = E_{\bar{\nu}} + (m_p - m_n)c^2 = E_{\bar{\nu}} - 1.293$ MeV. For an antineutrino energy of 2.5 MeV, the cross section becomes 12×10^{-44} cm^2.

Antineutrino capture was first observed by Reines, Cowan, and co-workers in 1956.[17] They set up a large and well-shielded liquid scintillation counter near a reactor. A reactor emits an intense stream of antineutrinos, in the Los Alamos experiment about $10^{13} \bar{\nu}/$cm^2-sec. A few of these are captured in the liquid and give rise to a neutron and a positron. These produce a characteristic signal, and the Los Alamos group was able to determine the cross section as

$$\sigma_{\text{exp}} = (11 \pm 4) \times 10^{-44} \text{ cm}^2.$$

To compare this number to the one expected from Eq. (11.76), the antineutrino spectrum must be known. It can be deduced from the beta spectrum of the fission fragments of ^{238}U,[18] and a cross section of about 10×10^{-44} cm^2 is computed, in good agreement with the actually observed value. The agreement is reassuring; it indicates that the low-energy features of the weak interaction theory are capable of describing neutrino reactions.

We next turn to *neutrino reactions at high energies*. The feasibility of such experiments was pointed out by Pontecorvo and by Schwartz.[19] The theoretical possibilities were first explored by Lee and Yang.[20] As so often in physics, the basic idea is simple, and it is sketched in Fig. 11.9:

17. F. Reines and C. L. Cowan, *Science* **124**, 103 (1956); *Phys. Rev.* **113**, 273 (1959); F. Reines, C. L. Cowan, F. B. Harrison, A. D. McGuire, and H. W. Kruse, *Phys. Rev.* **117**, 159 (1960).

18. R. E. Carter, F. Reines, J. J. Wagner, and M. E. Wyman, *Phys. Rev.* **113**, 280 (1959).

19. B. Pontecorvo, *Soviet Phys.* JETP **37**, 1751 (1959); M. Schwartz, *Phys. Rev. Letters* **4**, 306 (1960). For a fascinating personal account, see B. Maglich, ed., *Adventures in Experimental Physics*, Vol. α, World Science Communications, Princeton, N.J., 1972, p. 82.

20. T. D. Lee and C. N. Yang, *Phys. Rev. Letters* **4**, 307 (1960); *Phys. Rev.* **126**, 2239 (1962).

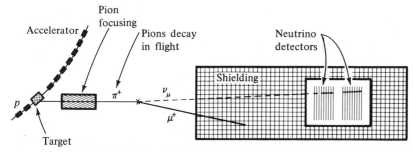

Fig. 11.9. Principle of high-energy muon neutrino experiments.

Protons from a high-energy accelerator strike a target and produce high-energy pions. Pions of one charge, for instance, π^+, are selected and focused into the desired direction. If no material is placed in their path, they decay in flight and create positive muons and muon neutrinos. In the c.m. of the pion, muon and neutrino are emitted with opposite momenta. Because of the large momentum of the decaying pion, in the laboratory, most of the decay products move forward in a small cone. The detector is placed in the resulting beam but is so well shielded that only the neutrinos can reach it. At first sight, neutrino experiments at high-energy accelerators appear to be hopeless because the neutrino flux is much smaller than at reactors. However, Eq. (11.75) predicts a rapid increase of the cross section with energy if the matrix element does not decrease too rapidly with energy. Lee and Yang used the conserved vector current hypothesis of Gell-Mann and Feynman to compute the expected cross sections; their result is shown in Fig. 11.10. The cross section increases very steeply up to laboratory neutrino energies of about 1 GeV and then levels off. The maximum cross section is of the order of 10^{-38} cm², about five orders of

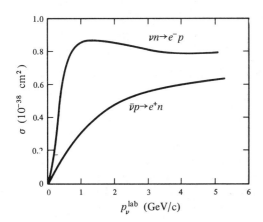

Fig. 11.10. Cross section for the reaction $vn \longrightarrow l^- p$, as predicted by T. D. Lee and C. N. Yang, *Phys. Rev. Letters*, **4**, 307 (1960).

magnitude larger than the one observed in the Los Alamos neutrino experiment. The larger cross section made it possible for the Columbia group to perform the memorable experiment that revealed the existence of two kinds of neutrinos (Section 7.4). In the present chapter we take the existence of muon neutrinos for granted and turn our attention to the behavior of the matrix element of H_w at high energies. We shall evaluate the cross section for the reaction $\nu_\mu N \rightarrow \mu^- N'$, where N and N' are spinless hypothetical nucleons. We shall discuss the modifications required to describe real nucleons later. The cross section for this reaction is given by Eq. (11.75) with small changes in notation. At high energies, the lepton mass can be neglected and E_μ can be replaced by $p_\mu c$. Equation (11.75) then reads

$$d\sigma_{\text{c.m.}}(\nu_\mu N \longrightarrow \mu^- N') = \frac{1}{4\pi^2 h^4 c^2} \frac{E}{W} p_\mu^2 |\langle \mu^- N' | H_w | \nu N \rangle|^2 \, d\Omega.$$

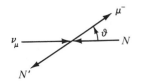

Fig. 11.11. "Elastic" reaction $\nu_\mu N \longrightarrow \mu^- N'$ in the c.m.

Here E is the energy of N' and W the total energy in the c.m. The reaction $\nu_\mu N \rightarrow \mu^- N'$ is shown in Fig. 11.11. In the c.m., all momenta have the same magnitude so that the square of the momentum transfer becomes

$$-q^2 = (\mathbf{p}_\nu - \mathbf{p}_\mu)^2 = 2p_\mu^2 (1 - \cos \vartheta), \tag{11.77}$$

where ϑ is the c.m. scattering angle. With Eq. (11.77), the solid-angle element $d\Omega = 2\pi \sin \vartheta \, d\vartheta$ can be written as

$$d\Omega = -\frac{\pi}{p_\mu^2} \, dq^2, \tag{11.78}$$

so that

$$d\sigma = \frac{-1}{4\pi h^4 c^2} \frac{E}{W} |\langle \mu^- N' | H_w | \nu N \rangle|^2 \, dq^2. \tag{11.79}$$

The central problem is now the matrix element. At low energies, where the structure of the particles can be neglected, we have already considered weak $0^+ \rightarrow 0^+$ transitions: The matrix element is given by Eq. (11.66), and the differential cross section in this case is

$$d\sigma = -\frac{G^2 \cos^2 \theta}{2\pi h^4 c^2} \frac{E}{W} \, dq^2. \tag{11.80}$$

The total cross section is obtained by integrating over dq^2. The minimum momentum transfer is $-4p_\mu^2$, the maximum as given by Eq. (11.77) is 0, and the integration from 0 to $-4p_\mu^2$ yields

$$\sigma_{\text{tot}} = \frac{2G^2 \cos^2 \theta}{\pi h^4 c^2} \frac{E}{W} p_\mu^2. \tag{11.81}$$

At high energies, $E \approx W/2$, and the total cross section increases as p_μ^2. It is physically unreasonable to expect that a cross section could exceed all bounds, and there must be some feature that *damps* the cross section at high energies. In the present example, the damping is provided by the structure of the hadron. We have so far assumed that it is a point particle. Equation (11.80) thus is analogous to the Rutherford cross section, Eq.

(6.17), or to the Mott cross section, Eq. (6.19), both also computed for point particles. In Chapter 6 the modification introduced by a finite particle size has been derived for a spinless particle. The particle structure is expressed through a form factor, and the cross section is multiplied by the absolute square of $F(q^2)$, as shown in Eq. (6.20). Similarly, the cross section for neutrino capture is multiplied by a *weak form factor*, and Eq. (11.80) is modified to read

$$d\sigma = -\frac{G^2 \cos^2 \theta}{2\pi\hbar^4 c^2} \frac{E}{W} |F_w(q^2)|^2 \, dq^2. \tag{11.82}$$

In principle, the weak form factor F_w can be determined from experiment. However, since such experiments are very difficult, theoretical guidance is welcome, and such guidance does exist. In Section 11.9, the conserved vector current hypothesis was discussed, and, by Eq. (11.71), conservation of the weak current is postulated. However, CVC goes a lot further than Eq. (11.71): Feynman and Gell-Mann postulated that the vector form factors appearing in the electromagnetic and in the weak currents must have the same form. For our simplified example CVC states that for the vector interaction

$$F_w(q^2) = F_{em}(q^2). \tag{11.83}$$

No spinless nucleons exist, and the form factor F_{em} for our specific example cannot be determined. However, we can assume that F_{em} has the same form as the form factors that appear in the nucleon structure. In particular we can identify F_{em} with G_D as given in Eq. (6.47): The weak cross section then becomes, with Eq. (11.83),

$$d\sigma = \frac{G^2 \cos^2 \theta}{2\pi\hbar^4 c^2} \frac{E}{W} \frac{dq^2}{(1 + |q^2|/q_0^2)^4}. \tag{11.84}$$

The total cross section is obtained by integration from 0 to $-4p_\mu^2$,

$$\sigma = \frac{G^2 \cos^2 \theta E q_0^2}{6\pi\hbar^4 c^2 W} \left\{ 1 - \frac{1}{(1 + 4p_\mu^2/q_0^2)^3} \right\}. \tag{11.85}$$

This expression displays the essential features of the theoretical cross sections shown in Fig. 11.10: At low energies, the term in the braces can be expanded, the result is identical to Eq. (11.81), and the cross section increases as p_μ^2. At very high energies, the term in the braces becomes unity, and the cross section is a constant.

The cross section given in Eq. (11.85) has been derived under unrealistic assumptions. Since it applies to a $0^+ \rightarrow 0^+$ transition, only the operator V and only one form factor enter. Real nucleons have spin $\frac{1}{2}$, and more than one form factor is needed to describe the cross section. It is assumed that the particle structure does not change the transformation character of the weak current. Since the weak current for point particles contains two operators, V and A, two types of form factors are introduced, vector and axial vector ones. Neutrino experiments at present do not yield sufficient information to allow an unambiguous determination of all form factors.

A semitheoretical approach is therefore chosen: The vector form factors are assumed to be identical to the electromagnetic form factors G_E and G_M that occur in the Rosenbluth formula. It is further assumed that one axial form factor dominates and that it has the same *form* as G_D, Eq. (6.47). Thus only one free parameter is left, $q_0^2 \equiv M_A^2 c^2$. Figure 11.12 presents

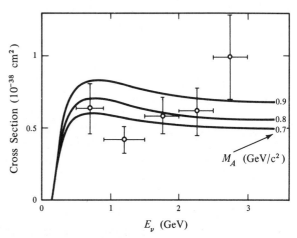

Fig. 11.12. Cross section for the reaction $\nu_\mu n \longrightarrow \mu^- p$, measured at CERN with a bubble chamber. Three theoretical fits are shown. [After I. Budagov et al., *Nuovo Cimento Lettere* **2**, 689 (1969).]

CERN data for the elastic scattering $\nu_\mu n \longrightarrow \mu^- p$. The theoretical curves are cross sections computed with three form factors, G_E, G_M, and G_A. G_E and G_M are given in Eq. (6.48) and G_A by Eq. (6.47), with $q_0^2 \equiv M_A^2 c^2$ and M_A as indicated in Fig. 11.12. The data show that the experimental results are compatible with these form factors and with an *axial mass* M_A that is not very different from the *vector mass* $M_V \equiv q_0/c = 0.71$ GeV/c^2.

So far the discussion of neutrino reactions has been restricted to the *elastic* case, where only one lepton and one hadron are present in the final state. Neutrinos can, however, produce a multitude of inelastic reactions, for instance,

$$\nu_\mu p \longrightarrow \mu^- \pi^+ p.$$

Consequently, the total cross section for neutrino scattering is bigger than the one discussed so far: Figure 11.13 indicates that the total cross section for both neutrino and antineutrino scatterings from protons increases approximately linearly with the neutrino laboratory energy up to the highest available energy, whereas the elastic cross section remains constant above about 1 GeV. How far will this linear increase continue?

It is likely that neutrino reactions will yield greatly improved data in the future because the accelerators will provide more intense and more energetic neutrino beams and because the detection systems will be "bigger

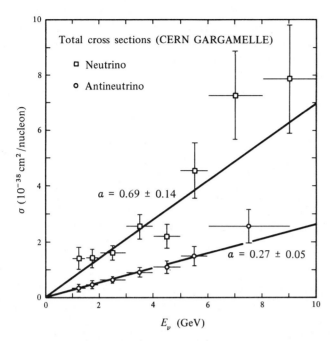

Fig. 11.13. *Weak Interaction:* Total cross sections for the scattering of neutrinos and antineutrinos from nucleons as a function of neutrino and antineutrino laboratory energy. Data are extracted from neutrino events in freon and propane bubble chambers. The lines are simple fits to $\sigma = \alpha E$, with α given in units of 10^{-38} cm²/GeV. (From D. H. Perkins, *XVI Intrn. Conf. on High Energy Physics*, Batavia, 1972).

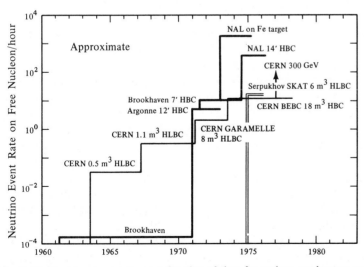

Fig. 11.14. Neutrino event rates as a function of time for various accelerators and detectors. [After C. H. Llewellyn Smith, *Phys. Rept.* **3C**, 261 (1972), with some updating.]

and better." Figure 11.14 presents a prediction first made by C. A. Ramm
in 1968,[21] and updated since then by C. H. Llewellyn-Smith.[21] It
shows the neutrino event rate extraplotated until 1980. It is interesting to
note that the most important discovery, the existence of two kinds of neutrinos, was made in 1961 when the rate was smallest. Hopefully, the much
higher neutrino energies available at NAL will lead to new discoveries of
similar significance.

11.11 Summary and Open Questions

In the present chapter we have shown that many weak phenomena can
be explained by one interaction, usually called the *universal Fermi interaction*. The universal weak Hamiltonian is given by Eq. (11.25) as

$$H_w = \frac{G}{\sqrt{2}\,c^2} \int d^3x\, J_w(\mathbf{x}) \cdot J_w^\dagger(\mathbf{x}),$$

where G is the weak coupling constant, given by Eq. (11.54). The weak
current consists of four parts, as shown in Eqs. (11.35) and (11.57):

$$J_w = J_w^e + J_w^\mu + \cos\theta J_w^0 + \sin\theta J_w^1.$$

The first two terms refer to the leptons: The electron and the muon appear
on equal footing, and this fact is called the electron-muon universality.
The hadron current branches into a $\Delta Y = 0$ and a $|\Delta Y| = 1$ stream, and
the Cabibbo angle θ, given by Eq. (11.58), determines the strength in each
branch. Each weak current is of the form of Eq. (11.47),

$$J_w = c\psi_{\text{final}}^*(V + A)\psi_{\text{initial}};$$

each contains a vector and an axial vector contribution. For leptons, at
energies available today, the form of the operators V and A is fully known.
The form of the hadron current is much less well understood, but the
following facts seem established: The vector current is conserved, and the
low-energy semileptonic decays involving only the vector interaction are
governed by the same coupling constant as the leptonic ones. At high energies, form factors appear, and these are expected to be proportional to
the electromagnetic form factors. No such guidance and proportionality
exists for the axial vector current since no axial electromagnetic current
occurs in nature. Nevertheless, the axial vector current seems to be partially conserved, and high-energy neutrino scattering experiments are compatible with the assumption that the axial form factor does not differ
greatly from the vector one.

Computations with the universal weak interaction have been performed
for many decays and reactions, and, wherever the difficulties introduced by

21. C. A. Ramm, *Nature* **217**, 913 (1968); C. H. Llewellyn-Smith, *Phys. Rept.* **3C**, 261 (1972).

the hadronic force can be handled, agreement with experiment is satis-factory. However, there are unsolved problems. We shall sketch some of these in order to show that much experimental and theoretical work remains to be done.

The Intermediate Boson (W). Despite many attempts, the W has not yet been seen. As long as the elusive particle is not found, theorists have a good time predicting its mass. We have, for instance, pointed out that the conserved vector current hypothesis implies equality of the electromagnetic and the weak vector form factors. This idea can be expanded by assuming that the weak and the electromagnetic interaction have the same origin. In particular it can be assumed that the weak and the electromagnetic coupling constants are of the same order of magnitude. With $g = e$ and Eq. (11.55), the rest energy of the W becomes $m_W c^2 \approx 100$ GeV. A more sophisticated approach gives the relation $g = e/2\sqrt{2}$, which leads to $m_W c^2 = 37.29$ GeV;[22] a different approach leads to a mass twice that value.[23] Only experiment can decide if such speculations are correct. If W is found and if its properties can be elucidated, then weak interaction theory will take on a more definite form. If it is not found, the search for alternative ways of explaining the propagation of the weak force becomes urgent.

The Form of the Weak Interaction. Is the current-current interaction the correct one? The Hamiltonian as described above explains many features of the weak interaction. A number of crucial predictions, however, have not yet been tested fully. Two, in particular, remain to be explored more deeply, namely diagonal (self-interaction) terms and neutral currents. Consider first the *diagonal terms*. The current-current interaction contains terms of the form $J_w^l \cdot J_w^{l\dagger}$ and $J_w^h \cdot J_w^{h\dagger}$. The former terms lead to processes of the type

$$\nu_e e^- \longrightarrow \nu_e e^-$$
$$\bar{\nu}_e e^- \longrightarrow \bar{\nu}_e e^-,$$

and the latter terms produce, for instance, a weak interaction between nucleons. Do such processes exist? Consider first neutrino-electron scatter-ing. Preliminary evidence for such processes has been obtained from the observation of bright white dwarf stars[24] Experiments with reactor[25]

22. T. D. Lee, *Phys. Rev. Letters* **26**, 801 (1971); J. Schechter and Y. Ueda, *Phys. Rev. D2*, 736 (1970).

23. S. Weinberg, *Phys. Rev. Letters* **19**, 1264 (1967) and *Phy. Rev. Letters*, **27**, 1688 (1971).

24. R. B. Stothers, *Phys. Rev. Letters* **24**, 538 (1970).

25. F. Reines and H. S. Gurr, *Phys. Rev. Letters* **24**, 1448 (1970); H. S. Gurr, F. Reines, and H. W. Sobel, *Phys. Rev. Letters* **28**, 1406 (1972); H. H. Chen, *Phys. Rev. Letters* **25**, 768 (1970).

and accelerator[26] neutrinos have not yet reached the necessary detection efficiencies and neutrino fluxes, but it is likely that the neutrino reactions will be seen within the next few years if the cross sections are as predicted by the universal Fermi interaction. The weak interaction between hadrons, described by $J_w^h \cdot J_w^{h\dagger}$, appears at first sight to be impossible to detect because the hadronic interaction swamps the weak one. However, the weak interaction has one characteristic feature not shared by the hadronic one: It does not conserve parity. The discovery of a parity-violating component of the force between two nucleons consequently amounts to an indirect, but convincing, proof for the existence of a weak force between nucleons. Such a component has been found in many experiments.[27] Unfortunately, these involve nuclear decays, and it is therefore difficult to make unambiguous comparisons between the experimental data and the theoretical predictions. Nevertheless, the existence of diagonal terms in the hadronic component of the weak current is by now well established.

The weak interaction in the form described so far does not contain *neutral currents;* the weak currents J_w conserve the lepton and muon number, but the electric charge changes by one unit. Examples are currents $v_e \rightarrow e^-$, $n \rightarrow p$, $e^+ \rightarrow \bar{v}_e$. As Fig. 11.4 shows, the exchanged W is then always charged. In principle, neutral weak currents can also exist; they would conserve the electric charge, and the corresponding intermediate boson would be neutral. Such currents would give rise to weak processes of the type

$$v_\mu e^- \rightarrow v_\mu e^- \qquad\qquad vp \rightarrow vp$$
$$K^0 \rightarrow e^+e^- \qquad\qquad K^+ \rightarrow \pi^+e^+e^-$$
$$\rightarrow \mu^+\mu^- \qquad\qquad \rightarrow \pi^+v\bar{v}.$$

These decays and reactions are allowed by all selection rules, but they can proceed (at least in first order) only if neutral weak currents exist. None of the processes has been observed so far at a rate compatible with sizable neutral weak currents. However, neutrino reactions at CERN and NAL have produced tentative evidence for neutral currents. If these findings are substantiated, then the weak Hamiltonian must be correspondingly changed.

Very High Energies. Scattering theory teaches that conservation of probability limits the maximum cross section in any given angular momentum state. For a point interaction, the scattering occurs with zero orbital angular momentum (*s* waves), and the so-called *unitarity limit* is

26. H. J. Steiner, *Phys. Rev. Letters* **24**, 746 (1970); D. C. Cundy, G. Myatt, F. A. Nezbrick, J. B. M. Pattison, D. H. Perkins, C. A. Ramm, W. Venus, and H. W. Wachsmuth, *Phys. Letters* **31B**, 478 (1970); C. H. Albright, *Phys. Rev.* **D2**, 1330 (1970).

27. L. B. Okun, *Comments Nucl. Particle Phys*, **1**, 181 (1967); E. M. Henley, *Comments Nucl. Particle Phys.* **4**, 206 (1970). For details and additional literature, see E. M. Henley, *Ann. Rev. Nucl. Sci.* **19**, 367 (1969); E. Fischbach and D. Tadić, *Phys. Rept.* **6C**, 123 (1973); M. Gari, *Phys. Rept.* **6C**, 317 (1973).

given by[28]

$$\sigma_{max} = \frac{4\pi\hbar^2}{p^2}. \tag{11.86}$$

Now consider neutrino-electron scattering. Neglecting spin, the c.m. cross section for neutrino-electron scattering at very high energies is given by Eq. (11.81) as

$$\sigma_{tot} = \frac{G^2 \cos^2 \theta}{\pi\hbar^4 c^2} p^2.$$

This relation is valid for spinless point particles and a zero-range nuclear force (contact interaction). For such a situation, scattering can occur only in s waves. The cross section increases with p^2, and it becomes equal to the unitarity limit, Eq. (11.86), at an energy given by

$$E^4_{crit} = (pc)^4 = \frac{4\pi^2(\hbar c)^6}{G^2}$$

or

$$E_{crit} \approx 600 \text{ GeV}.$$

More careful calculations give a critical energy of 300 GeV. While such a c.m. energy is far beyond present experiments, the basic problem is nevertheless present: What prevents the weak interaction of leptons from exceeding the unitarity limit? One possible, but not necessarily sufficient damping mechanism is the intermediate boson. If it exists, then the interaction will have very short, but not zero, range. Further experimental and theoretical work is needed to clear up this problem.

CP and *T* **Violation.** In Chapter 9, the operations *C*, *P*, and *T* were discussed. It turned out that *C* and *P* are violated in the weak interaction. It is straightforward to see that H_w, as described in the present chapter, violates *P* and *C*. However, the combined operation *CP* and time reversal, *T*, are conserved by H_w. As shown in Section 9.8, experiments with neutral kaons have indicated that *CP* and *T* are violated in the decay of the K^0. The violation is small ($\sim 10^{-3}$) but well established. At present it is not known what causes it and whether it is unique to the system of neutral kaons. Some clues point to a new and otherwise unobserved superweak interaction as the culprit.[29] Such an interaction can be weaker than the ordinary weak one by a factor of about 10^8 and still give rise to the observed *CP* violation.

There Is No Theory of Weak Interactions. Purists claim that the theory of weak interactions does not exist. True, the current-current interaction as presently used gives good agreement with many experiments and de-

28. Merzbacher, Eq. (11.62).

29. L. Wolfenstein, *Phys. Rev. Letters* **13**, 562 (1964); T. D. Lee and L. Wolfenstein, *Phys. Rev.* **138B**, 1490 (1965).

scribes a large body of data well. However, there is a major problem: All calculations are performed in lowest-order perturbation theory. Computations of higher-order weak processes lead to meaningless infinities. Experimentally, however, it is known that the rate of higher-order weak processes is extremely small. Consequently, the theory in the present form is not satisfactory. Various remedies have been proposed to cure the infinities and make the theory renormalizable. The most exciting of the cures achieves the goal by combining the weak and the electromagnetic interaction[23,30,31]. If correct, such a unification would rank with Maxwell's theory which unified the electric and the magnetic forces. At the present time, a large number of different models exist.[22,23,32] In all of these, heavy bosons (W) are assumed to exist. Moreover, the theories demand the existence of heavy leptons (particles with the quantum numbers of the electron or muon) and/or the presence of neutral currents. Verification or rejection of these theories will require more experimental data. The most direct evidence for a particular model would be the discovery of all the particles predicted by it. Since the masses can be very high, it may take considerable time before such a proof is feasible. Further studies of neutral currents, already discussed above, may also rule for or against some models. Finally it turns out that the renormalizable theories differ from the standard theory in the cross sections for the diagonal terms. Consequently there is hope that the neutrino-electron scattering experiments currently under way will provide some decision concerning the future of weak interactions.

11.12 References

Weak interactions have been extensively reviewed. We list here only a few books and articles. These contain extensive bibliographies so that further references can be traced without too much work. Very few articles cover the entire field of weak interactions; the emphasis is usually either on nuclear beta decay or on the weak interaction of elementary particles. We first list books and articles that stress nuclear beta decay.

C. S. Wu and S. A. Moszkowski in *Beta Decay*, Wiley-Interscience, New York, 1966, provide a balanced and lucid treatment of theory and experiment, which covers nuclear beta decay in depth but also contains discussions of particle decays. The level is not much above the level of this book, and it is recommended as the easiest place to find more information.

The next two books are more advanced: E. J. Konopinski, *The Theory*

30. A. Salam, *Elementary Particle Physics* (N. Svartholm, ed.), Almqvist and Wiksell, Stockholm, 1968. p. 367.

31. G.'t Hooft, *Nucl. Phys.* **B33**, 173 (1971), **B35**, 167 (1971).

32. T. D. Lee, *Phys. Today* **25**, 23 (April 1972). T. D. Lee and G. C. Wick, *Phys. Rev.* **D2**, 1033 (1970). H. Georgi and S. L. Glashow, *Phys. Rev.* **D6**, 2977 (1972). B. W. Lee, *Phys. Rev.* **D6**, 1188 (1972). J. Prentki and B. Zumino, *Nucl. Phys.* **B47**, 99 (1972).

of Beta Decay, Oxford University Press, London, 1966; H. F. Schopper, *Weak Interactions and Nuclear Beta Decay*, North-Holland, Amsterdam, 1966. Both books are thorough and complete, and they can be consulted if an answer cannot be found in Wu and Moszkowski.

An easy-to-read introduction to the physical ideas involved in nuclear beta decay, particularly in experiments that test parity nonconservation, is H. J. Lipkin, *Beta Decay for Pedestrian*, North-Holland, Amsterdam, 1962.

The experimental aspects of nuclear beta decay are discussed in Siegbahn and in O. M. Kofoed-Hansen and C. J. Christensen, "Experiments on Beta Decay," in *Encyclopedia of Physics*, Vol. XLI/2, (S. Flügge, ed.), Springer, Berlin, 1962.

Information on the weak interaction of particles can be found in two places: R. E. Marshak, Riazuddin, and C. P. Ryan, *Theory of Weak Interactions in Particle Physics*, Wiley-Interscience, New York, 1969; T. D. Lee and C. S. Wu, "Weak Interactions," *Ann. Rev. Nucl. Sci.* **15**, 381 (1965); **16**, 471 (1966). These two sources are hopelessly above the present level. However, you may find just the right answer or the right reference in one of these, even without understanding the rest.

The neutrino plays a crucial role in nearly all weak interactions, and many fascinating accounts of its history and its properties have been given. References are quoted and commented on in L. M. Lederman, "Resource Letter Neu-1, History of the Neutrino," *Am. J. Phys.* **38**, 129 (1970). Recent experiments are also discussed in A. W. Wolfendale, "Neutrino Physics," in *Essays in Physics* (G. K. T. Conn and G. N. Fowler, eds.), Academic Press, New York, 1970. Some of the unsolved problems in neutrino physics are reviewed in B. Pontecorvo, *Soviet Phys. Usp.* **14**, 235 (1971). Neutrino reactions are treated in detail in C. H. Llewellyn Smith, *Phys. Rept.* **3C**, 261 (1972).

Many of the influential papers on weak interactions, including a translation of Fermi's classic work, are reprinted in P. K. Kabir, *The Development of Weak Interaction Theory*, Gordon Breach, New York, 1963. Original papers, with introductory discussions, are also collected in C. Strachan, *The Theory of Beta Decay*, Pergamon, Elmsford, N. Y., 1969.

The renormalizable theories that unify the weak and the electromagnetic interaction are reviewed in B. W. Lee, *Proceedings of the XVI International Conference on High Energy Physics*, Vol. 4, National Accelerator Laboratory, 1973; E. S. Abers and B. W. Lee, *Phys. Rept.* C (1973).

PROBLEMS

11.1. Verify that the proton recoil energy can be neglected in the discussion of neutron beta decay.

11.2. Plot the phase–space distribution, Eq. (11.4), and check that a typical beta spectrum is well represented by it.

11.3. Discuss how the upper end of the beta spectrum and the Kurie plot are distorted if the neutrino has a finite rest mass.

11.4. Discuss the beta decay of the neutron:
(a) Sketch the measurement of the mean life.
(b) Discuss the measurement of the spectrum.
(c) Use Eqs. (11.9) and (11.10) to compute the value of f for the neutron decay. Assume that $F(-, 1, E) = 1$. Compare the resulting value of ft with the one given in Table 11.1.
(d) In what observables does parity-nonconservation show up in neutron decay? How can it be observed experimentally? Discuss the results of such measurements.

11.5. Beta spectra can be measured in a variety of instruments. Two that are often used are magnetic beta spectrometers and solid-state detectors.
(a) Discuss both methods. Compare momentum resolution and counting statistics for a given source strength.
(b) What are the advantages and disadvantages of either method?

11.6. Assume that the mass difference between the charged and the neutral pions is caused by the electromagnetic interaction. Compare the corresponding energy to the weak energy given in Eq. (11.14).

11.7. Verify the integration leading to Eq. (11.19).

11.8. List three nuclear beta decays, one with a very small, one with an average, and one with a very large ft value. Consider the spin and parities involved and discuss why the variation in ft is not an argument against the universal Fermi interaction.

11.9. Compute the ratio of lifetimes for the decays

$$\Sigma^+ \longrightarrow \Lambda^0 e^+ \nu \qquad \text{and} \qquad \Sigma^- \longrightarrow \Lambda^0 e^- \bar{\nu}.$$

Compare your value to the experimental ratio.

11.10. Verify all entries in Table 11.3.

11.11. Consider the branching (intensity) ratio

$$\frac{\pi \longrightarrow e\nu}{\pi \longrightarrow \mu\nu}.$$

(a) How were the two decay modes observed?
(b) Compute the branching ratio expected if the matrix elements for both decays are assumed to be equal. Compare the result with the experimental ratio.
(c) Discuss the helicities of the charged leptons emitted in the pion decay; assume that neutrinos and antineutrinos are fully polarized, as shown in Fig. 7.2. Sketch the helicities of the e^+ and e^-.
(d) Experiment indicates that the helicity of negative leptons emitted in beta decay is given by $-v/c$, where v is the lepton velocity. Use this fact, together with the result of part (c), to explain the low branching ratio that is experimentally found.

11.12. Why do positive muons in matter usually come to rest before they decay? Describe the processes involved and give approximate values for the characteristic times that enter the considerations. Why do negative muons behave differently?

11.13. Discuss the experimental determination of the polarization of the muon emitted in the decay of the pion.

11.14. Discuss the experimental determination of the electron spectrum in muon decay:
(a) Sketch a typical arrangement.
(b) How thin should the target be (in g/cm²) in order not to affect the spectrum appreciably?
(c) How does one guarantee that the spectrum observed is that of an unpolarized muon source?
(d) How can the spectrum at low electron momenta be found?

11.15. Use the spectrum of Fig. 11.6 to construct an approximate Kurie plot for the muon decay. Show that a simple phase-space spectrum does not fit the observed data.

11.16. List the reactions and decays that are described by the leptonic Hamiltonian, Eq. (11.36).

11.17. Show that the linear momentum and the angular momentum have the same transformation properties under ordinary rotations.

11.18. Verify that the weak current densities, Eqs. (11.41) and (11.42), conserve the baryonic, leptonic, and muonic numbers but that the electric charge number changes.

11.19. Show that the electron spectrum in Fig. 11.6 can be fitted with Eq. (11.52), after proper change of the variable.

11.20.
(a) Determine the value of E_{max} in Eq. (11.53). Assume that $m_e = 0$.
(b) Verify the result of the integration in Eq. (11.53).
(c) Use the value of the muon mean life listed in the Appendix to verify the value of G given in Eq. (11.54).

11.21. Verify Eq. (11.55).

11.22.
(a) What are the properties of W as predicted by the arguments in Sections 11.3 and 11.6?
(b) Discuss experiments that could give information about W.

11.23. Find some examples other than the ones given in Table 11.3 to demonstrate that hypercharge-changing weak decays are systematically slower than the corresponding hypercharge-conserving ones. Use your examples to find a value for the Cabibbo angle.

11.24. Verify that the wave functions of the neutrino and the electron, given in Eq. (11.60), are essentially constant over the nuclear volume.

11.25. Prove in detail that the integral containing $\mathbf{A}$ in Eq. (11.61) vanishes.

11.26. The computation of the lepton matrix element in Eq. (11.65) gives

$$|u_e^*(V_0 + A_0)u_{\bar{v}}|^2 = 2\left(1 + \frac{v}{c}\cos\theta_{ev}\right),$$

where v is the positron velocity and θ_{ev} the angle between positron and neutrino momenta.
(a) How can the positron-neutrino correlation be measured? Discuss the principle of the method and a typical experiment.
(b) Show that the observed positron-neutrino (and electron-antineutrino) correlations are in agreement with a $V - A$ interaction.

11.27. List superallowed $0^+ \longrightarrow 0^+$ transitions and show that their ft values are all closely identical.

11.28. High-energy neutrinos have been observed in bubble chambers (propane and hydrogen) and in spark chambers.
(a) Compare typical count rates.
(b) What are the advantages and the disadvantages of the various detectors?

11.29. Plot a few numerical values of the cross section equation (11.85) as a function of the neutrino momentum
(a) In the c.m.
(b) In the laboratory.
Compare your curve with the ones shown in Figs. 11.10 and 11.12.

11.30. Consider the weak current of hadrons, for instance, in the case $\Lambda^\circ \longrightarrow p$. Such a current satisfies the selection rule

$$\Delta S = \Delta Q,$$

where ΔS is the change of strangeness and ΔQ the change in charge.
(a) Give a few additional currents that have been observed and that satisfy this selection rule.
(b) Have currents with $\Delta S = -\Delta Q$ been observed? (The quantum numbers S and Q always refer to the hadrons.)
(c) Which selection rule does the universal Fermi interaction, as written in Section 11.11, satisfy?

11.31. Discuss the isospin selection rules that are satisfied by the weak interaction
(a) In nonstrange decays, and
(b) In decays involving a change of strangeness.

(c) What experiments can be used to test these selection rules?

11.32. Discuss the evidence for and against the existence of neutral currents.

11.33. Derive the unitarity limit, Eq. (11.86).

11.34. What experiments can be carried out to test the absence of $\Delta S \geq 2$ weak currents?

12

Hadronic Interactions

All good things must come to an end. The last two chapters have shown that the electromagnetic and the weak interactions of leptons are described by unified theories. Difficulties arise in the description of the electromagnetic and weak interactions of hadrons but there is good reason to believe that they can be blamed on the hadronic interaction. Apart from these complications, the two interactions appear to be universal: All particles are governed by the same law, and each interaction is characterized by one coupling constant. The situation with regard to the hadronic interactions is different in three aspects: (1) The hadronic interactions are so strong that perturbation theory works only badly or not at all. If a calculation gives a result that disagrees with observation, it is not always clear whether the underlying assumptions or the computational techniques are to be blamed. (2) The hadronic interaction is very complex. In the nucleon-nucleon interaction, for instance, almost every term allowed by general symmetry principles appears to be required to fit the experimental data. This situation can be contrasted to the other three interactions; nature uses only one term in the electromagnetic and the gravitational interactions; the first is a vector, and the second a tensor force. In the weak interaction, two terms appear, a vector and an axial vector one.

The complexity of the hadronic interaction may be an indication that we are not yet looking at the truly fundamental force but at secondary ones. (3) The hadronic interactions do not appear to be governed by one universal coupling constant. The electromagnetic interaction is characterized by the one constant e, the weak interaction by G, and the gravitational interaction by the universal gravitational constant. In the hadronic interactions, a number of constants turn up. Consider, for instance, Figs. IV. 1 and IV. 3. The strength of the interaction of the pion with the baryon is described by the constant $f_{\pi NN^*}$ in the first case, and by $f_{\pi NN}$ in the second one. The two constants are not identical. The interaction of the pion with pions is again characterized by another constant. Since many hadrons exist, a large number of coupling constants occur. The corresponding interactions are all called hadronic or strong because they all are about two orders of magnitude stronger than the electromagnetic one. However, they are not exactly alike. While some connections among the coupling constants can be derived by using symmetry arguments, these relations have not been tested accurately, and many constants appear at present to be unrelated. The situation resembles a jigsaw puzzle in which it is not known if all pieces are present and in which the shape of some pieces cannot be seen clearly.

Because of the difficulties just described, we cannot give a coherent and unified treatment of the hadronic force. A case-by-case description of all hadronic interactions would be only moderately more interesting than a telephone book. We shall therefore try to bring out the essential features of the hadronic interactions by concentrating on two forces, the pion-nucleon and the nucleon-nucleon ones. The first provides insight into the construction of an interaction from invariance arguments, and the second shows how the exchange of mesons leads to a hadronic force between two nucleons.

12.1 Range and Strength of the Hadronic Interactions

Some features are common to all hadronic interactions, and in this section we shall describe two of the most important ones, range and strength. Historically, much of the information on hadronic forces was gleaned from studying nuclei, and the force between nucleons therefore enters heavily into the discussion here.

Range. The early alpha-particle scattering experiments by Rutherford indicated that the nuclear force must have a range of at most a few fm. In 1933, Wigner pointed out that a comparison of the binding energies of the deuteron, the triton, and the alpha particle leads to the conclusion that nuclear forces must have a range of about 1 fm and be very strong.[1]

1. E. P. Wigner, *Phys. Rev.* **43**, 252 (1933).

The argument goes as follows. The binding energies of the three nuclides are given in Table 12.1. Also listed are the binding energies per particle and per "bond." The increase in binding energy cannot be due only to the increased number of bonds. However, if the force has a very short range, the increase can be explained: The larger number of bonds pulls the nucleons together, and they experience a deeper potential; the binding energies per particle and per bond increase correspondingly.

Table 12.1 Binding Energies of ^{2}H, ^{3}H, and ^{4}He

Nuclide	Number of bonds	Binding energy (MeV)		
		Total	Per particle	Per bond
^{2}H	1	2.2	2.2	2.2
^{3}H	3	8.5	2.8	2.8
^{4}He	6	28	7	4.7

Strength. The strength of a hadronic force is best described by a coupling constant. However, to extract a coupling constant from experimental data, a definite form of the hadronic Hamiltonian must be assumed. We shall do this in later sections. Here we compare the strength of the hadronic forces to that of the electromagnetic and the weak ones by giving the values for the total cross sections. The total absorption cross section for photons on nucleons at high energies is given in Fig. 10.21. The total cross section for scattering of neutrinos from nucleons is shown in Fig. 11.13. Two examples of hadronic processes are shown in Figs. 12.1 and 12.2. Finally, the various cross sections are compared in Fig. 12.3.

The relative strengths of the three interactions can be taken somewhat arbitrarily to be the ratio of cross sections at a few GeV; from Fig. 12.3 it

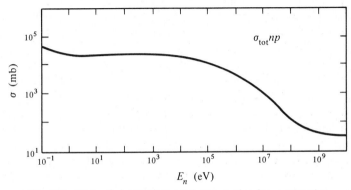

Fig. 12.1. Hadronic interaction: total cross section for *np* scattering.

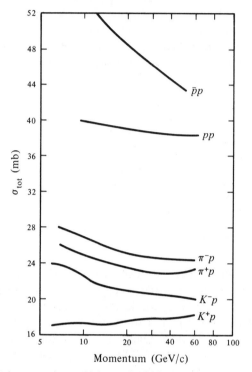

Fig. 12.2. Total cross sections at high energies. [After S. P. Denisor et al., *Phys. Letters* **36B**, 415 (1971).]

then follows that

$$\text{hadronic/electromagnetic/weak} \approx 1/10^{-2}/10^{-12}. \qquad (12.1)$$

Since the coupling constant of the electromagnetic interaction in dimensionless units is of the order of 10^{-2}, as indicated in Eq. (10.77), the corresponding coupling constant for the hadronic force is of the order of unity. Consequently, the perturbation approach is at best of limited use in the theory of hadronic interactions.

The fact that the absolute strength of the hadronic interactions is characterized by a coupling constant of the order of 1 can be seen in a different manner in Fig. 12.3. At the energies where the comparison of the coupling strengths was performed, namely at a few GeV, the hadronic cross section is of the order of the geometrical cross section of the proton, which is about 3 fm². If the proton were *transparent* to the incident hadrons, we would expect the cross section to be much smaller than the geometrical cross section. However, the size of the total cross section, of the order of a few fm², indicates that nearly every incident hadron that comes within

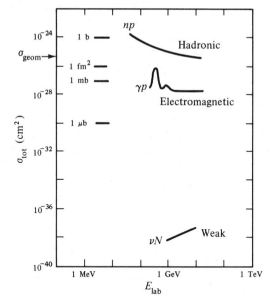

Fig. 12.3. Comparison of the total cross sections for hadronic, electromagnetic, and weak processes on nucleons. σ_{geom} indicates the geometrical cross section of a nucleon.

"reach" of a scattering center suffers an interaction. In this sense, the hadronic interaction is indeed strong. Even if it were much stronger, it could not scatter appreciably more.

12.2 The Pion-Nucleon Interaction—Survey

Since the early days of subatomic physics, explaining the nuclear forces has been one of the main goals. We have already pointed out in Section 5.8 that there was almost complete ignorance as to the nature of the nuclear force before Yukawa postulated the existence of a heavy boson in 1934.[2] Yukawa's revolutionary step did not solve the nuclear force problem completely because no calculation reproduced the experimental data well and because it was not even clear what properties the proposed quantum should have.[3] When the pion was discovered, identified with the Yukawa quantum, and found to be a pseudoscalar isovector particle, some of the uncertainties were removed, but it was still not possible to describe the nuclear force satisfactorily. Today we know that many more quanta exist and must be taken into account. Nevertheless, the pion and

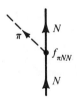

Pion emission

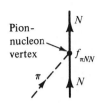

Pion absorption

Fig. 12.4. Pions can be emitted and absorbed singly. The strength of the pion-nucleon interaction is characterized by the coupling constant $f_{\pi NN}$.

2. H. Yukawa, *Proc. Phys. Math. Soc. Japan* **17**, 48 (1935).
3. W. Pauli, *Meson Theory of Nuclear Forces*, Wiley-Interscience, New York, 1946.

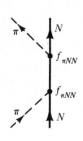

Pion-nucleon
scattering

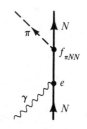

Photoproduction
of pions

Fig. 12.5. Typical diagrams
for pion-nucleon scattering
and for pion photoproduction.

its interaction with nucleons play a special role. First, the pion lives long
enough so that intense pion beams can be prepared and the interaction
of pions with nucleons can be studied in detail. Second, the pion is the
lightest meson; it is more than three times lighter than the next heavier one. In the energy range up to 500 MeV, the pion-nucleon interaction
can be studied without interference from other mesons. Moreover, because
the range of a force, $R = \hbar/mc$, is inversely proportional to the mass of
the quantum, the pion alone is responsible for the long-range part of the
nuclear force. In principle, the properties of the nuclear force beyond a
distance of about 1 fm can be compared with the theoretical predictions
without severe complications from other mesons. Experimentally and
theoretically, then, the pion-nucleon force plays the role of a test case,
and we shall therefore discuss some of the important aspects here.

Pions, being bosons, can be emitted and absorbed singly, as shown in
Fig. 12.4. The actual experimental exploration of the pion-nucleon force
is performed, for instance, through studies of pion-nucleon scattering and
of the photoproduction of pions. Two typical diagrams are shown in Fig.
12.5. In principle, 10 different pion-nucleon scattering processes can be
observed, but only the following three can be readily investigated:

$$\pi^+ p \longrightarrow \pi^+ p \tag{12.2}$$

$$\pi^- p \longrightarrow \pi^- p \tag{12.3}$$

$$\pi^- p \longrightarrow \pi^0 n. \tag{12.4}$$

The total cross sections for the scattering of positive and negative pions
have been displayed in Fig. 5.28. The cross sections for the elastic processes, Eqs. (12.2) and (12.3), and for charge exchange, Eq. (12.4), are

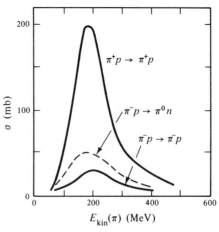

Fig. 12.6. Cross sections for the low-energy elastic and charge-exchange pion-proton
reactions.

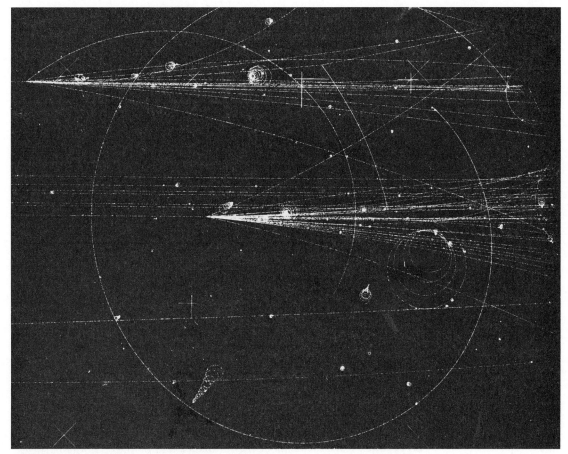

Photo 11. High-multiplicity event, produced by 300 GeV protons. (Courtesy National Accelerator Laboratory). (See p. 342.)

sketched in Fig. 12.6 up to a pion kinetic energy of about 500 MeV.[4]

The best-known photoproduction processes are

$$\gamma p \longrightarrow \pi^0 p \qquad (12.5)$$

$$\gamma p \longrightarrow \pi^+ n. \qquad (12.6)$$

The reaction γn can also be studied by using deuterium targets and subtracting the contribution of the proton. The cross sections for the processes in Eqs. (12.5) and (12.6) are shown in Fig. 12.7. The feature that dominates Eqs. (12.2)–(12.6) is the appearance of a resonance. In pion scattering, it

4. It is no longer meaningful to reproduce detailed cross sections with data points in textbooks because published compilations, available from the Lawrence Radiation Laboratory, Berkeley, and from CERN, Geneva, provide computer printouts that are complete and up-to-date. The pion-nucleon scattering data, for example, are contained in G. Giacomelli, P. Pini, and S. Stagni, "A Compilation of Pion-Nucleon Scattering Data," *CERN/HERA Report 69–1*, 1969.

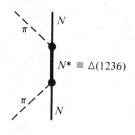

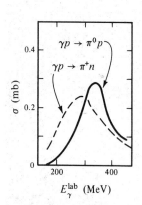

Fig. 12.7. Total cross sections for the photoproduction of neutral and charged pions from hydrogen, as a function of the incident photon energy.

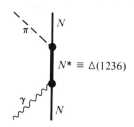

Fig. 12.8. Pion scattering and pion photoproduction at low energies are dominated by the formation of an excited nucleon, N^*, usually called $\Delta(1236)$.

occurs at a pion kinetic energy of about 170 MeV; in photoproduction, the photon energy at the peak is about 300 MeV. Despite this difference in kinetic energies, the peaks in pion scattering and pion photoproduction can be interpreted by one phenomenon, the formation of an excited nucleon state, N^*, as indicated in Fig. 12.8. The mass of this resonance particle is approximately given by $m_{N^*} \approx m_N + m_\pi + E_{\text{kin}}/c^2 = 1250 \text{ MeV}/c^2$ in pion scattering, and by $m_{N^*} \approx m_N + E_\gamma = 1240 \text{ MeV}/c^2$ in photoproduction. Proper computation, taking into account the recoil of the N^*, gives a mass of 1236 MeV/c^2 for both processes, and it is appealing to assume that they represent the same resonance. The discovery of this resonance, called $\Delta(1236)$, was already discussed in Section 5.11. The cross sections in Figs. 12.6 and 12.7 show that the interaction of pions with nucleons at energies below about 500 MeV is dominated by this resonance.

Isospin and spin of $\Delta(1236)$ can be established by simple arguments. Pion ($I = 1$) and nucleon ($I = \frac{1}{2}$) can form states with $I = \frac{1}{2}$ and $I = \frac{3}{2}$. If $\Delta(1236)$ had $I = \frac{1}{2}$, only two charge states of the resonance would occur. According to the Gell-Mann–Nishijima relation, Eq. (8.30), they would have the same electric charges as the nucleons, namely 0 and 1. These two resonances, Δ^0 (1236) and Δ^+ (1236), are indeed observed. In addition, however, the $\Delta^{++}(1236)$ appears in the process $\pi^+ p \longrightarrow \pi^+ p$, and Δ consequently must have $I = \frac{3}{2}$. The fourth member of the isospin multiplet, $\Delta^-(1236)$, cannot be observed with proton targets; deuteron targets permit the investigation of the reaction $\pi^- n \longrightarrow \pi^- n$, where Δ^- shows up. To establish the spin of $\Delta(1236)$, we note that the maximum cross section for the scattering of unpolarized particles is given by[5]

$$\sigma_{\max} = 4\pi\lambdabar^2 \frac{2J + 1}{(2J_\pi + 1)(2J_N + 1)} = 4\pi\lambdabar^2 (J + \tfrac{1}{2}). \qquad (12.7)$$

J, J_π, and J_N are the spins of the resonance and of the colliding particles, and $\lambdabar$ is the reduced pion wavelength at resonance. $4\pi\lambdabar^2$ at 190 MeV is about 100 mb, and $\sigma_{\max}$ is about 200 mb, so that $J + \frac{1}{2} \approx 2$ or $J = \frac{3}{2}$. To form a state with spin $\frac{3}{2}$ in pion-nucleon scattering, the incoming pions must carry one unit of orbital angular momentum. Pion-nucleon scattering at low energies occurs predominantly in p waves.

● The fact that pion-nucleon scattering at low energies occurs predominantly in the state $J = \frac{3}{2}$, $I = \frac{3}{2}$ (the so-called 3-3 resonance) can be verified by a spin-isospin phase-shift analysis. We shall not present the complete analysis here, but we shall outline its isospin part because it provides an example for the use of isospin invariance. We first note that experimental states are prepared with well-defined charges. Theoretically, however, it is more appropriate to use well-defined values of the total isospin. It is therefore necessary to express the experimentally prepared states in terms of eigenstates of I and I_3, denoted by $|I, I_3\rangle$. With the technique used in Problem 13.7, the following relations are established:[6]

5. The maximum cross section for the scattering of spinless particles with zero orbital angular momentum is given by Eq. (11.86). A particle with spin J is $(2J + 1)$-fold-degenerate. By assuming that Eq. (11.86) holds for each substate, Eq. (12.7) follows from Eq. (11.86).
   6. Merzbacher, Section 16.6.

$$|\pi^+ p\rangle = |\tfrac{3}{2}, \tfrac{3}{2}\rangle$$
$$|\pi^- p\rangle = \sqrt{\tfrac{1}{3}}\,|\tfrac{3}{2}, -\tfrac{1}{2}\rangle - \sqrt{\tfrac{2}{3}}\,|\tfrac{1}{2}, -\tfrac{1}{2}\rangle \qquad (12.8)$$
$$|\pi^0 n\rangle = \sqrt{\tfrac{2}{3}}\,|\tfrac{3}{2}, -\tfrac{1}{2}\rangle + \sqrt{\tfrac{1}{3}}\,|\tfrac{1}{2}, -\tfrac{1}{2}\rangle.$$

To describe pion-nucleon scattering, a scattering operator S is introduced. The operator S is not as frightening as it usually appears to the beginner, and all we have to know about it are two properties. (1) The scattering amplitude f for a collision $ab \longrightarrow cd$ is proportional to the matrix element of S,

$$f \propto \langle cd \,|\, S \,|\, ab \rangle.$$

The cross section is related to f by Eq. (6.10), or $d\sigma/d\Omega = |f|^2$. (2) The pion-nucleon force is hadronic and assumed to be charge-independent. Thus the Hamiltonian $H_{\pi N}$ must commute with the isospin operator,

$$[H_{\pi N}, \vec{I}] = 0.$$

Since pion-nucleon scattering occurs through the pion-nucleon force as shown in Fig. 12.5, the scattering operator can be constructed from $H_{\pi N}$. It therefore must also commute with $\vec{I}$,

$$[S, \vec{I}] = 0, \qquad (12.9)$$

and with I^2,

$$[S, I^2] = 0. \qquad (12.10)$$

Thus, if $|I, I_3\rangle$ is an eigenstate of I^2 with eigenvalues $I(I+1)$, so is $S\,|I, I_3\rangle$. Consequently, the state $S\,|I, I_3\rangle$ is orthogonal to the state $|I', I_3'\rangle$, and the matrix element $\langle I', I_3' \,|\, S \,|\, I, I_3\rangle$ vanishes unless $I' = I,\ I_3' = I_3$. Moreover, S does not depend on I_3, as is indicated by Eq. (12.9); the matrix element is independent of I_3 and can simply be written as $\langle I \,|\, S \,|\, I \rangle$. With the abbreviations

$$f_{1/2} = \langle \tfrac{1}{2} \,|\, S \,|\, \tfrac{1}{2} \rangle, \qquad f_{3/2} = \langle \tfrac{3}{2} \,|\, S \,|\, \tfrac{3}{2} \rangle$$

and with Eqs. (12.8), the matrix elements for the elastic and the charge exchange processes become

$$\langle \pi^+ p \,|\, S \,|\, \pi^+ p \rangle = f_{3/2}$$
$$\langle \pi^- p \,|\, S \,|\, \pi^- p \rangle = \tfrac{1}{3} f_{3/2} + \tfrac{2}{3} f_{1/2} \qquad (12.11)$$
$$\langle \pi^0 n \,|\, S \,|\, \pi^- p \rangle = \frac{\sqrt{2}}{3} f_{3/2} - \frac{\sqrt{2}}{3} f_{1/2}.$$

The matrix elements are complex numbers, and three reactions are not sufficient to determine $f_{1/2}$ and $f_{3/2}$. However, if the resonances shown in Fig. 12.6 occur in the $I = \tfrac{3}{2}$ state, then $f_{3/2}$ should dominate at the resonance energy. With $|f_{3/2}| \gg |f_{1/2}|$ and with $\sigma \propto |f|^2$, Eq. (12.11) predicts for the ratios of cross sections at resonance

$$\sigma(\pi^+ p \longrightarrow \pi^+ p)/\sigma(\pi^- p \longrightarrow \pi^- p)/\sigma(\pi^- p \longrightarrow \pi^0 n) = 9/1/2. \qquad (12.12)$$

The agreement of this prediction with experiment provides additional support for the hypothesis of charge independence of the pion-nucleon force. ●

12.3 The Form of the Pion-Nucleon Interaction

In this section, we shall construct a possible form for the Hamiltonian $H_{\pi N}$ at low pion energies by using invariance arguments and the properties of pions and nucleons. The pion is a pseudoscalar boson with isospin 1; consequently, the wave function $\vec{\Phi}$ of the pion is a pseudoscalar in ordinary space but a vector in isospace. The nucleon is a spinor in ordinary space and in isospace. The Hamiltonian $H_{\pi N}$ must be a scalar in ordinary and in isospace. In the nonrelativistic case (static limit), the nucleon recoil is neglected, and the building blocks available for the construction of $H_{\pi N}$ are

$$\vec{\Phi}, \qquad \vec{\tau}, \qquad \boldsymbol{\sigma}. \qquad (12.13)$$

Here, $\vec{\Phi}$ is the pion wave function, $\vec{\tau} = 2\vec{I}$ is related to the nucleon isospin operator, and $\boldsymbol{\sigma} = 2\mathbf{J}/\hbar$ is related to the nucleon spin operator. The Hamiltonian is a scalar in isospace if it is proportional to the scalar product of the two isovectors listed in (12.13),

$$H_{\pi N} \propto \vec{\tau} \cdot \vec{\Phi}.$$

$H_{\pi N}$ is a scalar in ordinary space if it is proportional to the scalar product of two vectors or two axial vectors. The list (12.13) contains only one axial vector, $\boldsymbol{\sigma}$, and a pseudoscalar, $\boldsymbol{\Phi}$. The easiest way to create a second axial vector is to form the gradient of $\vec{\Phi}$ so that

$$H_{\pi N} \propto \boldsymbol{\sigma} \cdot \nabla \vec{\Phi}.$$

Combining the ordinary and isoscalars gives

$$H_{\pi N} = F_{\pi N} \boldsymbol{\sigma} \cdot (\vec{\tau} \cdot \nabla \vec{\Phi}(\mathbf{x})), \qquad (12.14)$$

where $F_{\pi N}$ is a coupling constant. This Hamiltonian describes a point interaction: Pion and nucleon interact only if they are at the same point. To smear out the interaction, a weighting (source) function $\rho(\mathbf{x})$ is introduced; $\rho(\mathbf{x})$ can, for instance, be taken to be the nucleon probability density, $\rho = \psi^*\psi$. The function $\rho(\mathbf{x})$ falls rapidly to zero beyond about 1 fm and is normalized so that

$$\int d^3x \, \rho(\mathbf{x}) = 1. \qquad (12.15)$$

The Hamiltonian between a pion and an extended nucleon fixed at the origin of the coordinate system becomes

$$H_{\pi N} = F_{\pi N} \int d^3x \, \rho(\mathbf{x}) \boldsymbol{\sigma} \cdot (\vec{\tau} \cdot \nabla \vec{\Phi}(\mathbf{x})). \qquad (12.16)$$

This interaction is the simplest one that leads to single emission and absorption of pions. It is not unique; additional terms such as $F' \vec{\Phi}^2$ may be present. Moreover, it is nonrelativistic and therefore limited in its range of validity. However, at higher energies, where Eq. (12.16) is no longer valid, other particles and processes complicate the situation so that consideration of the pion-nucleon force alone becomes meaningless anyway.

The integral in Eq. (12.16) vanishes for a spherical source function $\rho(r)$ unless the pion wave function describes a p wave ($l = 1$). This prediction is in agreement with the experimental data described in the previous section.

The first successful description of pion-nucleon scattering and pion photoproduction was due to Chew and Low,[7] who used the Hamiltonian (12.16). Because of the angular momentum barrier present in the $l = 1$ state, the low-energy pion-nucleon scattering cross section (below about 100 MeV) can be computed in perturbation theory. At higher energies, the approach is more sophisticated, but it can be shown that the Hamiltonian (12.16) leads to an attractive force in the state $I = \frac{3}{2}$, $J = \frac{3}{2}$ and thus can explain the observed resonance.[8] At still higher energies, the nonrelativistic approach is no longer adequate.

The numerical value of the pion-nucleon coupling constant $F_{\pi N}$ is determined by comparing the measured and computed values for the pion-nucleon scattering cross section. It is customary not to quote $F_{\pi N}$ but rather the corresponding dimensionless and rationalized coupling constant, $f_{\pi NN}$. The dimension of $F_{\pi N}$ in Eq. (12.16) depends on the normalization of the pion wave function $\vec{\Phi}$. Since the pion should be treated relativistically, the probability density is normalized not to unity, but to E, where E is the energy of the state. This normalization gives the probability density the correct Lorentz transformation properties; the probability density is not a relativistic scalar, but transforms like the zeroth component of a four-vector. With this normalization, $\vec{\Phi}$ has the dimension of $E^{-1/2} L^{-3/2}$ and the dimensionless rationalized coupling constant has the value[9]

$$f_{\pi NN}^2 = \frac{m_\pi^2}{4\pi h^5 c} F_{\pi N}^2 = 0.080 \pm 0.005. \qquad \textbf{(12.17)}$$

When the pion was the only known meson, the subject of the pion-nucleon interaction played a dominant role in theoretical and experimental investigations. It was felt that a complete knowledge of this interaction would be the clue to a complete understanding of hadronic physics. However, attempts to explain, for instance, the nucleon-nucleon force and the nucleon structure in terms of the pion alone were never successful. Other mesons were postulated, and these and some unexpected ones were found. It became clear that the pion-nucleon interaction is not the only problem of interest and that an interaction-by-interaction approach would not necessarily solve the entire problem. Other methods, relying heavily on the analytical properties of the scattering amplitude, were introduced.[10] At present, the field is very complicated and far beyond a brief and low-

7. G. F. Chew, *Phys. Rev.* **95**, 1669 (1954); G. F. Chew and F. E. Low, *Phys. Rev.* **101**, 1570 (1956); G. C. Wick, *Rev. Modern Phys.* **27**, 339 (1955).

8. Detailed descriptions of the Chew-Low approach can be found in G. Källen, *Elementary Particle Physics*, Addison-Wesley, Reading, Mass., 1964; E. M. Henley and W. Thirring, *Elementary Quantum Field Theory*, McGraw-Hill, New York, 1962; and J. D. Bjorken and S. D. Drell, *Relativistic Quantum Mechanics*, McGraw-Hill, New York, 1964. While these accounts are not elementary, they contain more details than the original papers.

9. R. G. Moorhouse, *Ann. Rev. Nucl. Sci.* **19**, 301 (1969).

10. G. F. Chew, when asked in 1962 what he considered to be the most important discovery in high-energy physics in the last few years, answered "The complex plane."

brow description. Our discussion here is therefore limited; we shall not treat other interactions or proceed to higher energies but shall turn to the nucleon-nucleon force because it plays a role in nuclear *and* particle physics.

12.4 The Yukawa Theory of Nuclear Forces

We have stated at the beginning of Section 12.2 that Yukawa introduced a heavy boson for the explanation of nuclear forces in 1934. The fundamental idea thus antedates the discovery of the pion by years. The role of mesons in nuclear physics was not discovered experimentally; it was predicted through a brilliant theoretical speculation. For this reason we shall first sketch the basic idea of Yukawa's theory before expounding the experimental facts. We shall introduce the Yukawa potential in its simplest form by analogy with the electromagnetic interaction.

The interaction of a charged particle with a Coulomb potential has been discussed in Chapter 10. The scalar potential A_0 produced by a charge distribution $q\rho(\mathbf{x}')$ satisfies the wave equation[11]

$$\nabla^2 A_0 - \frac{1}{c^2}\frac{\partial^2 A_0}{\partial t^2} = -4\pi q\rho. \qquad (12.18)$$

If the charge distribution is time-independent, the wave equation reduces to the Poisson equation,

$$\nabla^2 A_0 = -4\pi q\rho. \qquad (12.19)$$

It is straightforward to see that the potential (10.45),

$$A_0(\mathbf{x}) = \int d^3x' \, \frac{q\rho(\mathbf{x}')}{|\mathbf{x} - \mathbf{x}'|}, \qquad (12.20)$$

solves the Poisson equation.[12] For a point charge q located at the origin, A_0 reduces to the Coulomb potential,

$$A_0(r) = \frac{q}{r}. \qquad (12.21)$$

When Yukawa considered the interaction between nucleons in 1934, he noticed that the electromagnetic interaction could provide a model but that it did not fall off rapidly enough with distance. To force a more rapid decrease, he added a term $k^2\Phi$ to Eq. (12.19):

$$(\nabla^2 - k^2)\Phi(\mathbf{x}) = 4\pi g\rho(\mathbf{x}). \qquad (12.22)$$

The electromagnetic potential A_0 has been replaced by the field $\Phi(\mathbf{x})$, and the strength of the field is determined by the hadronic source $g\rho(\mathbf{x})$, where

11. The inhomogeneous wave equation can be found in most texts on electrodynamics, for instance, in Jackson, Eq. (6.37). As in Chapter 10, our notation differs slightly from Jackson; here ρ is not a charge but a probability distribution.

12. See, for instance, Jackson, Section 1.7. The important step can be summarized in the relation $\nabla^2(1/r) = -4\pi\delta(\mathbf{x})$, where δ is the Dirac delta function.

g determines the strength and ρ is a probability density. The sign of the source term has been chosen opposite to the electromagnetic case.[3] The solution of Eq. (12.22) that vanishes at infinity is

$$\Phi(\mathbf{x}) = -g \int \frac{e^{-k|\mathbf{x}-\mathbf{x}'|}}{|\mathbf{x} - \mathbf{x}'|} \rho(\mathbf{x}')d^3x'. \qquad (12.23)$$

For a hadronic point source, placed at position $\mathbf{x}' = 0$, this solution becomes the *Yukawa potential*,

$$\Phi(r) = -g\frac{e^{-kr}}{r}. \qquad (12.24)$$

The constant k can be determined by considering Eq. (12.22) for the free case $[\rho(\mathbf{x}) = 0]$ and comparing it to the corresponding quantized equation. The substitution

$$E \longrightarrow i\hbar\frac{\partial}{\partial t}, \qquad \mathbf{p} \longrightarrow i\hbar\nabla, \qquad (12.25)$$

changes the energy-momentum relation,

$$E^2 = (pc)^2 + (mc^2)^2,$$

into the Klein-Gordon equation,

$$\left\{\frac{1}{c^2}\frac{\partial^2}{\partial t^2} - \nabla^2 + \left(\frac{mc}{\hbar}\right)^2\right\}\Phi(\mathbf{x}) = 0. \qquad (12.26)$$

For a time-independent field and for $\rho(\mathbf{x}) = 0$, comparison of Eqs. (12.26) and (12.22) yields

$$k = \frac{mc}{\hbar}. \qquad (12.27)$$

The constant k in the Yukawa potential is just the inverse of the Compton wavelength of the field quantum. The mass of the field quantum determines the range of the potential. We have thus regained the result already expressed in Section 5.8. In addition, we have found the radial dependence of the potential for the case of a point source. The simple form of the Yukawa theory thus provides a description of the hadronic potential produced by a point nucleon in terms of the mass of the field quantum. It "explains" the short range of the hadronic forces. Before delving deeper into meson theory, we shall describe in more detail what is known about the forces between nucleons.

12.5 Properties of the Nucleon-Nucleon Force

The properties of the forces between nucleons can be studied *directly* in collision experiments or *indirectly* by extracting them from the properties of bound systems, namely the nuclei. In the present section, we shall first discuss the properties of the nuclear force as deduced from nuclear characteristics and then sketch some of the results obtained in scattering experiments below a few hundred MeV.

From the observed characteristics of nuclei, a number of conclusions about the nuclear force, that is, the hadronic force between nucleons, can be drawn. The most important ones will be summarized here.

Attraction. The force is predominantly attractive; otherwise stable nuclei could not exist.

Range and Strength. As explained in Section 12.1, comparison of the binding energies of ^{2}H, ^{3}H, and ^{4}He indicates that the range of the nuclear force is of the order of 1 fm. If the force is represented by a potential with such a width, a depth of about 50 MeV is found (Section 14.2).

Charge Independence. As discussed in Chapter 8, the hadronic force is charge-independent. After correction for the electromagnetic interaction, the *pp*, *nn*, and *np* forces between nucleons in the same states are identical.

Saturation. If every nucleon interacted attractively with every other one, there would be $A(A - 1)/2$ distinct interacting pairs. The binding energy would be expected to be proportional to $A(A - 1) \approx A^2$, and all nuclei would have a diameter equal to the range of the nuclear force. Both predictions, binding energy proportional to A^2 and constant nuclear volume, disagree violently with experiment for $A > 4$. For most nuclei, the volume and the binding energy are proportional to the mass number A. The first fact is expressed in Eq. (6.34); the second one will be discussed in Section 14.1 Consequently, the nuclear force exhibits saturation: One particle attracts only a limited number of others; additional nucleons are either not influenced or are repelled. A similar behavior occurs in chemical bonding and with van der Waals' forces. Saturation can be explained in two ways; through exchange forces[13] or through strongly repulsive forces at short distances (hard core).[14] Exchange forces lead to saturation in chemical binding, and hard cores account for it in classical liquids. In the hadronic case, the decision between the two cannot be made by considering nuclear properties, but scattering experiments indicate that both contribute. We shall return to both phenomena later.

The next two properties require a somewhat longer discussion; after stating the properties, they will be treated together.

Spin Dependence. The force between two nucleons depends on the orientation of the nucleon spins.

Noncentral Forces. Nuclear forces contain a noncentral component.

The two properties follow from the quantum numbers of the deuteron and from the fact that it has only one bound state. The deuteron consists

13. W. Heisenberg, *Z. Physik* **77**, 1 (1932).
14. R. Jastrow, *Phys. Rev.* **81**, 165 (1951).

of a proton and a neutron. Spin, parity, and magnetic moment are found to be

$$J^\pi = 1^+$$

$$\mu_d = 0.85742\mu_N. \tag{12.28}$$

The total spin of the deuteron is the vector sum of the spins of the two nucleons and of their relative orbital angular momentum,

$$\mathbf{J} = \mathbf{S}_p + \mathbf{S}_n + \mathbf{L}.$$

The even parity of the deuteron implies that L must be even. There are then only two possibilities for forming total angular momentum 1, namely $L = 0$ and $L = 2$. In the first case, shown in Fig. 12.9(a), the two nucleon spins add up to the deuteron spin; in the second, shown in Fig. 12.9(b), orbital and spin contributions are antiparallel. In the s state, where $L = 0$, the expected magnetic moment is the sum of the moments of the proton and the neutron, or

$$\mu(s \text{ state}) = 0.879634\mu_N.$$

The actual deuteron moment deviates from this value by a few percent,

$$\frac{\mu_d - \mu(s)}{\mu_d} = -0.026. \tag{12.29}$$

The approximate agreement between μ_d and $\mu(s)$ implies that the deuteron is predominantly in an s state, with the two nucleon spins adding up to the deuteron spin. If the nuclear force were spin-independent, proton and neutron could also form a bound state with spin 0. The absence of such a bound state is evidence for the *spin dependence* of the nucleon-nucleon force. The deviation of the actual deuteron moment from the s-state moment can be explained if it is assumed that the deuteron ground state is a superposition of s and d states. Part of the time, the deuteron has orbital angular momentum $L = 2$. Independent evidence for this fact comes from the observation that the deuteron has a small, but finite, *quadrupole* moment. The electric quadrupole moment measures the deviation of a charge distribution from sphericity. Consider a nucleus with charge Ze to have its spin $\mathbf{J}$ point along the z direction, as shown in Fig. 12.10. The charge density at point $\mathbf{r} = (x, y, z)$ is given by $Ze\rho(\mathbf{r})$. The classical quadrupole moment is *defined* by

$$Q = Z \int d^3r\,(3z^2 - r^2)\rho(\mathbf{r}) = Z \int d^3r\, r^2(3\cos^2\theta - 1)\rho(\mathbf{r}). \tag{12.30}$$

For a spherically symmetric $\rho(\mathbf{r})$, the quadrupole moment vanishes. For a cigar-shaped (oblate) nucleus, the charge is concentrated along z, and Q is positive. The quadrupole moment of a disk-shaped (prolate) nucleus is negative. As defined here, Q has the dimension of an area and is given in cm², or barns (10^{-24} cm²), or fm². In an external inhomogeneous electric field, a nucleus with quadrupole moment acquires an energy that depends

(a)
$L = 0$
s state

(b)
$L = 2$
d state

Fig. 12.9. The two possible ways in which spin and orbital contribution can form a deuteron of spin 1.

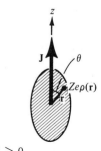

$Q > 0$

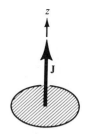

$Q < 0$

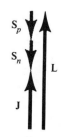

Fig. 12.10. Oblate and prolate nuclei, with spins pointing in the z direction. The nuclei are assumed to be axially symmetric; z is the symmetry axis.

on the orientation of the nucleus with respect to the field gradient.[15] This interaction permits the determination of Q; for the deuteron, a nonvanishing value was found.[16] The present value is

$$Q_d = 0.282 \text{ fm}^2. \qquad (12.31)$$

s states are spherically symmetric and have $Q = 0$. The nonvanishing value of Q_d thus verifies the conclusion drawn from the nonadditivity of the magnetic moments: The deuteron ground state must possess a d-state admixture. (See also Section 6.9, in particular Fig. 6.29.)

The presence of a d-state component implies that the nuclear force cannot be purely central, because the ground state in a central potential is always an s state; the energies of states with $L \neq 0$ are pushed higher by the centrifugal potential. The *noncentral force* giving rise to the deuteron quadrupole moment is called the *tensor force*. Such a force depends on the angle between the vector joining the two nucleons and the deuteron spin. Figure 12.11 shows two extreme positions. Since the deuteron quadrupole moment is positive, comparison of Figs. 12.10 and 12.11 indicates that

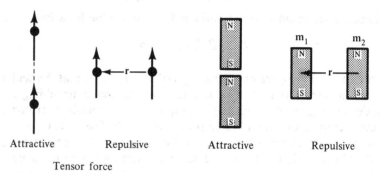

Attractive   Repulsive   Attractive   Repulsive

Tensor force

Fig. 12.11. The tensor force in the deuteron is attractive in the cigar-shaped configuration and repulsive in the disk-shaped one. Two bar magnets provide a classical example of a tensor force.

the tensor force must be attractive in the oblate and repulsive in the prolate configuration. A simple and well-known example of a classical tensor force is also shown in Fig. 12.11. Two bar magnets, with dipole moments $\mathbf{m}_1$ and $\mathbf{m}_2$, attract each other in the cigar-shaped arrangement but repel each other in the disk-shaped one. The interaction energy between two dipoles is well known[17]; it is

$$E_{12} = \frac{1}{r^3}(\mathbf{m}_1 \cdot \mathbf{m}_2 - 3(\mathbf{m}_1 \cdot \hat{\mathbf{r}})(\mathbf{m}_2 \cdot \hat{\mathbf{r}})). \qquad (12.32)$$

The vector $\mathbf{r}$ connects the two dipoles; $\hat{\mathbf{r}}$ is a unit vector along $\mathbf{r}$. In analogy

15. Careful discussions of the quadrupole moment are given in E. Segrè, *Nuclei and Particles,* Benjamin, Reading, Mass., Section 6.8; and Jackson, Section 4.2.

16. J. M. B. Kellog, I. I. Rabi, and J. R. Zacharias, *Phys. Rev.* **55**, 318 (1939).

17. Jackson, Section 4.2.

to this expression, a tensor operator is introduced to describe the non-central part of the force between two nucleons. This operator is defined by

$$S_{12} = 3(\boldsymbol{\sigma}_1 \cdot \hat{\mathbf{r}})(\boldsymbol{\sigma}_2 \cdot \hat{\mathbf{r}}) - \boldsymbol{\sigma}_1 \cdot \boldsymbol{\sigma}_2, \qquad (12.33)$$

where $\boldsymbol{\sigma}_1$ and $\boldsymbol{\sigma}_2$ are the spin operators for the two nucleons [Eq. (11.50)]. E_{12} and S_{12} have the same dependence on the orientation of the two components. S_{12} is dimensionless; the term $\boldsymbol{\sigma}_1 \cdot \boldsymbol{\sigma}_2$ makes the value of S_{12} averaged over all angles equal to zero and thus eliminates components of the central force from S_{12}.

The arguments given so far show that the properties of nuclei allow many conclusions concerning the nucleon-nucleon interaction. However, it is hopeless to extract the strength and the radial dependence of the various components of the nuclear force from nuclear information. *Collision experiments* with nucleons are required for a complete elucidation of the nucleon-nucleon interaction. Here we shall show that collision experiments provide evidence for *exchange* and *spin-orbit* forces.

Exchange Forces. The existence of exchange forces is readily apparent in the angular distribution (differential cross section as a function of the scattering angle) of np scattering at energies of a few hundred MeV. The expected angular distribution can be obtained with the help of the Born approximation. This approximation is reasonable here because the kinetic energy of the incident nucleon is much larger than the depth of the potential. The particle therefore crosses the potential region rapidly and barely feels the interaction. The differential cross section for a scattering process is given by Eqs. (6. 10) and (6.13) as

$$\frac{d\sigma}{d\Omega} = |f(\mathbf{q})|^2,$$

where

$$f(\mathbf{q}) = -\frac{m}{2\pi\hbar^2} \int V(\mathbf{x}) e^{i\mathbf{q}\cdot\mathbf{x}/\hbar}\, d^3x. \qquad (12.34)$$

Here $V(\mathbf{x})$ is the interaction potential and $\mathbf{q} = \mathbf{p}_i - \mathbf{p}_f$ is the momentum transfer. For elastic scattering in the c.m., $p_i = p_f = p$, and the magnitude of the momentum transfer becomes

$$q = 2p \sin \tfrac{1}{2}\theta.$$

The maximum momentum transfer is given by $q_{\max} = 2p$. At low energies, $2pR/\hbar \ll 1$, where R is the nuclear force range. Equation (12.34) then predicts isotropic scattering. At higher energies, where $2pR/\hbar \gg 1$, the situation is different. For forward scattering, at a sufficiently small scattering angle θ, q is small, and the cross section will remain large. For backward scattering, $q \approx q_{\max} = 2p$, the exponent in Eq. (12.34) oscillates rapidly, and the integral becomes small. The predicted behavior, isotropy at low energies and forward scattering at higher energies, is shown in Fig. 12.12. The two features do not depend on the Born approximation; they

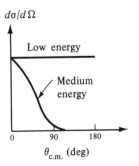

$d\sigma/d\Omega$

Fig. 12.12. Predicted differential cross section for np scattering at low and medium energies. The curves follow from the first Born approximation using an ordinary potential.

are more general. Low-energy scattering in a short-range potential is always isotropic, and the high-energy scattering usually acquires a diffractionlike character where small angles (low momentum transfers) are preferred.

Experiments at low energies indeed give an isotropic c.m. differential cross section. Even at a neutron energy of 14 MeV, the angular distribution is isotropic, as displayed in Fig. 12.13(a).[18] At higher energies, however, the behavior is very different from the one sketched in Fig. 12.12. An early measurement at a neutron energy of 400 MeV is reproduced in Fig. 12.13(b).[19] The differential cross section displays a pronounced peak in the backward direction. Such a behavior cannot be understood with an ordinary potential that leaves the neutron a neutron and the proton a proton. It is evidence for an *exchange force* that changes the incoming neutron into a proton through the exchange of a charged meson with the target proton. The forward-moving nucleon now has become a proton, and the recoiling target proton a neutron. In effect, then, the neutron is observed in the backward direction after scattering.

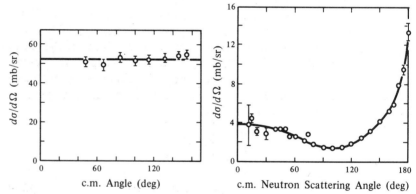

Fig. 12.13. Observed differential cross sections for *np* scattering. (a) The angular distribution at 14-MeV neutron energy is isotropic. [J. C. Alred et al., *Phys. Rev.* **91**, 90 (1953).] (b) At a neutron energy of 400 MeV, a pronounced backward peak is present. [A. J. Hartzler et al., *Phys. Rev.* **95**, 591 (1954).]

Spin-Orbit Force. The existence of a spin-orbit interaction can be seen in scattering experiments involving either polarized particles or polarized targets.[20] The idea underlying such experiments can be explained with a simple example, the scattering of polarized nucleons from a spinless target nucleus, for instance, ^{4}He or ^{12}C. Assume that the nucleon-nucleus force is

18. J. C. Alred, A. H. Armstrong, and L. Rosen, *Phys. Rev.* **91**, 90 (1953).

19. A. J. Hartzler, R. T. Siegel, and W. Opitz, *Phys. Rev.* **95**, 591 (1954).

20. Detailed information about such experiments can be found in the proceedings of the various conferences on polarization phenomena of nucleons, for instance, H. H. Barshall and W. Haeberli, eds., *Proceedings of the 3rd International Symposium on Polarization Phenomena in Nuclear Reactions*, Univ. of Wisconsin, Madison, 1970. See also G. G. Ohlsen, *Rept. Progr. Phys.* **35**, 399 (1972).

attractive; it then gives rise to trajectories as shown in Fig. 12.14(a). Assume further that the two incoming protons are fully polarized, with spins pointing "up," perpendicular to the scattering plane. Proton 1, scattered to the right, has an orbital angular momentum $\mathbf{L}_1$ with respect to the nucleus that is pointing "down." Proton 2, scattered to the left, has its orbital angular momentum $\mathbf{L}_2$ "up." Assume that the nuclear force consists of two terms, a central potential, V_c, and a *spin-orbit potential* of the form $V_{LS}\mathbf{L}\cdot\boldsymbol{\sigma}$,

$$V = V_c + V_{LS}\mathbf{L}\cdot\boldsymbol{\sigma}. \tag{12.35}$$

Figure 12.14(b) implies that the scalar product $\mathbf{L}\cdot\boldsymbol{\sigma}$ has opposite sign for nucleons 1 and 2. Consequently, the total potential V is larger for one nucleon than for the other, and more polarized nucleons will be scattered to one side than to the other. Experimentally, such left/right asymmetries are observed[20] and provide evidence for the existence of a spin-orbit force.

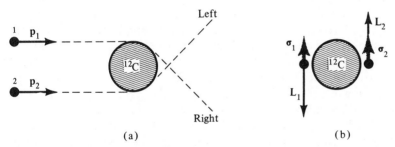

Fig. 12.14. Scattering of polarized protons from a spinless nucleus. (a) The trajectories in the scattering plane. (b) The spins and the orbital angular momenta of nucleons 1 and 2.

The information obtained in the present section can be summarized by writing the potential energy between two nucleons 1 and 2 as

$$V_{NN} = V_c + V_{sc}\boldsymbol{\sigma}_1\cdot\boldsymbol{\sigma}_1 + V_T S_{12} + V_{LS}\mathbf{L}\cdot\tfrac{1}{2}(\boldsymbol{\sigma}_1 + \boldsymbol{\sigma}_2). \tag{12.36}$$

Here $\boldsymbol{\sigma}_1$ and $\boldsymbol{\sigma}_2$ are the spin operators of the two nucleons and $\mathbf{L}$ is their relative orbital angular momentum,

$$\mathbf{L} = \tfrac{1}{2}(\mathbf{r}_1 - \mathbf{r}_2) \times (\mathbf{p}_1 - \mathbf{p}_2). \tag{12.37}$$

V_c in Eq. (12.36) describes the ordinary central potential energy, V_{sc} is the spin-dependent central term discussed above. V_T gives the tensor force; the tensor operator S_{12} is defined in Eq. (12.33). V_{LS} characterizes the spin-orbit force introduced in Eq. (12.35). V_{NN} in Eq. (12.36) is nearly the most general form allowed by invariance laws.[21]

Charge independence of the hadronic force implies invariance under

21. S. Okubo and R. E. Marshak, *Ann. Phys.* **4**, 166 (1958). Actually one term allowed by invariance arguments, the quadratic spin-orbit term, is missing in Eq. (12.36).

rotation in isospin space. The two isospin operators $\vec{I}_1$ and $\vec{I}_2$ of the two nucleons can only occur in the combinations

$$1 \quad \text{and} \quad \vec{I}_1 \cdot \vec{I}_2.$$

Thus each coefficient V_i in V_{NN} can still be of the form

$$V_i = V_i' + V_i'' \vec{I}_1 \cdot \vec{I}_2, \tag{12.38}$$

where V' and V'' can be functions of $r \equiv |\mathbf{r}_1 - \mathbf{r}_2|$, $p = \frac{1}{2}|\mathbf{p}_1 - \mathbf{p}_2|$, and $|\mathbf{L}|$.

The coefficients V_i must be determined from experiment. An incredible amount of pp and np collision experiments have already been performed.[22][23] In addition to total cross sections and angular distributions, collisions with polarized projectiles and polarized targets have been studied. The way to obtain the coefficients V_i from the experimental cross sections is, at least in principle, straightforward: A trial potential of the form of Eq. (12.36) is inserted into the Schrödinger equation, which is then solved with the help of a computer. From the solution, the theoretical cross section is computed. The coefficients in the trial potential are varied until agreement between calculation and observation is obtained. At the present time, a number of different choices fit the experiments about equally well.[24] The essential features of V_{NN} are common to the various fits. In particular, all require the presence of all terms listed in Eq. (12.36). The coefficients V_i depend on the spin and the parity of the state in which the interaction occurs. The force is strongly attractive in even-parity states (L even) and weak in odd-parity states (L odd). A strongly repulsive force appears at a radius of about 0.5 fm in most (or all) states. At large radii, V_{NN} falls off as predicted by the Yukawa potential, Eq. (12.24). At radii smaller than about 1 fm, the various fits differ. This fact is not surprising: Nucleons with a kinetic energy of 350 MeV have a reduced de Broglie wavelength of about 1 fm and are therefore not well suited for probing details smaller than about 1 fm. Moreover, the concept of a nonrelativistic potential may no longer be adequate for describing the nucleon-nucleon force at such distances.

One choice, introduced by Hamada and Johnston,[25] makes use of the insensitivity of the experimental data to the radial dependence of the V_i at small radii: The core is taken to have the same radius in all states, and it is assumed to become infinite and repulsive at a radius

$$r(\text{hard core}) = 0.48 \text{ fm}. \tag{12.39}$$

Figure 12.15 shows one component of the Hamada-Johnston potential. We have singled out this potential here because it has proved very useful in the calculation of nuclear properties, for instance, binding energies, in terms of the basic nuclear force. It appears to encompass the essential physical features of the nuclear force.

22. M. H. MacGregor, *Phys.* Today **22**, 21 (Oct. 1969).
23. *Rev. Modern Phys.* **39**, 495 (1967).
24. P. Signell, *Advan. Nucl. Phys.* **2**, 223 (1969).
25. T. Hamada and I. D. Johnston, *Nucl. Phys.* **34**, 382 (1962).

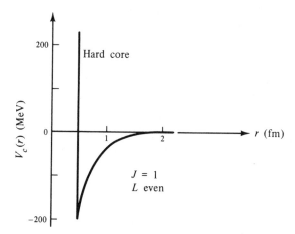

Fig. 12.15. Hamada-Johnston potential for the central part of the spin-triplet even-parity state.

12.6 Meson Theory of the Nucleon-Nucleon Force

● In the previous section, the nucleon-nucleon force has been described by a phenomenological potential energy V_{NN}. In this section we shall show that part of V_{NN} can be understood on the basis of pion exchange and that there is good evidence that the entire V_{NN} can be explained with the known mesons.

In Section 12.4, the Yukawa potential was introduced in analogy to electromagnetism by finding the solution to a Poisson equation with a mass term. In the present section we shall establish the expression for the interaction energy between two nucleons. We begin with the simplest case, where the interaction is mediated by the exchange of a *neutral scalar* meson. The emission and absorption of such a meson is described by an interaction Hamiltonian. For the pseudoscalar case, the corresponding Hamiltonian $H_{\pi N}$ has been discussed in Section 12.3. The Hamiltonian, H_s, for the scalar interaction can be obtained by similar invariance arguments: Φ is now a scalar in ordinary and in isospin space, and the simplest expression for the energy of interaction between a scalar meson and a fixed nucleon characterized by a source function $\rho(\mathbf{x})$ is

$$H_s = g \int d^3x \; \Phi(\mathbf{x})\rho(\mathbf{x}). \tag{12.40}$$

Between emission and absorption, the meson is free. The wave function of a free spinless meson satisfies the Klein-Gordon equation, Eq. (12.26). In the time-independent case, it reads

$$\left[\nabla^2 - \left(\frac{mc}{\hbar}\right)^2\right]\Phi(\mathbf{x}) = 0. \tag{12.41}$$

Together with Hamilton's equations of motion[26], Eqs. (12.40) and (12.41) lead to

$$\left[\nabla^2 - \left(\frac{mc}{\hbar}\right)^2\right]\Phi(\mathbf{x}) = 4\pi g\rho(\mathbf{x}). \tag{12.42}$$

26. A brief derivation is given in W. Pauli, *Meson Theory of Nuclear Forces*, Wiley-Interscience, New York, 1946. The elements of Lagrange and Hamiltonian mechanics can be found in most texts on mechanics. The application to wave functions (fields) is described in E. M. Henley and W. Thirring, *Elementary Quantum Field Theory*, McGraw-Hill, New York, 1962, p. 29, or F. Mandl, *Introduction to Quantum Field Theory*, Wiley-Interscience, New York, 1959, Chapter 2.

This expression is identical to Eq. (12.22). In Section 12.4, we constructed it by starting from the corresponding one in electromagnetism and adding a mass term. Here it follows logically from the wave equation for the scalar meson together with the simplest form for the interaction Hamiltonian. The solution to Eq. (12.42) has already been given in Section 12.4. In particular, for a point nucleon at position $\mathbf{x} = 0$, it is the Yukawa potential, Eq. (12.24). The nucleon acts as a source of the meson field and

$$\Phi(\mathbf{x}) = -\frac{g}{r}e^{-kr}, \qquad r = |\mathbf{x}|, \qquad k = \frac{mc}{\hbar}, \tag{12.43}$$

is the field produced at $\mathbf{x}$ by a point nucleon sitting at the origin. The interaction energy between this and a second point nucleon at position $\mathbf{x}$ is found by inserting Eq. (12.43) into Eq. (12.40) and by using the fact that $\rho(\mathbf{x})$ now also describes a point nucleon. The interaction energy then becomes

$$V_s = -g^2 \frac{e^{-kr}}{r}. \tag{12.44}$$

The negative sign means attraction and two nucleons consequently attract each other if the force is produced by a neutral scalar meson.

Pions are pseudoscalar and not scalar particles. Is the discussion just given therefore completely academic? Table A4 in the Appendix shows that scalar neutral mesons exist; the ϵ with a mass around 700 MeV/c^2 qualifies. At distances of the order of 0.2 fm, a contribution from the potential energy V_s consequently should appear. Many other mesons with smaller masses exist, however, and must be considered. The next step is the contribution from a *neutral pseudoscalar meson*. Table A4 in the Appendix indicates that η, with a mass of 549 MeV/c^2, is such a particle. The interaction Hamiltonian is very similar to the one given in Eq. (12.16); for an isoscalar particle, this relation simplifies to

$$H_p = F \int d^3x\, \rho(\mathbf{x})\boldsymbol{\sigma}\cdot\nabla\Phi. \tag{12.45}$$

The free pseudoscalar meson is also described by the Klein-Gordon equation, Eq. (12.41), because it is not possible to distinguish between free scalar and pseudoscalar particles. For the meson field in the presence of a nucleon, Eqs. (12.45) and (12.41), together yield

$$\left[\nabla^2 - \left(\frac{mc}{\hbar}\right)^2\right]\Phi = 4\pi F\rho(x)\boldsymbol{\sigma}\cdot\nabla.$$

A partial integration moves the gradient operator onto the source function $\rho(\mathbf{x})$ so that $\Phi(\mathbf{x})$ satisfies

$$\left[\nabla^2 - \left(\frac{mc}{\hbar}\right)^2\right]\Phi = -4\pi F\boldsymbol{\sigma}\cdot\nabla\rho(\mathbf{x}). \tag{12.46}$$

This equation is solved as in Section 12.4. Inserting the solution into Eq. (12.45) then gives, for the potential energy due to the exchange of the neutral pseudoscalar meson between point nucleons A and B,

$$V_P = F^2(\boldsymbol{\sigma}_A\cdot\nabla)(\boldsymbol{\sigma}_B\cdot\nabla)\frac{e^{-kr}}{r}. \tag{12.47}$$

The differentiations can be performed, and the final result is[27]

$$V_P = F^2\left\{\frac{1}{3}\boldsymbol{\sigma}_A\cdot\boldsymbol{\sigma}_B + S_{AB}\left[\frac{1}{3} + \frac{1}{kr} + \frac{1}{(kr)^2}\right]\right\}k^2\frac{e^{-kr}}{r}, \tag{12.48}$$

where k is given in Eq. (12.43) and S_{AB} is the tensor operator defined in Eq. (12.33).

---

27. Details can be found in L. R. B. Elton, *Introductory Nuclear Theory*, 2nd ed., Saunders, Philadelphia, Section 10.3. V_p, as given in Eq. (12.48), is not complete; a term proportional to $\delta(\mathbf{r})$ is missing. The omission is unimportant because the short-range repulsion between nucleons makes the term ineffective.

V_p can be generalized immediately to the pion: The only modification is a factor $\vec{I}_A \cdot \vec{I}_B$ multiplying Eq. (12.48).

It is remarkable that the exchange of a pseudoscalar meson leads to the experimentally observed tensor force. Even before the pion was discovered and its pseudoscalar nature established, Eq. (12.48) was known and was taken as a hint as to the properties of the Yukawa quantum.[3] However, it turned out to be impossible to explain all features of the nucleon-nucleon force in terms of the exchange of pions only. Today we know the reason for the failure: The pion is only one of many mesons; it leads to the longest-range part of the nucleon-nucleon force, but for distances less than about 2 fm, the others must be taken into account.[28] Table A4 in the Appendix shows that scalar, pseudoscalar, and vector mesons occur below a mass of about 1 GeV/c^2. We have already discussed the potentials produced by scalar and pseudoscalar mesons. The potential that results from the exchange of vector mesons can be obtained similarly[3]. For example, the exchange of the ω meson gives rise to a repulsive core as indicated schematically in Fig. 12.15. Modern approaches to the potential problem are based on these physical ideas, but the mathematical techniques are more sophisticated.[29] The essential result can be summarized as follows: If all known mesons up to a mass of about 1 GeV/c^2 are introduced with their experimentally determined masses and coupling constants (where known), and if the exchange of two pions is taken into account,[30] the experimentally deduced nucleon-nucleon potential energy V_{NN} can be reproduced satisfactorily. As stated earlier, the potential concept loses validity above an energy of about 500 MeV. It is, however, possible to compare the results of scattering experiments directly with calculations; it turns out that meson theory can explain the nucleon-nucleon scattering data up to about 3 GeV.[31] At least up to such energies, the fundamental idea of Yukawa, namely the explanation of the nucleon—nucleon force through the exchange of heavy hadronic quanta, is confirmed. ●

12.7 Hadronic Processes at High Energies

Early explorers of the earth faced an uncertain fate. They did not know if they would fall off into the unknown when they reached the end of the disk-shaped world. Bounds were placed on the possible disasters when it was realized that the earth was approximately a sphere. Further exploration led to more bounds, and the presently existing topographic maps leave little room for major surprises. The situation in high-energy physics resembles that of the early explorers. The immediate neighborhood, the hadronic interaction at energies below, say, 30 GeV, is reasonably well explored experimentally. Much remains to be explained, but it is possible that no major new feature will emerge in this energy region in future experiments. At higher energies, however, a new world may be waiting for us. Experiments, with the scarce cosmic rays and at the ISR at CERN, provide some glimpses into the ultrahigh-energy region, but it is very likely

28. The description of the nucleon-nucleon force in terms of mesons is reviewed in many papers. Much information can be found in M. J. Moravcsik and H. P. Noyes, *Ann. Rev. Nucl. Sci.* **11**, 95 (1961), and in the papers in *Progr. Theoret. Phys.* (*Kyoto*) *Suppl. 39* (1967).

29. M. H. Partovi and E. L. Lomon, *Phys. Rev.* **D2**, 1999 (1970); R. A. Bryan and A. Gersten, *Phys. Rev.* **D6**, 341 (1972); M. J. Moravcsik, *Rept. Progr. Phys.* **35**, 587 (1972).

30. M. Chemtob, J. W. Durso, and D. O. Riska, *Nucl. Phys.* **B38**, 141 (1972).

31. T. Ueda, *Phys. Rev. Letters* **26**, 588 (1971).

that much more will be learned in the next few years. In this section, we shall sketch three aspects of ultrahigh-energy collisions.

Inelastic Collisions.[32] Most of the discussions so far have been restricted to elastic collisons. These are dominant at low energies. As the energy increases, more and more particles can be created. At ultrahigh energies, the interaction of two nucleons can indeed be a spectacular event. One such event, produced by a 300 GeV proton, is shown on p. 325. The experimental data, obtained at various high-energy accelerators and in cosmic ray studies, display the following prominent features: (i) Small transverse momenta. The pp elastic differential cross sections reproduced in Fig. 6.23 decrease exponentially with t: collision events with large perpendicular momentum transfer are rare. The reluctance of particles to transfer momentum perpendicular to its motion persists in inelastic events. The number of particles produced falls off very rapidly as a function of p_T, the momentum transverse to the incident beam. The average value of p_T is of the order of 0.3 GeV/c and nearly independent of the incoming energy. (ii) Low multiplicity. The multiplicity, the number n of secondary particles, can be compared with the maximum allowed by energy conservation. By this criterion, n increases only slowly with energy. The average

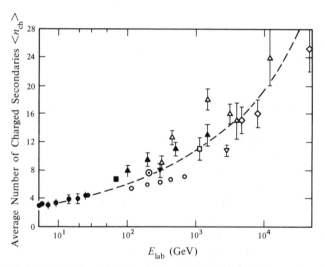

Fig. 12.16. Multiplicity, $\langle n_{\mathrm{ch}} \rangle$, of charged secondaries in pp collisions as a function of laboratory energy. The dashed curve represents $\langle n_{\mathrm{ch}} \rangle = 1.97 E_{\mathrm{lab}}^{1/4}$. (After Carruthers and Minh Duong-van.)[33]

32. Textbooks contain little information concerning ultrahigh-energy reactions. Most reviews are contained in the proceedings of the various high-energy conferences. A recent survey from which more references can be found is W. R. Frazer, L. Ingber, C. H. Metha, C. H. Poon, D. Silverman, K. Stowe, P. D. Ting, and H. J. Yesian, *Rev. Modern Phys.* **44**, 284 (1972). See also J. D. Jackson, *Rev. Modern Phys.* **42**, 12 (1970), and R. P. Feynman, *Phys. Rev. Letters* **23**, 1415 (1969). D. Horn, *Phys. Reports* **4C**, 2 (1972).
33. P. Carruthers and Minh Duong-van, *Phys. Letters* **41B**, 597 (1972).

multiplicity of charged secondaries, $\langle n_{ch} \rangle$, is shown in Fig. 12.16 as a function of the laboratory energy of the incident proton. The dashed curve represents the relation[33]

$$\langle n_{ch} \rangle = \text{const. } E_{lab.}^{1/4} \tag{12.49}$$

(iii) Poisson-like distributions. The cross sections for the production of events with n prongs are shown for two energies in Fig. 12.17.[34] The distributions are similar in shape, but broader than Poisson distributions, Eq. (4.3), of the same mean.

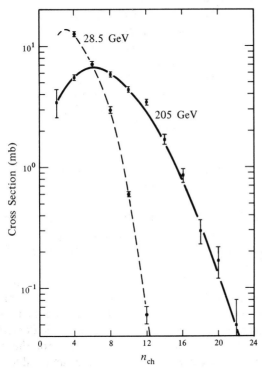

Fig. 12.17. Cross sections for the production of n charged particles in pp collisions at 28.5 GeV and 205 GeV.[34] The values at $n = 2$ (at 205 GeV) does not include the elastic cross section.

(iv) Pionization. Most of the secondaries are pions. In the c.m. of the colliding protons, nearly all the pions have small momenta. The formation of such a pion cloud is called pionization.

The theoretical description of very-high energy collisions is still in its infancy. One problem is the numerical value of the multiplicity: It is too big to allow a simple description in terms of well-defined particle exchanges and too small to be treated well by statistical theories. The numerous models

34. W. H. Sims et al., *Nucl. Phys.* **B41**, 317 (1972); G. Charlton et al., *Phys. Rev. Letters* **29**, 515 (1972).

invented so far[32][33] are consequently still crude and the experimental data are not yet precise and complete enough to allow focusing on one model.

High-Energy Theorems (Asymptotia). As Fig. 12.16 shows, processes at ultrahigh energies can be extremely complex. It is nevertheless possible to extrapolate lower-energy data to predict features of cross sections that should emerge as the total energy in the c.m., W, tends toward infinity. This energy region is usually called *asymptotia*, and it is not yet clear if and where this strange land begins.

In Section 9.8, we stated that the *TCP* theorem can be proved with very general arguments. These are based on axiomatic quantum field theory, which is an extension of quantum mechanics into the relativistic region. This theory can also be used to derive theorems on high-energy collisions.[35] Quantum field theory lies far outside the scope of this book, but we shall state two theorems because they are typical of the results that can be expected from this approach. The first theorem follows rigorously from quantum field theory,[35] and it gives an upper bound on the total cross section as $s = W^2$ tends to infinity:

$$\sigma_{\text{tot}} < \text{const.}(\log s)^2. \tag{12.50}$$

This bound was discovered by Froissart[36] and it limits the rise of the total cross section with increasing energy regardless of the type of interaction involved. An example of a cross section that increases with increasing energy is shown in Fig. 6.24, namely the *pp* total cross section at values of s greater than about 1000 GeV^2. It is not yet clear whether this increase follows the maximum rate allowed by the Froissart bound, Eq. (12.50), or is much slower. Will this increase continue indefinitely or will the cross section flatten out again? The answer to this question and the physical meaning of the increase is still unclear. The second theorem follows from quantum field theory if it is additionally assumed that the total cross sections at asymptotic energies become constants. The Pomeranchuk theorem[37] then predicts that the total cross sections for the particle-target and antiparticle-target collisions approach the same value as the energy tends toward infinity:

$$\frac{\sigma_{\text{tot}}(\bar{A} + B)}{\sigma_{\text{tot}}(A + B)} \longrightarrow 1 \qquad \text{in asymptotia.} \tag{12.51}$$

In a simplified geometrical interpretation, the Pomeranchuk theorem can be understood as follows: As the energy approaches infinity, so many reactions are possible that the collision can almost be thought of as one between two totally absorbing black disks. The cross section is thus essentially geometric (the radii of the two objects are not well defined, but

35. R. J. Eden, *Rev. Modern Phys.* **43**, 15 (1971).
36. M. Froissart, *Phys. Rev.* **123**, 1053 (1961).
37. I. Ia. Pomeranchuk, *Soviet Phys. JETP* **7**, 499 (1958).

we are only providing a qualitative argument). Since the geometrical structures of the positive and negative pions are identical (the charge is certainly not important), the cross sections for $\pi^+ p$ and $\pi^- p$ would be expected to be identical. The fact that $\pi^+ p$ can only be in an isospin state $I = \frac{3}{2}$ whereas $\pi^- p$ can scatter in both $I = \frac{3}{2}$ and $\frac{1}{2}$ is of no importance because there is a huge (infinite) number of possible final states in both cases. The same argument can be made, for instance, for $\bar{p}p$ and pp scatterings. The experimental data appear to bear out the Pomeranchuk theorem. Figure 12.18 shows some results from Serpukhov. The relevant cross sections indeed tend toward a common constant value, and the differences $\Delta\sigma$ tend towards zero.

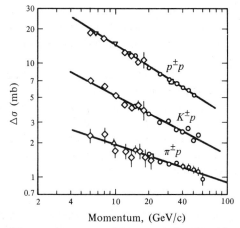

Fig. 12.18. The differences between particle-target and antiparticle-target cross sections, measured at Serpukhov by Gorin et al., tend towards zero with increasing energy.

Scale Invariance.[38] Where is asymptotia? At the present time, this question is not settled, but some insight can be obtained with simple arguments. Consider first a world in which only the electron and positron exist. The bound system in such a world is positronium, an "atom" in which an electron and a positron revolve around the common c.m. The energy levels of positronium are given by the Bohr formula,

$$E_n = -\alpha^2 m_e c^2 \frac{1}{(2n)^2}, \qquad n = 1, 2, \ldots, \qquad (12.52)$$

where $\alpha = e^2/\hbar c$ is the fine structure constant. Apart from the factor $(2n)^{-2}$, the energy levels are determined by two factors, α^2 and $m_e c^2$. The first describes the strength of interaction, and the second sets the *scale*. At energies of the order of, or smaller than, the scale energy $m_e c^2$, the

38. T. D. Lee, *Phys. Today* **25**, 23 (April 1972); R. Jackiw, *Phys. Today* **25**, 23 (Jan. 1972); J. D. Bjorken, *Phys. Rev.* **179**, 1547 (1969). M. S. Chanowitz and S. D. Drell, *Phys. Rev. Letters* **30**, 807 (1973).

physical phenomena are dominated by the existence of discrete energy levels. At energies large compared to $m_e c^2$, asymptotia has been reached in the positronium world, and physical phenomena satisfy simple laws in which m_e does not appear. Consider Bhabha scattering,

$$e^+ e^- \longrightarrow e^+ e^-. \tag{12.53}$$

The total cross section for electron-positron scattering in asymptotia can depend only on W, the total c.m. energy, and on the strength factor α^2 but not on m_e. The cross section has the dimension of an area, and the only possible form *not* containing m_e is

$$\sigma = \text{const.} \frac{\alpha^2}{W^2}, \qquad \text{in asymptotia.} \tag{12.54}$$

This form expresses *scale invariance:* It is not possible from the measured cross section to determine the mass of the colliding particles. Of course, this invariance is only approximate; it would be exact only if the colliding particles had zero mass.

Now consider $e^+ e^-$ scattering in the real world. Equation (12.54) is valid at c.m. energies greater than a few MeV. At energies of a few hundred MeV, deviations begin to occur, and a peak appears at $W = 760$ MeV, as indicated in Fig. 10.17. The deviation and the observed resonance reveal that m_e is not the only mass that sets a scale but that higher-mass particles exist, in this case the pions and their resonances. In addition to Bhabha scattering, processes such as

$$e^+ e^- \longrightarrow \text{hadrons} \tag{12.55}$$

become possible, and σ depends on the masses of the various hadrons. The departure of the total cross section from the form of Eq. (12.54) indicates that a new basic energy scale has appeared. The energy scale is now given by

$$E_h = m_h c^2, \tag{12.56}$$

where m_h is the mass of a suitably chosen hadron. Usually, m_h is taken to be the nucleon mass, $m_h = m_N$. What could have been considered asymptotia for Bhabha scattering has turned out to be nothing but a transition region. However, the game can now be replayed. At energies large compared to the new scale energy, E_h, we again expect independence of the total cross section on the hadron masses. Dimensional arguments then show that σ_{tot} must again be of the form of Eq. (12.54):

$$\sigma_{\text{tot}} = \text{const.} \frac{\alpha^2}{W^2}, \qquad \text{for } W \gg m_h c^2. \tag{12.57}$$

The constant can be different from the one given in Eq. (12.54), but the energy dependence is thé same. Will Eq. (12.57) now be valid up to infinite energies, or is the new, the hadronic, asymptotia again only a transition region? The answer to this question is not yet known, but some remarks can be made. In Section 11.11, it was pointed out that an appealing speculation leads to the introduction of an intermediate boson (W) with a mass of about 40 GeV/c^2. If particles with such a mass exist, then it is likely that the present hadronic asymptotia is again only a transition region and that a new fundamental energy scale appears at around 40 GeV. Subatomic physics, as we know it today, then would only be a beginning, and a revolution in hadronic physics could well occur. Only experiments can give answers to this fundamental question.

12.8 References

The literature covering the field of strong interactions is immense. Most texts and reviews, and particularly the original theoretical papers, are rather sophisticated. In the following we list some reviews and books from which, with some effort, information can be extracted even at the level assumed here.

The book *Pion-Nucleon Scattering* by R. J. Cence, Princeton University Press, Princeton, N. J., 1969, presents the essential experimental data and provides the necessary theoretical background for a discussion of the data.

A concise and elegant introduction to the theoretical treatment of the pion-nucleon interaction is given in J. D. Jackson, *The Physics of Elementary Particles*, Princeton University Press, Princeton, N. J. 1958. The pion-nucleon interaction is also discussed at length in the texts by S. De Benedetti, *Nuclear Interactions*, Wiley, New York, 1964 and H. Muirhead, *The Physics of Elementary Particles*, Pergamon, Elmsford, N. Y., 1965.

An interesting introduction to nuclear forces and a collection of some of the pioneering papers can be found in D. M. Brink, *Nuclear Forces*, Pergamon, Elmsford, N. Y., 1965.

The nucleon-nucleon interaction is discussed in the following books and reviews:

G. Breit and R. D. Haracz, "Nucleon-Nucleon Scattering," in *High-Energy Physics* (E. H. S. Burhop, ed.), Academic Press, New York, 1967.

M. J. Moravcsik, *The Two Nucleon Interaction*, Oxford University Press, Inc., London, 1963.

P. Signell, "The Nuclear Potential," *Advan. Nucl. Phys.* **2**, 223 (1969).

R. Wilson, *The Nucleon-Nucleon Interaction*, Wiley-Interscience, New York, 1963.

A good introduction to the meson theory of nuclear forces is given in L.R.B. Elton, *Introductory Nuclear Theory*, Saunders, Philadelphia, 1966.

High-energy scattering is treated in R.J. Eden, *High-Energy Collisions of Elementary Particles*, Cambridge University Press, Cambridge, 1967.

PROBLEMS

12.1.
(a) List the 10 possible pion-nucleon scattering processes.
(b) Which of these are related by time-reversal invariance?
(c) Express all cross sections in terms of $M_{3/2}$ and $M_{1/2}$.

12.2. Sketch an experimental arrangement used to study pion-nucleon scattering.
(a) How is the total cross section observed?
(b) How is the charge exchange reaction determined?

12.3. Use the observed cross sections to show that the peaks of the first resonance in pion-nucleon and in photonucleon reactions occur at the same mass of the Δ. Take recoil into account.

12.4. Treat the pion-nucleon scattering at the first resonance classically: Compute the classical distance from the center of the nucleon at which a pion with angular momentum $l = 0, 1, 2, 3$ (in units of $\hbar$) will strike. Which partial waves will contribute significantly according to this argument? Use a parity argument to rule out the values $l = 0$ and $l = 2$.

12.5. Justify Eq. (12.7) by a crude (nonrigorous) argument.

12.6. Verify the expansions (12.8).

12.7. Consider $H_{\pi N}$, Eq. (12.16). Assume a spherical source function $\rho(r)$. Assume the pion wave function to be a plane wave. Show that only the p-wave part of this plane wave leads to a nonvanishing integral.

12.8. Consider Fig. 5.28. The second and third resonances in the $\pi^- p$ system have no counterpart in the $\pi^+ p$ system. What is the isospin of these resonances?

12.9.
(a) Do conservation laws permit terms in the pion-nucleon interaction that are quadratic in the pion wave function $\vec{\Phi}$?
(b) Repeat part (a) for terms cubic in $\vec{\Phi}$.

12.10. Use second-order nonrelativistic perturbation theory and the two diagrams in Fig. P. 12.10 to compute the low-energy pion-nucleon scattering cross section. Compare to the experimental data.

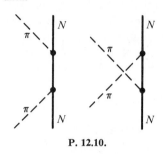

P. 12.10.

12.11. Use the decay $\Delta \longrightarrow \pi N$ to compute a very approximate value for the coupling constant $f_{\pi N \Delta}$. Compare to $f_{\pi NN}$.

12.12. Assume that particles of 1-GeV kinetic energy are produced at the center of a lead nucleus. Estimate the fraction of particles that escape from the nucleus without interaction if the particles interact
(a) Hadronically.
(b) Electromagnetically.
(c) Weakly.

12.13. Show that the Coulomb potential, Eq. (12.21), solves the Poisson equation, Eq. (12.19).

12.14. Show that the Yukawa potential, Eq. (12.24), is a solution of Eq. (12.22).

12.15. Assume attractive spherically symmetric nuclear forces with a range R and point nucleons. Show that the most stable nucleus has a diameter about equal to the force range R. (*Hint*: Consider the total binding energy, the sum of the kinetic and the potential energy, as a function of the nuclear diameter. The nucleus is in its ground state; the nucleons obey Fermi statistics. The arguments in Chapter 14 may be helpful.)

12.16. *Deuteron—Experimental.* Describe how the following deuteron characteristics have been determined:
(a) The binding energy.
(b) The spin.
(c) The isospin.

(d) The magnetic moment.

(e) The quadrupole moment.

12.17. Show that the ground state of a two-body system with central force must be an s state, that is, have orbital angular momentum zero.

12.18. *Deuteron—Theory.* Treat the deuteron as a three-dimensional square well, with depth $-V_0$ and range R.

(a) Write the Schrödinger equation. Justify the value of the mass used in the Schrödinger equation.

(b) Assume the ground state to be spherically symmetric. Find the ground-state wave function inside and outside the well. Determine the binding energy in terms of V_0 and R. Show that B fixes only the product $V_0 R^2$.

(c) Sketch the ground-state wave function. Estimate the fraction of time that the neutron and proton spend outside each other's force range. Why does the deuteron not disintegrate when the nucleons are outside the force range?

12.19. Dineutrons and diprotons, that is, bound states consisting of two neutrons or two protons, are not stable. Explain why not in terms of what is known about the deuteron.

12.20. Evidence for a bound state consisting of an antiproton and a neutron has been found, and the binding energy of this $\bar{p}n$ system is 83 MeV. [L. Gray, P. Hagerty, and T. Kalogeropoulos, *Phys. Rev. Letters* **26**, 1491 (1971).] Describe this system by a square well with radius $b = 1.4$ fm and depth V_0. Compute V_0 and compare the numerical value with that of the deuteron.

12.21. Antideuterons have been observed. How were they identified? [D. E. Dorfan et al., *Phys. Rev. Letters* **14**, 1003 (1965); T. Massam et al., *Nuovo Cimento* **39**, 10 (1965).]

12.22. Verify that a cigar-shaped nucleus, with the nuclear symmetry axis parallel to the z axis, has a positive quadrupole moment.

12.23. Show that the quadrupole moment of a nucleus with spin $\frac{1}{2}$ is zero.

12.24. Show that the quadrupole moment of the deuteron is "small," i.e., that it corresponds to a small deformation.

12.25. The lowest-lying singlet state of the neutron-proton system, with quantum numbers $J = 0$, $L = 0$, is sometimes called the *singlet deuteron*. It is not bound, and scattering experiments indicate that it occurs just a few keV above *zero* energy; it is just slightly *unbound*. Assume that the singlet state occurs at zero energy, and find the relation between well depth and well radius. Assume equal singlet and triplet well radii, and show that the singlet well depth is smaller than the triplet one.

12.26. Show that the tensor operator, Eq. (12.33), vanishes if it is averaged over all directions $\hat{r}$.

12.27. Prove that the operator $L = \frac{1}{2}(\mathbf{r}_1 - \mathbf{r}_2) \times (\mathbf{p}_1 - \mathbf{p}_2)$ [Eq. (12.37)] is the orbital angular momentum of the two colliding nucleons in their c.m.

12.28. Show that hermiticity of V_{NN}, Eq. (12.36), demands that the coefficients V_i be real.

12.29. Show that translational invariance implies that the coefficients V_i in Eq. (12.36) can depend only on the relative coordinate $\mathbf{r} = \mathbf{r}_1 - \mathbf{r}_2$ of the two colliding nucleons and not on $\mathbf{r}_1$ or $\mathbf{r}_2$ separately.

12.30. Galilean invariance demands that the transformation

$$\mathbf{p}_i' = \mathbf{p}_i + m\mathbf{v}$$

leaves the V_i in Eq. (12.36) unchanged. Show that this condition implies that V_i can depend only on the relative momentum $\mathbf{p} = \frac{1}{2}(\mathbf{p}_1 - \mathbf{p}_2)$.

12.31. Show that the spin operators $\boldsymbol{\sigma}_1$ and $\boldsymbol{\sigma}_2$ satisfy the relations

$$\sigma_x^2 = \sigma_y^2 = \sigma_z^2 = 1$$
$$\sigma_x \sigma_y + \sigma_y \sigma_x = 0$$
$$\boldsymbol{\sigma}^2 = 3$$
$$(\mathbf{a} \cdot \boldsymbol{\sigma})^2 = a^2$$
$$(\boldsymbol{\sigma}_1 \cdot \boldsymbol{\sigma}_2)^2 = 3 - 2\boldsymbol{\sigma}_1 \cdot \boldsymbol{\sigma}_2.$$

12.32. Show that the following eigenvalue equations hold:

$$\boldsymbol{\sigma}_1 \cdot \boldsymbol{\sigma}_2 |t\rangle = -1 |t\rangle$$
$$\boldsymbol{\sigma}_1 \cdot \boldsymbol{\sigma}_2 |s\rangle = -3 |s\rangle.$$

Here $|s\rangle$ and $|t\rangle$ are the spin eigenstates of the two-nucleon system: $|s\rangle$ is the singlet and $|t\rangle$ is the triplet state.

12.33. Show that the operator

$$P_{12} = \tfrac{1}{2}(1 + \boldsymbol{\sigma}_1 \cdot \boldsymbol{\sigma}_2)$$

exchanges the spin coordinates of the two nucleons in the two-nucleon system.

12.34. At which energy in the laboratory system does pp scattering become inelastic, i.e., can pions be produced?

12.35. Show that Hamilton's equations of motion, together with Eqs. (12.40) and (12.41), lead to Eq. (12.42).

12.36. Verify Eq. (12.47).

12.37. Show that Eq. (12.48) follows from Eq. (12.47).

12.38.
(a) Compute the expectation value of the single-pion exchange potential energy in the s states of two nucleons.

(b) Compute the effective force in any even angular momentum state with spin 1 and with spin 0.

12.39. Explain why the $\bar{p}p$ and the $\bar{p}n$ cross sections are much larger than the pp and the pn ones. [J. S. Ball and G. F. Chew, *Phys. Rev.* **109**, 1385 (1958).]

12.40. Verify Eq. (12.52).

12.41. Show that dimensional analysis leads to Eq. (12.53). Determine the dimension of the constant.

12.42. Show that the total cross section for the scattering of neutrinos and nucleons in asymptotia is given by

$$\sigma_{\text{tot}} = \text{const.}\ G^2 W^2,$$

where G is the weak coupling constant and W the total energy in the c.m. Compare this result with experiment.

PART **V**

Models

Atomic physics is very well understood. A simple model, the Rutherford model, describes the essential structure: A heavy nucleus gives rise to
a central field, and the electrons move primarily in this central field. The
force is well known. The equation describing the dynamics is the Schrödinger equation or, if relativity is taken into account, the Dirac equation.
Historically, this satisfactory picture is not the end result of one single line
of research, but it is the confluence of many different streams of discoveries, streams that at one time appeared to have nothing in common. The
Mendeleev table of elements, the Balmer series, the Coulomb law, elec-
trolysis, black-body radiation, cathode rays, the scattering of alpha particles, and Bohr's model all were essential steps and milestones. What is
the situation with regard to particles and nuclei? We have described the

Photo V.I. Sky and Water I, (1938). From *The Graphic Work of M. C. Escher*, Hawthorn Books, New York. (Courtesy of M. C. Escher Foundation, Gemeente Museum, The Hague.) Compare this illustration to Figure 13.3.

elementary particle zoo and the nature of the forces. Are the known facts sufficient to build a coherent picture of the subatomic world? The answer is no. The theoretical description of *nuclei* is in fairly good shape: There exist successful models, and most aspects of the structure and the interaction of nucleons and nuclei can be described reasonably well. However, most of these models do not start from first principles. They involve the known properties of the nuclear forces but focus on simple modes of motion. Much remains to be done until nuclear theory is as complete and as free from assumptions as atomic physics. The *particle* situation is even less satisfactory, and it is closer to the state of atomic theory at the time of Mendeleev than at the time of Bohr, Schrödinger, and Heisenberg. Some properties of the particle zoo can be explained rather well by introducing subunits with definite properties (quarks), but no such subunits have yet been seen directly and they may not exist.

In the following chapters, we shall briefly outline the quark model of particles and some of the most successful nuclear models. The discussion in these chapters is restricted to *hadrons;* leptons, the most mysterious of the particles, are omitted. They are not understood at all.

13

Quarks and Regge Poles

"Consider all substances; can you find among them any enduring "self"? Are they not all aggregates that sooner or later will break apart and be scattered?"

The Teaching of Buddha

The number of particles is at least as large as the number of elements. To find out how progress in understanding the particle zoo could occur, it may be a good idea to take a look at the history of chemistry and atomic physics. The discovery of the periodic table of elements was an essential cornerstone for the development of a systematic chemistry. Rutherford's model of the atom brought a first understanding of the atomic structure, and it formed the basis on which the periodic system of elements could be explained. Quantum mechanics then provided a deeper understanding of Bohr's atom and of the periodic system. Progress in atomic theory thus started from the empirical observation of regularities, proceeded via a model, and it came to a conclusion with the discovery of the dynamical equations.

The time delay between recognizing regularities and explaining them fully was long. The Balmer formula was proposed in 1885; the Schrödinger equation made its appearance 40 years later. The periodic table of elements was discovered in 1869; its explanation in terms of the exclusion principle came 55 years later. Where do we stand in particle physics? As we shall see in the following sections, impressive regularities have been found and partially explained, but a deep and complete understanding still eludes us. We do not even know if all the keys are on hand or if they remain to be discovered.

13.1 The Ur-Particles

In 1949, Fermi and Yang noticed that a pion has the same quantum numbers as the s_0 state of a nucleon-antinucleon pair ($A = 0, J^\pi = 0^-$, $S = 0, q = 0$ or ± 1) and suggested that the pion could be considered as a bound state of the nucleon-antinucleon pair with a very large binding energy.[1] After the discovery of strange particles, Sakata pointed out that the model of Fermi and Yang could be extended to take into account the quantum number S by adding the lambda to the set of primary objects.[2] The three particles p, n, and Λ are sometimes called sakatons.

The triplet of sakatons suffices to "build" all particles, but some problems exist. The three fundamental particles would once appear as building blocks and once as a combined system, indicating a certain lack of symmetry. More important, however, particles can be constructed that are not observed in nature. Additional reasons exist to prefer a different set of primary objects. The most important one is the success of SU(3), and a word about this subject is in order here. SU(3) is a mathematical group, the "special unitary group of order 3." Gell-Mann, and independently Ne'eman, showed that this group can be used with enormous success to explain the systematics of particles.[3] We shall not treat the group theoretical approach here but shall remark that the discussion in the present chapter can be understood in terms of SU(3). A second important reason for assuming a new set of fundamental objects is the observation of a possible structure of the nucleons. If nucleons are indeed "made up" from some constituents, then it is logical to use such constituents as building blocks for all particles.

In the present section we shall introduce family relationships among particles by returning to earlier concepts. While these concepts are hard to test by direct experiments they form a useful guide to an understanding of the more formal aspects of particle models.

In Sections 8.2 and 8.5 we discussed symmetry breaking in an external magnetic field and by the electromagnetic interaction. As a reminder, we repeat three cases in Fig. 13.1. In the case of the proton, the two magnetic substates $M = \frac{1}{2}$ and $M = -\frac{1}{2}$ are degenerate in the absence of an external magnetic field. The magnetic field then lifts the degeneracy, and the differ-

1. E. Fermi and C. N. Yang, *Phys. Rev.* **76**, 1739 (1949).
2. S. Sakata, *Progr. Theoret. Phys.* **16**, 636 (1956).
3. See M. Gell-Mann and Y. Ne'eman, *The Eightfold Way*, Benjamin, Reading, Mass., 1964.

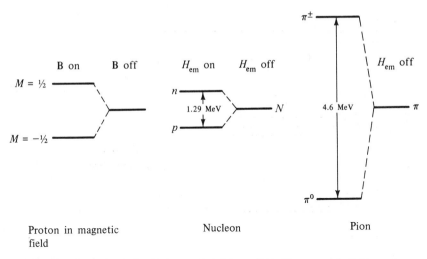

$M = \frac{1}{2}$

$M = -\frac{1}{2}$

$\pi^{\pm}$

n

1.29 MeV N 4.6 MeV π

p

π^0

Proton in magnetic field   Nucleon   Pion

Fig. 13.1. Mass (energy) splitting produced by a field. The magnetic field can be switched off; the two magnetic sublevels of the proton then become degenerate. The electromagnetic interaction, however, can be switched off only in a gedanken-experiment.

ent magnetic sublevels have different masses. The states are labeled by the values of the spin J and the magnetic quantum number M. In the case of the nucleon and the pion π, the electromagnetic interaction cannot be switched off, but it is assumed that the various members of the isomultiplet would have the same mass *if H_{em} were absent*. The various particle states can be labeled by I and I_3, the values of the isospin and its third component. These ideas have been discussed in detail in Chapter 8.

The question now arises: If part of the strong interaction could be switched off, would additional simplicity result? To answer the question, we look at particles with the same spin and parity within a reasonable mass range. To estimate the *reasonable mass range*, we note that the mass splitting due to the electromagnetic interaction is of the order of a few MeV, as indicated in Fig. 13.1. Since the hadronic interaction is about 100 times stronger than the electromagnetic one, a mass splitting of the order of a few hundred MeV can be expected. Since the pion is the lightest hadron, it is tempting to look first at the low-lying 0^- bosons. Table A4 in the Appendix lists nine such particles below 1 GeV: three pions, two kaons, two antikaons, the eta, and the eta-prime (or X^0). These particles are shown to the left in Fig. 13.2. In nature, only the positive and negative members of the same isomultiplet are degenerate, and all other particles possess different masses. If the weak interaction is switched off, the very small splitting between K^0 and $\overline{K^0}$ disappears. If in addition H_{em} is turned off, the neutral and the charged members of the same isospin multiplet become degenerate. Finally, it is assumed that all nine pseudoscalar mesons become degenerate if part of the hadronic interaction is turned off. We call the resulting nine-

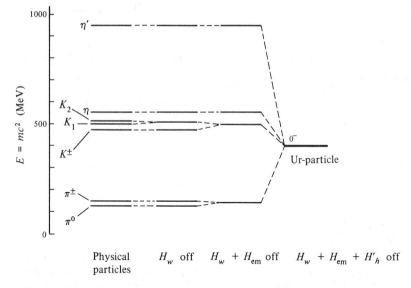

Fig. 13.2. The nine pseudoscalar mesons with mass below 1 GeV. At the left the masses are given as they occur in nature. Going to the right, first the weak interaction is switched off, then the electromagnetic interaction, and finally part of the hadronic interaction. The mass splittings caused by H_w and H_{em} are exaggerated. The position of the ur-particle is unknown.

fold degenerate pseudoscalar state the pseudoscalar ur-particle.[4] The mass of the ur-particle is determined by the part of the hadronic interaction that has not been switched off. According to Fig. 13.2, the ur-particle 0^- gives rise to a native family of nine different particles.

Closer inspection shows that three other ur-particles can be discerned in the region below about 1 GeV. The characteristics of the four ur-particles are summarized in Table 13.1. The mass or rest energy of the ur-particle is of course unknown; we have placed it approximately in the middle of the masses of the native particles. The set of anti-ur-particles

Table 13.1 Ur-Particles. The four lowest-lying ur-particles are listed. They give rise to a total of 36 particles. The rest energy is arbitrarily taken as the central energy of the multiplet.

Spin-parity, J^π	Rest energy (GeV)	Type	Members of the multiplet	Number of members
0^-	0.4	Boson	$\pi K \bar{K} \eta \eta'$	9
1^-	0.8	Boson	$\rho K^* \bar{K}^* \omega \phi$	9
$\frac{1}{2}^+$	1.2	Fermion	$N \Lambda \Sigma \Xi$	8
$\frac{3}{2}^+$	1.4	Fermion	$\Delta \Sigma^* \Xi^* \Omega$	10

4. Ur means original or primitive in German.

coincides with the set of the ur-particles for bosons, but the sets are distinct for fermions, with opposite baryon number.

The crucial question is now: Is this scheme useless, or can it be brought into more precise form? Does it then yield new predictions? To make the classification more quantitative, we turn to a discussion of quarks.

13.2 Quarks

We have stated above that all particles can be constructed from the three sakatons, neutron, proton, and the lambda. In 1964, Gell-Mann, and independently Zweig, suggested a different baryonic triplet, consisting of three hypothetical particles of remarkable properties.[5] Gell-Mann called his particles *quarks*, after *Finnegan's Wake*;[6] Zweig called his particles *aces*. The name quark won. At present, the quark model explains many features of elementary particles, but puzzles remain. However, the model is easy to grasp and gives a good insight into the structure of the hadron division of the elementary particle zoo.

To introduce quarks we ask: What are the simplest building blocks from which all elementary particles can be constructed? What are the quantum numbers of these building blocks, and how many do we need? The following discussion of these questions is not unique, but other quantum number assignments lead to equivalent physical predictions. The essential steps in building a quark scheme are

1. Quarks must be fermions; only with fermion building blocks can fermions *and* bosons be constructed.

2. In analogy to the original Fermi-Yang idea it is assumed that bosons are quark-antiquark pairs:

$$\text{boson} = (q\bar{q}). \tag{13.1}$$

Any fermion that is not one of the original quarks must be built from at least three quarks. It is assumed that the low-lying fermions are constructed from three quarks:

$$\text{fermion} = (qqq). \tag{13.2}$$

It is not excluded that particles with more constituents exist; bosons could, for instance, contain two quark-antiquark pairs, $(qq\bar{q}\bar{q})$, and fermions could be built from five quarks $(qqqq\bar{q})$. We restrict our considerations to the two simplest systems, $(q\bar{q})$ and (qqq).

3. To account for the nonstrange particles of charge 0 and ± 1, at least two quarks are needed. (Since quarks are assumed to be fermions, the statement "two quarks" implies the existence of two corresponding antiquarks which are different from the quarks.) The two nonstrange

5. M. Gell-Mann, *Phys. Letters* **8**, 214 (1964); G. Zweig, *CERN Report 8182/Th. 401* (1964) (unpublished).

6. James Joyce, *Finnegan's Wake*, Viking, New York, 1939, p. 383.

quarks could either both be isosinglets or be the two members of an isodoublet. With two $I = 0$ particles it is impossible to build $I = 1$ mesons, such as the pions. Consequently, the two nonstrange quarks must be members of an isodoublet. To construct strange mesons and strange baryons, at least one strange quark is needed. The smallest number of quark species required to build strange and nonstrange hadrons is *three*. We shall indeed find that three quarks and three antiquarks are sufficient to build all elementary particles.

The strange quark appears in only one form, and it must be an isosinglet. The quark triplet consists of an $I = \frac{1}{2}$ nonstrange doublet and an $I = 0$ strange singlet. The isospin and strangeness quantum numbers of the three quarks match those of the proton, the neutron, and the lambda, and we denote the quarks by q_p, q_n, and q_λ. By convention, the strangeness of q_λ is taken to be -1, just as for Λ^0.

4. In analogy to p, n, and Λ^0, the three quarks are taken to have identical space-time quantum numbers A, J, and π. Equations (13.1) and (13.2) then give

$$A(q) = -A(\bar{q}) = \tfrac{1}{3}. \tag{13.3}$$

Since quarks are fermions, their spins must be half-integer. For simplicity, the lowest value is taken and the parity is assumed to be positive:

$$J(q) = \tfrac{1}{2}, \qquad \pi(q) = +. \tag{13.4}$$

5. Assuming the Gell-Mann-Nishijima relation, Eq. (8.30), to be valid for quarks, the electric charges become[7]

$$q(q_p) = \tfrac{2}{3}e, \qquad q(q_n) = q(q_\lambda) = -\tfrac{1}{3}e. \tag{13.5}$$

The quantum numbers of the three quarks are summarized in Table 13.2. The prominent features are the fractional baryonic number and the

Table 13.2 PROPERTIES OF QUARKS

Quark	Quantum numbers						
	J	A	S	Y	I	I_3	q/e
q_p	$\frac{1}{2}$	$\frac{1}{3}$	0	$\frac{1}{3}$	$\frac{1}{2}$	$\frac{1}{2}$	$\frac{2}{3}$
q_n	$\frac{1}{2}$	$\frac{1}{3}$	0	$\frac{1}{3}$	$\frac{1}{2}$	$-\frac{1}{2}$	$-\frac{1}{3}$
q_λ	$\frac{1}{2}$	$\frac{1}{3}$	-1	$-\frac{2}{3}$	0	0	$-\frac{1}{3}$

fractional electric charges. These properties elevate the quarks to a unique position in the particle zoo. If they are ever found, they are certainly not going to be mixed up easily with other particles.

The quantum numbers listed in Table 13.2 lead to some definite conclusions:

7. With a little care there will be no confusion between the electric charge q and the quarks q.

1. Quarks cannot decay completely into known particles, since such decays would violate baryon and charge conservation.

2. Quarks can decay weakly into one another. The type of decay that can occur depends on the masses of the quarks. If, for instance, the strange quark q_λ is the heaviest, likely decays are

$$q_\lambda \longrightarrow q_p\pi^-, \qquad q_n\pi^0, \qquad q_n\gamma, \qquad q_pe^-\bar{\nu}, \qquad \dots \qquad \textbf{(13.6)}$$

If q_p and q_n have unequal masses, the heavier one can decay into the lighter one. The lightest quark is always stable.

13.3 Hunting the Quark

Do quarks really exist? Considerable effort has been spent by many experimental groups since 1964 to find quarks, but no conclusive positive evidence has yet been uncovered. The search is difficult because the mass of the quarks is not known. Fortunately the fractional electric charges would make quark signatures in careful experiments unambiguous.

In principle, quarks can be produced by high-energy protons through reactions of the type

$$pN \longrightarrow NNq\bar{q} + \text{bosons}$$
$$pN \longrightarrow Nqqq + \text{bosons}. \qquad \textbf{(13.7)}$$

The thresholds of these reactions depend on the mass m_q of the quarks; the magnitudes of the cross sections are determined by the forces between the hadrons and the quarks. Since neither the forces nor the quark masses are known, the search is an uncertain affair. If quarks are not found, one never knows if it is because they do not exist, because their mass is too high, or because the production cross section is too low.

The high energies required to produce massive particles are available in the biggest accelerators, in high-energy colliding beams, and in cosmic rays. Moreover, if quarks exist, and if the world were created in a "big bang," it is likely that quarks were produced during the very first stage when the temperature was exceedingly high. Some of these original quarks could still be around.

Quarks can be hunted at accelerators and in cosmic rays. Moreover, since at least one quark must be stable, they should have accumulated in the earth's crust, in meteorites, or in moon rocks. Quarks can be distinguished from other particles either by their fractional charge or by their mass. If the mass is studied, stability is taken as an additional criterion. If the charge is used as signature, the idea is simple. Equation (3.2) shows that the energy loss of a particle in matter is proportional to the square of its charge. A quark of charge $e/3$ would produce one ninth of the ionization of a singly charged particle of the same velocity. If the particle is relativistic, it produces approximately minimum ionization (see Fig. 3.5). A relativistic quark of charge $e/3$ would therefore show only one ninth of the minimum ionization, and it should have a very different appearance than an ordinary charged particle. A quark with charge $2e/3$ would yield four ninths of the standard ionization.

In reality, experiments are more complicated because it is difficult to find faint tracks. We shall not discuss any of the various experiments here because all the reliable ones have produced negative results.[8] If quarks are found, the excitement will be so enormous that the relevant experiment will be well advertised.

13.4 Mesons as Bound Quark States

According to Eq. (13.1), mesons are considered to be bound states of a quark-antiquark pair. The two spin-$\frac{1}{2}$ quarks q_p and $\bar{q}_n$ can be in a $^1S_0$ or a $^3S_1$ state, if we assume that their relative orbital angular momentum is zero.[9] The intrinsic parity of a fermion-antifermion pair is negative, and the two possibilities give

$$^1S_0 \qquad J^\pi = 0^- \qquad \text{pseudoscalar mesons}$$

$$^3S_1 \qquad J^\pi = 1^- \qquad \text{vector mesons.}$$

The interaction energies of these bound systems must be extremely large. The binding energy is given by

$$B = (2m_q - m_{\text{meson}})c^2. \qquad (13.8)$$

The best present experiments indicate that the quark mass must be larger than 5 GeV/c^2, if quarks indeed exist. The binding energy is thus of the order of 10 GeV. The possible combinations of the quark q and the antiquark $\bar{q}$ are

$$
\begin{array}{ccc}
q_p\bar{q}_p & q_n\bar{q}_p & q_\lambda\bar{q}_p \\[4pt]
q_p\bar{q}_n & q_n\bar{q}_n & q_\lambda\bar{q}_n \\[4pt]
q_p\bar{q}_\lambda & q_n\bar{q}_\lambda & q_\lambda\bar{q}_\lambda.
\end{array} \qquad (13.9)
$$

It is already evident that nine different mesons emerge, in agreement with the numbers listed in Table 13.1. However, the arrangement in Eq. (13.9) is not made according to quantum numbers, and comparison with the experimentally observed mesons is thus not obvious. In Table 13.1, the nine combinations are reordered according to the values of the hypercharge Y and the isospin component I_3. Table 13.2 is helpful in such rearrangements. The states in Table 13.3 can now be compared with the nine pseudoscalar

8. L. W. Jones, in *Symmetries and Quark Models* (R. Chand, ed.), Gordon and Breach, New York, 1970. G. Morpurgo, in *Subnuclear Phenomena*, 1969 Intern. Sch. of Physics "Ettore Majorana" (A. Zichichi ed). Academic Press, New York. 1970, p. 640.

9. The ordinary spectroscopic notation is used where the capital letter gives the orbital angular momentum, the subscript indicates the value of the total angular momentum, and the left superscript is equal to $2S + 1$, where S is the spin. $^3S_1$ thus denotes a state with $l = 0$, $J = 1$, $S = 1$, $2S + 1 = 3$.

Table 13.3 Reordering the $q\bar{q}$ States According to Hypercharge Y and Isospin Component I_3

	$I_3 = -1$	$-\frac{1}{2}$	0	$\frac{1}{2}$	1
$Y = \begin{cases} 1 \\ 0 \\ -1 \end{cases}$	$q_n\bar{q}_p$	$q_n\bar{q}_\lambda$ $q_\lambda\bar{q}_p$	$q_p\bar{q}_p, q_n\bar{q}_n, q_\lambda\bar{q}_\lambda$	$q_p\bar{q}_\lambda$ $q_\lambda\bar{q}_n$	$q_p\bar{q}_n$

and the nine vector mesons. For the *pseudoscalar mesons*, arranging these in the same scheme gives

$$\begin{array}{ccc} & K^0 & K^+ \\ \pi^- & \pi^0\eta^0\eta' & \pi^+ \\ & K^- \quad \overline{K^0} & \end{array} \qquad (13.10)$$

and for the *vector mesons*

$$\begin{array}{ccc} & K^{*0} & K^{*+} \\ \rho^- & \rho^0\omega^0\phi^0 & \rho^+. \\ & K^{*-} \quad \overline{K^{*0}} & \end{array} \qquad (13.11)$$

In both cases, the assignments for the six states in the *outer ring* are unambiguous. The three states in the center, however, have the same quantum numbers Y and I_3. How are the states $q_p\bar{q}_p$, $q_n\bar{q}_n$, and $q_\lambda\bar{q}_\lambda$ related to the corresponding mesons with $Y = I_3 = 0$? Since any linear combination of states $q_p\bar{q}_p$, $q_n\bar{q}_n$, and $q_\lambda\bar{q}_\lambda$ has the same quantum numbers, it is not possible to identify one quark combination with one meson. To get more information, we summarize the properties of the nonstrange neutral mesons in Table 13.4, which shows that it will be straightforward to find the quark

Table 13.4 Nonstrange Neutral Mesons

Meson	$I(J^\pi)$	Rest energy (MeV)	Meson	$I(J^\pi)$	Rest energy (MeV)
π^0	$1(0^-)$	135	ρ^0	$1(1^-)$	770
η^0	$0(0^-)$	549	ω^0	$0(1^-)$	784
η'	$0(0^-)$	958	ϕ^0	$0(1^-)$	1019

content of the neutral pion and the neutral rho: These two particles are members of isospin triplets. Knowing the quark assignment of the other members of the isotriplet should help. Consider, for instance, the three rho mesons:

$$\rho^+ = q_p\bar{q}_n, \qquad \rho_0 = ?, \qquad \rho^- = q_n\bar{q}_p.$$

The charged members of the rho do not contain a contribution from the strange quark q_λ in their wave function. The neutral rho forms an isospin triplet with its two charged relatives and thus should also not contain a strange component. Of the three products listed in the $I_3 = 0$, $Y = 0$ entry in Table 13.3, only the first two can appear, and the wave function must have the form

$$\rho^0 = \alpha q_n \bar{q}_n + \beta q_p \bar{q}_p.$$

Normalization and symmetry give

$$|\alpha|^2 + |\beta|^2 = 1, \qquad |\alpha| = |\beta|,$$

or

$$\alpha = \pm\beta = \frac{1}{\sqrt{2}}$$

If we were to add up two ordinary spin-$\frac{1}{2}$ particles to get a spin-1 system, it would be easy to select the correct sign: The linear combination must be an eigenfunction of J^2, with eigenvalue $j(j + 1)\hbar^2 = 2\hbar^2$. This condition determines that the sign is positive.[10] The situation here is different, because we are dealing with particle-antiparticle pairs and the antiparticle introduces a minus sign. We shall not justify the appearance of this minus sign because it will not occur in any measurable quantity in our discussion. The wave functions of the three rho mesons in terms of their quark constituents are

$$\rho^+ = q_p \bar{q}_n$$

$$\rho^0 = \frac{q_n \bar{q}_n - q_p \bar{q}_p}{\sqrt{2}} \tag{13.12}$$

$$\rho^- = q_n \bar{q}_p.$$

These quark combinations also apply to the pions; the difference between the rho and the pion lies in the ordinary spin. The rho is a vector meson ($J^\pi = 1^-$), while the pion is a pseudoscalar meson ($J^\pi = 0^-$). The other neutral mesons will be discussed in Section 13.6.

13.5 Baryons as Bound Quark States

Three quarks form a baryon. Since quarks are fermions, the overall wave function of the three quarks must be antisymmetric; the wave function must change sign under any interchange of two quarks:

$$|q_1 q_2 q_3\rangle = -|q_2 q_1 q_3\rangle. \tag{13.13}$$

To explain that the wave function of the three quarks must be antisymmetric, the ideas expounded in Chapter 8 are generalized. There, with the introduction of isospin, proton and neutron were considered to be two states of the same particle. The total wave function, including isospin, of a two-nucleon system then must be antisymmetric under exchange of the

10. Park, Eq. (6.43); Merzbacher, Eq. (16.85); G. Baym, *Lectures in Quantum Mechanics*, Benjamin, Reading, Mass., 1969, Chapter 15.

two nucleons. Here it is assumed that the three quarks are three states of the same particle, and Eq. (13.13) is then the expression of the Pauli principle. The simplest situation arises when the three quarks have no orbital angular momentum and have their spins parallel.[11] The resultant baryon then has spin $\frac{3}{2}$ and positive parity. As in the case of the mesons, it is straightforward to find the quantum numbers of the various quark combinations. Consider, for instance, the combination $q_p q_p q_p$:

$$q_p q_p q_p: \quad A = 1, \quad Y = 1, \quad I_3 = \tfrac{3}{2}, \quad q = 2e, \quad J^\pi = (\tfrac{3}{2})^+.$$

These are just the quantum numbers of the Δ^{++}, the doubly charged member of $N^*_{3/2}(1236)$. The three quarks q_p, q_n, and q_λ can be combined to form 10 combinations, and particles exist for all 10. The quark combinations and the corresponding baryons are shown in Fig. 13.3. Also indicated are the rest energies of the isomultiplets. Since there are 10 particles, the array is called the $(\tfrac{3}{2})^+$ *decimet* (or decuplet). The similarity to Escher's "Sky and Water I" on p. 352 is impressive, particularly if it is noted that the decimet of the antiparticles exists also.

Three spin-$\frac{1}{2}$ fermions can also be coupled to form a state with spin $\frac{1}{2}$ and positive parity. Table 13.1 indicates that only eight members of the $(\tfrac{1}{2})^+$ family are known. The eight particles and the corresponding quark combinations are shown in Fig. 13.4.

Two questions are raised by the comparison of existing particles and quark combinations in Fig. 13.4: (1) Why are the corner particles $q_p q_p q_p$, $q_n q_n q_n$, and $q_\lambda q_\lambda q_\lambda$ present in the $(\tfrac{3}{2})^+$ decimet but absent in the $(\tfrac{1}{2})^+$ octet? (2) Why does the combination $q_p q_n q_\lambda$ appear twice in the octet but only once in the decimet? Both questions have a straightforward answer:

1. No antisymmetric state with spin $\frac{1}{2}$ and zero angular momentum can be formed from three identical fermions. (Try!) The "corner particles"

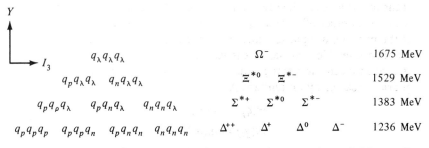

Fig. 13.3. Quarks and the $(\tfrac{3}{2})^+$ decimet. The states and the particles are arranged so that the x axis gives I_3, and the y axis Y. The rest energies are given at the right.

11. The lack of total orbital angular momentum does not mean that there is no orbital angular momentum between pairs of quarks. For three identical quarks, for instance, the spatial wave function must be antisymmetric. An example of an antisymmetric spatial wave function with total orbital angular momentum $L = 0$ is

$$\psi(\mathbf{r}_1, \mathbf{r}_2, \mathbf{r}_3) = (r_{12}^2 - r_{13}^2)(r_{23}^2 - r_{21}^2)(r_{32}^2 - r_{31}^2)\Phi(\mathbf{r}_1, \mathbf{r}_2, \mathbf{r}_3),$$

where $\Phi(\mathbf{r}_1, \mathbf{r}_2, \mathbf{r}_3)$ is a symmetric function, and $\mathbf{r}_{ij} \equiv \mathbf{r}_i - \mathbf{r}_j$.

$$q_p q_\lambda q_\lambda \qquad\qquad q_n q_\lambda q_\lambda \qquad\qquad\qquad \Xi^0 \qquad \Xi^- \qquad\qquad 1314 \text{ MeV}$$

$$\begin{array}{c} q_p q_n q_\lambda \\[2pt] q_p q_p q_\lambda \qquad\qquad\qquad q_n q_n q_\lambda \qquad \Sigma^+ \quad \Sigma^0 \quad \Sigma^- \qquad 1192 \text{ MeV} \\[2pt] q_p q_n q_\lambda \qquad\qquad\qquad\qquad\qquad \Lambda^0 \qquad\qquad 1115 \text{ MeV} \end{array}$$

$$q_p q_p q_n \qquad\qquad q_p q_n q_n \qquad\qquad\qquad p \qquad n \qquad\qquad 939 \text{ MeV}$$

Fig. 13.4. The $(\tfrac{1}{2})^+$ baryon octet and the corresponding quark combinations. The rest energies of the isomultiplets are given at the right.

in the $(\tfrac{1}{2})^+$ octet are therefore forbidden by the Pauli principle, Eq. (13.13), and indeed are not found in nature.

2. If the z component of each quark spin is denoted with an arrow, a state with $L = 0$ and $J_z = +\tfrac{1}{2}$ can be formed in three different ways:

$$q_p\!\uparrow q_n\!\uparrow q_\lambda\!\downarrow, \qquad q_p\!\uparrow q_n\!\downarrow q_\lambda\!\uparrow, \qquad q_p\!\downarrow q_n\!\uparrow q_\lambda\!\uparrow. \qquad\qquad \textbf{(13.14)}$$

From these three states, three different linear combinations can be formed that are orthogonal to each other and have a total spin J. Two of these combinations have spin $J = \tfrac{1}{2}$ and one has spin $J = \tfrac{3}{2}$. The one combination with $J = \tfrac{3}{2}$ turns up in the decimet; the two others are members of the octet.

13.6 The Hadron Masses

A remarkable regularity appears if the masses of the particles are plotted against their *quark content*. In the last two sections we have found definite assignments of quark combinations to most of the hadrons that comprise the set of the four ur-particles listed in Table 13.1. A careful look at the mass values of the various states shows that the mass depends strongly on the number of lambda quarks. In Fig. 13.5, the rest energies of most of the particles are plotted, and the number of strange quarks is indicated for each level. The masses of the various states can be understood if it is assumed that the nonstrange quarks are equally heavy but that the strange quark is heavier by an amount Δ:

$$\begin{aligned} m(q_p) &= m(q_n) \\ m(q_\lambda) &= m(q_p) + \Delta. \end{aligned} \qquad\qquad \textbf{(13.15)}$$

Figure 13.5 implies that the value of Δ is of the order of a few hundred MeV/c^2. The fact that the observed levels are not all equally spaced is not surprising. The mass of a meson built from quarks q_1 and $\bar{q}_2$ is given by

$$m = m(q_1) + m(\bar{q}_2) - \frac{B}{c^2}.$$

It is too much to hope that the binding energy B is the same for all mesons and baryons. B will very likely depend on the nature of the forces between

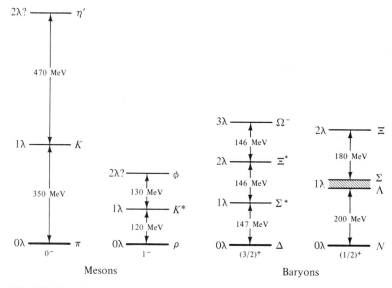

Fig. 13.5. Particle rest energies. Each level is labeled with the number of λ quarks that it contains.

quarks and on the state of the quarks. Figure 13.5 therefore provides only a crude value for the mass difference Δ.

A few observations follow directly from the simple arguments made so far. The first one concerns Ω^-. When Gell-Mann first introduced strangeness he conjectured that a particle with strangeness -3 should exist and he called it Ω^-. A few years later he introduced the group SU(3) into particle physics and predicted the mass of Ω^-.[3] The prediction can be easily understood by looking at Fig. 13.3. Once Gell-Mann had written down all particles except Ω^-, the top of the pyramid followed logically. Figure 13.5 shows that the energy differences between the three lower layers of the pyramid are about 146 MeV. Consequently, the top of the pyramid should be about 146 MeV above the rest energy of Ξ^* and that was where Ω^- was found.[12]

The second observation concerns the postulated ur-particles. Figure 13.5 can be interpreted as indicating that the splitting between the various levels shown in the first step from the right in Fig. 13.2 is caused by the fact that the force between a strange and a nonstrange hadron differs from that between two strange or two nonstrange ones. The mass of the ur-particle would be given by a hadronic force that does not distinguish between strange and nonstrange particles. In any case, in the quark model, the various particles belonging to one family are assumed to have similar

---

12. The first Ω^- was probably seen in a cosmic-ray experiment in 1954 [Y. Eisenberg, *Phys. Rev.* **96**, 541 (1954)]. The unambiguous discovery, however, occurred in 1964 [Barnes et al., *Phys. Rev. Letters* **12**, 204 (1964)]. See also W. P. Fowler and N. P. Samios, *Sci. Amer.* **211**, 36 (April 1964).

hadronic properties. If the force giving rise to the splitting could be switched off, all particles would collapse into one degenerate state. The *unitary multiplets* listed in Table 13.1 are consequently generalizations of isospin multiplets.

The third observation leads us back to the problem of the neutral mesons. This problem was only partially solved in Section 13.4. In Eq. (13.12) the quark composition of the ρ^0 was given, but ω^0 and ϕ^0 were left without assignment. Figure 13.5 implies that ϕ^0, which is about 130 MeV above K^*, contains two strange quarks:

$$\phi^0 = q_\lambda \bar{q}_\lambda. \tag{13.16}$$

The state function of ω^0 can now be found by setting

$$\omega^0 = c_1 q_p \bar{q}_p + c_2 q_n \bar{q}_n + c_3 q_\lambda \bar{q}_\lambda \tag{13.17}$$

The state representing ω_0 should be orthogonal to the states representing ρ^0 and ϕ^0. With Eqs. (13.16) and (13.12), the state of ω^0 then becomes

$$\omega^0 = \frac{1}{\sqrt{2}} (q_p \bar{q}_p + q_n \bar{q}_n), \tag{13.18}$$

and the mass of ω^0 should satisfy

$$m_{\omega^0} \approx m_{\rho^0}. \tag{13.19}$$

This prediction is in approximate agreement with reality.

13.7 Bootstrap and Regge Recurrences

Even if quarks are real particles and are discovered and their properties are elucidated, some questions will remain unanswered: Are quarks the ultimate building blocks, or are they again constructed from other entities? If they are composites, does the game continue indefinitely? If they are not, how can their existence be justified and understood? One fundamental approach to this problem has been suggested by Chew and Frautschi[13]: They assume that all *hadrons* are equally fundamental. Each hadron is assumed to "be made up" from all others so that it is impossible to say which ones are elementary and which ones composite. Gell-Mann called this picture *nuclear democracy*. It is assumed that such a model leads to self-consistency conditions and that these are such that the masses of all hadrons and their coupling constants are the unique result of the self-consistency requirement, or bootstrapping.[14]

To explain the idea of bootstrapping, consider a world in which only pions and rhos exist. The rho meson was discussed a number of times, particularly in Sections 5.3 and 10.7. It is a short-lived particle that decays

---

13. G. F. Chew and S. C. Frautschi, *Phys. Rev. Letters* **7**, 394 (1961).

14. Bootstrapping was probably first introduced by Hieronymus Karl Friedrich, Freiherr von Münchhausen (Oxford, 1786), who got out of a tight spot by lifting himself and his horse by his bootstraps. The French claim priority for Baron de Crac.

into two pions, and it can be considered as a resonant state of two pions. In our hypothetical world, it can be stable, but is still has the same quantum numbers as the real rho.

The force between two pions is assumed to be produced through the exchange of a virtual rho, as shown in Fig. 13.6(a). In a crude way, the force can be described as a potential with constant depth V_0 and radius R_0. The radius R_0 is of course given by the Compton wavelength of the rho,

$$R_0 = \frac{\hbar}{m_\rho c}.$$

(a) Pion-pion force

If the potential V_0, characterized by the coupling constant, g_ρ, is weak nothing special happens. However, for a square well of radius R_0 and depth V_0, a bound state occurs if

$$V_0 R_0^2 \geq \frac{\pi^2 \hbar^2}{4m_\pi}$$

or

$$V_0 \gtrsim \frac{\pi^2 \hbar^2}{4m_\pi R_0^2} = \frac{\pi^2}{4} m_\rho c^2 \left(\frac{m_\rho}{m_\pi}\right).$$

(b) Real rho

Similarly, for a sufficiently large g_ρ and relative angular momentum unity, a resonance (and for even stronger g_ρ a bound state) will occur. The resonance is just the rho, as indicated in Fig. 13.6(b). The pion-pion force, mediated by a virtual rho, has bootstrapped itself to form a real rho. The detailed calculation shows indeed that the force arising from the rho exchange is attractive in the $J = 1$, $I = 1$ state, and this state has the quantum numbers of the rho. In a complete self-consistent theory, the strength of the interaction and the mass and width of the rho should emerge. The extension of the approach to all observed particles is so difficult and complex that it is impossible to say if the bootstrap approach is indeed the correct way to look at the elementary particle zoo. We shall not pursue this line further but shall turn to a different aspect of the same problem, namely the description of particles as Regge recurrences.

Fig. 13.6. (a) The force between two pions is assumed to be produced entirely by rho mesons. (b) The force has become so strong that two pions bind to produce a rho.

In the search for a perfect nuclear democracy, Regge recurrences play an important role. We shall therefore sketch some relevant ideas, and we start by discussing the energy levels of the three-dimensional harmonic oscillator. Since these energy levels will reappear in the nuclear shell model in Chapter 15, the harmonic oscillator is treated here in more detail than would otherwise be necessary.[15] The physical facts are simple, but the complete mathematics is somewhat involved; only the parts needed here and in Chapter 15 are given.

---

15. The one-dimensional harmonic oscillator is treated, for instance, in Eisberg, Section 8.6. The three-dimensional oscillator can be found in Messiah, Section 12.15, or in detail in J. L. Powell and B. Crasemann, *Quantum Mechanics*, Addison-Wesley, Reading, Mass., 1961, Section 7.4.

A particle attracted toward a fixed point by a force proportional to the distance r' from the point has a potential energy

$$V(r') = \tfrac{1}{2}\kappa r'^2. \tag{13.20}$$

The Schrödinger equation for such a three-dimensional harmonic oscillator is

$$\nabla^2\psi + \frac{2m}{\hbar^2}\left(E - \frac{1}{2}\kappa r'^2\right)\psi = 0. \tag{13.21}$$

With the substitutions

$$\kappa = m\omega^2, \qquad r' = \left(\frac{\hbar}{m\omega}\right)^{1/2} r, \qquad E = \frac{1}{2}\hbar\omega\lambda, \tag{13.22}$$

the Schrödinger equation reads

$$\nabla^2\psi + (\lambda - r^2)\psi = 0. \tag{13.23}$$

Since the harmonic oscillator is spherically symmetric, it is advantageous to write the Schrödinger equation in spherical polar coordinates, r, θ, and and φ. In these coordinates, the operator ∇^2 becomes

$$\nabla^2 = \frac{1}{r^2}\frac{\partial}{\partial r}\left(r^2\frac{\partial}{\partial r}\right) - \frac{1}{r^2\hbar^2}L^2, \tag{13.24}$$

where L^2 is the operator of the square of the total angular momentum,

$$L^2 = -\hbar^2\left[\frac{1}{\sin\theta}\frac{\partial}{\partial\theta}\left(\sin\theta\frac{\partial}{\partial\theta}\right) + \frac{1}{\sin^2\theta}\frac{\partial^2}{\partial\varphi^2}\right]. \tag{13.25}$$

An ansatz of the form

$$\psi = R(r)\,Y_l^m(\theta, \varphi) \tag{13.26}$$

solves Eq. (13.23). Here, the Y_l^m are spherical harmonics given in Table A8 in the Appendix. Y_l^m is an eigenfunction of L^2 and L_z [compare Eq. (5.7)],

$$L^2 Y_l^m = l(l+1)\hbar^2 Y_l^m$$
$$L_z Y_l^m = m\hbar Y_l^m. \tag{13.27}$$

The radial wave function $R(r)$ satisfies

$$\frac{1}{r^2}\frac{d}{dr}\left(r^2\frac{dR}{dr}\right) + \left(\lambda - r^2 - \frac{l(l+1)}{r^2}\right)R = 0. \tag{13.28}$$

This equation can be solved in a straightforward way[15] and the results of interest here can be summarized as follows.[16] Equation (13.28) has acceptable solutions only if

$$E_N = (N + \tfrac{3}{2})\hbar\omega, \tag{13.29}$$

where N is an integer, $N = 0, 1, 2, \ldots$. The potential and the energy levels

---

16. Various definitions of the quantum numbers are in use. Our notation agrees with A. Bohr and B. R. Mottelson, *Nuclear Structure*, Vol. I. Benjamin, Reading, Mass. 1969. p. 220.

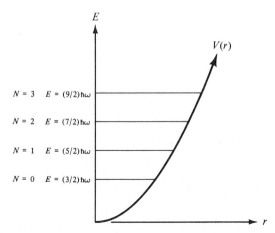

Fig. 13.7. Three-dimensional harmonic oscillator and its energy levels.

are shown in Fig. 13.7. The complete wave function is given by

$$\psi_{Nlm} = \left(\frac{2}{r}\right)^{1/2} \Lambda_k^{l+1/2}(r^2) Y_l^m(\theta, \varphi), \qquad k = \frac{1}{2}(N - l), \quad \textbf{(13.30)}$$

where $\Lambda(r^2)$ is a Laguerre function. It is related to the more familiar
Laguerre polynomials $L_k^\alpha(r)$ by

$$\Lambda_k^\alpha(r^2) = \left[\Gamma(\alpha + 1)\binom{k + a}{k}\right]^{-1/2} e^{-r^2/2} r^\alpha L_k^\alpha(r^2). \qquad \textbf{(13.31)}$$

At first, these functions appear terrifying. However, they become docile if
one simply looks up their properties and behavior in one of the many books
on mathematical physics[17]. The radial wave functions of the first three
levels ($N = 0, 1, 2$) are shown in Fig. 13.8. What is the physical meaning
of the indices N, l, and m? N has already been defined in Eq. (13.29); it
labels the energy levels. Equation (13.27) shows that l is the orbital angular
momentum quantum number; it is restricted to values $l \leq N$. For each
value of l, the magnetic quantum number m can assume the $2l + 1$ values
from $-l$ to l. The parity of each state is given by Eq. (9.10) as

$$\pi = (-1)^l.$$

States of even and odd parity exist, and consequently the possible orbital
angular momenta for a state with quantum number N are given by

$$
\begin{array}{llll}
N \text{ even} & \pi \text{ even} & l = 0, 2, \ldots, N \\
N \text{ odd} & \pi \text{ odd} & l = 1, 3, \ldots, N.
\end{array}
\qquad \textbf{(13.32)}
$$

The degeneracy of each level N can now be obtained by counting: The
possible angular momenta are determined by Eq. (13.32); each angular

17. For example, P. M. Morse and H. Feshbach, *Methods of Theoretical Physics*, Mc-
Graw-Hill, New York, 1953, Section 12.3, Eq. (12.3.37).

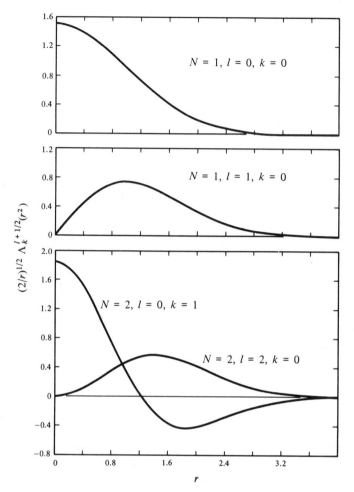

Fig. 13.8. Normalized radial wave functions $(2/r)^{1/2}\Lambda$ for the three-dimensional harmonic oscillator. The distance r is measured in units of $(\hbar/m\omega)^{1/2}$.

momentum contributes $2l + 1$ substates, and the total degeneracy becomes

$$\text{degeneracy} = \tfrac{1}{2}(N + 1)(N + 2). \tag{13.33}$$

The radial wave function $R(r) = (2/r)^{1/2}\Lambda$ is characterized by the number, n_r, of its nodes. It is customary to exclude nodes at $r = 0$ and include nodes at $r = \infty$ in counting. The examples in Fig. 13.8 then show that

$$n_r = 1 + k = 1 + \tfrac{1}{2}(N - l). \tag{13.34}$$

This relation is valid for all radial wave functions $R(r)$.

After this long preparation we return to our goal, connecting the properties of the harmonic oscillator to particle models. A state of a particle can be characterized by its mass (energy) and its angular momentum. For

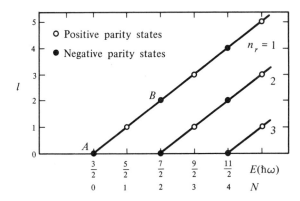

Fig. 13.9. Plot of the angular momentum versus energy for the states of the three-dimensional harmonic oscillator.

the harmonic oscillator, a plot of the angular momentum, l, versus energy is shown in Fig. 13.9. The states in Fig. 13.9 can be ordered into families in a variety of ways: States with equal values of N, of l, or of n_r can be connected. In Fig. 13.9, the last possibility has been chosen, and the result is a series of straight-line trajectories that rise with increasing energy. The straightness is a property of the harmonic oscillator; if a different potential shape is chosen, the trajectories will in general no longer be straight, but the general appearance remains. Why have levels with equal n_r rather than equal l been connected? The quantum numbers l and n_r have a different physical origin. We can, in principle, take a quantum mechanical system and spin it with various values of its angular momentum without changing its internal structure. The quantum number l describes the behavior of the system under rotations in space, and it can be called an *external quantum number*. The number of radial nodes, however, is a property of the structure of the state, and n_r (like intrinsic parity) can be called an *internal quantum number*. In this sense the states on one trajectory have a similar structure. Actually, the particles lying on a given trajectory can be further subdivided: States with the same parity recur at intervals $\Delta l = 2$.

The behavior of the harmonic oscillator states as expressed by Fig. 13.9 can be interpreted as saying that all states are equally fundamental. There is no essential difference between ground state and excited states, and no state appears as a fundamental building block. State B in Fig. 13.9 is considered to be state A recurring with a higher angular momentum. The question now arises: Do particles show a similar behavior if the particle mass is plotted versus particle spin for particles that have the same internal quantum numbers? Indeed, pronounced regularities appear,[18] and we present one example in Fig. 13.10, namely the hyperons with

18. V. D. Barger and D. B. Cline, *Phenomenological Theories of High Energy Scattering, An Experimental Evaluation*, Benjamin, Reading, Mass., 1969.

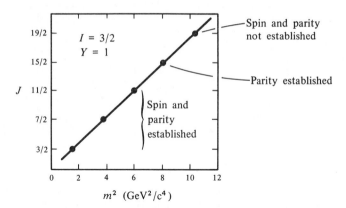

Fig. 13.10. Plot of the spin versus (mass)² for isospin-$\frac{3}{2}$, hypercharge-1, positive-parity hyperons. (Chew-Frautschi plot.)

isospin $\frac{3}{2}$, hypercharge 1, and positive parity. The plot becomes most impressive if the spin is plotted versus the square of the particle mass. The appearance of a family is clear, and the similarity to Fig. 13.9 is evident. The higher-mass particles are called Regge recurrences of the lowest-mass state. As in the case of the harmonic oscillator, the situation is interpreted by saying that such plots provide evidence for subatomic democracy: All states appear on an equal footing, and none is more fundamental than another. We shall finish the discussion of the particle models with this observation but also note that Reggiology, the science (or art) of the representation of particles in an angular momentum plane, only starts here.

13.8 Outlook and Problems

We have only scratched the surface of particle models. The detailed discussion goes far deeper and involves not only the assignment of particles to quark combinations but includes discussion of particle properties, such as decays and form factors, and of scattering at high energies. The quark model is very successful. While it does not explain all facts, it manages to correlate a great number of observations. The success leads to a number of questions; a few of these are fundamental enough so that they should be listed here:

1. Do quarks really exist? If they do, what are their properties and interactions? If they are not found, why not? This question is particularly intriguing to experimental physicists who would like to put their hands, or at least their eyes, on a real quark. Theorists profess relaxed detachment and claim that theoretical quarks are just as valuable to them as real ones. Only experiment will be able to decide the question, but if quarks are not

discovered, an explanation for the success of the quark model must be found on some other basis.

2. The basis of our discussion was the quark model with assignment boson $= (q\bar{q})$ and fermion $= (qqq)$. Moreover, only states with orbital angular momentum $L = 0$ were considered. Why is this simple model so successful? Do states with $L \neq 0$ or with assignments

$$\text{boson} = (q\bar{q}q\bar{q}), \qquad \text{fermion} = (qqqq\bar{q})$$

exist? If not, why not? If yes, where are they and what are their properties?

3. The properties outlined in the previous paragraph raise questions concerning the character of the forces between quarks. A number of serious puzzles exist. One of the most difficult ones can be stated as follows: In Section 13.5, it was pointed out that three identical quarks, for instance $q_\lambda q_\lambda q_\lambda$, give rise to a hadron with spin $\frac{3}{2}$, in this case the Ω^-. Total angular momentum $L = 0$ is assumed; since the spin state of the quarks in Ω^- is symmetric, the space state must be antisymmetric. For most forces, the lowest space state is symmetric (s state), and while it is possible to form an antisymmetric space state with $L = 0$,[19] it is difficult to invent forces that yield such a state as ground state. The question of the forces between quarks, and the possible existence of strong three-body forces among them, is still unsolved. The dynamical equations that govern the motion of quarks inside hadrons are unknown.

4. Is there an alternative way to explain all results obtained from the quark model? It appears as if most of the results can also be obtained by other theories. However, the quark model is the one that can most easily be explained in nontechnical terms. It has nearly always turned out that particles that were postulated on a reasonably firm basis have also turned up in the actual zoo, even if the delay between prediction and experimental discovery has been excruciatingly long in some instances. We can hope that the work at the ISR at CERN and at NAL will answer at least some of the questions concerning quarks.

5. We have only briefly touched on the subject of Regge poles, but an enormous number of open problems exist here also. Do Regge poles really describe all particles? Will a bootstrap approach ultimately explain all of hadron physics?

6. All remarks in the present chapter apply to hadrons. No model exists that "explains" the leptons. Attempts to introduce leptonic quarks have failed, and it is still unclear if hadrons and leptons are related. It is also unknown if the presently known leptons exhaust all possibilities or if leptons exist that are heavier than the muon.

7. How are quarks related to partons? Since the existence of neither particle is established, the question is somewhat premature. If one or both of the hypothetical entities are found, it will be easier to answer. If they do

19. R. H. Dalitz, in *High Energy Physics*, École d'Été de Physique Théorique (C. DeWitt and M. Jacob, eds.), Les Houches, 1965, Gordon & Breach, New York, 1965.

not turn up, the relevant experiments will at some point have to be described in a unified picture.

13.9 References

The following two articles provide very easy and readable introductions to the basic ideas of the quark model:

Ya. B. Zel' dovich, *Soviet Phys. Usp.* **8**, 489 (1965).

L. Brown, *Phys. Today* **19**, 44 (Febr. 1966).

On a considerably higher level, the quark model is discussed in three books:

B. T. Feld, *Models of Elementary Particles*, Ginn/Blaisdell, Waltham, Mass., 1969.

J. J. J. Kokkedee, *The Quark Model*, Benjamin, Reading, Mass., 1969.

D. B. Lichtenberg, *Unitary Symmetry and Elementary Particles*, Academic Press, New York, 1970.

Recent results are summarized in G. Morpurgo, "A Short Guide to the Quark Model," *Ann. Rev. Nucl. Sci.* **20**, 105 (1970). This article also contains an up-to-date list of references, including citations of reviews and theoretical and experimental papers.

Unitary symmetry is closely connected to the quark model. A good first introduction is

H. J. Lipkin, *Lie Groups for Pedestrians*, North-Holland, Amsterdam, 1965.

The original papers on unitary symmetry are collected in M. Gell-Mann and Y. Ne'eman, *The Eightfold Way*, Benjamin, Reading, Mass., 1964.

No low-level introduction to Regge poles is known to us. At an advanced level, a large number of books and reviews exist. Relatively easy to read are

R. J. Eden, *High Energy Collisions of Elementary Particles*, Cambridge University Press, Cambridge, 1967.

R. Omnès and M. Froissart, *Mandelstam Theory and Regge Poles, An Introduction for Experimentalists*, Benjamin, Reading, Mass., 1963.

A list of references and a summary of the experimental information is given in V. D. Barger and D. B. Cline, *Phenomenological Theories of High Energy Scattering, An Experimental Evaluation*, Benjamin, Reading Mass., 1969. See also D. V. Shirkov, *Soviet Phys. Usp.* **13**, 599 (1971).

Bootstrapping is treated in G. F. Chew, *Phys. Rev.* **D4**, 2330 (1971), and G. F. Chew, *The Analytic S Matrix*, Benjamin, Reading, Mass, 1966.

PROBLEMS

13.1. Assume that the nonstrange quarks q_p and q_n are stable. Describe their fate upon entering a solid. What will the ultimate fate of either one be, and where do you expect them to come to rest?

13.2. Describe the possible ways to search for quarks at accelerators. How are quarks distinguished from other particles? What limits the mass of the quark that can be found?

13.3. Would quarks be seen in a Millikan-type (oil drop) experiment? Estimate the lower limit of the concentration that can be observed in an ordinary oil droplet experiment. How can the approach be improved?

13.4. Use the quark model to compute the ratio of the magnetic moment of the proton to that of the neutron.

13.5. Use a simple potential well with range given by the exchange of a vector meson to justify that the nonrelativistic treatment of the quarks in the simple quark model is reasonable.

13.6. Justify that only one baryon state can be formed from three identical quarks with $L = 0$; verify that this state corresponds to a particle with spin $\frac{3}{2}$.

13.7.
(a) Prove that the square of the sum of two angular momentum operators, **J** and **J'**, can be written as

$$(\mathbf{J} + \mathbf{J'})^2 = J^2 + J'^2 + 2\mathbf{J}\cdot\mathbf{J'}$$
$$= J^2 + J'^2 + 2J_z J'_z + J_+ J'_- + J_- J'_+,$$

where

$$J_{\pm} = J_x \pm iJ_y, \qquad J'_{\pm} = J'_x \pm iJ'_y$$

are raising and lowering operators with properties as given in Eq. (8.27).
(b) Consider the two quark states

$$|\alpha\rangle = |q_p\uparrow\rangle|q_n\downarrow\rangle$$

$$|\beta\rangle = |q_p\downarrow\rangle|q_n\uparrow\rangle,$$

where, for instance, $|q_p\uparrow\rangle$ denotes a *p*-quark with spin up ($J_z = \frac{1}{2}$) and $|q_n\downarrow\rangle$ denotes an *n*-quark with spin down ($J'_z = -\frac{1}{2}$). Use the result of part (a) to find the linear combinations of the states $|\alpha\rangle$ and $|\beta\rangle$ that correspond to values $J_{\text{tot}} = 1$ and $J_{\text{tot}} = 0$ of the total angular momentum quantum number of the two quarks.

13.8. Verify Eqs. (13.18) and (13.19). Why should the various particle states be orthogonal to each other?

13.9. Apply the argument that leads to Eq. (13.19) to the neutral pseudoscalar mesons. Try to find possible explanations why the agreement with experiment is much less satisfactory than for the vector mesons.

13.10. Instead of the assignments made in Table 13.2, one could choose

	J	A	S	I	I_3
q_p'	$\frac{1}{2}$	1	0	$\frac{1}{2}$	$\frac{1}{2}$
q_n'	$\frac{1}{2}$	1	0	$\frac{1}{2}$	$-\frac{1}{2}$
q_λ'	$\frac{1}{2}$	1	-1	0	0

(a) What is q/e and Y for each quark in this case?
(b) Mesons would be constructed from $q'\bar{q}'$, as before, but baryons would be constructed from $q'q'\bar{q}'$, where $\bar{q}'$ is an antiquark. Would this assignment work? Explain any difficulties that are encountered.
(c) Why is this model not used?
[S. Sakata, Progr. Theoret. Phys. **16**, 686 (1956)].

13.11. If a real quark is ever seen, how would it be kept in captivity? To what uses could it be put?

13.12.
(a) Show that "normal" quark configurations

for bosons, $B = (q\bar{q})$, must satisfy the conditions

$$|S| \leq 1, \qquad |I| \leq 1, \qquad \left|\frac{q}{e}\right| \leq 1.$$

(b) Have "exotic" mesons, i.e., mesons that do not satisfy these conditions, been found?

13.13. Verify Eq. (13.23).

13.14. Show that L^2, Eq. (13.25), is indeed the operator of the square of the orbital angular momentum.

13.15. Show that $R(r)$ satisfies Eq. (13.28).

13.16. Prove Eq. (13.33).

13.17. Prepare a plot similar to Fig. 13.9 for the energy levels of the hydrogen atom.

13.18. Find two more examples of Regge plots (Fig. 13.10) and compare the slopes of the trajectories.

14

Liquid Drop Model, Fermi Gas Model

We have seen in Chapter 12 that the description of nuclear forces is not complete and that some ambiguities remain even in the energy range below about 350 MeV. Moreover, with the best forces, computation of nuclear properties ab initio are extremely difficult and tax the best computers. The force is very complicated, and nuclei are many-body problems. It is therefore necessary with most nuclear problems to simplify the approach and use specific nuclear models combined with simplified nuclear forces.

Nuclear models generally can be divided into independent particle models (IPM) in which the nucleons are assumed, in lowest order, to move nearly independently in a common nuclear potential and strong interaction (collective) models (SIM) in which the nucleons are strongly coupled to each other. The simplest SIM is the liquid drop model; the simplest IPM is the Fermi gas model. Both of these will be treated in this chapter. In the following two chapters we shall discuss the shell model (IPM) in which nucleons move nearly independently in a static spherical potential determined by the nuclear density distribution, and the collective model (SIM) in which collective motions of the nucleus are considered. The unified model combines features of the shell and of the collective model: The nucleons are assumed to move nearly independently in a common, slowly changing, nonspherical potential, and excitations of the individual nucleons and of the entire nucleus are considered.

14.1 The Liquid Drop Model

One of the most striking facts about nuclei is the approximate constancy of nuclear density: The volume of a nucleus is proportional to the number A of constituents. The same fact holds for liquids, and one of the early nuclear models, introduced by Bohr[1] and von Weizsäcker,[2] was patterned after liquid drops; nuclei are considered to be nearly incompressible liquid droplets of extremely high density. The model leads to an understanding of the trend of binding energies with atomic number, and it also gives a physical picture of the fission process. We shall sketch the simplest aspects of the liquid drop model in the present section.

In Section 5.3, nuclear mass measurements were introduced, and in Section 5.4 some basic features of nuclear ground states were mentioned. In particular, Fig. 5.20 represents a plot of the stable nuclei in a NZ plane. We return to the nuclear masses here, and we shall describe their behavior in more detail than in Chapter 5. Consider a nucleus consisting of A nucleons, Z protons, and N neutrons. The total mass of such a nucleus is somewhat smaller than the sum of the masses of its constituents because of the binding energy B which holds the nucleons together. For bound states, B is positive and represents the energy that is required to disintegrate the nucleus into its constituent neutrons and protons. B is given by

$$\frac{B}{c^2} = Zm_p + Nm_n - m_{\text{nuclear}}(Z, N). \tag{14.1}$$

Here, $m_{\text{nuclear}}(Z, N)$ is the mass of a nucleus with Z protons and N neutrons. It has become customary to quote *atomic* and not *nuclear* masses. The atomic mass includes the masses of all the electrons. The unit of atomic mass has been defined to be one twelfth of the mass of the atom ^{12}C; it is called mass unit and abbreviated u. In terms of MeV and g, u is given by

$$1 \text{ u} = 931.481 \text{ MeV} = 1.66043 \times 10^{-24} \text{ g}. \tag{14.2}$$

In terms of the atomic mass $m(Z, N)$, the binding energy can be written as

$$\frac{B}{c^2} = Zm_H + Nm_n - m(Z, N). \tag{14.3}$$

A small term due to atomic binding effects is neglected in Eq. (14.3); m_H is the mass of the hydrogen atom. The difference between the atomic rest energy $m(Z, N)c^2$ and the mass number times u is called the mass excess (or mass defect),

$$\Delta = m(Z, N)c^2 - A \text{ u}. \tag{14.4}$$

1. N. Bohr, *Nature* **137**, 344 (1936).
2. C. F. von Weizsäcker, *Z. Physik* **96**, 431 (1935).

Values of the mass excess are listed in Table A6 in the Appendix. Comparison between Eqs. (14.3) and (14.4) shows that $-\Delta$ and B measure essentially the same quantity but differ by a small energy. Tables usually list Δ because it is the quantity that follows from mass-spectroscopic measurements. The average binding energy per nucleon, B/A, is plotted in Fig. 14.1. The binding energy curve exhibits a number of interesting features:

1. Over most of the range of stable nuclei, B/A is approximately constant and of the order of 8–9 MeV. The constancy is another expression of the saturation of nuclear forces discussed in Section 12.5. If all nucleons inside a nucleus would be pulled into each other's force range, the total binding energy would be expected to increase proportionally to the number of bonds or approximately proportionally to A^2. B/A would then be proportional to A.

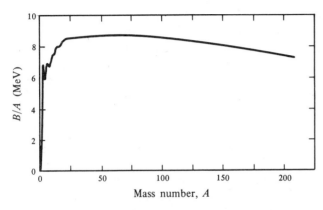

Fig. 14.1. Binding energy per particle for nuclei.

2. B/A reaches its maximum in the region of iron ($A \approx 60$). It drops off slowly toward large A and more steeply toward small A. This behavior is crucial for nuclear power production: If a nucleus of, say, $A = 240$ is split into two parts with $A \approx 120$, the binding of the two parts is stronger than that of the original nuclide, and energy is set free. This process is responsible for energy production in fission. At the other end, if two light nuclides are fused, the binding of the fused system will be stronger, and energy will again be set free. This energy release is the base for energy production in fusion.

The regularity and near-constancy of the binding energy B/A as a function of mass number A suggests that it should be possible to express the nuclear masses by a simple formula. The first semiempirical mass formula was obtained by von Weizsäcker, who noted that the constant average binding energy per particle and the constant nuclear density suggested a liquid drop model.[2] The primary fact needed to arrive at a mass formula is the tendency of B/A to be approximately constant for $A \gtrsim 50$. The binding energy per particle for an infinite nucleus without surface thus should have a constant value, a_v, the binding energy of nuclear matter. Since there are A particles in the nucleus, the *volume contribution* E_v to the binding energy is

$$E_v = +a_v A. \tag{14.5}$$

Nucleons at the *surface* have fewer bonds and the finite size of a real nucleus leads to a contribution E_s to the energy that is proportional to the surface area and decreases the binding energy,

$$E_s = -a_s A^{2/3}. \tag{14.6}$$

Volume and surface terms correspond to a liquid drop model. If only these two terms were present, isobars would be stable regardless of the value of N and Z. Figure 5.20, however, demonstrates that only nuclides in a narrow band are stable. For lighter nuclides, the self-conjugate isobars ($N = Z$ or $A = 2Z$) are the most stable ones, whereas heavier stable isobars have $A > 2Z$. These features are explained by two additional terms, a symmetry term and the Coulomb energy.

The *Coulomb energy* is caused by the repulsive electrical force acting between any two protons; this energy favors isobars with a neutron excess. For simplicity we assume that the protons are uniformly distributed throughout a spherical nucleus of radius $R = R_0 A^{1/3}$; with Eq. (8.37), the Coulomb energy becomes

$$E_c = -a_c Z^2 A^{-1/3}. \tag{14.7}$$

The fact that only nuclides in a small band are stable is explained by another term, the symmetry energy. The effect of the *symmetry energy* is best seen if the mass excess Δ is plotted versus Z for all isobars characterized by a given value of A. As an example, such a plot is shown in Fig. 14.2 for $A = 127$. The figure appears like a cross section through a deep valley; the isobar at the bottom is the only stable one, and the ones clinging to the steep sides tumble down toward the bottom of the valley, usually by emission of electrons or positrons. The isobars with $A = 127$ are not an isolated case; as Table A6 in the Appendix indicates, the mass excesses for all other isobars also are shaped like cross sections through a valley. Figure 5.20 can therefore be brought into a more informative form by adding a third dimension to the plot: the binding energy or the mass excess. Such a plot is analogous to a topographic map, and Fig. 14.3 presents the contour map of the binding energy in an NZ plane. Figure 14.2 is the cross section through the valley at the position indicated in Fig. 14.3. The sides of the valley are steep, and it is consequently not possible experimentally to explore the valley to the "top" because the nuclei are too shortlived. This fact is expressed in Fig. 14.3 by dotting the contour lines approximately where they can no longer by studied experimentally.

The symmetry energy arises because the exclusion principle makes it more expensive in energy for a nucleus to have more of one type of nu-

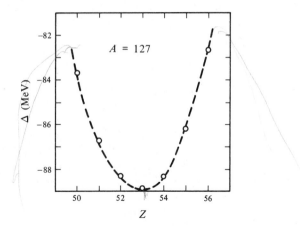

Fig. 14.2. Mass excess Δ as a function of Z for $A = 127$.

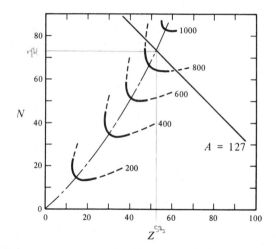

Fig. 14.3. The binding energy B is plotted in form of a contour map in an $N = Z$ plane. The *energy valley* appears clearly; it forms a canyon in the $N = Z$ plane. The numbers on the contour lines give the total binding energy in MeV.

cleon than the other. In the following section we shall derive an approximate expression for the symmetry energy; it is of the form

$$E_{\text{sym}} = -a_{\text{sym}}\frac{(Z - N)^2}{A}. \tag{14.8}$$

Collecting the terms gives the Bethe-Weizsäcker relation for the binding energy of a nucleus (Z, N),

$$B = a_v A - a_s A^{2/3} - a_{\text{sym}}(Z - N)^2 A^{-1} - a_c Z^2 A^{-1/3}. \tag{14.9}$$

The binding energy per particle becomes

$$\frac{B}{A} = a_v - a_s A^{-1/3} - a_{\text{sym}}\frac{(Z - N)^2}{A^2} - a_c Z^2 A^{-4/3}. \tag{14.10}$$

The constants in these relations are determined by fitting the experimentally observed binding energies; a typical set is

$$a_v = 15.6 \text{ MeV}$$
$$a_s = 17.2 \text{ MeV}$$
$$a_{\text{sym}} = 22.5 \text{ MeV}$$
$$a_c = 0.70 \text{ MeV}.$$

(14.11)

With these values, the general trend of the curves shown in Figs. 14.1 and 14.2 is reproduced well. Of course, finer details are not given, and relations with many more terms are employed when small deviations from the smooth behavior are studied.[3] Two remarks concerning the binding energy relation are in order. (1) Here, we have assumed that the coefficients in Eq. (14.9) are adjustable parameters to be determined by experiment. In a more thorough treatment of nuclear physics, the coefficients are derived from the characteristics of nuclear forces. In particular the calculation of the most important coefficient, a_v, has occupied theoretical physicists for a long time because it is intimately related to the properties of *nuclear matter*. Nuclear matter is the state of matter that would exist in an infinitely large nucleus. The closest approximation to nuclear matter presumably exists in neutron stars (Chapter 18). (2) The Bethe-Weizsäcker relation can be used to explore the stability properties of matter by extrapolating to regions that are not well known. Such studies are important, for instance, in the investigation of very heavy artificial elements, in the treatment of nuclear explosions, and in astrophysics.

14.2 The Fermi Gas Model

The semiempirical binding-energy relation obtained in the previous section is based on treating the nucleus like a liquid drop. Such an analogy is an oversimplification, and the nucleus has many properties that can be explained more simply in terms of independent-particle behavior rather than in terms of the strong-interaction picture implied by the liquid drop model. The most primitive independent-particle model is obtained if the nucleus is treated as a degenerate Fermi gas of nucleons. The nucleons are assumed to move freely, except for effects of the exclusion principle, throughout a sphere of radius $R = R_0 A^{1/3}$, $R_0 = 1.3$ fm. The situation is represented in Fig. 14.4 by two wells, one for neutrons and one for protons. Free neutrons and free protons, far away from the wells, have the same energy, and the zero levels for the two wells are the same. However, the two wells have different shapes and different depths because of the Cou-

3. See, for instance, J. Wing and P. Fong, *Phys. Rev.* **136**, B923 (1964); N. Zeldes, M. Gronau, and A. Lev, *Nucl. Phys.* **63**, 1 (1965); or G. T. Garvey, W. J. Gerace, R. L. Jaffe, I. Talmi, and I. Kelson, *Rev. Modern Phys.* **41**, No. 4, Part II (1969).

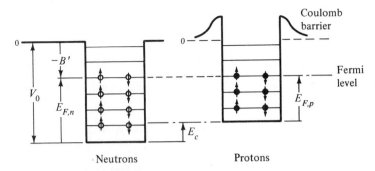

Fig. 14.4. Nuclear square wells for neutrons and protons. The well parameters are adjusted to give the observed binding energy B'.

lomb energy, Eq. (8.37): The bottom of the proton well is higher than the bottom of the neutron well by an amount E_c, and the proton potential sports a *Coulomb barrier*. Protons that try to enter the nucleus from the outside are repelled by the nuclear charge; they must either "tunnel" through the barrier or have enough energy to pass over it.

The wells contain a finite number of levels. Each level can be occupied by two nucleons, one with spin up and one with spin down. It is assumed that the *nuclear temperature* is so low that the nucleons occupy the lowest states available to them. Such a situation is described by the term *degenerate Fermi gas*. The nucleons populate all states up to a maximum kinetic energy equal to the Fermi energy E_F. The total number, n, of states with momenta up to p_{max} follows from Eq. (10.26), after integration over d^3p, as

$$n = \frac{V p_{max}^3}{6\pi^2 \hbar^3}. \tag{14.12}$$

Each momentum state can accept *two* nucleons so that the total number of one species of nucleons with momenta up to p_{max} is $2n$. If neutrons are considered, then $2n = N$, the number of neutrons, and N is given by

$$N = \frac{V p_N^3}{3\pi^2 \hbar^3}. \tag{14.13}$$

Here p_N is the maximum neutron momentum, and V is the nuclear volume. With $V = 4\pi R^3/3 = 4\pi R_0^3 A/3$, the maximum neutron momentum follows from Eq. (14.13) as

$$p_N = \frac{\hbar}{R_0} \left(\frac{9\pi N}{4A} \right)^{1/3}. \tag{14.14}$$

Similarly, the maximum proton momentum is obtained as

$$p_Z = \frac{\hbar}{R_0} \left(\frac{9\pi Z}{4A} \right)^{1/3}. \tag{14.15}$$

The appropriate value of the Fermi energy can be found by considering self-conjugate nuclei for which $N = Z$. Equation (14.14), after inserting the numerical values and using the nonrelativistic relation between energy and momentum, then yields

$$E_F = \frac{p_F^2}{2m} \approx 50 \text{ MeV}. \tag{14.16}$$

The average kinetic energy per nucleon can also be calculated, and it gives

$$\langle E \rangle = \frac{\int_0^{p_F} E d^3p}{\int_0^{p_F} d^3p} = \frac{3}{5}\left(\frac{p_F^2}{2m}\right) \approx 30 \text{ MeV}. \tag{14.17}$$

With Eqs. (14.14) and (14.15) the total average kinetic energy becomes

$$\langle E(Z, N) \rangle = N\langle E_N \rangle + Z\langle E_Z \rangle = \frac{3}{10m}(Np_N^2 + Zp_Z^2)$$

or

$$\langle E(Z, N) \rangle = \frac{3}{10m}\frac{\hbar^2}{R_0^2}\left(\frac{9\pi}{4}\right)^{2/3}\frac{(N^{5/3} + Z^{5/3})}{A^{2/3}}. \tag{14.18}$$

Equal masses for proton and neutron and equal radii for the proton and neutron wells have been assumed. Moreover, neutrons and protons move independently of each other. The interaction between the various particles has been replaced by the boundary of the nucleus, represented by the potential well.

For a given value of A, $\langle E(Z, N) \rangle$ has a minimum for equal numbers of protons and neutrons, or $N = Z = A/2$. To study the behavior of $\langle E(Z, N) \rangle$ around this minimum we set

$$Z - N = \epsilon$$

$$Z + N = A \qquad \text{(fixed)}$$

or $Z = \frac{1}{2}A(1 + \epsilon/A)$, $N = \frac{1}{2}A(1 - \epsilon/A)$, and assume that $(\epsilon/A) \ll 1$. With

$$(1 + x)^n = 1 + nx + \frac{n(n-1)}{2}x^2 + \cdots,$$

and after reinserting $Z - N$ for ϵ, Eq. (14.18) becomes, near $N = Z$,

$$\langle E(Z, N) \rangle = \frac{3}{10m}\frac{\hbar^2}{R_0^2}\left(\frac{9\pi}{8}\right)^{2/3}\left\{A + \frac{5}{9}\frac{(Z - N)^2}{A} + \cdots\right\}. \tag{14.19}$$

The first term is proportional to A, and it contributes to the volume energy. The leading-order deviation is of the form assumed in Eq. (14.8) for the symmetry energy, and the coefficient of $(Z - N)^2/A$ can be evaluated numerically:

$$\frac{1}{6}\left(\frac{9\pi}{8}\right)^{2/3}\frac{\hbar^2}{mR_0^2}\frac{(Z - N)^2}{A} \approx 11 \text{ MeV}\frac{(Z - N)^2}{A}. \tag{14.20}$$

The evaluation has produced the expected form for the symmetry energy, but the coefficient is only about half as big as a_{sym} in Eq. (14.11). We shall now briefly describe where the missing contribution to the symmetry energy comes from.[4]

In the discussion leading to Eq. (14.19) it was tacitly assumed that the potential depth V_0 (Fig. 14.4) does not depend on the neutron excess $(Z - N)$. This assumption is not very good because the average interaction between like nucleons is less than it is between neutron and proton, mainly because of the exclusion principle. The Pauli principle weakens the interaction between like particles by forbidding some of the two-body states, while the interaction between neutron and proton is allowed in all states. The change in the potential well depth has been determined, and it is of the order[5]

$$\Delta V_0 \text{ (in MeV)} \approx (30 \pm 10)\frac{(Z - N)}{A}. \qquad (14.21)$$

This decrease in depth of the potential well accounts for the missing contribution to the symmetry energy.[6]

14.3 References

The literature on nuclear structure up to 1964 is described and listed in M. A. Preston, "Resource Letter NS-1 on Nuclear Structure." This resource letter and a number of informative reprints are collected in *Nuclear Structure—Selected Reprints*, American Institute of Physics, New York, 1965.

All texts on nuclear physics contain sections or chapters on nuclear models. A modern account of nuclear structure at about the same level as the present book, but much more complete, is B. L. Cohen, *Concepts of Nuclear Physics*, McGraw-Hill, New York, 1971. This book also contains a good list of additional references. Much of the material on nuclear models either has been known for many years or the basic ideas date back a number of years. Two books were essential in teaching nuclear physics to most of the present-day physicists: E. Fermi, *Nuclear Physics*, notes complied by J. Orear, A. H. Rosenfeld, and R. A. Schluter, University of Chicago Press, Chicago, 1950, and J. M. Blatt and V. F. Weisskopf, *Theoretical Nuclear Physics*, Wiley, New York, 1952. While outdated, these two books still are gold mines; it is often a good idea to turn to them to find a difficult point explained lucidly.

The authoritative text on nuclear structure is A. Bohr and B. R. Mottelson, *Nuclear Structure*, Vol. I., Benjamin, Reading, Mass., 1969. This book is not easy to read. However, after studying a particular aspect first in some lower-level book, reading the same material in Bohr-Mottelson gives additional insight.

4. K. A. Brueckner, *Phys. Rev.* **97**, 1353 (1955).
5. F. G. Perey, *Phys. Rev.* **131**, 745 (1963).
6. B. L. Cohen, *Am. J. Phys.* **38**, 766 (1970).

A derivation of the semiempirical mass formula based on a nucleon-nucleon interaction is given in J. P. Wesley and A. E. S. Green, *Am. J. Phys.* **36**, 1093 (1968).

We have pointed out in the introduction to the present chapter that nuclear physics is still a long way from explaining nuclear structure fully in terms of the forces between nucleons. For futher information concerning this problem, see S. M. Austin and G. M. Crawley, eds., *The Two-Body Force in Nuclei*, Plenum, New York, 1972.

PROBLEMS

14.1. Estimate the magnitude of the correction that must be applied to Eq. (14.3) in order to take into account atomic binding effects.

14.2. Find the relation between the binding energy B and the mass excess Δ. Can either quantity be used if, for instance, the stability of isobars is studied?

14.3. Discuss the decays of the nuclides shown in Fig. 14.2.

14.4. Use the Bethe-Weizsäcker relation to estimate the position in Fig. 14.2 of the $A = 127$ isobars with $Z = 48, 49, 57,$ and 58. How would these isobars decay? With what decay energies? Estimate very crudely the lifetimes that you would expect.

14.5. Prepare a plot similar to Fig. 14.2 for the $A = 90$ isobars. Show that *two* parabolic curves appear. Explain why. How could the appearance of two such curves be introduced into the binding energy relation?

14.6. Consider possible decays $(A, Z) \rightarrow (A', Z')$. Write down criteria that involve the corresponding atomic masses $m(A, Z)$ and that indicate when a nucleus (A, Z) is stable against
(a) Alpha decay.
(b) Electron decay.
(c) Positron decay.
(d) Electron capture.

14.7. Derive Eq. (14.7) and find an expression for the coefficient a_c in terms of R_0. Compute a_c and compare with the empirical value quoted in Eq. (14.11).

14.8. Use Figs. 14.1–14.3 to find approximate values for the coefficients in the Bethe-Weizsäcker relation. Compare with the values in Eq. (14.11).

14.9. Verify that nucleons in the ground state of a nucleus indeed form a degenerate Fermi gas, i.e., occupy the lowest available levels, at all temperatures obtainable in the laboratory. At what temperature (approximately) would a fair fraction of nucleons be excited?

14.10. What would the ratio Z/A be for a nucleus if the exclusion principle were inoperative?

14.11. Consider a nucleus with $A = 237$. Use the semiempirical mass formula to:
(a) Find Z for the most stable isobar.
(b) Discuss the stability of this nuclide for various likely decay modes.

14.12. Symmetric fission is the splitting of a nucleus (A, Z) into two equal fragments $(A/2, Z/2)$. Use the Bethe-Weizsäcker relation to derive a condition for fission instability:
(a) Find the dependence on Z and A.
(b) For what values of A is fission possible for nuclides lying along the line of stability (Fig. 5.20)?
(c) Compare the result obtained in part (b) with reality.
(d) Compute the energy released in the fission of ^{238}U and compare with the actual value.

14.13.
(a) Consider isobars with A odd. How many

stable isobars would you expect for a given value of A? Why?

(b) Consider isobars with N and Z even. Explain why more than one even stable isobar can occur. Discuss an actual example.

14.14. Verify Eq. (14.19).

14.15. B/A gives the *average* binding energy of a nucleon in a nucleus. The separation energy is the energy required to remove the nucleon that is easiest to remove from a nucleus.

(a) Give an expression for the separation energy in terms of binding energies.

(b) Use Table A6 in the Appendix to find the neutron separation energies for ^{113}Cd and ^{114}Cd.

14.16. Compare the ratio of the binding energy to the mass of the system for atoms, nuclei, and elementary particles. (Assume that elementary particles are built from heavy quarks.)

14.17. Use the dependence of the potential depth V_0 on $N - Z$, as expressed by Eq. (14.21), to compute the corresponding contribution to the symmetry energy.

14.18. Discuss the symmetry energy due to
(a) An ordinary central force.
(b) A space and spin exchange potential (Heisenberg force). If s_1 and s_2 are the spins of particles 1 and 2, this potential is given by

$$V\psi(\mathbf{r}_1, s_1; \mathbf{r}_2, s_2) = f(r)\psi(\mathbf{r}_2, s_2; \mathbf{r}_1, s_1),$$

where $\mathbf{r} = \mathbf{r}_1 - \mathbf{r}_2$.

15

The Shell Model

The liquid drop and the Fermi gas models represent the nucleus in very crude terms. While they account for gross nuclear properties, they cannot explain specific properties of excited nuclear states. In Section 5.10 we have given some aspects of the nuclear energy spectrum, and we have also pointed out that progress in atomic physics was tied to an unraveling of the atomic spectra. In atomic physics, solid-state physics, and quantum electrodynamics, unraveling began with the independent-particle model. It is therefore not surprising that this approach was tried early in nuclear theory also. Bartlett, and also Elsasser,[1] pointed out that nuclei display particularly stable configurations if Z or N (or both) is one of the *magic numbers*

$$2, \quad 8, \quad 20, \quad 28, \quad 50, \quad 82, \quad 126. \qquad (15.1)$$

The main evidence at that time consisted of the number of isotopes, alpha-particle emission energies, and elemental abundance. Elsasser tried to understand this stability in terms of the neutrons and protons moving independently in a single particle potential well, but he was unable to account for the stability of N or Z at 50 and 82 and N at 126. Scant attention was paid to his work for two reasons. One was that the model had no apparent

---

1. J. H. Bartlett, *Phys. Rev. 41*, 370 (1932); W. M. Elsasser, *J. Phys. Radium* **4**, 549 (1933); **5**, 625 (1934).

theoretical basis. Unlike atoms, nuclei have no fixed center, and the short range of nuclear forces seems to imply that one cannot use a smooth average potential to represent the actual potential felt by a nucleon. The second reason was the meager experimental evidence available.

However, the evidence for the existence of magic numbers continued to increase. As in the case of atoms, such magic numbers try to tell us that some kind of *shells* exist in nuclei. Finally, in 1949, the magic numbers were explained in terms of single-particle orbits by Maria Goeppert Mayer[2] and by J. H. D. Jensen.[3] The crucial element was the realization that spin-orbit forces are essential for an understanding of the closed shells at 50, 82, and 126. Moreover, the suggestion was made that the Pauli principle strongly suppresses collisions between nucleons and thereby provides for nearly undisturbed orbits for the nucleons in nuclear matter.[4]

The naive shell model assumes that nucleons move independently in a spherical potential. The assumptions of independence and sphericity are oversimplifications. Interactions between nucleons are present that cannot be described by an average central potential, and the nuclear shape is known to not always be spherical. The shell model can be refined by taking some of the *residual interactions* into account and by studying orbits in a deformed well.

In the following sections we shall first exhibit some of the experimental evidence for the existence of magic numbers. We shall then discuss shell closures and the single-particle shell model and finally sketch some refinements.

15.1 The Magic Numbers

In this section, we shall review some experimental evidence for the fact that nuclides with either Z or N equal to one of the magic numbers 2, 8, 20, 28, 50, 82, or 126 are particularly stable. Of course, these numbers are now well explained by the shell model but the adjective *magic* is so descriptive that it is retained.

In Fig. 15.1, the *relative abundance* of different even-even nuclides is plotted as a function of the atomic number A, for $A > 50$. Nuclides with $N = 50$, 82, and 126 form three pronounced peaks.

Clear evidence for the magic numbers comes from the *separation energies of the last nucleons*. To explain the concept, consider atoms. The separation energy or ionization potential is the energy needed to remove the least tightly bound (the *last*) electron from a neutral atom. The separation energies of the elements are shown in Fig. 15.2. The atomic shells are

2. M. G. Mayer, *Phys. Rev.* **74**, 235 (1948); **75**, 1969 (1949); **78**, 16 (1950).

3. O. Haxel, J. H. D. Jensen, and H. Suess, *Phys. Rev.* **75**, 1766 (1949); *Z. Physik* **128**, 295 (1950).

4. E. Fermi, *Nuclear Physics*, University of Chicago Press, Chicago, 1950; V. F. Weisskopf, *Helv. Phys. Acta* **23**, 187 (1950); *Science* **113**, 101 (1951).

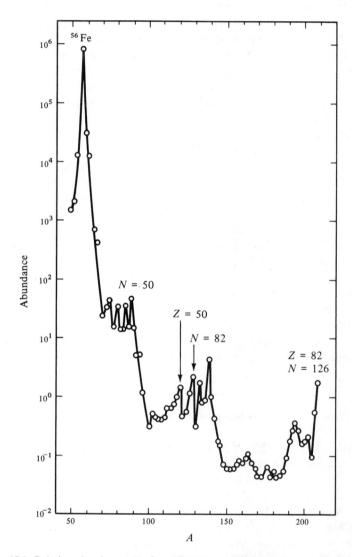

Fig. 15.1. Relative abundance, H, for different even-even nuclides, plotted as a function of A. The abundances are measured relative to Si, with $H(\text{Si}) = 10^6$. [Based on A.G.W. Cameron, "A New Table of Abundance of the Elements in the Solar System," *Origin and Distribution of the Elements* (L. H. Arens, ed.), Pergamon Press, New York, 1968, p. 125.]

responsible for the pronounced peaks: If the last electron fills a major shell, it is particularly tightly bound, and the separation energy reaches a peak. The next electron finds itself outside a closed shell, has very little to hold onto, and can be removed easily. The nuclear quantity that is analogous to the ionization potential is the separation energy of the last nucleon. If, for

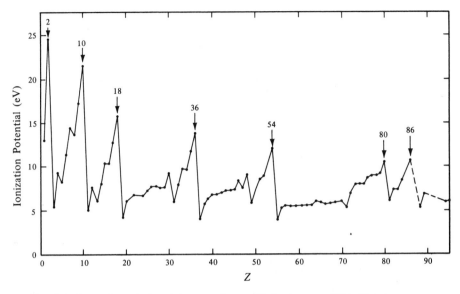

Fig. 15.2. Separation energies of the neutral atoms (ionization potentials). (Based on data from C. E. Moore, "Ionization Potentials and Ionization Limits Derived from the Analyses of Optical Spectra," *NSRDS—NBS 34,* 1970.)

instance, a neutron is removed from a nuclide (Z, N), a nuclide $(Z, N - 1)$ results. The energy needed for removal is the difference in binding energies between these two nuclides,

$$S_n(Z, N) = B(Z, N) - B(Z, N - 1). \tag{15.2}$$

An analogous expression holds for the proton separation energy. With Eqs. (14.3) and (14.4), the separation energy can be written in terms of the mass excesses,

$$S_n(Z, N) = m_n c^2 - u + \Delta(Z, N - 1) - \Delta(Z, N) \tag{15.3}$$

or with the numerical values of the neutron mass and the atomic mass unit

$$S_n(Z, N) = 8.07 \text{ MeV} + \Delta(Z, N - 1) - \Delta(Z, N).$$

The mass excess is given in Table A6 in the Appendix, and the separation energy can be computed quickly. The result can be presented in two different ways: either Z can be kept fixed, or the neutron excess $N - Z$ can be kept constant. The first situation is easier to visualize: We start with a certain nuclide, continue adding neutrons, and record the energy with which each one is bound. Such a plot is shown in Fig. 15.3 for the isotopes of cerium, $Z = 58$. Two effects are apparent, an even-odd difference and a closed-shell discontinuity. The even-odd behavior indicates that neutrons are more tightly bound when N is even than when N is odd. The same holds for protons. This fact, together with the empirical observation that all even-even nuclei have spin zero in their ground states, shows that an extra attractive interaction occurs when two like particles pair off to zero angular

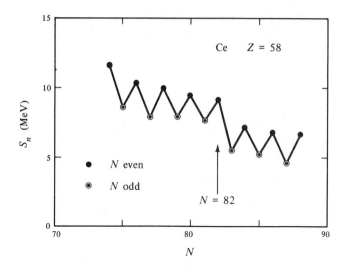

Fig. 15.3. Separation energy of the last neutron for the isotopes of cerium.

momentum. This *pairing interaction* is important for understanding nuclear structure in terms of the shell model, and we shall explain it later. Here we note that a similar effect occurs in superconductors where two electrons of opposite momenta and spins form a Cooper pair.[5] From Fig. 15.3, it follows that the pairing energy is of the order of 2 MeV in cerium. Once this pairing is corrected for, for instance by only considering isotopes with even N, the second effect, namely the influence of the closed shell at $N = 82$, stands out. Neutrons after a closed shell are less tightly bound by about 2 MeV than just before the closed shell. With the data in Table A6 in the Appendix, figures similar to Fig. 15.3 can be prepared for other regions, and shell closure at all magic numbers can be observed.

Closed shells should be spherically symmetric, have a total angular momentum of zero, and be especially stable. The stability of closed shells can be seen from the energies of the first excited states; a pronounced

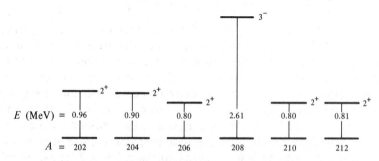

Fig. 15.4. Ground and first-excited states of the even-A isotopes of Pb.

5. See, for instance, G. Baym, *Lectures on Quantum Mechanics*, Benjamin, Reading, Mass., 1969, Chapter 8.

stability means that it will be hard to excite a closed shell, and consequently the first excited state should lie especially high. An example of this behavior is given in Fig. 15.4 where the ground states and first excited states of the Pb isotopes with even A are shown. ^{208}Pb, with $N = 126$, has an excitation energy that is nearly 2 MeV larger than that of the other isotopes. Furthermore, unlike all the other isotopes for which the spins and parities of the first excited states are 2^+, that of ^{208}Pb is 3^-. The closed shell affects not only the energy of the first excited state but also its spin and parity.

15.2 The Closed Shells

The first task in the construction of the shell model is the explanation of the magic numbers. In the independent-particle model it is assumed that the nucleons move independently in the nuclear potential. Because of the short range of the nuclear forces, this potential resembles the nuclear density distribution. To see the resemblance explicitly, we consider a two-body force of the type

$$V_{12} = V_0 f(\mathbf{r}_1 - \mathbf{r}_2),\qquad(15.4)$$

where V_0 is the central depth of the potential and f describes its shape. The function f is assumed to be smooth and of very short range. A crude estimate of the strength of the central potential acting on nucleon 1 in the nucleus can be obtained by averaging over nucleon 2. Such an averaging represents the action of all nucleons (except 1) on 1. Averaging is performed by multiplying V_{12} by the density distribution of nucleon 2 in the nucleus, $\rho(\mathbf{r}_2)$,

$$V(1) = V_0 \int d^3r_2\, f(\mathbf{r}_1 - \mathbf{r}_2)\rho(\mathbf{r}_2).$$

If f is of sufficiently short range, $\rho(\mathbf{r}_2)$ can be approximated by $\rho(\mathbf{r}_1)$, and $V(1)$ becomes

$$V(1) = CV_0\rho(\mathbf{r}_1),$$

$$C = \int d^3r\, f(\mathbf{r}).\qquad(15.5)$$

The potential seen by a particle is indeed proportional to the nuclear density distribution. The density distribution, in turn, is approximately the same as the charge distribution. The charge distribution of spherical nuclei was studied in Section 6.5, and it was found that it can be represented in a first approximation by the Fermi distribution, Fig. 6.6. It would therefore be appropriate to start the investigation of the single-particle levels by using a potential that has the form of a Fermi distribution but is attractive. The Schrödinger equation for such a potential cannot be solved in closed form. For many discussions the realistic potential is consequently replaced by one that can be treated easily, either a square well or a harmonic oscillator potential. We have encountered the latter in Section 13.7,

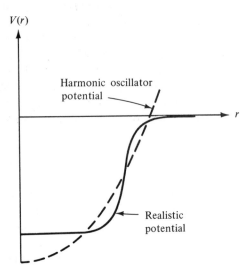

Fig. 15.5. The more realistic potential resembling the actual nuclear density distribution is replaced by a harmonic oscillator potential.

and we can now use the relevant information with very minor changes. The nuclear potential and its approximation by the harmonic oscillator are shown in Fig. 15.5.

Consider first the harmonic oscillator whose energy levels are shown in Fig. 13.7. The group of degenerate levels corresponding to one particular value of N is called an oscillator shell. The degeneracy of each shell is given by Eq. (13.33). In the application to nuclei, each level can be occupied by two nucleons, and consequently the degeneracy is given by $(N+1)(N+2)$. In Table 15.1 the oscillator shells, their properties, and the total number of levels up to the shell N are listed. The orbitals are denoted by a number and a letter; $2s$, for instance, means the second level with an orbital angular momentum of zero.

Table 15.1 shows that the harmonic oscillator predicts shell closures

Table 15.1 OSCILLATOR SHELLS FOR THE THREE-DIMENSIONAL HARMONIC OSCILLATOR

N	Orbitals	Parity	Degeneracy	Total number of levels
0	$1s$	+	2	2
1	$1p$	−	6	8
2	$2s, 1d$	+	12	20
3	$2p, 1f$	−	20	40
4	$3s, 2d, 1g$	+	30	70
5	$3p, 2f, 1h$	−	42	112
6	$4s, 3d, 2g, 1i$	+	56	168

at nucleons numbers 2, 8, 20, 40, 70, 112, and 168. The first three agree
with the magic numbers, but after $N = 2$, the real shell closures differ from
the predicted ones. One of the two conclusions is forced on us: Either the
agreement of the first three numbers is fortuitous or an important feature
is still missing. Of course by now it is well understood that the second conclusion is correct. To introduce the missing feature, we turn again to the
level diagram.

The energy levels of the harmonic oscillator are degenerate for two
different reasons. Consider, for example, the level with $N = 2$, which contains the orbitals $2s$ and $1d$. The $2s$ state has $l = 0$, and it can accept two
particles because of the two possible spin states. Rotational symmetry
gives the d state ($l = 2$) a $(2l + 1)$-fold degeneracy, and, considering the
two spin states, this degeneracy leads to $2(4 + 1) = 10$ states. The fact
that the $2s$ and the $1d$ state have the same energy is a feature peculiar to
the harmonic oscillator. It is somewhat unfortunate that the harmonic
oscillator, which is otherwise so straightforward to understand, possesses
this dynamical degeneracy. What happens to the degeneracy in a more
realistic potential, such as the one shown in Fig. 15.5? The wave functions
of the harmonic oscillator in Fig. 13.8 indicate that particles in states with
higher angular momenta are more likely to be found at larger radii than in
states with small or zero orbital angular momenta. Figure 15.5 shows that
the Fermi potential has a flat bottom and for identical central depth is thus
deeper at large radii than the oscillator potential. Consequently, the states

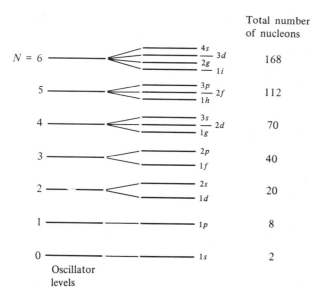

Fig. 15.6. Oscillator shells. At the left are the harmonic oscillator levels. If the accidental degeneracy in each oscillator shell is lifted by a change in potential shape, a level
diagram as given at the right appears. The total number of nucleons that can be placed
into the well up to, and including, the particular shell are also given.

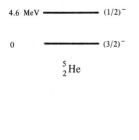

4.6 MeV ——————— $(1/2)^-$

0 ——————— $(3/2)^-$

$_2^5$He

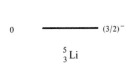

7 MeV ——————— $(1/2)^-$

0 ——————— $(3/2)^-$

$_3^5$Li

Fig. 15.7. Lowest energy levels of ^{5}He and ^{5}Li. Actually, the states have very short half-lives and consequently very large widths. Since the widths are not germane to our arguments, they are not shown.

with higher angular momenta see a deeper potential in the realistic case, the degeneracy will be lifted, and the high-l states will move to lower energies. The lifting of the degeneracy by this feature can be shown explicitly for the square well, and the result is given in Fig. 15.6. The number of nucleons in each shell remains unchanged and the magic numbers 50, 82, and 126 are still not explained.

So far, the energy levels have been labeled only with n and l, but the nucleon spin has been neglected. A nucleon in a state with orbital angular momentum l can give rise to two states, with total angular momenta $l \pm \frac{1}{2}$. As an example, consider the oscillator shell $N = 1$. A nucleon in state $1p$ can have total angular momentum $\frac{1}{2}$ or $\frac{3}{2}$, and the corresponding states are denoted with $1p_{1/2}$ and $1p_{3/2}$. In the central harmonic oscillator potential and in the square well potential, these states are degenerate. The situation is altered by *spin-dependent* forces. Consider, for instance, the lowest energy levels of ^{5}He and ^{5}Li, as given in Fig. 15.7. The ground states of these nuclides have spin $\frac{3}{2}$ and negative parity, and the first excited states spin $\frac{1}{2}$ and negative parity. These quantum numbers are explained by considering ^{5}He(^{5}Li) as a closed shell core of ^{4}He plus one neutron (proton). In ^{4}He, the $1s$ levels for neutrons and protons are filled, and it is the first doubly magic nucleus. The next nucleon, neutron or proton, must go into one of the $1p$ levels, either $1p_{1/2}$ or $1p_{3/2}$. The spins of the observed levels (Fig. 15.8) tell us that the $1p_{3/2}$ level has the lower energy. If the nucleon outside the closed shell, the so-called valence nucleon, is lifted to the next higher level, the first excited state of ^{5}He results. The spin and parity values, $(\frac{1}{2})^-$, of this state indicate that it is a $1p_{1/2}$ single-particle level. The degeneracy of the $1p_{1/2}$ and $1p_{3/2}$ levels is lifted in actual nuclei, and the energy splitting is of the order of a few MeV in the light nuclei. This conclusion can be tentatively generalized by assuming that the degeneracy between the levels $l + \frac{1}{2}$ and $l - \frac{1}{2}$ is always lifted in real nuclei, as shown in Fig. 15.9.

The splitting between states $l + \frac{1}{2}$ and $l - \frac{1}{2}$ is now known to be caused primarily by the interaction between the nucleon spin and its orbital an-

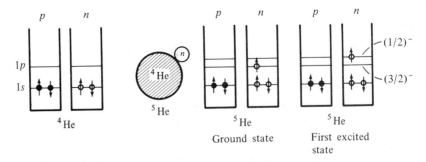

Fig. 15.8. Occupation of the nucleon energy levels in ^{4}He, ^{5}He, and ^{5}He*. For simplicity the Coulomb interaction has been neglected, and the neutron and proton wells have been drawn identically. Moreover, only the two lowest energy levels are shown.

gular momentum. Such a spin-orbit force is well known in atomic physics,[6] but it was not expected that it would be so strong in nuclei. We shall
return to the spin-orbit force in the next section but shall show here that
the magic numbers can be explained if its effects are taken into account.
A nucleon, moving in the central potential of the nucleus with orbital
angular momentum $\mathbf{l}$, spin $\mathbf{s}$, and total angular momentum $\mathbf{j}$,

$$\mathbf{j} = \mathbf{l} + \mathbf{s}, \tag{15.6}$$

acquires an additional energy

$$V_{ls} = C_{ls}\mathbf{l}\cdot\mathbf{s}. \tag{15.7}$$

We must find the effect of this potential-energy operator on a state $|\alpha; j, l, s\rangle$.
Here α denotes all quantum numbers other than j, l, and s. (The reason
that j, l, and s can be specified simultaneously is that states of $l = j \pm \frac{1}{2}$
have opposite parities, and parity is conserved in the hadronic force.) With
the square of Eq. (15.6), the operator $\mathbf{l}\cdot\mathbf{s}$ is written as

$$\mathbf{l}\cdot\mathbf{s} = \tfrac{1}{2}(j^2 - l^2 - s^2). \tag{15.8}$$

The actions of the operators j^2, l^2, and s^2 on $|\alpha; j, l, s\rangle$ are given by Eq.
(5.7) so that

$$\mathbf{l}\cdot\mathbf{s}|\alpha; j, l, s\rangle = \tfrac{1}{2}\hbar^2\{j(j + 1) - l(l + 1) - s(s + 1)\}|\alpha; j, l, s\rangle. \tag{15.9}$$

For a nucleon, with spin $s = \frac{1}{2}$, only two possibilities exist, namely $j = l + \frac{1}{2}$ and $j = l - \frac{1}{2}$, and for these Eq. (15.9) yields

$$\mathbf{l}\cdot\mathbf{s}|\alpha; j, l, \tfrac{1}{2}\rangle = \begin{cases} \tfrac{1}{2}\hbar^2 l |\alpha; j, l, \tfrac{1}{2}\rangle & \text{for } j = l + \tfrac{1}{2} \\ -\tfrac{1}{2}\hbar^2(l + 1)|\alpha; j, l, \tfrac{1}{2}\rangle & \text{for } j = l - \tfrac{1}{2} \end{cases} \tag{15.10}$$

The energy splitting ΔE_{ls}, shown in Fig. 15.9, is proportional to $l + \frac{1}{2}$:

$$\Delta E_{ls} = (l + \tfrac{1}{2})\hbar^2 C_{ls}. \tag{15.11}$$

The spin-orbit splitting increases with increasing orbital angular momentum l. It consequently becomes more important for heavier nuclei,
where higher l values appear. For a given value of l, the level with higher
total angular momentum, $j = l + \frac{1}{2}$, lies lower, and it has a degeneracy of
$2j + 1 = 2l + 2$. The upper level, with $j = l - \frac{1}{2}$, is $2l$-fold degenerate.
With these remarks, shell closure at the magic numbers can be understood.
Consider Fig. 15.6. The total number of nucleons up to the oscillator shell
$N = 3$ is 40; the correct magic number is 50. The $1g_{9/2}$ state has a degeneracy of 10, as shown in Fig. 15.10. This level is depressed by the spin-orbit
interaction so that it intrudes into the $N = 3$ oscillator shell, and the total
number of nucleons adds up to 50, the correct magic shell closure. Similarly, the $1h_{11/2}$ state has a degeneracy of 12; depressed and added to the
$N = 4$ oscillator shell, it produces the number 82. The $1i_{13/2}$, depressed
into the $N = 5$ shell, adds 14 nucleons and produces the magic number

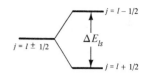

Fig. 15.9. Splitting of the states with a given value of l into two states. The spin-orbit interaction depresses the state with total angular momentum $j = l + \frac{1}{2}$ and raises the one with $j = l - \frac{1}{2}$.

6. Eisberg, Section 11.4; H. A. Bethe and R. Jackiw, *Intermediate Quantum Mechanics*,
2nd ed., Benjamin, Reading Mass., 1968, Chapter 8; Park, Chapter 14; G. P. Fisher, *Am. J. Phys.* **39**, 1528 (1971).

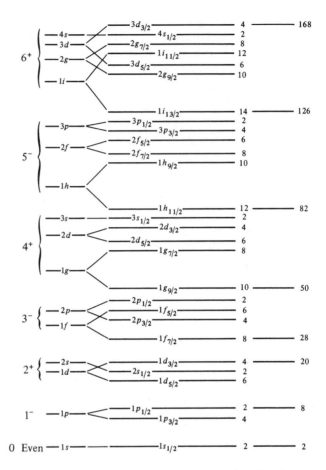

Fig. 15.10. Approximate level pattern for nucleons. The number of nucleons in each level and the cumulative totals are shown. The oscillator grouping is shown at the left. Neutrons and protons have essentially the same level pattern up to 50. From then on, some deviations occur. Low neutron angular momenta are more favored than low proton angular momenta.

126. The situation is summarized in Fig. 15.10, where the level pattern is shown. The details differ slightly for protons and neutrons and this fact will be important in Section 19.1. The situation can be summarized by saying that a sufficiently strong spin-orbit interaction which is attractive in the states $j = l + \frac{1}{2}$ can account for the experimentally observed shell closures.

15.3 The Spin-Orbit Interaction

In the previous section it was shown that a spin-orbit interaction, of the form of Eq. (15.7), can produce the experimentally observed shell closures, provided the constant C_{ls} is sufficiently large. Is the evidence from nuclear properties in agreement with what is known about the nucleon-nucleon potential? In Section 12.5 it was shown that the nucleon-nucleon potential energy represented in Eq. (12.36) contains a spin-orbit term,

$$V_{LS}\mathbf{L}\cdot\mathbf{S}.\tag{15.12}$$

Here $\mathbf{L} = \frac{1}{2}(\mathbf{r}_1 - \mathbf{r}_2) \times (\mathbf{p}_1 - \mathbf{p}_2)$ is the relative orbital angular momentum of the two nucleons and $\mathbf{S} = \mathbf{s}_1 + \mathbf{s}_2 = \frac{1}{2}(\boldsymbol{\sigma}_1 + \boldsymbol{\sigma}_2)$ is the sum of the spins. Such a term in the nucleon-nucleon force will produce a term

$$V_{ls} = C_{ls}\mathbf{l}\cdot\mathbf{s}$$

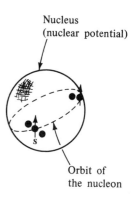

Nucleus
(nuclear potential)

Orbit of
the nucleon

Fig. 15.11. Nucleon with orbital angular momentum $\mathbf{l}$ and spin $\mathbf{s}$ moving in the nuclear potential.

in the *nuclear* potential. Here, $\mathbf{l}$ is the orbital angular momentum of the nucleon that moves in the nuclear potential and $\mathbf{s}$ is its spin. To see the connection, we consider an orbit as shown in Fig. 15.11. In the interior of the nucleus, where the nuclear density is constant, there are an equal number of nucleons on either side of the orbit within reach of the nuclear force. The spin-orbit interaction consequently averages out. Near the surface, however, nucleons are only on the interior side of the orbit, the relative orbital angular momentum $\mathbf{L}$ in Eq. (15.12) always points in the same direction, and the two-body spin-orbit interaction gives rise to a term of the form of Eq. (15.7). To make this argument more precise, the spin-orbit interaction energy [Eq. (15.12)] between two nucleons, 1 and 2, is written as

$$V(1, 2) = \frac{1}{2}V_{LS}(r_{12})(\mathbf{r}_1 - \mathbf{r}_2) \times (\mathbf{p}_1 - \mathbf{p}_2)\cdot(\mathbf{s}_1 + \mathbf{s}_2).\tag{15.13}$$

If particle 1 is the nucleon under consideration, an estimate of the nuclear spin-orbit potential can be obtained by averaging $V(1, 2)$ over nucleon 2,

$$V_{ls}(1) = \text{Av}\int d^3r_2\, \rho(\mathbf{r}_2)V(1, 2),\tag{15.14}$$

where Av indicates that we must average over the spin and the momentum of nucleon 2, and where $\rho(\mathbf{r}_2)$ is the probability density of nucleon 2. After inserting $V(1, 2)$ from Eq. (15.13), $V_{ls}(1)$ becomes

$$V_{ls}(1) = \frac{1}{2}\int d^3r_2\, \rho(\mathbf{r}_2)V_{LS}(r_{12})(\mathbf{r}_1 - \mathbf{r}_2) \times \mathbf{p}_1\cdot\mathbf{s}_1;\tag{15.15}$$

the average of all other terms is zero. The nuclear density at position $\mathbf{r}_2$ can be expanded in a Taylor series about $\mathbf{r}_1$ because of the short range of the spin-orbit force:

$$\rho(\mathbf{r}_2) = \rho(\mathbf{r}_1) + (\mathbf{r}_2 - \mathbf{r}_1)\cdot\nabla\rho(\mathbf{r}_1) + \cdots.\tag{15.16}$$

After inserting the expansion into $V_{ls}(1)$, the integral containing the factor $\rho(\mathbf{r}_1)$ vanishes. The remaining integral can be computed; under the assumption that the range of the nucleon spin-orbit interaction is small compared to the nuclear surface thickness, which is the only region wherein $\nabla\rho$ is appreciable, it gives

$$V_{ls}(1) = C\frac{1}{r_1}\frac{\partial\rho(r_1)}{\partial r_1}\mathbf{l}_1\cdot\mathbf{s}_1,\tag{15.17}$$

where

$$C = -\frac{1}{6} \int V_{LS}(r)r^2 d^3r. \qquad \textbf{(15.18)}$$

The nucleon-nucleon spin-orbit interaction leads to a spin-orbit interaction for a nucleon moving in the average nuclear potential. As Eq. (15.17) shows, the interaction vanishes where the density is constant, and it is strongest at the nuclear surface. Numerical estimates with Eqs. (15.17) and (15.18) give the correct order of magnitude of V_{ls}.

15.4 The Single-Particle Shell Model

The simplest atomic system is hydrogen because it consists of only one electron moving in the field of a heavy nucleus. Next in simplicity are the alkali atoms which consist of a closed atomic shell plus one electron. In a first approximation they are treated by assuming that the one valence electron moves in the field of the nucleus shielded by the closed shells of electrons which form a spherically symmetric system with zero angular momentum. The entire angular momentum of the atom is provided by the valence electron (and the nucleus). In nuclear physics, the two-body system (deuteron) has only one bound state and does not provide much insight. In analogy to the atomic case, the next simplest cases then are nuclei with closed shells plus one valence nucleon (or nuclides with closed shells minus one nucleon). To discuss such nuclides we first return to closed shells.

What are the quantum numbers of nuclides with closed shells? In the shell model, protons and neutrons are treated independently. Consider first a subshell with a given value of the total angular momentum j, for instance, the proton subshell $1p_{1/2}$ (Fig. 15.10). There are $2j + 1 = 2$ protons in this subshell. Since protons are fermions, the total wave function must be antisymmetric. The spatial wave function of two protons in the same shell is symmetric, and consequently the spin function must be antisymmetric. Only one totally antisymmetric state can be formed from two protons, but a state described by *one* wave function only must have spin $J = 0$. The same argument holds for any closed subshell or shell of protons or neutrons: Closed shells always have a total angular momentum of zero. The parity of a closed shell is even because there are an even number of nucleons filling it.

Ground-state spin and parity of nuclides with closed shells plus or minus a single particle are now straightforward to predict. Consider first a single proton outside a closed shell. Because the closed shell has zero angular momentum and even parity, angular momentum and parity of the nucleus are carried by the valence proton. Angular momentum and parity of the proton can be read off from Fig. 15.10. The corresponding level diagram for neutrons is very similar. A first example was already given in Fig. 15.8 from which we deduced that the ground state assignment of ^{5}He should be

$p_{3/2}$, or spin $\frac{3}{2}$ and negative parity. A few additional examples are shown in Table 15.2. The agreement between predicted and observed values of spins and parities is complete. The quantum numbers of nuclei with a complete shell minus a single particle can also be obtained from Fig. 15.10. Such a *single-hole state* can be described in the language used in Section 7.5 for antiparticles; the hole appears as an antiparticle, and Eq. (7.42) tells us that the angular momentum of the state must be the same as that of the missing nucleon. Similarly, the parity of the hole state must be the same as that of the missing nucleon state.[7] These properties of holes also follow from the remark that a hole, together with the particle that can fill it, couple together to give $J = 0^+$ for the closed shell. As a simple example, consider ⁴He, shown in Fig. 15.8. Removing one neutron from ⁴He gives ³He. The removed neutron was in an $s_{1/2}$ state; the absence is indicated by the symbol $(s_{1/2})^{-1}$. The corresponding spin-parity assignment of ³He is $(\frac{1}{2})^+$, in agreement with experiment. Assignments for other single-hole nuclides can easily be given, and they also agree with the experimental values.

Next we turn to *excited states*. In the spirit of the extreme single-particle model, they are described as excitations of the valence nucleon alone; it moves into a higher orbit. The core (closed shell) is assumed to remain undisturbed. Up to what energies can such a picture be expected to hold? Figures 15.3 and 15.4 indicate that the pairing energy is of the order of about 2 MeV. At an excitation energy of a few MeV it is therefore possible that the valence nucleon remains in its ground state but that a pair from the core is broken up and that one of the nucleons of the pair is promoted to the next higher shell. It is also possible that a pair is excited to the next higher shell. In either case, the resulting energy level is no longer describable by the single-particle approach. It is consequently not surprising to find "foreign" levels at a few MeV. Two examples are shown in Fig. 15.12, both doubly magic nuclei plus one valence nucleon. In the case of ⁵⁷Ni, the single-particle shell-model assignments hold up to about 1 MeV, but

Table 15.2 Ground-State Spins and Parities as Predicted by the Single-Particle Shell Model and as Observed

Nuclide	Z	N	Shell-model assignment	Observed spin and parity
¹⁷O	8	9	$d_{5/2}$	$\frac{5}{2}^+$
¹⁷F	9	8	$d_{5/2}$	$\frac{5}{2}^+$
⁴³Se	21	22	$f_{7/2}$	$\frac{7}{2}^-$
²⁰⁹Pb	82	127	$g_{9/2}$	$\frac{9}{2}^+$
²⁰⁹Bi	83	126	$h_{9/2}$	$\frac{9}{2}^-$

7. A detailed discussion of hole states and of the particle-hole conjugation is given in A. Bohr and B. R. Mottelson, *Nuclear Structure*, Benjamin, Reading, Mass., 1969. See Vol. I, p. 312 and Appendix 3B.

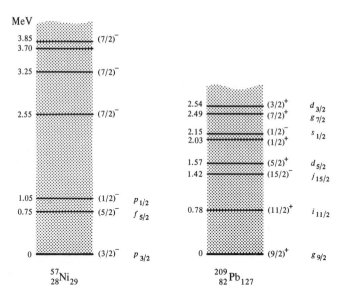

Fig. 15.12. Excited states in ⁵⁷Ni and ²⁰⁹Pb. The states that allow an unambiguous shell-model assignment are labeled with the corresponding quantum numbers.

above 2.5 MeV, foreign states appear. The foreign states are not really foreign. While they cannot be described in terms of the extreme single-particle shell model, they can be understood in terms of the general shell model, through excitations from the core. In the case of ²⁰⁹Pb, the first such state appears at 2.15 MeV. The estimate based on Figs. 15.3 and 15.4 that core excitation will play a role at about 2 MeV is verified.

We have discussed only two properties of nuclei that are well described by the single-particle model, spin and parity of ground states and the level sequence and quantum numbers of the lowest excited states. There are other features that are explained by the extreme single-particle model, for instance the existence of very-long-lived first excited states in certain regions of N and Z, the so-called islands of isomerism. However, the model applies only to a restricted class of nuclei—namely those with only one nucleon outside a closed shell—and an extension to more general conditions is necessary.

15.5 Generalization of the Single-Particle Model

The extreme single-particle shell model, discussed in the previous section, is based on a number of rather unrealistic assumptions: The nucleons move in a spherical fixed potential, no interactions among the particles are taken into account, and only the last odd particle contributes to the level properties. These restrictions are removed in various steps and to various degrees of sophistication; we briefly outline some of the extensions.

1. All particles outside the closed major shells are considered. The angular momenta of these particles can be combined in various ways to get the resulting angular momentum. The two main schemes are the Russell-Saunders, or *LS*, coupling, and the *jj* coupling. In the first, the orbital angular momenta are assumed to be weakly coupled to the spins; spin and orbital angular momenta of all nucleons in a shell are added separately to get the resulting $\mathbf{L}$ and $\mathbf{S}$: $\Sigma_i \mathbf{l}_i = \mathbf{L}$, $\Sigma_i \mathbf{s}_i = \mathbf{S}$. The total orbital angular momentum $\mathbf{L}$ and the total spin $\mathbf{S}$ of all nucleons in a shell are then added to form a given $\mathbf{J}$. In the *jj* coupling scheme, the spin-orbit force is assumed to be stronger than the residual force between individual nucleons so that the spin and the angular momentum of each nucleon are added first to give a total angular momentum $\mathbf{j}$; these $\mathbf{j}$'s are then combined to the total $\mathbf{J}$. In most nuclei, the empirical evidence indicates that the *jj* coupling is closer to the truth; in the lightest nuclei ($A \lesssim 16$), the coupling scheme appears to be intermediate between the *LS* and the *jj* coupling.

2. Residual forces between the particles outside the closed shells are introduced. That such residual interactions are needed can be seen in many ways. Consider, for instance, ^{69}Ga. It has three protons in state $2p_{3/2}$ outside the closed proton shell. These three protons can add their spins to get values of $J = \frac{1}{2}, \frac{3}{2}, \frac{5}{2}, \frac{7}{2}, \frac{9}{2}$. In the absence of a residual interaction, these states are degenerate. Experimentally, one state is observed to be lowest—quite often the state $J = j(= \frac{3}{2}$ in this case). There must be an interaction that splits these degenerate states. In principle, one should derive the residual interaction as what remains after the nucleon-nucleon interaction is replaced by an average single-nucleon potential. In practice, such a program is too difficult, and the residual interaction is determined empirically. However, many of the features of the residual interaction can be understood on theoretical ground. Consider as an example the *pairing force* described in Section 15.1. We have pointed out there that two like nucleons prefer to be in an antisymmetric spin state, with spins opposed and with a relative orbital angular momentum of zero ($^1S_0$). If the residual force has a very short range and is attractive, this behavior can be understood immediately. Consider for simplicity a zero-range force. The two nucleons can take advantage of such a residual attraction only when they are in a relative *s* state; the exclusion principle then forces their spins to be opposed, as is observed in reality. Although the true nuclear forces are not of such short range (indeed there is a repulsion at about 0.5 fm), the net effect is unchanged. The energy gained by the action of the *pairing force* is called *pairing energy*, and it is found empirically to be of the order of $12A^{-1/2}$ MeV. The pairing energy leads to an understanding of the energies of the first excited states of even-even nuclei: A pair must be broken, and the corresponding first excited state lies roughly 1–2 MeV above the ground state.

3. It is known that many nuclides are permanently deformed and hence cannot be described properly by a spherical potential. For such nuclei, the potential in which the single particles move is assumed to be non-

spherical.[8] This deformed-shell or Nilsson model will be described in Section 16.4.

When the three restrictions discussed here are removed, the shell model can describe many states very well. However, there still remain features that it cannot explain. We have already mentioned the existence of levels that must be ascribed to core excitations, and such excitations must be taken into account. More revealing, however, are some systematic deviations from properties predicted by the shell model. The two most pronounced ones are quadrupole moments that are much larger than expected and electric quadrupole transitions that are much faster than calculated. These features are most pronounced far away from closed shells, and they point to the existence of collective degrees of freedom that we have not yet considered. We shall turn to the collective model in the following chapter.

15.6 Isobaric Analog Resonances

"Since 1964 isospin has become an industry."

D. H. Wilkinson in *Isospin in Nuclear Physis*,
D. H. Wilkinson, ed., North-Holland,
Amsterdam, 1969, Ch. 1.

So far we have discussed states in a given nuclide, without considering neighboring isobars. In Section 8.7 we proved that the charge independence of nuclear forces leads to the assignment of an isospin I to a nuclear state; as long as the Coulomb interaction can be neglected, such a state will show up in $2I + 1$ isobars. Such isobaric analog states have even been found in medium and heavy nuclei[9,10] and are receiving a great deal of attention because of their value for nuclear structure studies.[11]

To describe analog states, we consider the isobars (Z, N) and $(Z + 1, N - 1)$. The energy levels in the absence of the Coulomb interaction are shown in Fig. 15.13. The difference in energy between the two ground states can be computed from the symmetry term in the semiempirical mass formula. Equation (14.8) gives

$$\Delta_{\text{sym}} = E_{\text{sym}}(Z + 1, N - 1) - E_{\text{sym}}(Z, N) = -4a_{\text{sym}} \frac{N - Z - 1}{A},$$

or

$$\Delta_{\text{sym}} \text{ (in MeV)} = -90 \frac{N - Z - 1}{A}. \tag{15.19}$$

8. S. G. Nilsson, *Kgl. Danske Videnskab. Selskab, Mat.-fys. Medd.* **29**, No. 16 (1955).

9. J. D. Anderson and C. Wong, *Phys. Rev. Letters* **7**, 250 (1961); J. D. Anderson, C. Wong, and T. W. McClure, *Phys. Rev.* **126**, 2170 (1962).

10. J. D. Fox, C. F. Moore, and D. Robson, *Phys. Rev. Letters* **12**, 198 (1964).

11. H. Feshbach and A. Kerman, *Comments Nucl. Particle Phys.* **1**, 69 (1967); M. H. Macfarlane and J. P. Schiffer, *Comments Nucl. Particle Phys.* **3**, 107 (1969). D. Robson, *Science* **179**, 133 (1973).

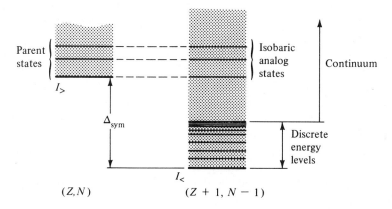

Fig. 15.13. Energy level diagram for the isobars (Z, N) and $(Z + 1, N - 1)$ in the absence of the Coulomb interaction.

The volume and surface terms are equal for the isobars, and thus the ground state of the isobar with higher Z lies lower by the amount Δ_{sym}. For the pair ^{209}Pb and ^{209}Bi, for instance, Δ_{sym} is about 19 MeV.

In the absence of the Coulomb interaction, isospin is a good quantum number. As stated in Section 8.7, the isospin of a nuclear ground state assumes the smallest allowed value. The isospin of the ground state of the isobar (Z, N) is thus given by Eq. (8.34) as

$$I_> = \frac{N - Z}{2}, \tag{15.20}$$

whereas for the isobar $(Z + 1, N - 1)$, the assignment is

$$I_< = \frac{N - Z}{2} - 1 = I_> - 1. \tag{15.21}$$

Because of charge independence, the levels of the *parent nucleus* (Z, N) also appear with the same energy in the isobar $(Z + 1, N - 1)$. These analog states are shown in Fig. 15.13. At this point, a crucial difference between light and heavy nuclei appears. To appreciate it we return to Fig. 5.27, Table 5.7, and Eq. (15.3) and note that nuclei have discrete levels (bound states) up to excitation energies of about 8 MeV. Above about 8 MeV, emission of nucleons becomes possible, and the spectrum is continuous. In *light nuclei*, where the symmetry energy is small, the isobaric analogs of the ground state and of low-lying excited states of the parent nucleus lie in the discrete part of the spectrum and consequently are *bound states*. An example is shown in Fig. 8.5 where the 0^+ state in ^{14}C is the parent state, and the first excited state in ^{14}N is the isobaric analog state. In *heavy nuclei*, the situation is as shown in Fig. 15.13: The symmetry energy is larger than the energy at which the continuum begins, and the analog states lie in the continuum. Nevertheless, in the absence of the Coulomb interaction, the analog states would remain bound, as can be seen as follows. Decay be neutron emission will lead to a neutron and a nucleus $(Z + 1, N - 2)$. The isospin of the ground state and the low-lying excited states of the nuclide $(Z + 1, N - 2)$ is given by $I = \frac{1}{2}(N - Z - 3)$ $= I_> - \frac{3}{2}$. Isospin conservation forbids the decay of the analog state with $I = I_>$ into a state with $I_> - \frac{3}{2}$ and a neutron. In the absence of the Coulomb interaction, the threshold for proton emission is so high that a decay of the analog state by proton emission is not possible.

Switching on the Coulomb interaction produces two effects: The analog levels move to higher energies than the parent states, and in general they are no longer bound, even in lighter nuclei, but become *resonances*. Consider first the shift of the energy levels. The Coulomb energy is different for the two isobars (Z, N) and $(Z + 1, N - 1)$. With Eq. (14.7), the relative shift becomes

$$\Delta_c = a_c \frac{2Z + 1}{A^{1/3}}. \tag{15.22}$$

For the pair ^{209}Pb and ^{209}Bi, Δ_c is about 19 MeV. Consequently the Coulomb energy shift roughly cancels the symmetry energy shift, and the energy levels are as shown in Fig. 15.14.

The effect of the Coulomb interaction on the decay properties of the isobaric analog states is twofold. Isospin is no longer completely conserved, and the decay of the analog state via neutron emission becomes possible. Moreover, the threshold of proton emission is lowered so that the analog states can also decay by proton emission. If the widths of these isobaric analog resonances were very large, say many MeV, it would be extremely difficult to observe them, and they would not be very interesting. However, the resonances are actually quite narrow, with widths less than about 200 keV. Before discussing the reasons for the narrowness, we shall show for the example given in Fig. 15.14 how the isobaric analog resonances are observed. Conceptually the simplest way to reach the isobaric

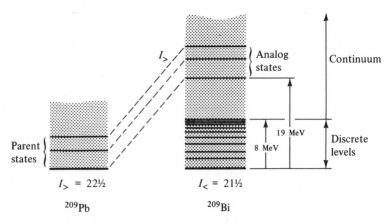

Fig. 15.14. Energy level diagram for the isobars (Z, N) and $(Z + 1, N - 1)$, for $A = 209$, $Z = 82$.

analog resonances in ^{209}Bi is through charge exchange reactions, for instance,

$$p + {}^{209}\text{Pb} \longrightarrow n + {}^{209}\text{Bi*}$$

or[12]

$$\pi^+ + {}^{209}\text{Pb} \longrightarrow \pi^0 + {}^{209}\text{Bi*}.$$

In both cases, one starts with the parent nucleus and changes a neutron into a proton. (Actually, ^{209}Pb is unstable and has a half-life of about 3 hr. The experiments discussed here would therefore be difficult to perform. The ideas, however, are unaffected by the radioactivity of ^{209}Pb.) Indeed, isobaric analog resonances were first observed in (p, n) reactions.[9] Here we shall discuss another approach in which the initial state is not the parent nucleus. Consider the reaction[13]

$$p + {}^{208}\text{Pb} \longrightarrow {}^{209}\text{Bi*} \longrightarrow p' + {}^{208}\text{Pb*}. \qquad \textbf{(15.23)}$$

The isospin of ^{208}Pb is $I_0 = 22$. Assuming isospin invariance, the reaction $p + {}^{208}$Pb can therefore lead to excited states with isospin $21\frac{1}{2}$ and $22\frac{1}{2}$ in ^{209}Bi. As Fig. 15.14 shows, the analog resonances have an isospin $22\frac{1}{2}$ and thus can be produced with reaction (15.23). They appear dramatically when the energy of the scattered proton, p', is selected so that the final nucleus ^{208}Pb* is in a particular state.[14] We shall discuss the reason for this fact below. The experiment proceeds as follows: The cross section for inelastic proton scattering is measured as a function of the energy of the incident proton. The energy of the scattered proton is selected so that the final nucleus is always in the same excited state. Results are shown in Fig. 15.15. At the left, the energy levels of ^{209}Pb are given. The cross sections at the right exhibit the appearance of the isobaric analog resonances corresponding to the various parent states in ^{209}Pb.

● An understanding of the results shown in Fig. 15.15 can be obtained by considering the analog resonances in more detail. The change from a parent state to the analog resonance corresponds to changing a neutron to a proton. Such a transformation is expressed with the raising operator I_+, Eq. (8.26):[15]

$$I_+ | \text{parent} \rangle = \text{const.} | \text{analog} \rangle. \qquad \textbf{(15.24)}$$

Since the ground state and the low-lying excited states of the parent nucleus are characterized by $I_> = |I_3|$ or $I_3 = -I_>$, Eq. (15.24) can be written as

$$I_+ | I_>, -I_> \rangle = (2I_>)^{1/2} | \text{analog} \rangle, \qquad \textbf{(15.25)}$$

where the constant has been taken from Eq. (8.27). To progress further, a model is needed. ^{209}Pb, the parent state, consists of a doubly magic core plus one valence

12. J. Alster, D. Ashery, A. I. Yavin, J. Duclos, J. Miller, and M. A. Moinester, *Phys. Rev. Letters* **28**, 313 (1972).

13. S. A. A. Zaidi, J. L. Parish, J. G. Kulleck, C. F. Moore, and P. von Brentano, *Phys. Rev.* **165**, 1312 (1968).

14. P. von Brentano, W. K. Dawson, C. F. Moore, P. Richard, W. Wharton, and H. Wieman, *Phys. Letters* **26B**, 666 (1968).

15. As stated in Chapter 8, the convention used here is the opposite of the one customary in nuclear physics, where our I_+ is written as I_-.

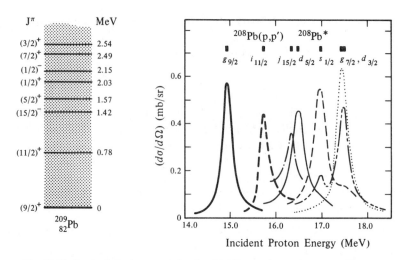

Fig. 15.15. At the left, the energy levels of ²⁰⁹Pb are shown (compare Fig. 15.12). At the right, the corresponding analog resonances in ²⁰⁹Bi* are apparent in the reaction $p + {}^{208}\text{Pb} \longrightarrow p' + {}^{208}\text{Pb*}$. [P. von Brentano et al., *Phys. Letters* **26B**, 666 (1968).] The curves give the cross section for the inelastic proton scattering leading to a specific final state as a function of the incident proton energy. Note that there is a one-to-one correspondence between the resonances and the parent states in ²⁰⁹Pb.

nucleon (Table 15.2). It is customary to picture it schematically as in Fig. 15.16: The filled proton and neutron shells are shown as blocks; the neutron block is larger because the core of ²⁰⁹Pb, namely ²⁰⁸Pb, contains 44 more neutrons than protons. The single neutron is depicted above the filled neutron shell. In the shell-model approximation, the wave function of ²⁰⁹Pb can thus be written as

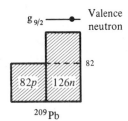

Fig. 15.16. Schematic representation of the ²⁰⁹Pb nucleus. The core is indicated by the blocks.

$$|{}^{209}\text{Pb}\rangle = |\text{core}\rangle\,|\text{single neutron}\rangle \equiv |{}^{208}\text{Pb}\rangle\,|n\rangle, \qquad (15.26)$$

where the valence neutron is in the $g_{9/2}$ state, as shown in Fig. 15.12. The raising operator can be decomposed into two parts

$$I_+ = I_+^c + I_+^{\text{sp}}, \qquad (15.27)$$

where c indicates core and sp single particle. With Eqs. (15.25)–(15.27), the analog state becomes

$$|\text{analog}\rangle = (2I_>)^{-1/2}\,(I_+^c + I_+^{\text{sp}})|{}^{209}\text{Pb}\rangle$$
$$= (2I_>)^{-1/2}\{|{}^{208}\text{Pb}\rangle I_+^{\text{sp}}|n\rangle + (I_+^c|{}^{208}\text{Pb}\rangle)|n\rangle\}. \qquad (15.28)$$

Equation (8.27) gives

$$I_+^{\text{sp}}|n\rangle = |p\rangle, \qquad (15.29)$$

where the proton is in the same state as the neutron was, namely $g_{9/2}$. The term $I_+^c|{}^{208}\text{Pb}\rangle$ describes a state in which one neutron from the core has been changed into a proton, leaving behind a hole in the otherwise filled neutron levels. The Pauli principle does not allow two protons to occupy the same state, and only neutrons above the 82-line shown in Fig. 15.16 participate. The original core corresponds to ²⁰⁸Pb; changing a neutron into a proton results in an excited state of ²⁰⁸Bi. The isospin quantum number of ²⁰⁸Pb is $I_> - \tfrac{1}{2}$ so that Eq. (8.27) gives

$$I_+^c|{}^{208}\text{Pb}\rangle = (2I_> - 1)^{1/2}|{}^{208}\text{Bi*}\rangle. \qquad (15.30)$$

With Eqs. (15.29) and (15.30), the analog state becomes

$$|\text{analog}\rangle = \left(\frac{1}{2I_>}\right)^{1/2}|{}^{208}\text{Pb} + p\rangle + \left(\frac{2I_> - 1}{2I_>}\right)^{1/2}|{}^{208}\text{Bi*} + n\rangle. \qquad (15.31)$$

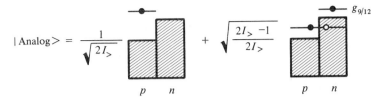

$$|\text{Analog}\rangle = \frac{1}{\sqrt{2I_>}} \quad + \quad \sqrt{\frac{2I_> - 1}{2I_>}}$$

Fig. 15.17. Representation of the analog state of a single-particle shell-model state.

In terms of diagrams like Fig. 15.16, the analog resonance is shown in Fig. 15.17. Figure 15.17 and Eq. (15.31) permit a discussion of some of the points stated without justification. First, consider the decay of the isobaric analog resonance. It can take place either by proton or by neutron emission. Figure 15.17 suggests that the following modes can occur:

$$\text{analog} \begin{cases} \longrightarrow {}^{208}\text{Pb} + p \\ \longrightarrow {}^{208}\text{Pb*} + p \\ \longrightarrow {}^{208}\text{Bi*} + n \end{cases}$$

The first mode comes from the first term in Fig. 15.17, whereas the other two are contributed by the second term. Consider, first, *neutron emission*. As justified above, it is forbidden if isospin invariance holds and can occur only through isospin impurities caused by the Coulomb interaction. For two reasons these impurities are small.[16] First, a constant electric field leads to a splitting of levels with different values of I_3 but does not induce transitions between states of different I. In a heavy nucleus, the electric field over most of the nuclear volume is nearly constant, as shown by Fig. 6.7, and transitions occur essentially only near the nuclear surface. Second, the excess neutrons in a nucleus have no Coulomb interaction. Since their number, $N - Z$, is large, they *dilute* the isospin impurity introduced by the protons. Generally, then, the width of the isobaric analog states due to neutron emission is much smaller than expected at first. Proton emission is allowed by isospin selection rules. However, the decay via the first term in Fig. 15.17 is reduced by $(2I_>)^{-1}$, the square of the expansion coefficient. Since $I_>$ is large in heavy nuclei, this reduction is considerable. The proton resulting from the second term has less energy than that from the first, and its decay is also slowed down. These factors all conspire to reduce the decay rate of isobaric analog resonances, and they explain the narrowness of the observed resonances.

The analog resonances in ${}^{209}\text{Bi*}$ can be understood in more detail by noting that the second term in Fig. 15.17 can be looked at as consisting of an excited state of ${}^{208}\text{Pb}$ plus a proton. Said differently, the state $|{}^{208}\text{Bi*} + n\rangle$ in Eq. (15.31) can be expanded in terms of states $|{}^{208}\text{Pb}^i + p^i\rangle$, where ${}^{208}\text{Pb}^i$ is the ith excited state of ${}^{208}\text{Pb}$ with the same spin and parity as the parent state and p^i is the proton emitted in the transition to that state. With such an expansion, Eq. (15.31) becomes

$$|\text{analog}\rangle = a_0 |{}^{208}\text{Pb}^0 + p^0\rangle + a_1 |{}^{208}\text{Pb}^1 + p^1\rangle + \cdots. \tag{15.32}$$

This equation expresses the utilty of experiments with analog resonances for nuclear structure studies: In principle, the coefficients a_i can be determined experimentally by measuring the decay rate of a particular analog state to the excited state ${}^{208}\text{Pb}^i$. The wave function of the analog state can thus be investigated and compared with the predictions based on specific nuclear models. Equation (15.32) also elucidates Fig. 15.15: Assume that for a particular analog resonance, the coefficient a_k is especially large. The analog resonance can then be observed most easily by looking at the reaction $p + {}^{208}\text{Pb}$

16. A. M. Lane and J. M. Soper, *Nucl. Phys.* **37**, 633 (1962); L. A. Sliv and Yu. I. Kharitonov, *Phys. Letters* **16**, 176 (1965); A. Bohr and B. R. Mottelson, *Nuclear Structure*, Benjamin, Reading, Mass., 1969. Vol. I, Section 2.1.

$\longrightarrow p^k + {}^{208}\mathrm{Pb}^k$. The cross section for this particular inelastic channel will be especially large, and the analog resonance will stand out clearly. The curves in Fig. 15.17 were obtained in this manner. ●

15.7 References

A very careful and easily readable introduction to the shell model, including a thorough discussion of the experimental material, is given by the two founders of the model: M. Goeppert Mayer and J. H. D Jensen, *Elementary Theory of Nuclear Shell Structure*, Wiley, New York, 1955.

The modern aspects of the shell model and a critical review of many experimental aspects is given in Chapter 3 of A. Bohr and B. R. Mottelson, *Nuclear Structure* Vol. 1. W. A. Benjamin, Reading, Mass. 1969.

An elementary approach that differs considerably from the one given in the present chapter is given in Chapter 4 of B. L. Cohen, *Concepts of Nuclear Physics*, McGraw-Hill, New York, 1971.

An extensive treatment of the experimental and the theoretical foundations of the shell model as it existed about 15 years ago is J. P. Elliott and A. M. Lane, "The Nuclear Shell Model," in *Encyclopedia of Physics*, Vol. 39 (S. Flügge, ed.), Springer, Berlin, 1957.

A concise review of recent results is H. J. Mang and H. A. Weidenmüller, *Ann. Rev. Nucl. Sci.* **18**, 1 (1968).

The mathematical problems that appear in the shell model are expounded in detail in A. de-Shalit and I. Talmi, *Nuclear Shell Theory*, Academic Press, New York, 1963.

Short introductions to isobaric analog states are given in H. Feshbach and A. Kerman, *Comments Nucl. Particle Phys.* **1**, 66 (1967), and M. H. Macfarlane and J. P. Schiffer, *Comments Nucl. Particle Phys.* **3**, 107 (1969). Considerably more details are given in a number of reviews collected in D. H. Wilkinson, ed., *Isospin in Nuclear Physics*, North-Holland, Amsterdam, 1969.

The theory of analog resonances is summarized and reviewed, at an advanced level, in N. Auerbach, J. Hüfner, A. K. Kerman, and C. M. Shakin, *Rev. Modern Phys.* **44**, 48 (1972).

PROBLEMS

15.1. Use the mass excesses given in Table A6 in the Appendix to discuss the evidence for shell closure as obtained from proton separation energies:

(a) Plot the proton separation energies for some nuclides across the magic numbers while keeping N constant.

(b) Repeat part (a) but keep $N - Z$ constant.

15.2. Discuss additional evidence for the exist-

ence of magic numbers by considering the following properties:
(a) The number of stable isotopes and isotones.
(b) Neutron absorption cross sections.
(c) The excitation energies of the first excited states of even-even nuclides.
(d) Beta decay energies.

15.3. Add the following spin-spin term to the two-body force, Eq. (15.4):

$$\boldsymbol{\sigma}_1 \cdot \boldsymbol{\sigma}_2 V_0' g(\mathbf{r}_1 - \mathbf{r}_2).$$

Assume that g is smooth and very short range. Show that this term gives no contribution to $V(1)$, Eq. (15.5), for closed shell nuclei. Show that the term can be neglected for a nucleus with one particle outside a closed shell.

15.4. Study the level sequence in the infinite three-dimensional square well. Compare the sequence with that obtained from the harmonic oscillator and given in Fig. 15.6.

15.5. Discuss additional evidence for the existence of a strong spin-orbit term in the nucleon-nucleus interaction by considering the scattering of protons from ^{4}He.

15.6. Verify Eq. (15.10).

15.7. Verify Eqs. (15.17) and (15.18).

15.8.
(a) Estimate the A dependence of the spin-orbit force.
(b) What is the strength of the two-body spin-orbit force needed to obtain the empirical nuclear spin-orbit splitting? Compare to ^{5}He, ^{5}Li splitting.
(c) What is the sign of the two-body spin-orbit force that gives the correct nuclear spin-orbit term?

15.9. Verify the step from Eq. (15.14) to (15.15). Prove that the terms that are not shown in Eq. (15.15) average to zero.

15.10. Find the spin and parity assignment for the following single-hole nuclear ground states: ^{15}O, ^{15}N, ^{41}K, ^{115}In, ^{207}Pb. Compare your predictions with the measured data.

15.11. Compare the first few excited states of the nuclides ^{15}N, ^{17}O, and ^{39}K with the prediction of the single-particle shell model. Discuss spin, parity, and level ordering.

15.12. Use the single-particle model to calculate for odd-mass nuclei the magnetic moments as a function of spin for
(a) Z odd, and
(b) N odd.
(c) Compare the result with experimental values.

15.13. What isospin value would you expect for the ground state of an odd-mass nuclide (Z, N) in the single-particle shell model?

15.14. Use the single-particle shell model to explain why the *islands of isomerism* exist. (Traditionally, a long-lived excited nuclear state was called an isomer.) In particular, explain why the nuclide ^{85}Sr has an excited state, at 0.225 MeV, with a half-life of about 70 min.

15.15. Discuss direct nuclear reactions, for instance, $(p, 2p)$, in the shell model and show in the case of a particular example (for instance $p\ ^{16}\text{O} \rightarrow 2p\ ^{15}\text{N}$) that the shell structure is readily apparent in the differential cross section. [For example, see Th. A. Maris, P. Hillman, and H. Tyrèn, *Nucl. Phys.* **7**, 1 (1958).]

15.16. Explain the reaction mechanism for exciting analog states in (d, n) reactions. Find an example in the literature.

15.17. The force acting on a nucleon incident on a nucleus can be represented by a single-particle *optical potential*. Such a potential can contain a term

$$C\,\vec{I}\cdot\vec{I}'f(\mathbf{r}),$$

where $\vec{I}$ is the isospin of the incident nucleon and $\vec{I}'$ that of the target nucleus.
(a) Show that such a term is allowed.
(b) Explain how such a term permits excitation of isobaric analog resonances in (p, n) and (n, p) reactions, among others. Are these reactions (either or both) still "allowed" if the electromagnetic interaction is switched off?
(c) Estimate the magnitude and the mass number dependence of the constant C.

15.18. Consider the state of a proton, with small excitation energy, in a heavy nucleus. Explain why the application of the charge-lowering operator, I_-, to such a state gives zero.

16
Collective Model

Although the shell model describes the magic numbers and the properties of many levels very well, it has a number of failures. The most outstanding one is the fact that many quadrupole moments are much larger than those predicted by the shell model.[1] It was shown by Rainwater that such large quadrupole moments can be explained within the concept of a shell model if the closed-shell core is assumed to be deformed.[2] Indeed, if the core is ellipsoidal it acquires a quadrupole moment proportional to the deformation. A deformation of the core is evidence for many-body effects, and collective modes of excitation are possible. The appearance of such modes is not surprising. Lord Rayleigh investigated the stability and oscillations of electrically charged liquid drops in 1877,[3] and Niels Bohr and F. Kalckar showed in 1936 that a system of particles held together by their mutual attraction can perform collective oscillations.[4] A classical example of such collective effects is provided by plasma oscillations.[5] The existence of large nuclear quadrupole moments

1. C. H. Townes, H. M. Foley, and W. Low, *Phys. Rev.* **76**, 1415 (1949).
2. J. Rainwater, *Phys. Rev.* **79**, 432 (1950).
3. J. W. S. Rayleigh, *The Theory of Sound*, Vol. II, Macmillan, New York, 1877, §364.
4. N. Bohr, *Nature* **137**, 344 (1936); N. Bohr and F. Kalckar, *Kgl. Danske Videnskab. Selskab. Mat.-fys. Medd.* **14**, No. 10 (1937).
5. Feynman Lectures, II-7-5ff; Jackson, Chapter 10.

provides evidence for the possibility of collective effects in nuclei. Beginning about 1950, Aage Bohr and Ben Mottelson started a systematic study of collective motions in nuclei;[6] over the years, they and their collaborators have improved the treatment so that today the model combines the desirable features of shell and collective models and is called the *unified nuclear model*.

The salient facts can be discussed most easily by describing two extreme situations. *Closed shell* nuclei are spherically symmetric and not deformed. The primary collective motions of such nuclei are surface oscillations, like the surface waves on a liquid drop. For small oscillations, harmonic restoring forces are assumed, and equally spaced vibrational levels result. *Far away* from closed shells, the nucleons outside the core polarize the core, and the nucleus can acquire a *permanent deformation*. The entire deformed nucleus can rotate, and this type of collective excitation leads to the appearance of rotational bands. The deformed nucleus acts as a nonspherical potential for the much more rapid single-particle motion; the energy levels of a single particle in such a potential can be investigated, and the result is the *Nilsson model*,[7] already mentioned at the end of the previous chapter.

We shall begin the discussion in the present chapter with deformations and rotational excitations because these two features are easiest to understand and give the most spectacular effects.

16.1 Nuclear Deformations

As early as 1935 optical spectra revealed the existence of nuclear quadrupole moments.[8] We have encountered the quadrupole moment in Section 12.5, and we have seen there that it measures the deviation of the shape of the nuclear charge distribution from a sphere. The existence of a quadrupole moment hence implies nonspherical (deformed) nuclei. For the discussion of nuclear models, the sign and magnitude of the deformation are important. As we shall see below, the quadrupole moments far away from closed shells are so large that they cannot be due to a single particle and thus cannot be explained by the naive shell model. The discrepancy is particularly clear around $A \approx 25$ (Al, Mg), $150 < A < 190$ (lanthanides), and $A > 220$ (actinides).

The classical definition of the quadrupole moment has already been given in Eq. (12.30) as

$$Q = Z \int d^3r \, (3z^2 - r^2)\rho(\mathbf{r}). \tag{16.1}$$

6. A. Bohr, *Phys. Rev.* **81**, 134 (1951); A. Bohr and B. R. Mottelson, *Kgl. Danske Videnskab. Selskab. Mat-fys. Medd.* **27**, No. 16 (1953).
7. S. G. Nilsson, *Kgl. Danske Videnskab. Selskab. Mat.-fys. Medd.* **29**, No. 16 (1955).
8. H. Schüler and T. Schmidt, *Z. Phys.* **94**, 457 (1935).

Note that the quadrupole moment as defined here has the dimension of an area. In some publications, an additional factor e is introduced in the definition of Q. For estimates, Q is computed for a homogeneously charged ellipsoid with charge Ze and semiaxes a and b. With b pointing along the z axis, Q becomes

$$Q = \tfrac{2}{5}Z\,(b^2 - a^2). \tag{16.2}$$

If the deviation from sphericity is not too large, the average radius $\bar{R} = \frac{1}{2}(a + b)$ and $\Delta R = b - a$ can be introduced. With $\delta = \Delta R / \bar{R}$, the quadrupole moment becomes

$$Q = \tfrac{4}{5}Z R^2 \delta. \tag{16.3}$$

Quantum mechanically, the probability density $\rho(r)$ is replaced by $\psi^*_{m=j}\psi_{m=j}$. Here j is the spin quantum number of the nucleus and $m = j$ indicates that the nuclear spin is taken to point along the z direction. Thus

$$Q = Z \int d^3r\,\psi^*_{m=j}(3z^2 - r^2)\psi_{m=j}. \tag{16.4}$$

It is customary to introduce a reduced quadrupole moment,

$$Q_{\text{red}} = \frac{Q}{ZR^2}. \tag{16.5}$$

For a uniformly charged ellipsoid, Eq. (16.3) shows that the reduced quadrupole moment is approximately equal to the deformation parameter δ:

$$Q_{\text{red}}\ (\text{ellipsoid}) = \tfrac{4}{5}\delta. \tag{16.6}$$

After these preliminary remarks we turn to some experimental evidence. Figure 16.1 displays the reduced quadrupole moments as a function of the number of odd nucleons (Z or N); it shows that the nuclear deformation is very small near the magic numbers but assumes values as large as 0.4 between shell closures. The large deformations are all positive. Equation (16.1) then indicates that these nuclei are elongated along their symmetry axes; they are cigar-shaped (prolate).

The first question is now: Can the observed deformations be explained by the shell model? In the single-particle shell model, the electromagnetic moments are due to the last nucleon; the core is spherically symmetric and does not contribute to the quadrupole moment. The situation for a single proton and a single proton hole are sketched in Fig. 16.2. To compute the quadrupole moment arising from the single particle, a single-particle wave function, for instance, as in Eq. (13.30), is inserted into Eq. (16.4); the result is

$$Q_{\text{sp}} = -\langle r^2 \rangle \frac{2j - 1}{2(j + 1)}. \tag{16.7}$$

Here, j is the angular momentum quantum number of the single particle and $\langle r^2 \rangle$ is the mean-square radius of the single-nucleon orbit. With

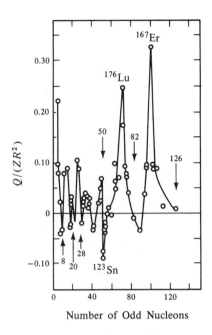

Fig. 16.1. Reduced quadrupole moment plotted versus the number of odd nucleons (Z or N). Arrows indicate the positions of closed shells, where $Q = 0$.

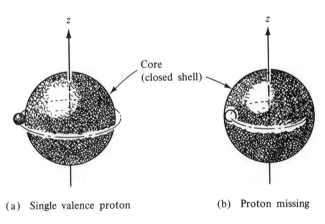

(a) Single valence proton   (b) Proton missing

Fig. 16.2. Quadrupole moment produced by a closed shell core plus (a) a single proton and (b) a single-proton hole.

$\langle r^2 \rangle \approx R^2$, the reduced quadrupole moment for a single proton becomes approximately

$$Q^p_{\text{red. sp}} \approx -\frac{1}{Z}. \tag{16.8}$$

A single neutron, in first order, produces no quadrupole moment. However, its motion affects the proton distribution by shifting the c.m., and the corresponding value is

$$Q_{sp}^n \approx \frac{Z}{A^2} Q_{sp}^p. \tag{16.9}$$

For single-hole states, relations similar to Eqs. (16.7) and (16.9) hold, but the sign is positive.

Even a quick glance at Fig. 16.1 shows that many of the observed quadrupole moments are far larger than the estimates given in Eqs. (16.8) and (16.9). A more detailed comparison for four specific cases is given in Table 16.1. For the estimates of the predicted single-particle quadrupole moments, $\langle r^2 \rangle$ has been taken equal to the square of the half-density radius c, given in Eq. (6.35). The values in the table show that in the case of a doubly magic nucleus plus a proton, the single-particle estimate agrees reasonably well with the actual quadrupole moment. In the other cases, the observed values are very much larger than the predicted ones. In the case of ^{175}Lu even the sign is wrong. The features shown in Table 16.1 for a few typical cases hold true when more nuclides are considered. The naive single-particle shell model cannot explain the observed large quadrupole moments.

Table 16.1 COMPARISON OF OBSERVED AND PREDICTED SINGLE-PARTICLE QUADRUPOLE MOMENTS

Nuclide	Z	N	Character	j	Q_{obs} (fm^2)	Q_{sp} (fm^2)	Q_{obs}/Q_{sp}
^{17}O	8	9	Doubly magic + 1 neutron	$\frac{5}{2}$	−2.6	−0.1	20
^{39}K	19	20	Doubly magic + proton hole	$\frac{3}{2}$	+5.5	+5	1
^{175}Lu	71	104	Between shells	$\frac{7}{2}$	+560	−25	−20
^{209}Bi	83	126	Doubly magic + 1 proton	$\frac{9}{2}$	−35	−30	1

How can the large quadrupole moments be explained? As stated earlier, the crucial step to a solution of the puzzle was taken by Rainwater. In the naive shell model it is assumed that the closed shells do not contribute to the nuclear moments: The core is assumed to be spherical. Rainwater suggested that the core of nuclides with large quadrupole moments is not spherical but permanently *deformed* by the valence nucleons. Since the core contains most of the nucleons and hence also most of the electric charge, even a small deformation produces a sizable quadrupole moment. An estimate of the deformation necessary to produce a certain reduced quadrupole moment can be obtained from Eq. (16.6). In the case of ^{17}O, for instance, a deformation of only $\delta = 0.07$ is needed to obtain the observed value.

The nuclear deformation can be understood by starting from a closed shell nuclide. As discussed in Chapter 15, the short-range pairing force makes such a nucleus spherical, with zero angular momentum. The addition of nucleons outside the closed shell tends to polarize the core through the long-range attractive part of the nuclear force. If only one nucleon is outside the core, the distortion is of the order of $1/A$. Since there are about Z electric charges in the core, such a distortion leads to an induced quadrupole moment of the order of $(Z/A)Q_{\rm sp}$. The distortion is about the same for neutrons as for protons, and nuclei with one neutron outside a closed shell thus should have a quadrupole moment of the same sign and about the same magnitude as odd-proton nuclides. The quadrupole moment of ^{17}O, listed in Table 16.1, can consequently be understood in a crude way. When more nucleons are added outside the closed shell, the polarization effect is enhanced, and the observed quadrupole moments can be explained.

The existence of a nuclear deformation makes itself felt not only in the static quadrupole moments but also in a number of other properties. We shall discuss two in the following sections: the appearance of a rotational spectrum and the behavior of shell-model states in a deformed potential.

16.2 Rotational Spectra of Spinless Nuclei

In the previous section we have shown that considerable evidence for the existence of permanently deformed nuclei exists. A nuclear deformation implies that the orientation of such a nucleus in space can be determined and can be described by a set of angles. This possibility leads to a prediction[9]. There exists an uncertainty relation between an angle, φ, and the corresponding orbital angular momentum operator, $L_\varphi = -i\hbar \, (\partial/\partial\varphi)$,

$$\Delta\varphi\,\Delta L_\varphi \gtrsim \hbar. \tag{16.10}$$

Because the angle can be determined to a certain extent, the corresponding angular momentum cannot be restricted to one sharp value, but a number of angular momentum states must exist. Such angular momentum states have been observed in many nuclides. They are called *rotational states*, and their physical characteristics will be discussed in more detail below. A particularly beautiful example of a rotational spectrum is shown in Fig. 16.3. A large number of similar spectra have been found in other nuclides.

The levels of ^{170}Hf in Fig. 16.3 show remarkable regularities: All levels have the same parity, the spin increases in units of 2, and the spacing

9. A. K. Kerman, "Nuclear Rotational Motion," in *Nuclear Reactions*, Vol. I, (P. M. Endt and M. Demeur, eds.), North-Holland, Amsterdam, 1959. The uncertainty relation Eq. (16.10), which underlies the discussion here, gives rise to interesting problems and arguments. If such arguments surface, read M. M. Nieto, *Phys. Rev. Letters* **18**, 182 (1967), and P. Carruthers and M. M. Nieto, *Rev. Modern Phys.* **40**, 411 (1968).

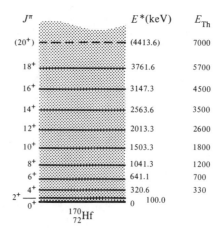

J^π		E^*(keV)	E_{Th}
(20^+)		(4413.6)	7000
18^+		3761.6	5700
16^+		3147.3	4500
14^+		2563.6	3500
12^+		2013.3	2600
10^+		1503.3	1800
8^+		1041.3	1200
6^+		641.1	700
4^+		320.6	330
2^+		100.0	
0^+		0	

$^{170}_{72}$Hf

Fig. 16.3. Rotational spectrum of the strongly deformed nuclide ^{170}Hf. [After F. S. Stephens, N. L. Lark and R. M. Diamond, *Nucl. Phys.* **63**, 82 (1965).] The levels were observed in the reaction ^{165}Ho(^{11}B, $6n$)^{170}Hf. The values E_{Th} are taken from Eq. (16.14), assuming that $E_2 = 100$ keV.

between adjacent levels increases with increasing spin. These properties are very different from those of shell-model states, discussed in Chapter 15. Moreover, ^{170}Hf is an even-even nucleus. We expect that in its ground state all nucleons have their spins paired. The energy needed to break a pair is of the order of 2 MeV (Fig. 15.3), much larger than the energy of the first excited state of ^{170}Hf. The levels therefore do not involve the breaking of a pair. ^{170}Hf is not an exception. Figure 16.4 shows the energies of the lowest 2^+ states of even-even nuclides. With very few exceptions, almost all of which occur for magic nuclei, these are the first excited states. The figure indicates clearly that the excitation energies far away from closed shells are much smaller than the pairing energy.

We shall now show that levels of the type shown in Fig. 16.3 can be explained by collective rotations of deformed nuclei. For simplicity, we assume the deformed nucleus to be axially symmetric (spheroidal), as shown in Fig. 16.5. A Cartesian system of axes, 1, 2, and 3, is fixed in the nucleus, with 3 being chosen as the nuclear symmetry axis. Axes 1 and 2 are equivalent. Naively it could be expected that such a nucleus could rotate about its symmetry axis as well as about any axis perpendicular to it. However, rotation about the symmetry axis is quantum mechanically not a meaningful concept. This fact can be seen as follows: Denote the angle about the symmetry axis 3 by ϕ. Axial symmetry implies that the wave function, ψ, is independent of ϕ,

$$\frac{\partial \psi}{\partial \phi} = 0.$$

R_3, the operator of the component of the orbital angular momentum along the 3 axis, is given by $R_3 = -i\hbar(\partial/\partial\phi)$. Axial symmetry thus implies that the component of the orbital angular momentum along the symmetry

Fig. 16.4. Three-dimensional model of the energies of the first excited 2⁺ states of even-even nuclei. The dependence of the excitation energy of these on proton and neutron number can be seen here. The magic numbers are indicated with heavy bars in the Z-N plane. The model shows that the excitation energies are very small between magic numbers, very large at the magic numbers. (Courtesy Gertrude Scharff-Goldhaber, Brookhaven National Laboratory; based on work published beginning with *Physics* **18**, 1105 (1952) and *Phys. Rev.* **90**, 587 (1953), and including data up to 1967.)

axis is zero: No collective rotation about the symmetry axis can occur. Rotation about any axis perpendicular to the symmetry axis, however, can lead to observable results. For simplicity we first assume a deformed nucleus with zero intrinsic angular momentum and consider rotations about axis 1 (Fig. 16.5). If the nucleus possesses a rotational angular momentum **R**, the energy of rotation is given by

$$H_{\text{rot}} = \frac{R^2}{2\mathcal{I}}, \qquad (16.11)$$

where $\mathcal{I}$ is the moment of inertia about axis 1. Translation into quantum mechanics yields the Schrödinger equation

$$\frac{R_{\text{op}}^2}{2\mathcal{I}}\psi = E\psi. \qquad (16.12)$$

We have already encountered the operator R_{op}^2 in Chapter 13; we called it L^2 there, and it is given by Eq. (13.25). According to Eq. (13.27), the eigenvalues and eigenfunctions of R_{op}^2 are given by

$$R_{\text{op}}^2 Y_J^M = J(J+1)\hbar^2 Y_J^M, \qquad J = 0, 1, 2, \ldots, \qquad (16.13)$$

where Y_J^M is a spherical harmonic (Table A8 in the Appendix). The

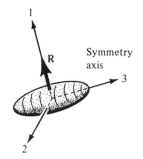

Fig. 16.5. Permanently deformed axially symmetric nucleus. **R** is the rotational angular momentum discussed in the text.

parity of Y_J^M is given by Eq. (9. 10) as $(-1)^J$. The spinless nucleus assumed here is invariant against reflection in the 1–2 plane. Since the spherical harmonics of odd J have odd parity, they change sign under such a reflection and are not admissible eigenfunctions. Only even values of J are allowed; with Eq. (16.12), the rotational energy eigenvalues of the nucleus become

$$E_J = \frac{\hbar^2}{2\mathcal{J}}J(J+1), \qquad J = 0, 2, 4, \dots. \tag{16.14}$$

The spin assignments of the levels in Fig. 16.3 agree with these values. If the energy of the first excited state is taken as given, the energies of the higher levels follow from Eq. (16.14) as

$$E_J = \tfrac{1}{6}J(J+1)E_2. \tag{16.15}$$

The values of E_J for ^{170}Hf predicted by this relation are given in Fig. 16.3. The general trend of the experimental spectrum is reproduced, but the computed values are all higher than the observed ones. The deviation can be explained by a centrifugal stretching of the nucleus. Taking stretching into account, the ratios observed for ^{170}Hf can be explained.[10]

Through Eq. (16.14), the energies of the rotational levels are described in terms of a moment of inertia, $\mathcal{J}$. The experimental value of this parameter for a particular nucleus can be obtained from the observed excitation energies, and this value can then be compared with that computed for a model. Two extreme models suggest themselves, rigid and irrotational motions. For a uniform rigid spherical body, of radius R_0 and mass Am, the moment of inertia is given by

$$\mathcal{J}_{\text{rigid}} = \tfrac{2}{5} Am R_0^2. \tag{16.16}$$

Rigid

In the other extreme, the nuclear rotation is considered as a wave traveling around the nuclear surface; the nuclear shape rotates and the nucleons oscillate. The moment of inertia is given by

$$\mathcal{J}_{\text{irrot}} = \tfrac{2}{5} Am(\Delta R)^2, \tag{16.17}$$

or

$$\mathcal{J}_{\text{irrot}} = \mathcal{J}_{\text{rigid}}\delta^2. \tag{16.18}$$

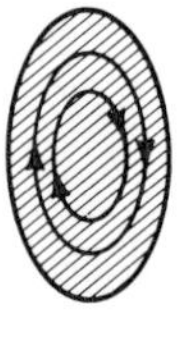

Irrotational

Fig. 16.6. Rigid and wavelike (irrotational) rotations. The two rotations are seen from a coordinate system that rotates with the nucleus. For the rigid rotation, the velocities vanish. For the irrotational motion, the streamlines form closed loops. The particles circulate opposite to the rotation of the entire nucleus.

Here $\delta = \Delta R/R_0$ is the deformation parameter already encountered in Eq. (16.3). The streamline picture for the two types of rotation, seen from a rotating coordinate system, are given in Fig. 16.6.[11] The empirical values of the moment of inertia lie between the two extremes. The nucleus is certainly not a rigid rotator, but the flow is also not completely irrotational.

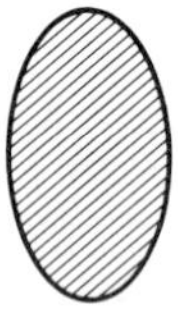

Finally, we come to a conceptual problem: A favorite examination question in quantum mechanics is to ask for a proof that a particle with spin J less than 1 cannot have an observable quadrupole moment. Yet

10. A. S. Davydov and A. A. Chaban, *Nucl. Phys.* **20**, 499 (1960); R. M. Diamond, F. S. Stephens, and W. J. Swiatecki, *Phys. Rev. Letters* **11**, 315 (1964).
11. The two models can be appreciated by playing with a hard-boiled and a raw egg.

we have assumed that a spinless nucleus, as in Fig. 16.5, possesses a permanent deformation. How does this assumption agree with the theorem just mentioned? The solution to the problem lies in a distinction between the *intrinsic quadrupole moment* and the *observed quadrupole moment*.[12] A spinless nucleus can have a permanent deformation (intrinsic quadrupole moment), and its effect can be seen in the existence of rotational levels and also in the rates of transitions leading to and from the $J = 0$ level. However, the quadrupole moment cannot be observed directly because the absence of a finite spin does not permit singling out a particular axis. In any measurement, an average over all directions is involved, and the permanent deformation appears only as a particularly large skin thickness.

16.3 Rotational Families

Deformed nuclei with spin zero in their ground state give rise to a rotational band, with spin-parity assignments 0^+, 2^+, ... Since many deformed nuclei with spins different from zero exist, the treatment of rotations must be extended to this more general case. The situation then becomes considerably more complicated, and we shall only treat the simplest situation, namely a nucleus consisting of a deformed, axially symmetric, spinless core and one valence nucleon, and we shall neglect the interaction between the intrinsic and the collective (rotational) motion. We assume that the valence nucleon does not affect the core so that it behaves like the deformed spinless nucleus treated in the previous section. The core then gives rise to a rotational angular momentum **R** perpendicular to the symmetry axis, 3, so that $R_3 = 0$. The valence nucleon produces an angular momentum **j**; **R** and **j** are shown in Fig. 16.7(a); they add up

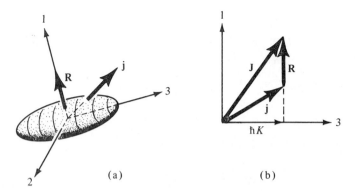

(a)   (b)

Fig. 16.7. (a) The deformed core gives rise to a collective angular momentum **R**; the valence nucleon produces an angular momentum **j**. (b) **R** and **j** add up to the total nuclear angular momentum **J**. The eigenvalue of the component of **J** along the symmetry axis 3 is denoted by $\hbar K$.

12. K. Kumar, *Phys. Rev. Letters* **28**, 249 (1972).

to the total nuclear angular momentum $\mathbf{J}$:

$$\mathbf{J} = \mathbf{R} + \mathbf{j}. \tag{16.19}$$

The total angular momentum $\mathbf{J}$ and its component, J_3, along the nuclear symmetry axis are conserved, and they satisfy the eigenvalue equations

$$
\begin{aligned}
J_{\text{op}}^2 \psi &= J(J+1)\hbar^2 \psi \\
J_{3,\text{op}} \psi &= K\hbar\psi.
\end{aligned}
\tag{16.20}
$$

Because $R_3 = 0$, the eigenvalue of $j_{3,\text{op}}$ is also given by $\hbar K$.

If, as assumed, the state of the valence nucleon is not affected by the collective rotation, then it is to be expected that each state of the valence nucleon can form the base (head) of a separate rotational band. In the following we shall compute the energy levels of these bands. The Hamiltonian is the sum of the rotational energy and the energy of the valence nucleon,

$$H = H_{\text{rot}} + H_{\text{nuc}},$$

or, with Eqs. (16.11) and (16.19),

$$H = \frac{R_{\text{op}}^2}{2\mathscr{I}} + H_{\text{nuc}} = \frac{1}{2\mathscr{I}}(\mathbf{J}_{\text{op}} - \mathbf{j}_{\text{op}})^2 + H_{\text{nuc}}.$$

The physical meaning becomes clearer if the Hamiltonian is written as the sum of three terms,

$$
\begin{aligned}
H &= H_R + H_p + H_c \\
H_R &= \frac{1}{2\mathscr{I}}(J_{\text{op}}^2 - 2J_{3,\text{op}}j_{3,\text{op}}) \\
H_p &= H_{\text{nuc}} + \frac{1}{2\mathscr{I}}j_{\text{op}}^2 \\
H_c &= -\frac{1}{\mathscr{I}}(J_{1,\text{op}}j_{1,\text{op}} + J_{2,\text{op}}j_{2,\text{op}}).
\end{aligned}
\tag{16.21}
$$

The third term resembles the classical Coriolis force, and it is called the Coriolis, or rotation-particle coupling, term. It can be neglected except for the special case $K = \frac{1}{2}$.[13] The second term is independent of the rotational state of the nucleus, and its contribution to the energy can be found by solving

$$H_p \psi = E_p \psi.$$

The first term describes the energy of the rotational motion. With Eq. (16.20), the energy eigenvalues of this term are given by

$$E_R = \frac{\hbar^2}{2\mathscr{I}}[J(J+1) - 2K^2], \qquad J \geq K. \tag{16.22}$$

The total energy is then[13]

$$E_{J,K} = \frac{\hbar^2}{2\mathscr{I}}[J(J+1) - 2K^2] + E_p. \tag{16.23}$$

---

13. For the treatment of the case $K = \frac{1}{2}$, see references 6 or 9.

This relation describes a sequence of levels, similar to the one given in Eq. (16.14) for spinless nuclei. Following the terminology in molecular physics, the sequence belonging to a particular value of K is called a *rotational band*, and the state with lowest spin is called the *band head*. Characteristic differences exist between the case $K = 0$ and $K \neq 0$:

1. The spins for the case $K = 0$ are the even integers, while the spins for $K \neq 0$ are given by

$$J = K, K + 1, K + 2, \ldots, \qquad K \neq 0. \qquad (16.24)$$

2. The ratios of excitation energies above the band head are not given by Eq. (16.15). For instance, the ratio of excitation energies of the second to the first excited state is not $\frac{10}{3}$, but

$$\frac{E_{K+2,K} - E_{K,K}}{K_{K+1,K} - E_{K,K}} = 2 + \frac{1}{K + 1}. \qquad (16.25)$$

The value of the component K can be determined from this ratio. As an example of the appearance of rotational bands in an odd-A nucleus, the level diagram of ^{249}Bk is shown in Fig. 16.8. The energy levels are drawn at the left, with spins and parities. Three bands can be distinguished; their band heads have assignments $K = (\frac{7}{2})^+$, $(\frac{3}{2})^-$, and $(\frac{5}{2})^+$. The level sequences satisfy Eq. (16.24), and the energies are reasonably well described by Eq. (16.23). The values of K follow unambiguously from Eq. (16.25).

The rotational families can be represented as trajectories in an angular momentum plot, just as was done for the harmonic oscillator levels in

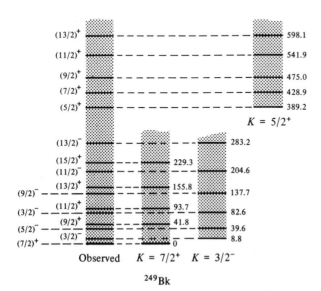

Fig. 16.8. Energy levels of ^{249}Bk. All observed energy levels up to an excitation energy of about 600 keV are given at the left. The levels fall into three rotational bands; these are shown at the right. All energies are in keV.

Fig. 13.9 and for some hyperons in Fig. 13.10. Such a plot is shown in Fig. 16.9 for the three families that have emerged from the ^{249}Bk decay scheme of Fig. 16.8. The states on one trajectory have the same internal structure and are distinct only in their collective rotational motion.

So far we have discussed nuclear deformations and the resulting rotational structure of energy levels. While the treatment has been superficial and many complications and justifications have been omitted, the most important physical ideas have emerged. In the following sections, two more aspects of collective motions must be taken up—the influence of the nuclear deformation on shell-model states (the Nilsson model) and collective vibrations.

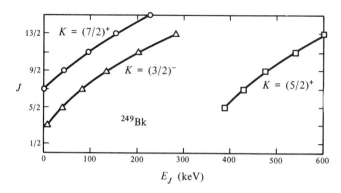

Fig. 16.9. Angular momentum plot for the three rotational families of ^{249}Bk displayed as energy levels in Fig. 16.8.

16.4 One-Particle Motion in Deformed Nuclei (Nilsson Model)

In Chapter 15, the nuclear shell model is treated; in the previous section, nuclei are considered as collective systems that can rotate. These two models are prototypes of two extreme and opposite points of view. Is there a way to weld the two models into one? In the present section, we shall describe the first step to a unified picture, namely the Nilsson model.[7] This model considers a deformed nucleus as consisting of independent particles moving in a *deformed well*. In Chapter 15, shell-model states in a spherical well were treated. As justified in Section 15.2, the average potential seen by nucleons resembles the nuclear density distribution. With Eqs. (13.20), (13.22), and (15.17), the spherical shell model potential can be written as

$$V(r) = \tfrac{1}{2}m\omega^2 r^2 - C\mathbf{l}\cdot\mathbf{s}. \tag{16.26}$$

The first term is the central potential, and the second the spin-orbit potential. The factor ω is related to the energy of an oscillator level (Fig. 13.7) through Eq. (13.29), $E = (N + \tfrac{3}{2})\hbar\omega$. The levels in the potential (16.26) are given, for instance, in Fig. 15.10; they are labeled by the quantum numbers N, l, and j. Because of rotational and parity invariance, the total angular momentum, j, and the orbital angular momentum, l (or the parity), of the nucleon are good quantum numbers, and N, l, and j are used to label the levels.

Since many nuclei possess large permanent deformations, as described in Section 16.1, nucleons do not always move in a spherical potential, and Eq. (16.26) must be generalized. A well-known generalization is due to Nilsson, who wrote, instead of Eq. (16.26),

$$V_{def} = \tfrac{1}{2}m[\omega_{\perp}^2(x_1^2 + x_2^2) + \omega_3^2 x_3^2] + C\mathbf{l}\cdot\mathbf{s} + Dl^2. \qquad (16.27)$$

This potential describes an axially symmetric situation—the one that applies to most deformed nuclei. The coordinates x_1, x_2, and x_3 are fixed in the nucleus: x_3 lies along symmetry the axis, 3 (Fig. 16.5). C determines the strength of the spin-orbit interaction. The term Dl^2 corrects the radial dependence of the potential: The oscillator potential differs markedly from the realistic potential near the nuclear surface, as shown in Fig. 15.5. States with large orbital angular momentum are most sensitive to this difference, and the term Dl^2, with $D < 0$, lowers the energy of these states. Nuclear matter is nearly incompressible: For a given form of the deformation, the coefficients $\omega_{\perp}$ and ω_3 are thus related. For a pure quadrupole deformation, discussed in the following section, the relation between the coefficients $\omega_{\perp}$ and ω_3 is expressed in terms of a deformation parameter ϵ:

$$\begin{aligned} \omega_3 &= \omega_0(1 - \tfrac{2}{3}\epsilon) \\ \omega_{\perp} &= \omega_0(1 + \tfrac{1}{3}\epsilon). \end{aligned} \qquad (16.28)$$

For $\epsilon^2 \ll 1$, $\omega_{\perp}^2$ and ω_3 satisfy

$$\omega_{\perp}^2\omega_3 = \omega_0^3, \qquad (16.29)$$

and this relation expresses the constancy of the nuclear volume on deformation. The parameter ϵ is connected to the deformation parameter δ introduced in Section 16.1 by

$$\delta = \epsilon(1 + \tfrac{1}{2}\epsilon). \qquad (16.30)$$

With Eqs. (16.3), (16.30), and (6.37), the intrinsic quadrupole moment can be written as

$$Q = \tfrac{4}{3}Z\langle r^2\rangle\epsilon(1 + \tfrac{1}{2}\epsilon). \qquad (16.31)$$

Equations (16.27) and (16.28) show that V_{def} is determined by four parameters, ω_0, C, D, and ϵ. Only ϵ depends strongly on the nuclear shape. For a given nuclide, ϵ is found by measuring Q and $\langle r^2 \rangle$. The first three parameters, ω_0, C, and D, are independent of the nuclear shape for $\epsilon^2 \ll 1$, and they are determined from the spectra and radii of spherical nuclei, where $\epsilon = 0$. Approximate values of these parameters are

$$\hbar\omega_0 \approx 41 A^{-1/3} \text{ MeV} \qquad (16.32)$$

and

$$C \approx -0.1\hbar\omega_0, \qquad D \approx -0.02\hbar\omega_0. \qquad (16.33)$$

The choice (16.27) of the potential V_{def} is not unique, and forms other than the one introduced by Nilsson have been studied extensively.[14] Since the salient features of the resulting spectra are unchanged, we restrict the discussion to the Nilsson model.

In the Nilsson model, as in the spherical single-particle model treated in Chapter 15, it is assumed that all nucleons except the last odd one are paired and do not contribute to the nuclear moments. To find the wave function and the energy of the last nucleon, the Schrödinger equation with the potential V_{def} is solved numerically with the help of a computer. A typical result for small A is shown in Fig. 16.10. For zero deformation the levels agree with the ones shown in Fig. 15.10, and they can be labeled with the quantum numbers N, j, and l. (N characterizes the oscillator shell and

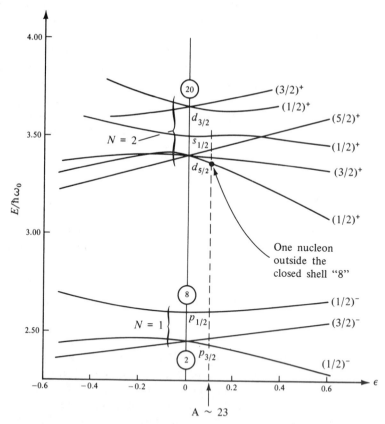

Fig. 16.10. Level diagram in the Nilsson model. The notation is explained in the text. Each state can accept two nucleons.

14. A detailed investigation of the single-particle levels of nonspherical nuclei in the region $150 < A < 190$ is given by W. Ogle, S. Wahlborn, R. Piepenbring, and S. Fredriksson, *Rev. Modern Phys.* **43**, 424 (1971).

is given in Table 15.1.) In this limit ($\epsilon = 0$), the states are $(2j + 1)$-fold degenerate. The deformation lifts the degeneracy, as can be seen from Fig. 16.10: State $p_{3/2}$ splits into two and state $d_{5/2}$ into three levels. A nucleon with total angular momentum j in the spherical case gives rise to $\frac{1}{2}(2j + 1)$ different energy levels, with K values $j, j - 1, j - 2, \ldots, \frac{1}{2}$. The factor $\frac{1}{2}$ describes a remaining twofold degeneracy which is caused by the symmetry of the nucleus about the 1–2 plane: The states K and $-K$ have the same energy (Fig. 16.11). A state with a given value of K can accommodate two nucleons of a given kind.

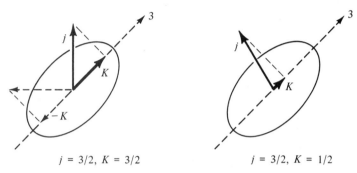

$$j = 3/2, \ K = 3/2 \qquad\qquad\qquad j = 3/2, \ K = 1/2$$

Fig. 16.11. In a nonspherical nucleus, the total angular momentum, j, of a nucleon is no longer a conserved quantity. Only its component, K, along the nuclear symmetry axis is conserved. A nucleon with spin j (in the spherical case) gives rise to K values, $j, j - 1, \ldots, \frac{1}{2}$. States K and $-K$ have the same energy.

Which *quantum numbers* describe the levels in a deformed potential? Rotational symmetry, except about the symmetry axis, is destroyed, and the angular momenta $\mathbf{j}$ and $\mathbf{l}$ are no longer conserved. Only two quantum numbers remain exact in the Nilsson model, the parity, $\pi = (-1)^N$, and the component K. (The fact that a nucleon with total angular momentum j can give rise to the various states K can be understood in the vector model: The angular momentum $\mathbf{j}$ precesses rapidly around the symmetry axis 3. Any component perpendicular to 3 is averaged to zero and has no effect.) A state is consequently denoted by K^π. Actually, three partially conserved quantum numbers are used to describe a given level further. We shall not need these *asymptotic* quantum numbers here.

As an application of the Nilsson model, we consider the ground states of some nuclides with a neutron or proton number around 11. Figure 16.1 shows that these nuclides are expected to have a deformation of the order of 0.1, and consequently the Nilsson model should be applicable. The relevant properties of a number of nuclides are summarized in Table 16.2. If it is assumed that the nuclides are described by the single-particle spherical shell model, their ground-state spin-parity assignment can be read from Fig. 15.10: Only the last odd nucleon is assumed to determine the moments. The listed nuclides have one or three nucleons outside the

Table 16.2 DEFORMED NUCLEI AROUND $A \approx 23$

Nuclide	Z	N	Q	$\delta \approx \epsilon$	Ground-state assignment		
					Exp.	Shell model	Nilsson model
^{19}F	9	10			$(\frac{1}{2})^+$	$(\frac{5}{2})^+$	$(\frac{1}{2})^+$
^{21}Ne	10	11	9 fm^2	0.09	$(\frac{3}{2})^+$	$(\frac{5}{2})^+$	$(\frac{3}{2})^+$
^{21}Na	11	10			$(\frac{3}{2})^+$	$(\frac{5}{2})^+$	$(\frac{3}{2})^+$
^{23}Na	11	12	14 fm^2	0.11	$(\frac{3}{2})^+$	$(\frac{5}{2})^+$	$(\frac{3}{2})^+$
^{23}Mg	12	11			$(\frac{3}{2})^+$	$(\frac{5}{2})^+$	$(\frac{3}{2})^+$

closed shell 8: According to Fig. 15.10, they should all have an assignment $(\frac{5}{2})^+$. In reality, the spins are different, even for ^{19}F, which has only one proton outside the magic number 8. The quadrupole moment has been measured for two of the listed nuclides, and $\langle r^2 \rangle$ can be taken from Eq. (6.34); Eq. (16.31) then provides the value of the deformation parameter δ ($\approx \epsilon$). In agreement with the estimate from Fig. 16.1, δ is of the order of 0.1. The value $\delta = 0.1$ is indicated in Fig. 16.10. Following this line the predicted assignments can be read: For one nucleon outside the closed shell 8, $(\frac{1}{2})^+$ is predicted. Three nucleons outside the shell lead to an assignment $(\frac{3}{2})^+$. As Table 16.2 shows, these values agree with experiment and demonstrate that the Nilsson model can explain at least some of the properties of deformed nuclei. (In all these assignments it is assumed that the even number of nucleons, for instance, the 10 neutrons in ^{19}F, remain coupled to zero.)

The prediction of ground-state moments is only one of the successes of the Nilsson model. It is also able to correlate a great many other observed properties of deformed nuclei.[15][16]

So far we have studied the motion of a single particle in a stationary deformed potential without regard to the motion of this well. The well is fixed in the nucleus. If the nucleus rotates, the potential rotates with it. In the previous section we have shown that the rotation of a deformed nucleus gives rise to a rotational band. Now the question arises: Is it correct to treat rotation and intrinsic motion separately, as was done in Eq. (16.21)? The separation is permissible if the motion of the particle in the deformed well is fast compared to the rotation of the well so that the particle traverses many orbits in one period of collective motion. In real nuclei, the condition is reasonably well satisfied because the rotational motion involves A nucleons and consequently is slower than the motion of the single valence nucleon. Nevertheless, for a realistic treatment, the

15. B. R. Mottelson and S. G. Nilsson, *Kgl. Danske Videnskab. Selskab. Mat-fys. Medd.* **1**, No. 8 (1959).

16. M. E. Bunker and C. W. Reich, *Rev. Modern Phys.* **43**, 348 (1971).

effect of the rotational motion on the intrinsic level structure, given by the term H_p in Eq. (16.21), must be taken into account.[17][18]

After asserting that intrinsic and rotational motion are indeed independent to a good approximation, we can return to the interpretation of the spectra of deformed nuclei. Since the nucleus can rotate in any state of the deformed nucleus, each intrinsic level (Nilsson level) is the band head of a rotational band. In other words, a rotational band is built onto each intrinsic level. Figure 16.8 gives an example of three bands, built on three different Nilsson states.

16.5 Vibrational States in Spherical Nuclei

So far we have discussed two types of nuclear states, rotational and intrinsic. The occurrence of different types of excitations is not peculiar to nuclei; diatomic molecules were known long ago to display three different types of excitations, intrinsic (electronic), rotational, and vibrational.[19] In a first approximation, the wave function of a given state can be written as

$$|\text{total}\rangle = |\text{intrinsic}\rangle |\text{rotation}\rangle |\text{vibration}\rangle. \qquad (16.34)$$

It turns out that nuclei are similar to molecules in that they, also, can have vibrational excitations.[6][20][21] In the present section, we shall describe some aspects of nuclear vibrations, restricting the treatment to spherical nuclei.

The simplest vibration corresponds to a density fluctuation about an equilibrium value, as shown in Fig. 16.12(a). Since such a motion carries no angular momentum, it is called the *monopole* mode. However, the compressibility of nuclei is extremely high, the energies of such monopole modes are thus very high, and none has been found yet.

Even an incompressible system can perform *shape oscillations*, without change of density. Such oscillations were first treated by Rayleigh[3], who observed: "The detached masses of liquid into which a jet is resolved do not at once assume and retain a spherical figure, but execute a series of vibrations, being alternately compressed and elongated in the direction of the axis of symmetry." The investigations of nuclear vibrations use much of the mathematical approach developed by Rayleigh, but, of course,

17. O. Nathan and S. G. Nilsson, in *Alpha-, Beta- and Gamma-Ray Spectroscopy*, Vol. 1 (K. Siegbahn, ed.), North-Holland, Amsterdam, 1965, p. 646.

18. A. K. Kerman, *Kgl. Danske Videnskab. Selskab Mat.-fys. Medd.* **30**, No. 15 (1956).

19. G. Herzberg, *Molecular Spectra and Molecular Structure*, Van Nostrand Rinehold, New York, 1950; L. D. Landau and E. M. Lifshitz, *Quantum Mechanics*, Pergamon, Elmsford, N.Y., 1958, Chapters 11 and 13.

20. N. Bohr and J. A. Wheeler, *Phys. Rev.* **56**, 426 (1939). D. L. Hill and J. A. Wheeler, *Phys. Rev.* **89**, 1102 (1953).

21. A. Bohr, *Kgl. Danske Videnskab. Selskab. Mat.-fys. Medd.* **26**, No. 14 (1952).

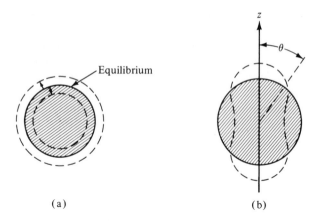

Fig. 16.12. (a) Monopole vibration. (b) Quadrupole vibration, $l = 2$, $m = 0$.

the oscillations are quantized. Before describing shape oscillations, we shall briefly outline how permanent nuclear *deformations* are expressed mathematically. After Rayleigh, the surface of a figure of arbitrary shape can be expanded as

$$R = R_0\left[1 + \sum_{l=0}^{\infty}\sum_{m=-l}^{l}\alpha_{lm}Y_l^m(\theta, \varphi)\right],$$ (16.35)

where the $Y_l^m(\theta, \varphi)$ are the spherical harmonics listed in Table A8 in the Appendix; θ and φ are polar angles with respect to an arbitrary axis and the α_{lm} are expansion coefficients. If the expansion coefficients are time-independent, Eq. (16.35) describes a permanent deformation of the nucleus. If α_{lm} is time dependent, then the term $l = 0$ is absent because nuclei are essentially incompressible. The term $l = 1$ corresponds to a displacement of the center-of-momentum and is not allowed, since no external force is acting on the system.[22] The lowest term of interest is thus $l = 2$, describing a *quadrupole deformation*. Since the salient features of nuclear collective vibrations appear in this mode, we restrict the following discussion to these terms. The nuclear radius then is written as

$$R = R_0\left[1 + \sum_{m=-2}^{2}\alpha_{2m}Y_2^m(\theta, \varphi)\right].$$ (16.36)

The quadrupole deformation is determined by the five constants α_{2m}. For a mode with $\alpha_{2m} = 0$, $m \neq 0$, the radius is (with Table A8 in the Appendix)

$$R(t) = R_0\left[1 + \alpha_{20}\left(\frac{5}{16\pi}\right)^{1/2}(3\cos^2\theta - 1)\right].$$ (16.37)

Such a deformation ($l = 2$, $m = 0$) is shown in Fig. 16.12(b).

---

22. The dipole vibration of protons against neutrons is allowed, however, because it leaves the nuclear c.m. unaffected. The giant dipole resonance that occurs in nuclei at excitation energies between 10 and 20 MeV is explained as being due to such dipole vibrations, and it is particularly clearly observed in electromagnetic processes.

Equation (16.36) describes a quadrupole deformation if the coefficients α_{2m} are constants. Shape vibrations are expressed through the time dependence of the expansion coefficients. To write the relevant Hamiltonian, we note first that for small oscillations about an equilibrium position, the motion can be treated as harmonic. For such harmonic motion, we saw in Section 13.7 that the kinetic energy is given by $\frac{1}{2}mv^2 = \frac{1}{2}m\dot{x}^2$, the potential energy by $\frac{1}{2}m\omega^2 r^2$, and the Hamiltonian by $H = \frac{1}{2}m\dot{x}^2 + \frac{1}{2}m\omega^2 r^2$. In the present situation, the dynamical variable is the deviation of the radius vector from its equilibrium value. This deviation is given by α_{2m} so that the Hamiltonian for an oscillating liquid drop, for $l = 2$ and for small deformation, has the form[3][21][23]

$$H = \frac{1}{2}B \sum_m |\dot{\alpha}_{2m}|^2 + \frac{1}{2}C \sum_m |\alpha_{2m}|^2. \tag{16.38}$$

Here B is the parameter corresponding to the mass and C is the potential energy parameter. H describes a five-dimensional harmonic oscillator because there are five independent variables α_{2m}. In analogy to Eq. (13.29), the energies of the quantized oscillator are given by

$$E_N = \left(N + \frac{5}{2}\right)\hbar\omega, \qquad \hbar\omega = \left(\frac{C}{B}\right)^{1/2}. \tag{16.39}$$

The angular dependence of the shape oscillations is described by the spherical harmonics Y_2^m, and we know from Eq. (13.27) that these are eigenfunctions of the total angular momentum with quantum number $l = 2$. The vibration carries an angular momentum of 2 and positive parity. Nuclear physicists have borrowed the expression *phonons* from their solid-state colleagues,[24] and the situation is described by saying that the phonon angular momentum is 2, and that one phonon is present in the first excited state, two phonons in the second excited state, and so forth. Since the ground states of even-even nuclei always have spin 0, the first excited vibrational states should have assignments 2^+. Two phonons have an energy $2\hbar\omega$ and they can couple to form states 0^+, 2^+, and 4^+. The states with spin 1 and 3 are forbidden by the requirement that the wave function of two identical bosons must be symmetric under exchange. The expected spectrum is sketched in Fig. 16.13.

Nuclei indeed show spectra with the characteristics predicted by the vibrational model.[25] They show up in even-even nuclides near closed shells. The degeneracy between the state 0^+, 2^+, and 4^+ is lifted by residual forces, and not all three members of the second excited state have been

23. A detailed derivation of Eq. (16.38) is given by S. Wohlrab, in *Lehrbuch der Kernphysik*, Vol. II (G. Hertz, ed.), Verlag Werner Dausien, 1961, p. 592.

24. C. Kittel, *Introduction to Solid State Physics*, 3rd ed., Wiley, New York, 1968, Chapter 5; J. M. Ziman, *Electrons and Phonons*, Clarendon Press, Oxford University, 1960. J. A. Reisland, *The Physics of Phonons*, Wiley, New York, 1973.

25. G. Scharff-Goldhaber and J. Weneser, *Phys. Rev.* **98**, 212 (1955).

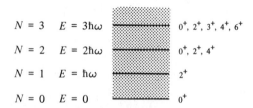

$N = 3$	$E = 3\hbar\omega$		$0^+, 2^+, 3^+, 4^+, 6^+$
$N = 2$	$E = 2\hbar\omega$		$0^+, 2^+, 4^+$
$N = 1$	$E = \hbar\omega$		2^+
$N = 0$	$E = 0$		0^+

Fig. 16.13. Vibrational states. The vibrational phonon carries an angular momentum 2 and positive parity. The states are characterized by the number, N, of phonons. The energy of the ground state has been set equal to zero.

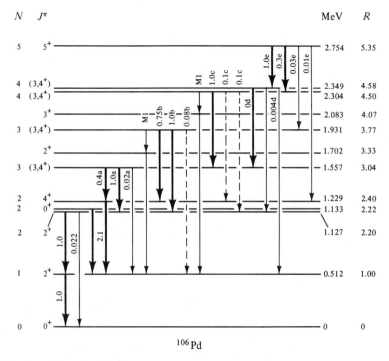

Fig. 16.14. Example of a vibrational spectrum. [After O. Nathan and S. G. Nilsson, in *Alpha, Beta- and Gamma-Ray Spectroscopy*, Vol. 1 (K. Siegbahn, ed.), North-Holland, Amsterdam, 1965.] The levels tentatively identified as vibrational are denoted by the number of phonons, N.

found in all cases. One example of a vibrational spectrum is shown in Fig. 16.14.

16.6 Nuclear Models—Concluding Remarks and Some Problems

In the last three chapters we have discussed the simplest aspects of nuclear models. In particular we have treated two extreme approaches, the shell and the collective models. The shell model is most successful near magic-number nuclei, the collective model for nuclei far removed from shell closures. The deformed shell model (Nilsson model) combines essential aspects of the two extremes. By considering the levels of nucleons in a deformed potential and collective excitations at the same time, the spectra of the low-lying states of most nuclei can be explained satisfactorily. However, such a unified model is not the final answer because it is phenomenological. What is really desired is a *microscopic* theory in which the observed features of the unified model are explained by the known properties of nuclear forces. Such a program is very ambitious, but considerable progress has been made in the past decade.

We will not describe the microscopic theory of nuclei in detail, but only make some pertinent remarks. The shell model potential is obtained by averaging the two-body force over the nucleus. However, when a nucleon is close to another one, there remain effects which cannot be included in the average potential. Such a short range residual force leads to pairing. The evidence for such pairing comes from the observation of an energy gap in the spectra of nuclei: The first intrinsic excitation of even-even heavy nuclei is at about 1 MeV (Fig. 15.4) whereas neighboring odd nuclei possess many levels below this energy. Even-even nuclei consequently exhibit an energy gap and this gap is taken as evidence for the pairing force:[26] Nucleons like to form pairs with angular momentum zero and the energy gap arises because to reach the first excited state requires a minimum energy corresponding to the break-up of such a pair. The pairing of nucleons bears a close resemblance to Cooper pairs[27] in superconductivity and it has been possible to use the tools and ideas develped to explain superconductivity[28] in nuclear physics.

There remains the question of the detailed nature of the residual force between nucleons. Does it have the characteristics of the free nucleon-nucleon force? Experiments so far are not conclusive but as the features of the residual force are being elucidated, it appears that their complexity and characteristics parallel those between free nucleons.[29] However, it is difficult to isolate residual force effects from average potential effects. For instance, two nucleons close together may behave like a quasi-bound state, in part due to the effects of the residual force. One might hope to study such nucleon-nucleon correlations in nuclei through (e, epp) reactions, but the observed correlations are influenced not only by the mutual force between the protons but also by the average effect of all the other nucleons in the nucleus. Nevertheless, short-range correlation effects have shown up in multiple scattering (Section 6.9) of high energy nucleons from nuclei.[30]

26. A. Bohr, B. R. Mottelson, and D. Pines, *Phys. Rev.* **110**, 936 (1958).
27. L. N. Cooper, *Phys. Rev.* **104**, 1189 (1956).
28. J. Bardeen, L. N. Cooper, and J. R. Schrieffer, *Phys. Rev.* **108**, 1175 (1957).
29. G. E. Brown, *Unified Theory of Nuclear Models and Forces*, 3d edition North-Holland, Amsterdam, 1971, Ch. 13; T. T. S. Kuo and G. E. Brown, *Nucl. Phys.* **85**, 40 (1966); M. Conze, H. Feldmeier and P. Manakos, *Phys. Letters* **43B**, 101 (1973).
30. V. G. Neudatchin and Yu. F. Smirnov, *Progress Nucl. Phys.* **10**, 273 (1969).

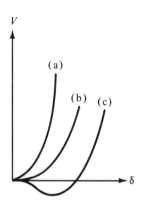

Fig. 16.15. Potential energy surfaces as a function of deformation (see Section 16.1). The three curves are (a) for a closed shell nucleus, (b) for a nucleus near a closed shell and (c) for a nucleus far from a closed shell. A permanent deformation occurs in the last case.

It is possible to understand the success of the spherical shell model for nuclei near closed shells and the transition to deformed nuclei for nuclei removed from magic numbers by considering the competition between the short-range pairing force and the longer range polarizing force. The latter is the force that nucleons outside a closed shell exert on those inside the shell. For a single nucleon outside a closed shell the polarizing effect is too small to deform the core. When two nucleons are present outside the closed shell, two competing effects occur: The pairing force tends to keep the nucleus spherical, whereas the polarizing force tries to deform the nucleus. When only a few nucleons are present outside a closed shell, the pairing force wins out, but the polarizing effects become dominant as more and more nucleons are added. This feature is shown schematically in Fig. 16.15.

The nuclear potential that acts on a nucleon can also be investigated with *hypernuclei*.[31] In such nuclei, one or sometimes two nucleons are replaced by hyperons, most often lambdas. Although the potential seen by the lambda is not identical to that which acts on a nucleon, the potentials are closely related; however, the lambda is unaffected by the Pauli exclusion principle. The study of hypernuclei is still in its infancy. This approach immediately suggests a host of other questions. Nucleons are presumably objects clothed with mesons. Are these meson clouds distorted in a nucleus, so that bound nucleons behave differently from free ones?[32] To what extent do excited nucleon states play a role in the nucleus? Recent evidence suggests that they may account for deviations of 1–5% in some features of the two-body and three-body problems.[33][34]

So far, we have only talked about two-body forces between nucleons. However, meson-theoretical considerations suggest that three-body forces should exist; such forces come into play only if *three* nucleons are close together. It is generally assumed that such forces play no major role in nuclei, but more theoretical and experimental work is needed before the absence or presence of three-body forces is established.[34]

A further nuclear structure problem parallels one of nucleons. Whereas low-lying excited states have been studied fairly thoroughly, highly excited states, except for the giant dipole resonance, remain to be investigated in detail. Consider first levels in the continuum. Are there monopole and quadrupole resonances at higher energies? How should such continuum states be explored and how described theoretically? There are indications

31. D. H. Davis and J. Sacton, in *High Energy Physics*, Vol. II. (E. H. S. Burhop, ed.), Academic Press, New York, 1967.

32. G. E. Brown, *Comments on Nuclear and Particle Physics* **5**, 6 (1972).

33. A. K. Kerman and L. S. Kisslinger, *Phys. Rev.* **180**, 1483 (1969); J. S. Vincent et al., *Phys. Rev. Letters* **24**, 236 (1970). H. Arenhövel and H. J. Weber, *Springer Tracts in Modern Physics* **65**, 58 (1972).

34. R. D. Amado, *Ann. Rev. Nucl. Sci.* **19**, 61 (1969); H. P. Noyes, G. E. Brown, and R. D. Amado in *International Confer. on Few Particle Problems in the Nuclear Interaction* (Los Angeles, 1972) North-Holland, Amsterdam, 1973.

of collective excited states at 8 to 12 MeV excitation energies, but little is yet known about these states.[35] Are there collective rotational and vibrational states built on single particle ones at these higher energies? Even highly-excited bound states are only partially explored. Of particular interest are the *yrast* levels.[36] An yrast level of a given nuclide, at a given angular momentum, is the level with least energy of that angular momentum.[37] Such levels play a crucial role in many nuclear reactions.[38] The yrast line, the line connecting the yrast levels of a given nuclide, shows how the moment of inertia changes as the rotational angular velocity of the nucleus varies.[39] Such investigations have become exciting in recent years because heavy ion accelerators permit the production of states with very high angular momenta. The properties of nuclear matter can thus be studied when it is being subjected to enormous rotational forces. To understand some of the experimental results, we note that the angular velocity and the moment of inertia of an axially symmetric rotor with angular momentum $\hat{J} = \hbar[J(J + 1)]^{1/2}$ are defined by[40]

$$\omega_{\text{rot}} = \frac{dE}{d\hat{J}} = \frac{dE}{\hbar d[J(J + 1)]^{1/2}} \approx \frac{dE}{\hbar dJ}, \qquad (16.40)$$

$$\mathcal{I} = \frac{\hat{J}}{\omega_{\text{rot}}} \approx \frac{\hbar J}{\omega_{\text{rot}}}. \qquad (16.41)$$

These two definitions together give

$$\mathcal{I} \approx \hbar^2 J \frac{dJ}{dE}. \qquad (16.42)$$

The yrast line of a nucleus gives E as a function of J, as for instance shown for rotational families in Fig. 16.9. From such a plot, Eqs. (16.40) and (16.42) permit determination of ω_{rot} and $\mathcal{I}$ for the yrast states. It has become customary to plot $2\mathcal{I}/\hbar^2$ versus the square of the rotational energy, $(\hbar\omega_{\text{rot}})^2$. The points on the plot are characterized by the values of the spin of the various yrast states. If nothing remarkable happens, then such a plot will indicate a smooth increase of the rotational energy with J, and a smooth increase of the moment of inertia with rotational energy. Such a

35. G. Chenevert et al., *Phys. Rev. Letters* **27**, 434 (1971); G. R. Satchler, *Nuclear Phys.* **A195**, 1 (1972); A. Bohr and B. R. Mottelson, *Nuclear Structure*, Vol. II, W. A. Benjamin, Reading, (to be published).

36. J. R. Grover, *Phys. Rev.* **157**, 832 (1967).

37. The origin of the word "yrast" is given by Grover:[36]

The English language seems not to have a graceful superlative form for adjectives expressing rotation. Professor F. Ruplin (of the Germanic Languages Department of the State University of New York, Stony Brook) suggested the use of the Swedish adjective *yr* for designating these special levels. This word derives from the same Old Norse verb *hvirfla* (to whirl) as the English verb *whirl*, and forms the natural superlative, *yrast*. It can thus be understood to mean "whirlingest," although literally translated from Swedish it means "dizziest" or "most bewildered."

38. J. R. Grover, *Phys. Rev.* **127**, 2142 (1962).

39. A. Johnson, H. Ryde, and S. A. Hjorth, *Nucl. Phys.* **A179**, 753 (1972).

40. Eqs. (16.41) and (16.42) are the rotational analogs of the relations $v = dE/dp$ and $m = p/v$.

behavior is indeed observed for many nuclei. In some nuclides, however, a dramatic departure from such a smooth picture has been discovered.[39] At some value of the spin J, the moment of inertia increases so rapidly that the rotational frequency actually decreases as higher spin states are reached. As an example, the yrast line for the even-spin states in ^{132}Ce are shown in Fig. 16.16. The yrast states up to $J = 18$ were found by using the reaction ^{16}O + ^{120}Sn $\longrightarrow 4n + {}^{132}$Ce.[41] At $J = 10$, a backbending occurs and the rotational frequency at $J = 14$ is about the same as at $J = 2$! The change in moment of inertia may be due to a phase transition between a superfluid and a normal state induced by the Coriolis force.[42] We have given this example here to show that new tools, like heavy ion accelerators, permit exploration of nuclei in new domains, such as that of very high angular momentum. Such extensions then can lead to new phenomena.

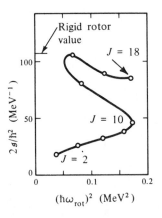

Fig. 16.16. Plot of the nuclear moment of inertia as a function of the square of the angular frequency. The rigid rotor value was calculated for the nucleus in its ground state ($\omega = 0$). [After O. Taras et al., *Phys. Letters* **41B**, 295 (1972).]

16.7 References

The theoretical treatment of nuclear models depends on the availability of complete and reliable information on nuclear spectroscopy and nuclear moments. The relevant experimental techniques are described in K. Siegbahn, ed., *Alpha-, Beta- and Gamma-Ray Spectroscopy*, North-Holland, Amsterdam, 1965, and in H. Kopfermann, *Nuclear Moments*, Academic Press, New York, 1958; N. F. Ramsey, "Nuclear Moments

41. O. Taras et al., *Phys. Letters* **41B**, 295 (1972).
42. B. R. Mottelson and J. G. Valatin, *Phys. Rev. Letters* **5**, 511 (1960). J. Krumlinde and Z. Szymanski, *Phys. Letters* **36B**, 157 (1971); **40B**, 314 (1972). But see also F. S. Stephens and R. S. Simon, *Nucl. Phys.* **A183**, 257 (1972). A. Molinari and T. Regge, *Phys. Lett.* **41B**, 93 (1972). A. Johnson and Z. Szymanski, *Phys. Rept.* **7C**, 182 (1973). R. A. Sorensen, *Rev. Modern Phys.* **45**, 353 (1973).

and Statistics," in *Experimental Nuclear Physics*, Vol. I (E. Segrè, ed.), Wiley, New York, 1953; and K. Alder and R. M. Steffen, *Ann. Rev. Nucl. Sci.* **14**, 403 (1964).

The authoritative work on the phenomenological description of the collective nuclear model is A. Bohr and B. R. Mottelson, *Nuclear Structure*, Vol. II, Benjamin, Reading, Mass. (to be published).

A careful and detailed description of the same aspects are also given in J. P. Davidson, *Collective Models of the Nucleus.*, Academic Press, New York, 1968.

Detailed comparisons between theoretical predictions and experimental data are given in Bohr and Mottelson, *Nuclear Structure*, Vol. II, and in B. R. Mottelson and S. G. Nilsson, *Kgl. Danske Videnskab. Selskab. Mat.-fys. Medd.* **1**, No. 8 (1959); M. E. Bunker and C. W. Reich, *Rev. Modern Phys.* **43**, 348 (1971); and W. Ogle, S. Wahlborn, R. Piepenbring, and S. Fredriksson, *Rev. Modern Phys.* **43**, 424 (1971).

The microscopic theory of nuclear models (quasi-particles, unified model, Hartree-Fock) is described in the following books and articles:

M. Baranger, "Theory of Finite Nuclei," in *Cargèse Lectures in Theoretical Physics* (M. Lévy, ed.), Benjamin, Reading, Mass., 1963. Concise. Good first introduction.

A. M. Lane, *Nuclear Theory*, Benjamin, Reading, Mass., 1964.

D. Nathan and S. G. Nilsson, "Collective Nuclear Motion and the Unified Model," in Siegbahn, Vol. I.

G. E. Brown, *Unified Theory of Nuclear Models and Forces*, 3rd ed., North-Holland, Amsterdam, 1971.

S. T. Belyaev, *Collective Excitations in Nuclei*, Gordon & Breach, New York, 1968.

A. B. Migdal, *Nuclear Theory: The Quasiparticle Method*, Benjamin, Reading, Mass., 1968.

D. J. Rowe, *Nuclear Collective Motion*, Methuen, London, 1970.

J. M. Eisenberg and W. Greiner, "Nuclear Theory," Vol. III, *Microscopic Theory of the Nucleus*, North-Holland, Amsterdam, 1972.

A complete discussion of nuclear models should also include *nuclear matter*. A readable first introduction is L. Gomes, J. D. Walecka, and V. F. Weisskopf, *Ann. Phys. (New York)* **3**, 241 (1958), reprinted in *Nuclear Structure*, selected reprints of the American Institute of Physics, New York, 1965.

A comprehensive review is H. A. Bethe, *Ann. Rev. Nucl.* Sci. **21**. 93 (1971).

PROBLEMS

16.1. Find the expression for the energy of interaction between a system with quadrupole moment Q and an electric field $\mathcal{E}$ with field gradient $\Delta \mathcal{E}$.

16.2. The electric quadrupole moment of a nucleus can be determined by using atomic beams.
(a) Describe the principle underlying the method.
(b) Sketch the experimental apparatus.
(c) What are the main limitations and sources of error?

16.3. Repeat Problem 16.2 for the method using optical hyperfine structure.

16.4. Quadrupole moments can also be determined by using nuclear quadrupole resonance and the Mössbauer effect. Answer the questions posed in Problem 16.2 for these two methods.

16.5. Verify Eq. (16.2).

16.6. The giant dipole resonance has a very different shape in spherical nuclei and in strongly deformed nuclei. Sketch typical resonances for the two cases. Explain the reason for the appearance of two peaks in deformed nuclei. How can the ground-state quadrupole moment be deduced from the positions of the two peaks? How are the relevant experiments performed? [F. W. K. Firk, *Ann. Rev. Nucl. Sci.* **20**, 39 (1970).]

16.7. How can the deformation of a nucleus be observed in an electron scattering experiment? [See, for instance, F. J. Uhrhane, J. S. McCarthy, and M. R. Yearian, *Phys. Rev. Letters* **26**, 578 (1971).]

16.8. Prepare a Z–N plot and indicate in this plot the regions where you expect spherical nuclei and where you expect large deformations. Plot the position of a few typical nuclides. [E. Marshalek, L. W. Person, and R. K. Sheline, *Rev. Modern Phys.* **35**, 108 (1963).]

16.9. Verify Eq. (16.7).

16.10. Show that the expectation value of the quadrupole operator in states with spins 0 and $\frac{1}{2}$ vanishes.

16.11. Discuss the transition rates for electric quadrupole transitions in strongly deformed nuclei:
(a) Find a particular example and compare the observed half-life with the one predicted by a single-particle estimate.

(b) How can the observed discrepancy be explained?

16.12. Coulomb excitation. Discuss:
(a) The physical process of Coulomb excitation, and
(b) The experimental approach.
(c) What information can be extracted from Coulomb excitation?
(d) Sketch the information that supports the assumption of collective states in strongly deformed nuclei. [K. Alder and A. Winther, *Coulomb Excitation*, Academic Press, New York, 1966; K. Alder et al., *Rev. Modern Phys.* **28**, 432 (1956).]

16.13. Verify the numbers in Table 16.1.

16.14. Compute the single-particle quadrupole moments for ^{7}Li, ^{25}Mg, and ^{167}Er. Compare with the observed values.

16.15.
(a) Draw the energy levels of ^{166}Yb, ^{172}W, and ^{234}U. Compare the ratios E_4/E_2, E_6/E_2, and E_8/E_2 with the ones predicted on the basis of rotation of a spherical nucleus.
(b) Repeat part (a) for ^{106}Pd and ^{114}Cd. Compare with the predictions of the vibrational model.

16.16. Assume ^{170}Hf to be a rigid body. Calculate, very approximately, the centrifugal force in state $J = 20$. What would happen to the nucleus if its mechanical properties were similar to those of steel? Support your conclusion with a crude calculation.

16.17. Verify the uncertainty relation equation (16.10).

16.18. Verify Eqs. (16.16) and (16.17).

16.19. Figure 16.6 shows the flow lines of particles for rigid and for irrotational motion in a rotating coordinate system. Draw the corresponding flow lines in a laboratory-fixed coordinate system.

16.20. Assume the moment of inertia, $\mathcal{I}$, in Eq. (16.14) to be a function of the energy E_J. Compute $\mathcal{I}(E_J)$ (in units of $\hbar^2$/MeV) for the rotational levels in ^{170}Hf, ^{184}Pt, and ^{238}U. Plot $\mathcal{I}(E_J)$ versus E_J and show that a linear fit

$\mathcal{I}_{\text{eff}} = c_1 + c_2 E_J$ reproduces the empirical data well.

16.21. Consider an even-even nucleus with equilibrium deformation δ_0 and spin $J = 0$ in its ground state. The energy in a state with spin J and deformation δ is the sum of a potential and a kinetic term,

$$E_J = a(\delta - \delta_0)^2 + \frac{\hbar^2}{2\mathcal{I}} J(J+1).$$

(a) Assume irrotational motion, $\mathcal{I} = b\delta^2$. Use the condition $(dE/d\delta) = 0$ to find the equation for the equilibrium deformation δ_{eq} in the state with spin J.
(b) Show for small deviations of the deformation from the ground-state deformation that the nucleus stretches and that the energies of the rotational states can be written as $E_J = AJ(J+1) + B[J(J+1)]^2$.
(c) Use this form of E_J to fit the observed energy levels of ^{170}Hf by determining the constants A and B from the two lowest levels. Then check how well the computed energies agree with the observed ones up to $J = 20$.

16.22. Consider an axially symmetric deformed core plus one valence nucleon (Fig 16.7). Why are J and K good quantum numbers, but not j?

16.23. Why are the states with odd J not excluded from the sequence (16.24)?

16.24. Discuss the rotational families of ^{249}Bk (Fig. 16.8):
(a) Check how well Eq. (16.23) fits the observed energy levels for each band.
(b) Show that K for each band can be found unambiguously from the lowest three levels of a band by using Eq. (16.25).

16.25. Compare the term H_c in Eq. (16.21) with the classical Coriolis force.

16.26. Use the slope of the trajectories in Fig. 16.9 and Eq. (16.23) for E_J to determine the moment of inertia as a function of J. Plot $\mathcal{I}$ versus J for the three families. Is stretching apparent?

16.27. Find another example for rotational families and prepare a plot similar to Fig. 16.9.

16.28. Find the energy levels of the anharmonic oscillator, described by the potential

$$V = \tfrac{1}{2}m[\omega_\perp(x_1^2 + x_2^2) + \omega_3^2 x_3^2].$$

16.29. Describe the complete labeling of Nilsson levels.

16.30. Verify Eq. (16.30).

16.31. Justify that the rotational and the intrinsic motion in deformed nuclei can be separated by finding approximate values for the time of rotation and the time a single nucleon needs to traverse the nucleus.

16.32. Discuss the level diagram of ^{165}Ho [M. E. Bunker and C. W. Reich, *Rev. Modern Phys.* **43**, 348 (1971)]:
(a) Find the various band heads and their rotational spectra.
(b) Plot the bands in a Regge plot.
(c) Use a Nilsson diagram to find the complete quantum number assignment for each band head.

16.33. Consider a completely asymmetric nucleus, with $\omega_1 > \omega_2 > \omega_3$. What is the spectrum of single particle levels in such a nucleus if $\omega_1/\omega_2/\omega_3 = \alpha/\beta/1$. (*Hint*: Use Cartesian coordinates.)

16.34. Compare molecular and nuclear spectra. Discuss the energies and energy ratios involved in the three types of excitations. Discuss the corresponding characteristic times. Sketch the essential aspects of the spectra.

16.35. Show that the term $l = 1$ in Eq. (16.35) corresponds to a translation of the nuclear c.m. Draw an example.

16.36. Find a relation between the coefficients α_{lm} and $\alpha^*_{l,-m}$ in Eq. (16.35) by using the reality of R and the properties of the Y_l^m given in Table A8 in the Appendix.

16.37. Use Eq. (16.35) to draw a deformed nucleus described by $\alpha_{30} \neq 0$, all other $\alpha = 0$.

16.38. Verify the solution (16.39).

16.39. Show that for an incompressible irrotational nucleus the semiempirical mass formula gives for the coefficients B and C in Eq. (16.38)

$$B^{-1} = \frac{3}{8\pi} AmR^2$$

$$C = 4R^2 a_s - \frac{6}{5}\frac{Z^2 e^2}{R}.$$

16.40. Show that vibrational motion implies the existence of excited vibrational states. (*Hint:* Consider the nuclear density and show that the density is always constant if only one state exists. Then consider a small admixture of an excited state.)

16.41. Discuss a plot of the energy ratio E_2/E_1 for even-even nuclei. Indicate where rotational and where vibrational spectra appear. Compare the corresponding excitation energies E_1.

16.42. Why can a state 3^+ turn up in the level $N = 3$, but not $N = 2$, in Fig. 16.13?

16.43. Consider nonazimuthally symmetric quadrupole deformations,

$$R = R_0\left(1 + \sum_m \alpha_{2m} Y_2^m\right)$$

$$\alpha_{20} = \beta \cos \gamma, \qquad \alpha_{22} = \alpha_{2,-2}$$

$$= \frac{1}{\sqrt{2}} \beta \sin \gamma.$$

(a) If $\gamma = 0$, what is $V(\beta)$ for a spherical harmonic oscillator?

(b) For a prolate nucleus and harmonic forces, what is $V(\beta)$?

(c) Consider harmonic γ vibrations for a prolate nucleus. What is the shape of the potential and the energy spectrum due to these vibrations?

16.44. What is the effect of an octupole term in Eq. (16.35) on

(a) The vibrational spectrum?

(b) Permanent deformations?

(c) The rotational spectrum?

16.45. Label the angular momenta and show the spacings of the first two excited states for nuclear octupole vibrations. (Take nuclear symmetries into account.)

Nuclear Science and Technology

Science is the best technology.

After Helmholtz[1]

I think the relation between application and basic science is like that between fish and water. Without water there can be no fish; without basic science there can be no application.

T. D. Lee

Our entire life is dominated, for better or for worse, by technology. Technology has been spawned by basic research. Electricity, nuclear power, solid-state electronics, the laser, and X rays are testimony to this causal relation; they are all by-products of fundamental research. It is unlikely the X rays would have been found in the search for a better tool to study broken bones or that lasers would have been discovered in an applied research project devoted to develop a better tool for laying straight rows of bricks.

1. C. S. Slichter, *Science in a Tavern*, University of Wisconsin Press, Madison, 1958, p. 51.

Subatomic physics has given rise to a wide range of uses in other fields. On the one extreme, the knowledge gained in subatomic physics is an essential component for the understanding of the fundamental problems in cosmology, creation of the elements, processes in neutron stars, and supernovae. At the other extreme, techniques developed in subatomic physics, such as very large superconductors or spark chambers, are used in technology. In the middle ground, knowledge and techniques such as nuclear fusion, fission, radioisotopes, and the Mössbauer effect are essential tools in many human endeavors, from power production to archaeology.

The following chapters should show the reader a few of the many ways in which subatomic physics interacts with other fields. Since the discussions are sketchy and condensed, we give enough references so that the interested student can delve deeper into any subject. Applications of subatomic physics in general are reviewed in a number of books.[2,3] The booklets in the series *Understanding the Atom* provide simple descriptions of the uses of nuclear physics.[4]

2. G. T. Seaborg and W. R. Corliss, *Man and Atom*, Dutton, New York, 1971.

3. L. C. L. Yuan, ed., *Elementary Particles, Science, Technology, and Society*, Academic Press, New York, 1971.

4. *Understanding the Atom*, a series of booklets, published by USAEC Division of Technical Information Extension, Oak Ridge, Tenn.

17

Nuclear Power

The (Human) race may date its development from the day of the discovery
of a method of utilizing atomic energy.

Rutherford

Rutherford uttered these words more than 50 years ago, long before
nuclear power was a reality. Much has happened in these 50 years, and
nuclear power has become a reality. Three different nuclear power sources
appear to be feasible; they are based on fission, on fusion, and on radioactivity. The first and the third are used; the future of fusion power is not
yet assured. In the present chapter we shall discuss the three power sources
and describe some applications.

17.1 The Fission Process

Soon after the discovery of the neutron in 1932, Fermi began systematic
studies of reactions induced by the bombardment of heavy nuclei with
neutrons.[1] Experiments with uranium gave puzzling results which were

1. E. Fermi, *Nature* **133**, 757 (1934).

only explained in 1939 by Hahn and Strassmann:[2] Uranium, upon bombardment by neutrons, produced nuclei such as barium that have a far smaller atomic number than uranium. Meitner and Frisch[3] quickly verified the result, suggested that the uranium nucleus, after neutron capture, must divide itself into two nuclei of roughly equal size, and borrowed the name *fission* from biology. They also pointed out the analogy between the fission process and the division of a small liquid drop into droplets. The theoretical understanding of the fission process was advanced in one giant step by Bohr and Wheeler.[4] Their paper provides much of the framework and the language for the treatment of the fission process to the present day.[5]

To describe the fission process in a simplified model, we consider the following fission reaction:

$$n + {}^{235}\text{U} \longrightarrow {}^{236}\text{U} \longrightarrow {}^{139}\text{La} + {}^{95}\text{Mo} + 2n. \qquad (17.1)$$

The ${}^{235}\text{U}$ nucleus captures a neutron, and the compound nucleus ${}^{236}\text{U}$ is formed. In its ground state, ${}^{236}\text{U}$ is essentially stable; it has a half-life of 2.4×10^7 y. As discussed in Section 16.5 and shown in Fig. 16.2, such a nucleus can perform vibrations about its equilibrium state without fissioning. However, when ${}^{235}\text{U}$ captures a neutron, the compound nucleus ${}^{236}\text{U}$ is highly excited, and the amplitude of vibrations can become so large that the nucleus separates into two parts. The Coulomb force between the two parts then drives them apart with considerable energy. The various stages of the fission process are shown in Fig. 17.1. However, not all the available energy goes into kinetic energy; some is stored as internal (excitation) energy of the two fragments. This energy is released primarily by evaporation of neutrons. The main products of the fission process are therefore two very roughly equal nuclei and a few neutrons.

We shall discuss some of the outstanding features of the fission process on the basis of the liquid drop model that we have just described. However, the model should not be taken too literally; the detailed treatment of the fission process is considerably more involved.[6-8] Nevertheless, the following semiquantitative analysis gives some feeling for the magnitudes involved. We first consider the *energy release* in fission. A crude estimate can be obtained from the binding energy curve, Fig. 14.1. It indicates that the

2. O. Hahn and F. Strassmann, *Naturwissenschaften* **27**, 89 (1939).

3. L. Meitner and O. R. Frisch, *Nature* **143**, 471 (1939).

4. N. Bohr and J. A. Wheeler, *Phys. Rev.* **56**, 426 (1939).

5. The history of fission reads like a first-class adventure story. The discovery was missed, for various reasons, by quite a few scientists. When it was finally confirmed, the news was carried to the United States by Niels Bohr, and it started feverish activities in many laboratories. The story has been told in a number of books, and we refer to the following ones for exciting reading: L. Fermi, *Atoms in the Family*, University of Chicago Press, Chicago, 1954; R. Moore, *Niels Bohr*, Knopf, New York, 1966; and O. R. Frisch and J. A. Wheeler, *Phys. Today* **20**, 43 (Nov. 1967).

6-8. I. Halpern, *Ann. Rev. Nucl. Sci.* **9**, 245 (1959); J. S. Fraser and J. C. D. Milton, *Ann. Rev. Nucl. Sci.* **16**, 379 (1966); L. Wilets, *Theories of Nuclear Fision*, Clarendon Press, Inc., Oxford, 1964.

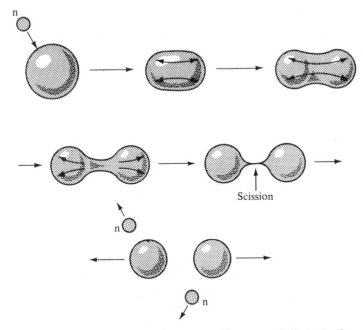

Fig. 17.1. Various stages of the fission process. The arrows indicate the flow of nuclear matter. *Scission* is the act of separation.

binding energy per particle, B/A, is about 7 MeV for $A = 250$ and about 8 MeV for $A = 125$. The medium-weight nuclei are more tightly bound than the heavier ones, and the energy release in a fission process $A = 250 \rightarrow 2A = 125$ is about 250 MeV. The Bethe-Weizsäcker relation leads to a more reliable estimate. Although fission occurs usually into two unequal parts, as shown, for instance, in Eq. (17.1), the separation into two equal parts can be used to obtain an estimate of the energy release. For a process $(A, Z) \rightarrow 2(A/2, Z/2)$, Eq. (14.9) predicts

$$Q = 2B\left(\frac{A}{2}, \frac{Z}{2}\right) - B(A, Z) = (1 - 2^{1/3})a_s A^{2/3} + (1 - 2^{-2/3})a_c Z^2 A^{-1/3},$$

or, with the values of the constants given in Eq. (14.11),

$$Q(\text{in MeV}) = -4.5A^{2/3} + 0.26Z^2 A^{-1/3}. \tag{17.2}$$

For ^{235}U, Q becomes about 180 MeV.

While the calculation just performed gives a reasonable value of Q, it is a static rather than dynamic one, based on the initial and final configuration. It fails to tell us for what nuclei fission can be expected, how much energy must be supplied to cause fission, and, more important for the dynamic considerations of a reactor, whether the energy Q goes into internal energy of the two fragments or into their kinetic energies. It is desirable that a sizable fraction goes into internal energy to produce neutrons, because neutrons are responsible for the chain reaction in reactors.

We begin the dynamical considerations by looking at small deformations of a nucleus that is initially spherically symmetric. It is then possible to use the spherical liquid drop model to study the condition that leads to instability for small deformations. If the binding energy decreases as the sphere is deformed, the spherical shape is stable; if it increases, then fission can occur. For a small deformation of a prolate nucleus, major and minor semiaxes are given by

$$a = R(1 + \epsilon)$$
$$b = R(1 - \tfrac{1}{2}\epsilon) \tag{17.3}$$

so that the volume remains constant. The surface energy, E_s, is assumed to be proportional to the surface area, and the generalization of Eq. (14.6) can be shown to be

$$E_s = -a_s A^{2/3} \left(1 + \tfrac{2}{5}\epsilon^2 + \cdots\right). \tag{17.4}$$

The Coulomb energy for an ellipsoid provides the generalization of Eq. (14.7),

$$E_c = -a_c Z^2 A^{-1/3} \left(1 - \tfrac{1}{5}\epsilon^2 + \cdots\right). \tag{17.5}$$

For small deformations, the total energy change ΔE in going from a sphere to an ellipsoid therefore depends quadratically on the deformation parameter ϵ,

$$\Delta E = \alpha\epsilon^2, \tag{17.6}$$

where

$$\alpha = \tfrac{1}{5}[a_c Z^2 A^{-1/3} - 2a_s A^{2/3}]. \tag{17.7}$$

The coefficient α is positive for

$$\frac{Z^2}{A} > \frac{2a_s}{a_c} = 49. \tag{17.8}$$

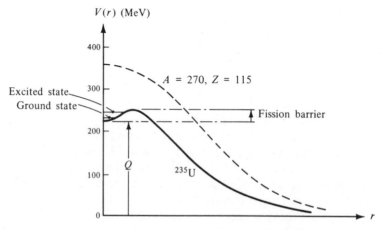

Fig. 17.2. The potential energy $V(r) = \text{const.} - B$ as a function of the separation, r, between fragments.

Equations (17.6) and (17.7) state that the Coulomb force tends to deform the nucleus from its spherical shape, whereas the surface tension tries to keep it spherical. The Coulomb force wins out if Eq. (17.8) is satisfied.

The considerations leading to Eqs. (17.6) and (17.7) are valid only for small deformations; they cannot provide us with information about the energy at larger ones. We can, nevertheless, guess at what can happen by also considering the potential energy of a fissioning nucleus after the fragments have separated and then smoothly interpolating in the region between small deformation and after scission. (Scission is the process of actual severance.) The resulting picture for the potential energy of a fission-ing nucleus as a function of the separation, r, of the centers of the two fission fragments is shown in Fig. 17.2. The potential energy is given by $V(r) = $ const. $- B$; it decreases as the binding energy, B, increases. The constant is so chosen that $V(\infty) = 0$. For zero separation, V is equal to the released energy Q, as given in Eq. (17.2). For small vibrations, the separation is proportional to the square of the deformation parameter, or with $r \approx 2\epsilon$,

$$V(r) = Q - \Delta E = Q - \frac{\alpha}{4}\left(\frac{r}{R}\right)^2, \qquad r \ll R. \qquad \textbf{(17.9)}$$

After scission, $V(r)$ is given by the Coulomb potential,

$$V(r) = \frac{Z_1 Z_2 e^2}{r}, \qquad \textbf{(17.10)}$$

where $Z_1 e$ and $Z_2 e$ are the charges of the two fragments. $V(r)$ is shown in Fig. 17.2 for $Z^2/A = 49$ and $Z^2/A = 36$. The first case applies to a hypo-thetical nucleus with $Z = 115$ and $A = 270$, and the second one, for in-stance, to ^{235}U. The behavior of $V(r)$ in the region not covered by Eqs. (17.9) and (17.10) has been drawn to connect smoothly to the regions of small and large r. Actually, the lower curve in Fig. 17.2 represents a very much oversimplified picture. Most nuclei that undergo fission are not spherical in their ground state; the minimum in the potential-energy curve then does not occur at $r = 0$. Moreover, evidence exists for a double-humped fission barrier in many nuclei.[9] We shall not treat these more complex situations but shall discuss spontaneous and induced fission with the help of the simple potentials shown in Fig. 17.2.

Consider first the upper curve in Fig. 17.2. The nucleus $Z = 115$, $A = 270$ would be unstable and would fission within characteristic nuclear times ($\approx 10^{-22}$ sec). Fission thus limits the range of stable or long-lived nuclides. Even at smaller values of Z and A, where $Z^2/A < 49$, *spontaneous fission* can occur:[10] Quantum mechanical tunneling through the fission barrier allows the decay into fission fragments. Although Eq. (17.8) holds only for small deformations, it can serve as a guide, and it suggests that the lifetime

9. W. J. Swiatecki and S. Bjørnholm, *Phys. Rept.* **4C**, 326 (1972).
10. G. N. Flerov and K. A. Pertshak, *Phys. Rev.* **58**, 89 (1940).

of a nucleus (A, Z) against spontaneous fission may depend on Z^2/A. Indeed, the general behavior of the lifetime shows a marked, approximately exponential, trend as a function of Z^2/A, as illustrated in Fig. 17.3.

If the fission barrier, shown in Fig. 17.2, becomes too high, spontaneous fission is no longer observed. Fission can still be *induced* by exciting the nucleus. To understand *neutron-induced fission*, we return to Fig. 15.3, which gives the separation energy of the last neutron. If, for instance, ^{235}U captures a slow neutron, the binding energy of the neutron, about 6 MeV, will be released to the nuclear system. The compound nucleus, ^{236}U, will consequently not be formed in its ground state but in a highly excited state. It is then much easier for the nucleus to overcome the fission barrier, and fission is likely to occur. Comparison of ^{235}U and ^{238}U reveals another fact: The compound nucleus that results when ^{235}U captures a neutron is ^{236}U; in the case $n + {}^{238}U$, the compound nucleus is ^{239}U. Figure 15.3 demonstrates that the separation energy is larger for even-even nuclei, such as ^{236}U, than for odd-A nuclei, such as ^{239}U. After slow neutron capture, more excitation energy will be available in $n + {}^{235}U$ (6.4 MeV) than in $n + {}^{238}U$ (4.8 MeV). Indeed, fission can be induced in the first case with thermal neutrons, whereas fast neutrons are needed in the second one.

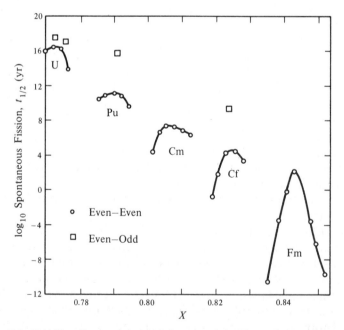

Fig. 17.3. Half-life of spontaneous fission as a function of x, where $x = (Z^2/A)/(Z^2/A)_{crit}$. $(Z^2/A)_{crit} \approx 49$ is given essentially by Eq. (17.8). (From R. Vandenbosch and J. R. Huizenga, *Nuclear Fission*, Academic Press, New York, 1973. Copyright © 1973 by Academic Press, New York.)

A number of characteristics of the fission process are crucial for the discussion of the *chain reaction* in nuclear reactors. In particular three properties must be known: the number of neutrons emitted per fission, the energy distribution of these neutrons, and the fission cross section as a function of neutron energy. For $n + {}^{235}U$, the average number of neutrons emitted is about 2.5. The energy distribution of these neutrons is shown in Fig. 17.4. The total and the fission cross section of ${}^{235}U$ is sketched in Fig. 17.5. The total cross section is larger than σ_f because processes other than fission, in particular elastic scattering and radiative capture, occur.

As a last point, we discuss the separation of the total fission energy into kinetic and internal fragment energy. If scission occurs for small deformations, then the kinetic energies of the fragments will be large since they have a considerable way to go down the potential energy curve, Fig. 17.2. On the other hand, if scission occurs for large deformations, then

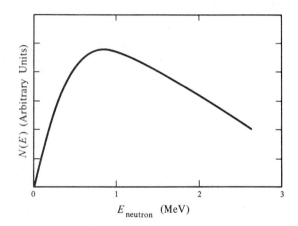

Fig. 17.4. Fission spectrum of $n + {}^{235}U$.

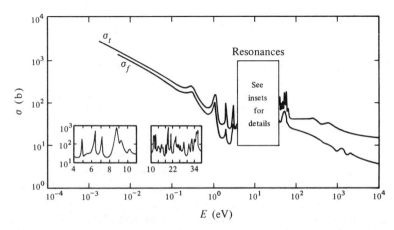

Fig. 17.5. Total (σ_t) and fission (σ_f) cross sections of ${}^{235}U$. (1 b $= 10^{-24}$ cm^2.)

more energy is stored in the stretched and distorted fragments and less is left to go into kinetic energy. A decision between these two extremes cannot be made on the basis of the simple liquid drop model. The average fragment distortion at scission is such that the fragments have enough excitation to release a few neutrons. The evidence is sketched in Fig. 17.6. As expected from Eq. (17.2), the average energy release increases with Z. Although the measured kinetic energies also increase with Z, the fractional increase is less rapid than for the average energy release. Thus, the percentage increase with Z of the energy stored as excitation energy of the fragments, which is the difference between the two curves, increases much faster. Since the energy stored in fission fragments is crucial for the working of nuclear reactors, Fig. 17.6 makes it clear why Ra and lighter nuclides cannot be used for such a purpose.

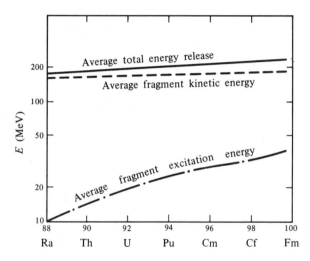

Fig. 17.6. Distribution of the energy in fission as a function of Z. The curves are approximate. (Courtesy of I. Halpern.)

17.2 Fission Reactors

Of all modern technologies, nuclear reactor technology is unique in having sprung up full blown almost overnight. Only four years separate the date of the discovery of fission (1938) and the date of the first chain reaction (1942).

A. M. Weinberg and E. P. Wigner[12]

The ideas underlying the workings of a nuclear reactor can be understood with the material presented in the previous section. The detailed computations and the actual engineering, however, are very complex and highly sophisticated. We shall only sketch the ideas here, and we refer to the literature given in Section 17.7 for further studies.

The salient components of a nuclear power plant are shown sche-matically in Fig. 17.7. The heat produced by fission in the reactor core and removed through a heat exchanger is used to drive turbines and thus is transformed into electricity. Of interest to us here are the processes in the reactor core.[11,12] To understand them we consider an arrangement of natural uranium and graphite as sketched in Fig. 17.8. Natural uranium consists of 99.3% ^{238}U and 0.7% ^{235}U. Only ^{235}U undergoes thermal fis-sion. Figures 17.4 and 17.5 show that fission produces mostly fast neutrons but is most efficiently induced by slow neutrons. To produce a viable chain reaction, the emitted fast neutrons must be slowed down to thermal ener-gies in the *moderator* through collisions with nuclei. The processes shown in Fig. 17.8 can be described by starting with one fission event: The fission of one ^{235}U nucleus produces on the average ν fast neutrons. Some of these produce fission before they are slowed down, and these events increase the number of fast neutrons by the fast fission factor, ϵ. Of the $\nu\epsilon$ neutrons entering the moderator, only a fraction p (resonance escape probability) survives thermalization; the rest will be captured in the moderator. Of the surviving $\nu\epsilon p$ neutrons, a fraction f, the thermal utilization factor, will be captured in the uranium. Of the $\nu\epsilon pf$ captured neutrons, a fraction $\sigma_f/\sigma_{\rm tot}$

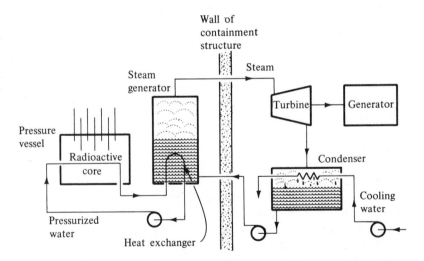

Fig. 17.7. Nuclear reactor. The fission of uranium in the reactor core is self-sustained through a chain reaction. The released energy converts to heat; the heat is carried away by the coolant which is pressurized water in the pile shown here. Through a heat exchanger, the power is transferred to a second steam system and used to drive a turbine.

11. The original description of the chain reacting pile is still worth reading: E. Fermi, *Science* **105**, 27 (Jan. 10, 1947), reprinted in E. Fermi, *Nuclear Physics*, notes by J. Orear, A. H. Rosenfeld, and R. A. Schluter, University of Chicago Press, Chicago, 1949, p. 208.

12. A clear description is also given in A. M. Weinberg and E. P. Wigner, *The Physical Theory of Neutron Chain Reactors*, University of Chicago Press, Chicago, 1958. Copyright © 1958 by the University of Chicago Press.

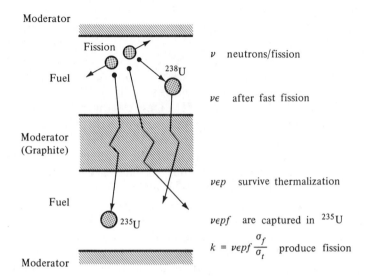

Moderator

Fission

Fuel

238U

ν neutrons/fission

$\nu\epsilon$ after fast fission

Moderator
(Graphite)

Fuel

235U

$\nu\epsilon p$ survive thermalization

$\nu\epsilon p f$ are captured in 235U

$k = \nu\epsilon p f \dfrac{\sigma_f}{\sigma_t}$ produce fission

Moderator

Fig. 17.8. Life of neutrons in a natural uranium reactor.

produces fission. One fission event consequently produces

$$k = \nu\epsilon p f \frac{\sigma_f}{\sigma_{\text{tot}}} \tag{17.11}$$

secondary fission events; k is called the reproduction factor. A chain reaction will take place if $k > 1$. Typical numbers of the various factors in a lattice consisting of natural uranium and graphite as moderator are $\nu = 2.47$, $\epsilon = 1.02$, $p = 0.89$, $f = 0.88$, $\sigma_f/\sigma_{\text{tot}} = 0.54$, and $k = 1.07$. The argument gives the reproduction factor for an infinite lattice; in a finite lattice, some neutrons will escape, and k_{eff} will be smaller than k.

The chain reaction is the basis of all nuclear reactors. Various types of reactors exist, and they are used as research tools, for the production of radioisotopes and for the production of power. We shall not discuss any of these reactors in more detail but shall add some remarks.

Power production may well be the most important function of reactors. The living standard is closely tied to the availability of inexpensive power; to raise the standard where it is most necessary, energy sources are required.[13] Hydroelectric, coal, oil, and gas energy exhaust natural resources that cannot be replenished. Can nuclear reactors provide the needed energy without damaging the environment and without exhausting irreplaceable supplies? The answer is far from clear, but some aspects are well understood. If all reactors use uranium, as described here, then the existing high-grade ores will be used up within a few decades. This problem

13. Energy, its role in human life, energy resources, and many other aspects of this most important resource are discussed in *Sci. Amer.* **224** (Sept. 1971).

can probably be surmounted by constructing *breeder reactors.*[14] Such reactors produce more fuel than they use, and we shall describe the essential idea.

Breeding occurs when more fissionable material is produced than is consumed. The principle was recognized early: Fermi and Zinn began to design a breeder reactor in 1944. Prerequisite for a breeder reaction is a fissionable nuclide that can be produced as the result of neutron capture and successive decays of a fertile nuclide. Consider as an example a reactor core consisting of ^{239}Pu, the fuel, and ^{238}U, the fertile nuclide. A fissioning ^{239}Pu nucleus will produce on the average 2.91 neutrons. One of these fast neutrons can produce fission in another ^{239}Pu nucleus, and a second can be captured in ^{238}U, giving rise to the sequence

$$n + {}^{238}\text{U} = {}^{239}\text{U} + \gamma$$

$$^{239}\text{U} \xrightarrow[\beta^-]{t_{1/2} = 25 \text{ min}} {}^{239}\text{Np} \xrightarrow[\beta^-]{t_{1/2} = 2.3d} {}^{239}\text{Pu}.$$

Neutron capture in the fertile nuclide produces the fissionable nuclide. In a well-designed breeding reactor, the amount of fissionable material can double in 7–10 y. At the present time, research and development in the field of breeder reactors are active, and it is likely that breeder reactors will form an important component of the energy resources within two decades.

17.3 Fusion and Fusion Power

Barring nuclear war and some general collapse of civilization, world power demands will probably be far higher than conventional extrapolations allow. We base this prophecy on three observations:
1. Much of the world is hungry.
2. Much of the world is poor.
3. Much of the world is polluted.

G. T. Seaborg and W. R. Corliss[15]

The age of fossil energy, barely begun, will very likely end soon. Three reasons are responsible: The amount of fossil fuel is limited,[16] it produces substantial pollution, and it is irreplaceable. Seen on the time scale of

---

14. G. T. Seaborg and J. L. Bloom, *Sci. Amer.* **223**, 13 (Nov. 1970); W. Häfele, D. Faude, E. A. Fischer, and H. J. Laue, *Ann. Rev. Nucl. Sci.* **20**, 393 (1970); A. M. Perry and A. M. Weinberg *Ann. Rev. Nucl. Sci.* **22**, 317 (1972); F. L. Culler and W. O. Harms, *Phys. Today* **25**, 28 (May 1972).

15. From the book *Man and Atom: Building a New World Through Nuclear Technology* by Glenn T. Seaborg and William R. Corliss. Copyright © 1971 by Glenn T. Seaborg and William R. Corliss. Published by E. P. Dutton & Co., Inc. New York, and used with their permission.

16. In units of Q, where $1Q \approx 10^{21}$ J, the total world reserve of coal is about $200Q$, and of natural gas and oil each about $10Q$. In 1960, the world energy consumption was about $0.1Q/y$; in 1975, it is expected to be about $0.4Q/y$.

history, the fossil-fuel reign will probably represent only a short interlude, as sketched in Fig. 17.9. Even if the quote above anticipates too high a power consumption, the coming void must be filled with other energy sources. High among the candidates is nuclear power, used in conjunction with superconducting power transmission[17] and with hydrogen as a secondary fuel.[18] Conventional fission reactors will very likely be important only for a limited time, but breeder reactors could be dominant for centuries. However, they possess two shortcomings: They produce a large amount of radioactive waste, and they lead to heat pollution. It is likely that both problems will be manageable,[19] but another power source, fusion, may well be cleaner and more efficient.

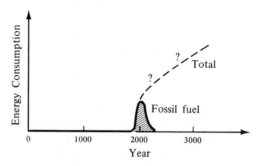

Fig. 17.9. Short reign of fossil fuel.

Fusion can be understood by looking at Fig. 14.1: Very light nuclei are much less tightly bound than the ones that are somewhat heavier; we have explained the reason in Section 12.1. If two light nuclei can be fused together, an energy Q will be produced in the form of kinetic energy. For a given reaction $ab \rightarrow cd + Q$, Q can be calculated easily with Eqs. (14.3) and (14.4) and with the values of the mass excess given in Table A6 in the Appendix. We list here some fusion reactions and Q (t denotes the triton, the nucleus of tritium, ^{3}H)

$$d\ d \longrightarrow {}^3\text{He} \quad n + 4.0 \text{ MeV} \tag{17.12}$$

$$d\ d \longrightarrow t \quad p + 3.25 \text{ MeV} \tag{17.13}$$

$$t\ d \longrightarrow {}^4\text{He} \quad n + 17.6 \text{ MeV} \tag{17.14}$$

$${}^3\text{He}\ d \longrightarrow {}^4\text{He} \quad p + 18.3 \text{ MeV}. \tag{17.15}$$

A fusion reaction is initiated if the two partners are brought within reach of their mutual hadronic attraction. To get so close, colliding nuclei must overcome (or tunnel through) their mutual long range electrostatic repulsion, the Coulomb barrier. The fusion reaction rate will therefore be van-

17. W. D. Metz, *Science* **178**, 968 (1972).
18. L. Lessing, *Fortune*, 138 (Nov. 1972). D. P. Gregory, *Sci. Amer.* **228**, 13 (Jan. 1973).
19. Modern pyramids, for instance, have been suggested as an elegant means for radioactive waste burial [C. Starr and R. P. Hammond, *Science* **177**, 744 (1972)].

ishingly small at energies below a few keV, but it increases rapidly with increasing kinetic energy of the reaction partners. The relevant cross sections are shown in Fig. 17.10. In an accelerator, the energies needed to initiate any fusion reaction can be obtained very easily. However, the main goal of fusion research is to establish a self-sustaining reaction. To obtain such a situation, the reaction mixture must be brought to such a temperature that the kinetic energy is sufficient to overcome the Coulomb barrier. To be specific, to achieve a particle energy of 10 keV, a temperature of about 10^8 K is required. By comparison, the temperature of the surface of the sun is about 6000 K. (In plasma physics it is customary to give the "temperature" in eV or keV; 1 keV corresponds to a temperature of 11.6×10^6 K.) At a temperature of 10 keV, the gas atoms are fully ionized and form a *plasma*. The number of fusion reactions per unit time and unit volume is then given by

$$R_{ab} = n_a n_b w_{ab}(T), \qquad (17.16)$$

where n_a and n_b are the number of particles a and b, respectively, per unit volume. The reaction probability $w_{ab}(T)$ is the product of cross section and relative velocity, averaged over the velocity distribution in the plasma,

$$w_{ab}(T) = \overline{\sigma_{ab} v_{ab}}; \qquad (17.17)$$

it depends only on the plasma temperature T. The energy released per unit volume and in a time τ becomes

$$W = R_{ab} Q_{ab} \tau = n_a n_b w_{ab}(T) Q_{ab} \tau. \qquad (17.18)$$

For $n_a = n_b = 10^{15}$ particles/cm³ and a temperature $T = 100$ keV, the *dt* reaction will produce about 10^3 W/cm³-sec.

A *fusion reactor* should yield more energy than is put in to heat and confine the plasma. Equations (17.16)–(17.18) imply that three conditions must be satisfied for a self-sustaining plasma reaction: The plasma must be raised to the required temperature, the plasma density must be adequate,

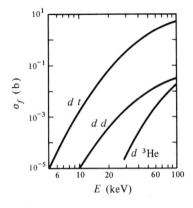

Fig. 17.10. Cross sections for the fusion reactions given in Eqs. (17.12)–(17.14), as a function of the deuteron energy.

and temperature and density must be maintained for a sufficiently long time. The energy input required to heat $n_a + n_b \approx 2n$ particles to a temperature T is given by $3nkT$, where k is the Boltzmann constant. A plasma reactor consequently requires that

$$n^2 w_{ab} Q_{ab} \tau > 3nkT. \tag{17.19}$$

Taking losses into account, this condition leads to Lawson's criterion[20] for the *dt* reaction

$$n\tau > 10^{14} \text{ sec-cm}^{-3} \qquad \text{for } T = 10 \text{ keV}. \tag{17.20}$$

Attaining and surpassing this criterion has been the goal of plasma physics for the past 20 y. Initially it appeared as if it would not be very difficult, and thermonuclear neutrons were indeed observed soon. However, nature had many surprises in store, and unexpected instabilities and losses turned up. What had been thought to be a fast run to a technological goal became an arduous path through the ever widening and deepening field of *plasma physics*. It now seems that the goal is in sight.[21] Two lines of work are possibly close to laboratory fusion reactors that will ultimately lead to economic power plants. Plasma at 10^8 K can clearly not be confined in any material vessel, and the two approaches are different in the way in which the plasma is confined and heated.

In *magnetic confinement experiments*, the plasma is held in suitably shaped electromagnetic fields, and it is heated electromagnetically. A large variety of field configurations has been explored, and the Russian Tokamak device, copied and modified in other countries, may be the front runner.[22] In the past few years, *laser fusion* has appeared on the horizon.[23] Here, a small pellet of deuterium and tritium is heated by powerful laser bursts. Confinement is through inertia: Heating must occur so fast that the fusion reaction occurs before the pellet has exploded. Existing pulsed lasers already have a power that exceeds for short times the total electrical power capacity of the United States. Pulses of 100 J in less than 1 nsec are routine, and much larger lasers are under construction. At present it is not clear if laser fusion will lead to economical reactors or if the real problems have not yet been encountered. In any case experts are confident that fusion power will be a reality before the end of this century.

17.4 Nuclear Explosives

In power reactors, the nuclear chain reaction is controlled so that the power output never exceeds a safe limit; only a very small fraction of the

20. J. D. Lawson, *Proc. Phys. Soc.* **B70**, 6 (1957).

21. D. J. Rose, *Science* **172**, 797 (1971); W. C. Gough and B. J. Eastlund, *Sci. Amer.* **224**, 50 (Feb. 1971).

22. B. Coppi and J. Rem, *Sci. Amer.* **227**, 65 (July 1972).

23. M. J. Lubin and A. P. Fraas, *Sci. Amer.* **224**, 21 (June 1971); W. D. Metz, *Science* **177**, 1180 (1972); J. Nuckolls, J. Emmett, and L. Wood, *Phys. Today* **26**, 46 (August 1973).

fuel is used every second. Nuclear explosives, in contrast, are so designed that the initial exponential increase in the number of reactions continues until the fuel is dispersed.

Nuclear explosives come in two forms, fission (*atom* bombs) and fission-fusion (*H* bombs). While details are classified, the basic facts are accessible,[24] and they can be explained with the help of Fig. 17.11. The fuel in a fission explosive consists of nearly pure ^{235}U or ^{239}Pu, both fissionable nuclides.[25] A small lump of such material cannot explode because too many neutrons escape. A large enough (*critical*) mass, however, can chain-react. An accidental initial neutron will start the fission events. The discussion of the chain reaction in pure ^{235}U is analogous to the one given for a system consisting of natural uranium and a moderator in Section 17.2. If N neutrons are present at a time t, their number will increase in the time dt by

$$dN = \alpha(t)N(t)dt.$$

Here, $\alpha(t)$ is a complicated function of geometry, material, and time. Typically, α is initially of the order of 10^8 sec^{-1}. As long as $\alpha(t)$ can be considered a constant, the neutron number will increase exponentially,

$$N(t) = N(0)e^{+t/\tau_g}, \tag{17.21}$$

where $\tau_g = 1/\alpha$ is called the generation time. The fission explosive shown in Fig. 17.11 works as follows: Before triggering, the fissile material is

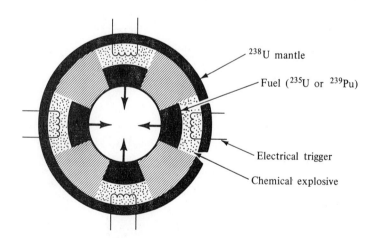

Fig. 17.11. Fission explosive.

24. S. Glasstone, ed., *The Effects of Nuclear Weapons*, Government Printing Office, Washington, D.C., 1964; H. F. York, "The Great Test-Ban Debate," *Sci. Amer.* **227**, 15 (Nov. 1972); H. L. Brode, *Ann. Rev. Nucl. Sci.* **18**, 153 (1968).

25. Large-scale isotope separation to obtain nearly pure ^{235}U was one of the key problems in the production of the first nuclear explosive. See H. D. Smyth, *Atomic Energy for Military Purposes*, Princeton University Press, Princeton, N.J., 1945.

divided into subcritical parts to prevent accidental explosion. Upon electrical triggering, chemical explosives shoot the subcritical parts into the center, where they form a critical mass. The chain reaction then starts and builds up exponentially, according to Eq. (17.21). After about 50 generations, or about 0.5 μsec, enough energy has been produced so that the device begins to explode. The neutron number will then decline and finally drop to zero. A *logarithmic* plot of the number of neutrons as a function of time and a *linear plot* of the corresponding power output (yield rate)[26] are shown in Fig. 17.12.

In a thermonuclear explosive, the fissile material is in the center, and it is surrounded by lithium deuteride. Explosion of the central core in the same manner as described above produces a very high temperature and a very high neutron flux and leads to a thermonuclear explosion in the outer part.

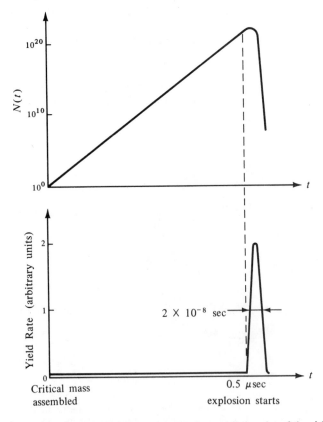

Fig. 17.12. Logarithmic plot of the neutron number and linear plot of the yield rate (energy released per unit time) as function of time for a fission explosive. [From H. A. Sandmeier, S. A. Dupree, and G. E. Hansen, *Nucl. Sci. Eng.* **48**, 343 (1972).]

26. H. A. Sandmeier, S. A. Dupree, and G. E. Hansen, *Nucl. Sci. Eng.* **48**, 343 (1972).

Scientific[27,28] *and engineering*[29] *applications* of nuclear explosives are based on two facts: intense neutron flux and large energy release. A typical device releases about 10^{24} neutrons in less than 10^{-7} sec and produces between 1 and 100 kilotons of energy during the same time. (1 kiloton = 1 kt = 10^{12} cal = 2.61×10^{31} eV.) We mention only two specific examples of applications: The elements Einsteinium (99) and Fermium (100) were first discovered in the debris of a large thermonuclear explosive in 1952. Nuclear reactions and levels can be studied because the neutron burst is so intense. Even at distances of a few hundred meters, the total neutron flux is still of the order of 10^{14} *n*/cm². Since the initial burst is sharp, as shown in Fig. 17.12, time-of-flight measurements allow separation of the neutrons according to energy.

17.5 Power from Radionuclides

Nuclear reactors are copious producers of radioactive nuclides. Fission products are neutron-rich and consequently decay predominantly through emission of electrons and antineutrinos. Using the intense neutron beams from reactors, alpha-radioactive nuclei can be produced. The energies of the electrons or of the alpha particles are typically of the order of a few MeV. By absorbing the emitted charged particles and transforming the heat so produced into electricity, reliable and long-lasting power sources can be constructed.[30,31] One basic arrangement is sketched in Fig. 17.13.

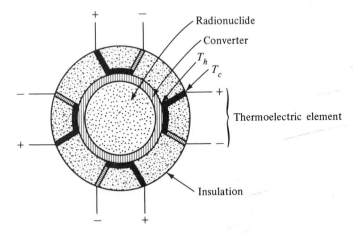

Fig. 17.13. Cross section through a cylindrical radionuclide power generator.

27. B. C. Diven, *Ann. Rev. Nucl. Sci.* **20**, 79 (1970).

28. H. C. Rodean, *Nuclear-Explosion Seismology*, AEC Critical Review Series, National Technical Information Service, Springfield, Va., 1971.

29. Proceedings, *Engineering with Nuclear Explosives*, Clearinghouse for Federal Scientific and Technical Information, Springfield, Va., 1970, 2 vols.

30. W. R. Corliss and D. G. Harvey, *Radioisotope Power Generation*, Prentice-Hall, Englewood Cliffs, N.J., 1964;

31. Y. Wang, ed., *Handbook of Radioactive Nuclides*, Chemical Rubber Co., Cleveland, 1969, pp. 559–568.

The fuel, a radionuclide, is placed in the center, so that all charged particles are absorbed inside the converter. Thermoelements, for instance Ge-Si couples, are embedded in an insulating mantle so that their junction is at the temperature, T_h, of the converter. The outer ends are at the colder temperature T_c. The maximum (Carnot) efficiency is then given by

$$\epsilon = \frac{T_h - T_c}{T_c}.$$

In practice, efficiencies of about 5% can be obtained. If a small steam engine (turbogenerator) is used, efficiencies of about 25% are reached.

The properties of the two most widely used radionuclides are summarized in Table 17.1. Power generators based on these two nuclides are successfully used in satellites, light buoys, weather stations, undersea acoustic beacons,[31] and heart pacemakers and even for a nuclear heart pump in a calf[32] This work may lead to a completely implantable artificial heart.[33]

Table 17.1 RADIONUCLIDES FOR POWER SOURCES[31]

Nuclide	Half-life (y)	Radiation	Power density (W/cm³)	
			theoretical	practical
^{90}Sr	27.7	β	1.1–1.7	0.85–1.5
^{238}Pu	87.5	α	4.8	3.6

17.6 Nuclear Propulsion

While man has been able to travel safely to the moon and back, the vast expanses of outer space still wait to be fully explored and possibly colonized. At the same time, *inner space*, the oceans, also remain to be studied more deeply, and they will conceivably also be "colonized" in the search for more growth space. Nuclear propulsion will very likely be crucial in these explorations. It is already an essential constituent in submarines; the nuclear-powered submarine *Nautilus* began operating in January 1955, and its exploration under the seas is the modern realization of Jules Verne's dream and prophesy. The icebreaker *Lenin* was the first nonmilitary nuclear-powered ship. The *Savannah*[34] served for many years as a cargo carrier. It is likely that, after a transition period, far more nuclear-powered ships will be constructed, because they possess advantages over conventional ships: The refueling period is much longer; a nuclear ship can sail 2–5 y without taking on fresh fuel. Fuel costs are

32. *The New York Times* 37 (March 21, 1972).
33. E. E. Fowler, *Isotopes Radiation Technol.* **9**, 253 (1972).
34. *National Geographic Magazine* **122**, 28 (Aug. 1962).

lower. Higher speeds can be obtained, and cargo capacity is larger.[35] At the present time, conventional ships are still more economical overall (capital costs, operations, fuel) than nuclear-powered ones. However, conventional ships are the product of a long evolution and cannot be improved much; nuclear ships are still in their infancy. For large merchant ships (80,000 shaft hp and larger), the prognosis is favorable for nuclear power to be economically competitive before 1980.[36]

Figure 17.14 shows a longitudinal section through the German nuclear ship *Otto Hahn*,[37] which is, like the *Savannah*, an experimental design. The fundamental aspects of a ship reactor are not different from a land-based one. What is different is the environment; shielding must be optimized to protect the crew but not overload the ship. Safety includes provisions against severe accidents in case of collision or other damage. We shall not discuss these problems here but shall turn to rockets.

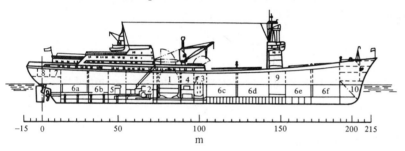

Fig. 17.14. German nuclear ship *Otto Hahn*. 1. Containment. 2. Engine room. 3. Service room. 4. Auxiliaries. 5. Auxiliary boiler. 6. a–f Shipping rooms. 7. Drinking water supply. 8. Steering servo unit. 9. Emergency engine. 10. Fore- and afterpeak. [After D. Bünemann et al., *Advan. Nucl. Sci. Technol.* **6**, 2 (1972).]

Nuclear rockets, compared to chemical ones, may permit more efficient use of power, longer travels, and larger payloads.[38] To understand the reason for the advantages, we briefly consider the mechanism of rocket propulsion. The thrust of a rocket with constant exhaust velocity, v, is given by Newton's law as

$$F = \frac{dp}{dt} = v\frac{dM}{dt}.$$ (17.22)

Here dM/dt is the mass of propellant ejected per unit time. If the propellant is a molecule with mass m and ejected at a temperature T, its energy is given by

$$E = \tfrac{1}{2}mv^2 = \tfrac{3}{2}kT,$$

35. A. W. Kramer, *Nuclear Propulsion for Merchant Ships*, Government Printing Office, Washington, D.C., 1962.

36. *The Wall Street Journal*, 13 (Oct. 20, 1972).

37. D. Bünemann, M. Kolb, H. Henssen, E. Müller, and W. Rossbach, *Advan. Nucl. Sci. Technol.* **6**, 2 (1972).

38. R. W. Bussard and R. D. DeLauer, *Fundamentals of Nuclear Flight*, McGraw-Hill, New York, 1965; R. S. Cooper, *Ann. Rev. Nucl. Sci.* **18**, 203 (1968); H. Löb, ed., *Nuclear Engineering for Satellites and Rockets*, K. Thiemig, Munich, 1970.

so that

$$v = \left(\frac{3kT}{m}\right)^{1/2}$$

and

$$F = \frac{dM}{dt}\left(\frac{3kT}{m}\right)^{1/2}. \tag{17.23}$$

In a rocket, the total propellant mass, $M = \int dM$ is given, as is the temperature T for which the system can be designed. Thus the only free variable in Eq. (17.23) is m, the mass of the propellant molecules. Here is where the nuclear rocket engine is superior: In a chemical engine, the designer is forced to use the end product of the energy-producing combustion as propellant. The burning of hydrogen and oxygen yields water, and it is ejected. In nuclear rockets, the energy is derived from the nuclear power plant, and the propellant can be chosen independently. Using hydrogen yields a velocity ratio

$$\frac{v_{\text{nuc}}}{v_{\text{chem}}} = \left(\frac{m_{H_2}}{m_{H_2O}}\right)^{1/2} \approx 3.$$

The propellant velocity in a nuclear rocket is three times higher than in a chemical one. In a nuclear rocket, then, hydrogen is pumped through a hot reactor core; it is heated there and ejected, thus driving the rocket.[39]

17.7 References

The literature covering nuclear power and power in general is reviewed in two excellent resource letters:

P. Michael and R. I. Schermer, "Resource Letter Rea-1 on Reactors," *Am. J. Phys.* **36**, 1 (1968).

R. H. Romer, "Resource Letter ERPEE-1 on Energy; Resources, Production, and Environmental Effects," *Am. J. Phys.* **40**, 805 (1972).

The Fission Process. The key papers and reports centering on the discovery of nuclear fission are collected in H. G. Graetzer and D. L. Anderson, *The Discovery of Nuclear Fission: A Documentary History*, Van Nostrand Reinhold, New York, 1971.

The experimental aspects of fission are summarized in E. K. Hyde, *Nuclear Properties of the Heavy Elements, III: Fission Phenomena*, Prentice-Hall, Englewood Cliffs, N.J., 1964.

Details of the fission process and the corresponding theories are reviewed in

J. S. Fraser and J. C. D. Milton, *Ann. Rev. Nucl. Sci.* **16**, 379 (1966).

39. For an exciting account of nuclear rocket propulsion, see the article of F. J. Dyson, in *Adventures in Experimental Physics*, Beta (B. Maglich, ed.), World Science Communications, Princeton, N.J., 1972.

I. Halpern, *Ann. Rev. Nucl. Sci.* **9**, 245 (1949).

L. Wilets, *Theories of Nuclear Fission*, Oxford University Press, Inc., Oxford 1964.

R. Vandenbosch and J. R. Huizenga, *Nuclear Fission*, Academic Press, New York, 1973.

The first of these is easiest to read; it is very descriptive and uses practically no formal theory.

Fission Reactors. The authoritative work on reactor physics, including the necessary nuclear physics background, is A. M. Weinberg and E. P. Wigner, *The Physical Theory of Neutron Chain Reactors*, University of Chicago Press, Chicago, 1958.

Reactor engineering is treated in books listed in "Resource Letter Rea-1," quoted above. In addition, the two following serial publications provide an enormous amount of information:

E. J. Henley and J. Lewins, eds., *Advances in Nuclear Science and Technology*, Academic Press, New York.

International Series of Monographs on Nuclear Energy, Pergamon, Elmsford, N.Y.

A modern book on advanced reactor theory is George I. Bell and S. Glasstone, *Nuclear Reactor Theory*, Van Nostrand Reinhold, New York, 1971.

Fusion and Fusion Power. The literature on *fusion and plasma physics* is increasing rapidly. Very readable first introductions are provided in the following two books:

A. S. Bishop, *Project Sherwood—The U.S. Program in Controlled Fusion*, Doubleday, Garden City, N.Y., 1960.

E. R. Hulme, *Nuclear Fusion*, Wykeham Publications, London, 1969.

Fusion is treated in more depth and detail in the following book and article:

L. A. Artsimovich, *Controlled Thermonuclear Reactions*, Van Nostrand Rinehold, New York, 1960.

R. F. Post, *Ann. Rev. Nucl. Sci.* **20**, 509 (1970).

A good way to remain reasonably up-to-date, without going through the original literature, is by reading B. D. Fried, ed., *Comments on Plasma Physics and Controlled Fusion*, Gordon & Breach, New York.

More thorough reviews are published in the journal *Nuclear Fusion* (quarterly, Vienna).

18
Nuclear Astrophysics

For millenia, the stars, sun, and moon have fascinated man; their properties have been subject to much speculation. Up to a very short time ago, however, observation of the heavens was restricted to the very small optical window between about 400 and 800 nm, and mechanics was the branch of physics most intimately involved in astronomy. In this century, the situation has changed dramatically and physics and astronomy have become much more closely intertwined. In this chapter, we shall sketch some of the areas in which subatomic physics and astrophysics are linked.

1. M. A. Ruderman and W. A. Fowler, "Elementary Particles," *Science, Technology and Society* (L. C. L. Yuan, ed.), Academic Press, New York, 1971, p. 72. Copyright © 1971 by Academic Press.

18.1 Cosmic Rays

*The planetary system is a gigantic laboratory where nature has been perform-
ing an extensive high-energy physics experiment for billions of years.*

T. A. Kirsten and O. A. Schaeffer[2]

We are constantly bombarded by energetic particles from outer space;
about 1 particle/sec passes through every cm² of the earth's surface. These
"rays" were discovered by Victor Hess in 1912 by observing the ionization
in an electrometer carried in a manned balloon; above 1000 m altitude, the
intensity began to increase and it doubled by 4000 m.[3] Since 1912, cosmic
rays have been studied extensively; their composition, energy spectrum,
spatial and temporal variation are being explored with ever-increasing
sophistication, and many theories concerning their origin have been proposed. Cosmic rays are one of the main components of the galaxy. This
assessment is based on the fact that the energy density of the cosmic rays in
our galaxy, about 1 eV/cm³, is of the same order of magnitude as the energy density of the magnetic field of the galaxy and of the thermal motion
of the interstellar gas.

Cosmic rays have been observed and studied at various altitudes, in
caverns deep underground, in mountaintop laboratories, with balloons at
altitudes up to 40 km, with rockets, and with satellites. Much of the future
of cosmic ray research will take place on extraterrestrial observation stations, on satellites, and possibly on the moon.

The radiation incident on the earth's atmosphere consists of nuclei,
electrons and positrons, photons and neutrinos. It is customary to call
only the charged particles cosmic rays. X-ray astronomy[4] has recently
led to spectacular discoveries,[5] but we shall not treat these here. Consider
first the fate of a cosmic ray proton of very high energy that strikes the top
of the earth's atmosphere. It will interact with an oxygen or nitrogen nucleus, and a cascade process will be initiated. A simplified scheme is shown
in Fig. 18.1 As discussed in Sections 12.7 and 6.9, the interaction will
produce a large number of hadrons; pions predominate, but antinucleons,
kaons, and hyperons also occur. These hadrons can again interact with
oxygen or nitrogen nuclei; the unstable ones can also decay weakly. The
decays result in electrons, muons, neutrinos, and photons (Chaper 11).
The photons can produce pairs; the muons decay, but because of the time
dilation [Eq. (1.9)], many penetrate into the earth's solid mantle before
doing so. Overall, a very-high energy proton can give rise to a large num-

2. T. A. Kirsten and O. A. Schaeffer, "Elementary Particles." *Science, Technology and
Society* (L. C. L. Yuan, ed.), Academic Press, New York, 1971, p. 76. Copyright © 1971 by
Academic Press.

3. V. F. Hess, *Physik. Z.* **13**, 1084 (1912).

4. H. Friedman, *Ann. Rev. Nucl. Sci.* **17**, 317 (1967); G. W. Clark, *Ann. Rev. Astronomy
Astrophys.* (to be published).

5. See the special issue on the X-ray star Cygnus X-3, *Nature* **239**, No. 95 (1972).

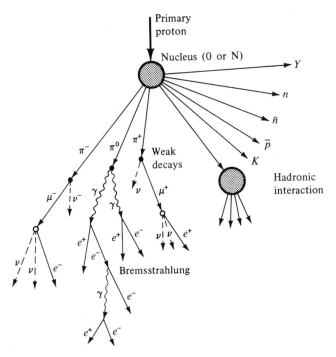

Fig. 18.1. An incident high-energy proton strikes the top of the atmosphere and produces a cascade shower.

ber of photons and leptons [Fig. 3.10]; such a cosmic-ray shower can cover an area of many km² on the surface of the earth.[6] We shall not discuss the phenomena in the atmosphere further, but shall turn to the primary radiation.

The *composition of the nuclear component* of the primary cosmic rays is shown in Fig. 18.2.[7] Also shown for comparison is the *universal* distribution, observed in the solar atmosphere and in meteorites. A few remarkable facts emerge from a comparison of the cosmic-ray and the universal data: (1) The elements Li, Be, and B are about 10^5 times more abundant in cosmic rays than universally. (2) The ratio ^{3}He/^{4}He is about 300 times larger in cosmic rays. (3) Very heavy nuclei are much more prevalent in cosmic rays. The first two facts can be explained by assuming that the cosmic rays have traversed about several g/cm² of matter between their source and the top of the earth atmosphere. In such an amount of matter, nuclear reactions produce the observed distribution. Since the interstellar density is about 10^{-25} g/cm³, the cosmic rays must have wandered around for 10^7–10^8 y. Two more facts have been established that may prove important for theories of the origin of cosmic rays: (4) So far,

6. G. Cocconi, "Extensive Air Showers," in *Encyclopedia of Physics*, Vol. 46.1, Springer, Berlin, 1961.

7. M. M. Shapiro and R. Silberberg, *Ann. Rev. Nucl. Sci.* **20**, 323 (1970); P. B. Price and R. L. Fleischer, *Ann. Rev. Nucl. Sci.* **21**, 295 (1971).

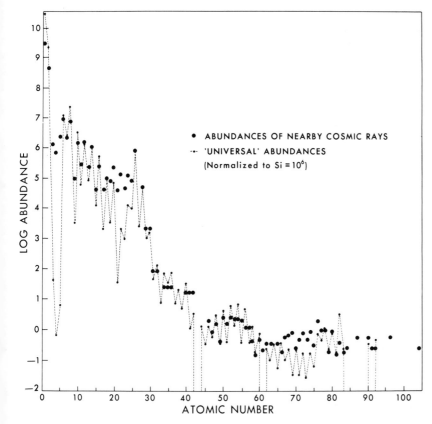

Fig. 18.2. Composition of the nuclear component of the primary cosmic rays. Shown for comparison is the universal reference distribution. [Courtesy P. B. Price, updated from P. B. Price and R. L. Fleischer, *Ann. Rev. Nucl. Sci.* **21**, 295 (1971).]

no antihadrons have been found in the primary cosmic rays.[8] (5) Electrons are about 1% as abundant as nuclei in the same energy interval; positrons form about 10% of the electron component.

The energy spectrum, the number of primary particles as a function of energy, has been measured over an enormous range. For the nuclear component, it is shown in Fig. 18.3; the data extend over about 14 decades in energy and 32 decades in intensity. The highest observed energy is 4×10^{21} eV or about 60 Joules.[9] Fig. 18.3 demonstrates that the cosmic ray spectrum does not have a thermal shape; it is not exponential, but decays more slowly. A good fit to the data, except at the lowest energies, is

$$I(E) \propto E^{-2.6}, \qquad (18.1)$$

where $I(E)$ is the intensity of the nuclear component with energy E. The

8. A. Buffington, L. H. Smith, G. F. Smoot, L. W. Alvarez, and M. A. Wahlig, *Nature* **236**, 335 (1972).

9. K. Suga, H. Sakuyama, S. Kawaguchi, and T. Hara, *Phys. Rev. Letters* **27**, 1604 (1971).

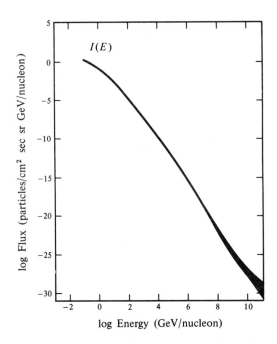

Fig. 18.3. Energy spectrum of the nuclear component of the primary cosmic rays.

electron spectrum is similar to Fig. 18.3 for energies above 1 GeV, but somewhat flatter below. Two more facts concerning the energy spectra are important for the discussion of the origin of cosmic rays. One is the isotropy of cosmic rays; the other is their constancy over a long period of time. Measurements in outer space indicate that the cosmic-ray flux is essentially isotropic; there exists the possibility that the flux from the center of our galaxy is about 1 % larger than average, but this fact has not yet been firmly established. The time dependence of the intensity over long periods has been studied by looking at the abundance of nuclides created in moon samples and meteorites. The cosmic ray intensity has been approximately constant over a period of about 10^9 y.

The experimental evidence discussed above implies that the *source of cosmic rays* must have the following properties:[10] It must produce cosmic rays with energies up to nearly 10^{22} eV and with a power spectrum, Eq. (18.1). The total produced energy must be of the order of 10^{49} ergs/y in our galaxy; the cosmic rays must be isotropic and constant during at least 10^9 y. The primary spectrum must include heavy elements up to about $Z = 100$ but less than about 1 % antihadrons.

No model has yet been proposed that explains all data uniquely and in a satisfactory way. Three of the most important questions still remain to be answered: (1) Where do cosmic rays come from? (2) How are they pro-

10. V. L. Ginzburg, *Sci. Amer.* **220**, 50 (February 1969); R. Cowsik and P. B. Price, *Phys. Today* **24**, 30 (Oct. 1971).

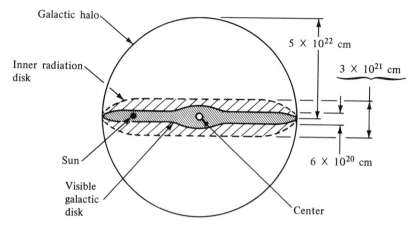

Galactic halo

Inner radiation disk

5×10^{22} cm

3×10^{21} cm

Sun

6×10^{20} cm

Visible galactic disk

Center

Fig. 18.4. Cross section through our galaxy.

duced? (3) How are they accelerated? A few remarks can be made on each of these problems.

1. The first question can be formulated more specifically by drawing a cross section through our galaxy, as in Fig. 18.4. Cosmic rays can be produced in the inner radiation disk, in the galactic halo, or they can flow into the galaxy from the outside.[11] It is not yet known where the bulk of cosmic rays originates, but most experts favor the galaxy.

2. At present, it is believed that supernovae and neutron stars can produce cosmic rays with the correct properties.[12] In our galaxy, a supernova appears about every 30 y and one supernova is believed to produce between 10^{51} and $10^{52.5}$ erg of energy. Supernovae consequently can provide the required 10^{49} erg/y.

3. It is possible that the sources emit cosmic rays with the energy spectrum Eq. (18.1). It is, however, also possible that nature uses the same technique as present day high-energy accelerators, acceleration in stages (Section 2.5). A mechanism for acceleration in interstellar space, collision of the particles with moving magnetic fields, has for instance been suggested by Fermi.[13]

18.2 Stellar Energy

And God said, Let there be light; and there was light.

Genesis

In the previous section we have indicated that the source of the most energetic radiation falling on the earth, the cosmic rays, is still shrouded

11. V. L. Ginzburg, *Soviet Phys. Usp.* **14**, 21 (1971).

12. R. M. Kulsrud, J. P. Ostriker, and J. E. Gunn, *Phys. Letters* **28**, 636 (1972).

13. E. Fermi, *Phys. Rev.* **75**, 12 (1949). Reprinted in *Cosmic Rays, Selected Reprints*, American Institute of Physics, New York.

in mystery. Another source of radiation, however, is identified, namely the sun. The mechanism of energy production in the sun is believed to be understood and we shall discuss it as an example of stellar power sources. In Section 17.3 we have implied that the construction of a terrestrial fusion reactor is difficult. The main difficulty is *containment*: A plasma with a temperature of about 10^8 K must be kept enclosed within a finite volume. Solid walls cannot withstand such a temperature and magnetic confinement is used. The magnetic field volume must be relatively small (a few m^3), or power and construction costs become prohibitive. The designer(s) of the sun has chosen a rather inelegant, but workable, solution: He has made the container very large, with a radius of about 7×10^{10} cm, with an outside temperature of about 6000 K, and with a central temperature of about 1.4×10^7 K. The fusion reaction then proceeds at a much lower rate than the one needed for terrestrial reactors. Nevertheless, total energy production is large because the volume is huge.

Before nuclear reactions were discovered, the energy production in the sun was unexplainable; no known source could provide sufficient energy, particularly because it was clear from geophysical studies that the sun must have had about the same temperature for at least 10^9 y. Among the first to recognize the nature of the energy-producing process was Eddington,[14] who showed that the fusion of four hydrogen atoms into one atom of He would release about 7 MeV/nucleon and thus provide millions of times more energy than a chemical reaction. However, one problem remained: Classically, fusion cannot occur because even at stellar temperatures protons do not have sufficient kinetic energy to overcome their mutual repulsion. Quantum mechanical tunneling, of course, permits reactions at much lower temperatures,[15] and specific reactions responsible for the stellar energy production were established.[16] The first sequence that was proposed is the carbon or *CNO cycle* in which a ^{12}C and 4p are transformed into an alpha particle and ^{12}C. The various steps in this cycle are:

$$^{12}\text{C } p \longrightarrow {}^{13}\text{N } \gamma$$

$$^{13}\text{N } \longrightarrow {}^{13}\text{C } e^+ \nu$$

$$^{13}\text{C } p \longrightarrow {}^{14}\text{N } \gamma$$

$$^{14}\text{N } p \longrightarrow {}^{15}\text{O } \gamma$$

$$^{15}\text{O } \longrightarrow {}^{15}\text{N } e^+ \nu$$

$$^{15}\text{N } p \longrightarrow {}^{12}\text{C } {}^4\text{He}. \tag{18.2}$$

14. A. S. Eddington, *Brit. Assoc. Advan. Sci. Rept. Cardiff*, 1920. In this talk, Eddington also said: "If, indeed, the subatomic energy in the stars is being freely used to maintain their great furnaces, it seems to bring a little nearer to fulfillment our dream of controlling this latent power for the well-being of the human race—or for its suicide."

15. R. Atkinson and F. Houtermans, *Z. Physik* **54**, 656 (1928).

16. H. A. Bethe, *Phys. Rev.* **55**, 434 (1939); C. F. Weizsäcker, *Physik. Z.* **39**, 633 (1938); H. A. Bethe and C. L. Critchfield, *Phys. Rev.* **54**, 248 (1938).

In this sequence, ^{12}C acts as a catalyst; it undergoes changes but it is not used up, it appears again in the final state. Thus the overall reaction is

$$4p \longrightarrow {}^4\text{He}.$$

The total energy release in this reaction can easily be found with the mass values given in Table A.6; it is

$$Q(4p \longrightarrow {}^4\text{He}) = 26.7 \text{ MeV}. \tag{18.3}$$

Of this energy, about 25 MeV heat the star, and the rest is carried off by the neutrinos. The CNO cycle dominates in *hot* stars; in cooler stars, particularly in the sun, the *pp cycle* is more important. The essential steps in the *pp* cycle are

$$\left.\begin{array}{c} pp \longrightarrow de^+\nu \\ \text{or} \\ ppe^- \longrightarrow d\nu \end{array}\right\} \quad dp \longrightarrow {}^3\text{He } \gamma \tag{18.4}$$

and

$$\begin{array}{c} {}^3\text{He } {}^3\text{He} \longrightarrow {}^4\text{He } 2p \\ \text{or} \\ {}^3\text{He } {}^4\text{He} \longrightarrow {}^7\text{Be } \gamma. \end{array} \tag{18.5}$$

In the first part of Eq. (18.5), the overall reaction $4p \longrightarrow {}^4\text{He} + 2e^+ + 2\nu$ has already been achieved. In the second part, ^{7}Be has been formed, and it, in turn, leads to ^{4}He through two sequences:

$$^7\text{Be } e^- \longrightarrow {}^7\text{Li } \nu; \quad {}^7\text{Li } p \longrightarrow 2 \, {}^4\text{He}$$

or

$$\tag{18.6}$$

$$^7\text{Be } p \longrightarrow {}^8\text{B } \gamma; \quad {}^8\text{B} \longrightarrow {}^8\text{Be*} \, e^+\nu; \quad {}^8\text{Be*} \longrightarrow 2 \, {}^4\text{He}.$$

The *pp* cycle has the same energy release, Eq. (18.3), as the CNO cycle. To compute the reaction rates, two very different input data are required. First, the temperature distribution in the interior of the sun must be known. The original work goes back to Eddington;[17] the improved versions[18] are claimed to be reliable, and the astrophysicists are willing to bet that the temperature of the sun at its center is about 14 million K. Second, the cross sections for the reactions listed above must be known at 14 million K. This temperature corresponds to kinetic energies of only a few keV, and the relevant cross sections are extremely small. A glance at Eqs. (18.4)–(18.6) shows that two types of reactions are involved, hadronic and weak ones. All reactions where neutrinos are involved are weak. The mean life of the decay $^8\text{B} \longrightarrow {}^8\text{Be*} \, e^+\nu$ has been measured. The two weak reactions in Eq. (18.4), however, are so slow that they cannot be measured in the laboratory; they must be computed using the weak Hamiltonian

17. A. S. Eddington, *Internal Constitution of Stars*, Cambridge University Press, Cambridge, 1926.

18. D. D. Clayton, *Principles of Stellar Evolution and Nucleosynthesis*, McGraw-Hill, New York, 1968; M. Schwarzschild, *Structure and Evolution of the Stars*, Princeton University Press, Princeton, N.J., 1958.

discussed in Chapter 11.[19] To find the cross sections for the hadronic reactions, values measured at higher energies are extrapolated down to a few keV. Most of the relevant very careful experiments have been performed at the California Institute of Technology under the leadership of W. A. Fowler.[20] Nuclear experimentalists and theorists are confident that their numbers are meaningful and cannot be changed much. Both the stellar-structure and the nuclear-physics aspects of solar energy production thus appear to be understood. We shall, however, return to this point in the next section.

18.3 Neutrino Astronomy

Classical astronomy is based on observations in the narrow band of visible light, from 400 to 800 nm. In the past few decades, this window has been enlarged enormously through radio astronomy on one hand and through X-ray and gamma-ray astronomy on the other. The charged cosmic rays provide another extension. However, all these observations have one limitation in common: They cannot look at the inside of stars, because the radiations are absorbed in a relatively small amount of matter (Chapter 3). Fortunately, there is one particle that presumably escapes even from the inside of a very dense star, the *neutrino*; and neutrino astronomy,[21,22] even though extremely difficult, promises to become an irreplacable tool in astrophysics. The properties that make the neutrino unique have already been treated in Sections 7.4 and 11.10:

1. The absorption of neutrinos and antineutrinos in matter is very small. For the absorption cross section, Eq. (11.76) gives

$$\sigma(\text{cm}^2) = 2.3 \times 10^{-44} \frac{p_e}{m_e c} \frac{E_e}{m_e c^2}, \tag{18.7}$$

where p_e and E_e are momentum and energy of the final electron in the reaction $\nu N \rightarrow eN'$. With Eqs. (6.8) and (6.9) it is then found that the mean free path of a 1-MeV neutrino in water is about 10^{21} cm. It far exceeds the linear dimensions of stars, which range up to 10^{13} cm. (See also Fig. 1.1.)

2. Neutrino and antineutrino can be distinguished by their reactions with matter.

Rather than discuss the possibilities of neutrino astronomy in general,

19. J. N. Bahcall and R. M. May, *Astrophys. J.* **155**, 501 (1969); J. N. Bahcall and C. P. Moeller, *Astrophys. J.* **155**, 511, (1969).

20. T. A. Tombrello, in *Nuclear Research with Low Energy Accelerators* (J. B. Marion and D. M. Van Patter, eds.), Academic Press, New York, 1967, p. 195. See also: *New Uses for Low-Energy Accelerators*, National Acad. Sciences, Washington, D. C., 1968. C. A. Barnes, in *Advances in Nuclear Physics*, Vol. 4 (M. Baranger and E. Vogt, eds.), Plenum Press, New York, 1971.

21. P. Morrison, *Sci. Amer.* **207**, 90 (Aug. 1962).

22. H. Y. Chiu, *Ann. Rev. Nucl. Sci.* **16**, 591 (1966).

we shall present the case of the *solar neutrinos*, because it is a mystery story whose ending is not yet known. In Eqs. (18.4)–(18.6), the steps in the *pp* cycle are described. Four reactions in this cycle produce neutrinos. These reactions are listed in Table 18.1, together with the neutrino energy and with their calculated relative intensities.[23] In 1964, it was pointed out

Table 18.1 NEUTRINO-PRODUCING REACTIONS, *pp* CYCLE

Reaction	Relative intensity	Neutrino energy
$pp \longrightarrow de^+ v_e$	0.9975	Spectrum, $E_{max} = 0.42$ MeV
$ppe^- \longrightarrow dv_e$	0.0025	Monoenerg., $E = 1.44$ MeV
$^7\text{Be } e^- \longrightarrow {}^7\text{Li } v_e$	0.86	Monoenerg., $E = 0.86$ MeV
$^8\text{B} \longrightarrow {}^8\text{Be* } e^+ v_e$	0.002	Spectrum, $E_{max} = 14.1$ MeV

by Bahcall and Davis[24] that it should be possible to verify directly the occurrence of the *pp* fusion reaction in the sun by detecting the solar neutrinos through Eq. (7.31):

$$v_e \, {}^{37}\text{Cl} \longrightarrow e^- \, {}^{37}\text{Ar}. \tag{18.8}$$

Electron neutrinos (but not antineutrinos) with an energy of more than 0.814 MeV can be captured by ^{37}Cl; the result is ^{37}Ar and an electron. As indicated in Eq. (18.7), the cross section for the process is extremely small. However, it should still be possible to detect the reaction, for four reasons: (1) The neutrino flux from the sun should be extremely large, of the order of 10^{11} neutrinos/cm²-sec on the surface of the earth; it then follows from Table 18.1 that the flux of energetic neutrinos is $\sim 2 \times 10^8$ neutrinos/cm²-sec. (2) Detectors can be made very large. (3) The ^{37}Ar can be observed in minute quantities because it is radioactive. It decays by electron capture,

$$e^- \, {}^{37}\text{Ar} \longrightarrow v_e \, {}^{37}\text{Cl}, \tag{18.9}$$

with a half-life of 35 days. The electron is usually captured from the K shell and leaves a hole there. The energy released when an electron from a higher shell drops into this hole is either emitted as an X ray or is used to eject an electron from an outer shell, called an Auger electron. The Auger electron has a well-defined energy, in this case 2.8 keV, and thus can be counted cleanly. (4) Ar is a noble gas. Consequently, it can be separated easily from chlorine and can be concentrated.

The experiment of Davis and collaborators is based on the four points listed above.[25] A tank containing 390,000 liters of C_2Cl_4 (tetrachloro-ethylene, a common cleaning fluid) is housed in a rock cavern in the Home-

23. J. N. Bahcall and R. K. Ulrich, *Astrophys. J.* **170**, 479 (1971).

24. J. N. Bahcall, *Phys. Rev. Letters* **12**, 300 (1964); R. Davis, Jr., *Phys. Rev. Letters* **12**, 303 (1964). J. N. Bahcall, *Sci. Amer.* **221**, 28 (July 1969).

25. R. Davis, Jr., D. S. Harmer, and K. C. Hoffman, *Phys. Rev. Letters* **20**, 1205 (1968).

stake gold mine in Lead, South Dakota. It is 1.5 km underground, in order to reduce the cosmic ray background. The abundance of the crucial isotope ^{37}Cl is 25%. A neutrino reaction with it produces ^{37}Ar. The radioactive argon is allowed to accumulate for several months and is then removed by flushing the tank with helium. The argon is separated from the helium by adsorption in a cold charcoal trap and it is then counted in a 0.5-cm^3 proportional counter. The result is expressed in solar neutrino units, pronounced "snew," where 1 SNU $= 10^{-36}$ events/sec-target atom.

The theoretically predicted neutrino counting rates are summarized in Table 18.2.[26]

Table 18.2 Theoretical Predictions for the Neutrino Counting Rate.
(1 SNU $= 10^{-36}$ neutrino captures/sec-target atom)

Assumption	Expected rate (SNU)
CNO-cycle-produced solar energy	35
Current "best" theory, pp cycle	6 ± 3
Lower limit consistent with theory	1.7 ± 0.3

The most recent result of Davis is[26,27]

$$\text{experimental rate} < 1 \text{ SNU.}$$

There is a serious disagreement between the experiment and the prediction. Solar neutrinos are missing and nobody knows why. Either the experiment is wrong, or some aspect of the solar theory must be modified, or some nuclear-physics input data must be altered radically, or some unexpected particle physics aspect (such as decaying neutrinos) has announced its existence.

18.4 Nucleosynthesis

In the beginning, God created the heavens and the earth.

Genesis

In the beginning, according to Gamow,[28] there was an extremely hot and highly compressed neutron ball, bathed in radiation. Because of its great internal energy, this primordial fireball or *ylem*[29] began to expand,

26. J. N. Bahcall, *Comments Nuclear Particle Phys.* **5**, 59 (1972).
27. *Phys. Today*, **25**, 17 (August 1972).
28. G. Gamow, *Phys. Rev.* **70**, 572 (1946); *Rev. Modern Phys.* **21**, 367 (1949). It should be well understood that nucleosynthesis as described here is a hypothesis. It fits most facts, but it is possible that it will be superseded by a different theory in the future.
29. Gamow introduced the word "ylem" which is an obsolete noun meaning "the primordial substance from which the elements were formed." In scientific slang, the initial event is called "big bang."

and neutrons decayed to protons. When, owing to the expansion, the temperature of the ylem dropped below about 10^9 K, deuterons that formed in the capture reaction

$$np \longrightarrow d\gamma$$

remained stable. Further neutron capture led to ^{3}H formation. The ^{3}H beta decayed to ^{3}He, which captured another neutron to form ^{4}He. (It is also possible to reach ^{4}He through the *pp* fission cycle discussed in Section 18.2.) Neutron capture, however, fails to account for elements beyond ^{4}He: $n + {}^4$He leads to ^{5}He which is unstable and decays back to ^{4}He. Further, the alpha particle capture reaction,

$$^4\text{He} \,^4\text{He} \longrightarrow {}^8\text{Be}, \tag{18.10}$$

leads to the highly unstable nuclide ^{8}Be which breaks up immediately into two alpha particles. Thus, some time after the big bang, the elements H and He existed in the expanding fireball, but production of heavier elements could not be achieved through neutron capture.

Although it was believed by some physicists[28] that other reactions in the expanding fireball could explain the formation of all elements this view was not widely shared. Since little information about the temperature of the ylem was available, it was difficult to eliminate the hypothesis. However, in 1965, Penzias and Wilson discovered a microwave background radiation.[30] This radiation essentially fits a blackbody spectrum with a temperature of about 2.7 K. It is interpreted as the radiation that is left over from the primordial fireball and thus provides some information about the conditions in the ylem. Big-bang nucleosynthesis was reexamined and one conclusion emerged firmly: For a present-day temperature of 2.7 K and universal density $\lesssim 10^{-28}$ g/cm^3 the observed abundance of heavy elements cannot be explained through big-bang synthesis.[31] Heavier elements apparently were produced later, after stars had already been formed. Nucleosynthesis, the explanation of the abundance of nuclear species, thus becomes intimately involved with problems of stellar structure and evolution.

In a star, gravity pressure tends to decrease the star's volume, while the pressure of the gas inside tends to oppose this reduction. Pressure and temperature inside a star are immense. In the sun, for instance, the pressure at the center is about 2×10^{10} atm and the temperature 14 MK. Under these circumstances, atoms will be completely ionized, resulting in a mixture of free electrons and bare nuclei. This mixture forms the "gas" mentioned above. The internal pressure is maintained by the nuclear reaction that provides the energy for the star's radiation. As long as this reaction proceeds, gravitational and internal pressure balance and the star will be

30. A. A. Penzias and R. W. Wilson, *Astrophys. J.* **142**, 419 (1965). See also G. B. Fields in "The Growth Points of Physics," *Rivista del Nuovo Cimento* **1**, 87, (1969), and P. J. E. Peebles and D. T. Wilkinson, *Sci. Amer.* **216**, 28 (June 1967).

31. R. V. Wagoner, W. A. Fowler, and F. Hoyle, *Astrophys. J.* **148**, 3 (1967).

in equilibrium. What will, however, happen when the fuel is used up? Or to give one example, what will happen to our sun when all hydrogen is used up and the pp cycle stops? At this point, the star will contract gravitationally and the central temperature and pressure increase. At some higher temperature new reactions occur, a new equilibrium will be reached, and new elements will be formed. There are thus alternate stages of nuclear burning and contraction. Burning may be quiescent as in the sun or explosive as in supernovae, and both are involved in the synthesis of heavier elements.

After the formation of ^{4}He, the next important step is the creation of ^{12}C. ^{8}Be, formed through the reaction Eq. (18.10), is unstable. However, if the density of ^{4}He is very high, measurable quantities of ^{8}Be are present in the equilibrium situation

$$^4He\,^4He \rightleftharpoons\, ^8Be^*.$$

Capture of an alpha particle can then occur,

$$^4He\,^8Be^* \longrightarrow\, ^{12}C. \tag{18.11}$$

This capture reaction is enhanced because the formation of ^{12}C proceeds mainly through a resonant capture to an excited state, $^{12}C^*$.

The formation of ^{16}O can occur through the CNO cycle, Eq. (18.2), but the dominant reaction is *helium burning*,

$$^4He\,^{12}C \longrightarrow\, ^{16}O\,\gamma. \tag{18.12}$$

This sequence can be repeated up the ladder of elements, and n and p capture reactions can form the elements that lie in between the alpha-like nuclides. However, fusion reactions, sometimes called *carbon burning*, appear to be vital in accounting for the abundance of the elements $20 \lesssim A \lesssim 32$. Such reactions,

$$
\begin{aligned}
^{12}C\,^{12}C &\longrightarrow\, ^{20}Ne\,\alpha \\
&\longrightarrow\, ^{23}Na\,p \\
&\longrightarrow\, ^{23}Mg\,n
\end{aligned}
\tag{18.13}
$$

require temperatures higher than about 10^9 K. Such temperatures occur only in some very massive stars and carbon burning thus is believed to occur predominantly in *exploding stars*. If it is assumed that the temperature in exploding stars in about 2×10^9 K, the abundance of elements produced appears to agree closely with observation, as is shown in Fig. 18.5. Similarly, *oxygen burning*,

$$
\begin{aligned}
^{16}O\,^{16}O &\longrightarrow\, ^{28}Si\,\alpha \\
&\longrightarrow\, ^{31}P\,p \\
&\longrightarrow\, ^{31}S\,n
\end{aligned}
\tag{18.14}
$$

can account for the abundance of the elements $32 \lesssim A \lesssim 42$, but requires

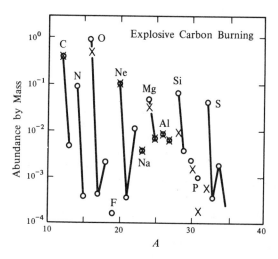

Fig. 18.5. Products of carbon burning in an exploding star. Circles represent solar-system abundances, calculated abundances are shown as crosses. Solid lines connect all stable isotopes of a given element. The assumed peak temperature is 2×10^9K, the density 10^5 g/cm³. [After W. D. Arnett and D. D. Clayton, *Nature* **227**, 780 (1970).]

a temperature of about 3.6×10^9 K. *Silicon burning* helps to explain the formation of the elements up to Ni.

As the formation of elements reaches iron, a new aspect appears. As Fig. 14.1 shows, the binding energy per nucleon reaches a maximum at the iron group. Beyond these elements, the binding energy per nucleon decreases. Hence the iron group cannot serve as fuel, and burning must cease once iron has been formed. This feature explains why the elements centered about Fe are more abundant than others.

Most elements beyond the iron group are probably formed through neutron and proton capture reactions. These processes continue as long as the stellar furnaces or explosions produce neutrons and protons. Once the power-producing reactions stop, further build-up of the heavy elements stops as well.

The synthesis of the elements described here bypasses the elements Li, Be, and B. The abundances of these elements shown in Fig. 18.2 can be accounted for by spallation reactions: Cosmic rays interact with heavy nuclei in the interstellar dust; these nuclei break into various parts, one of which is much lighter than the others. Very recent experimental observations indicate, however, that the Be and Li abundances are larger by about a factor 300 than given in Fig. 18.2. It is possible that reactions in exploding stars are responsible for these elements also.

We have only sketched the simplest ideas of nucleosynthesis. The correctness of these ideas can be examined only through detailed computations, involving nuclear physics and stellar evolution. Such investigations have arrived at encouraging results: Most of the salient abundance features can at least be qualitatively explained. However, further research is required before a complete understanding is in hand.[32-34]

18.5 Neutron Stars

> *Twinkle, twinkle, little star,*
> *How I wonder what you are,*
> *Up above the world so high,*
> *Like a diamond in the sky.*

In the previous section we have described the various burning processes in stars. These fusion reactions give rise to the elements; at the same time, they exhaust more and more of the nuclear fuel. What happens after the fuel is gone? According to present theory, the star can die one of four deaths; it can become a black hole, a white dwarf, a neutron star, or it can become completely disassembled. The ultimate fate is determined by the initial mass of the star. If this mass is less than about four solar masses, the star sheds mass until it becomes a white dwarf. If the initial mass is larger than about four solar masses, it may become a supernova which then results either in a neutron star, a black hole, or becomes completely disassembled. Black holes contract forever and they approach, but never reach, a radius of roughly 3 km and a density exceeding 10^{16} g/cm^3. Neutron stars have a radius of about 10 km and a central density exceeding that of nuclear matter, about 10^{14} g/cm^3. We shall limit the discussion here to neutron stars.[35]

The cross section of a typical neutron star, according to present theory, is shown in Fig. 18.6.[36] How did the star reach this terminal stage and why does it not collapse completely? The answer to these questions involves many fields, relativity, quantum theory, nuclear, particle, and solid state physics. Here we sketch some of the features of interest to subatomic physics.

Consider first *density and composition*. For a given stellar mass, the radius and the density distribution can be computed.[37] For a star with a radius of 10 km, the central density is of the order of 10^{14} to 10^{15} g/cm^3. The density thus increases from zero at the top of the atmosphere to a value larger than the density of nuclear matter at the center.

32. E. M. Burbidge, G. R. Burbidge, W. A. Fowler, and F. Hoyle, *Rev. Modern Phys.* **29**, 548 (1957). Reprinted in *Origin of the Elements*, Selected Reprints, American Institute of Physics, New York.

33. J. W. Truran, "Theories of Nucleosynthesis," in *Symposium on Cosmochemistry*, Cambridge, Mass., 1972.

34. W. D. Arnett and D. D. Clayton, *Nature*, **227**, 780 (1970); D. D. Clayton, *Comments Astrophys. Space Phys.* **3**, 13 (1971).

35. M. A. Ruderman, *Sci. Amer.*, **224**, 24 (Feb. 1971).

36. G. Baym, *Computational Solid State Physics*, (F. Herman, N. W. Dalton, and T. R. Koehler, eds.), Plenum, New York, 1972.

37. G. A. Baym, *Neutron Stars*, Nordita, Copenhagen, 1970.

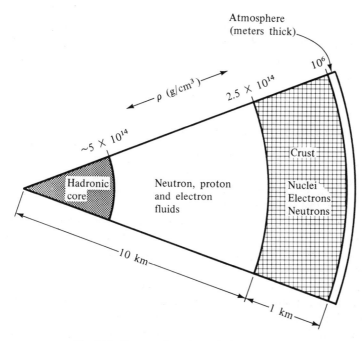

Fig. 18.6. Cross section of a typical neutron star.

From a knowledge of the density, the composition at a given depth can be inferred. In the early stages of stellar evolution, where the temperature is high, hadronic processes are dominant. These are fast compared to the times characteristic of stellar evolution and the evolving star will remain in nuclear equilibrium. At later times, when the weak interactions become important, equilibrium will no longer hold, but the star will still try to reach the energetically most favorable state. Each time a proton beta decays into a neutron, or vice versa, a neutrino is emitted; it leaves the star and carries away energy and entropy.

The absolute ground state of matter as a function of pressure involves atomic and subatomic physics. At zero pressure, the answer is given in Fig. 14.1, or Table A.6 and Eq. (14.3): The most stable atom at zero pressure is ^{56}Fe. In order to get the lowest solid-state energy, the iron atoms are arranged in a lattice. The outermost layer of a neutron star is thus expected to be mainly ^{56}Fe. Deeper inside the star, the pressure increases. At a density of about 10^4 g/cm^3, the atomic cores begin to touch, the atoms become fully ionized, and the electrons are no longer bound to given atoms. The behavior of these electrons is important for the evolution of the star and it can be understood with the Fermi gas model treated in Section 14.2. We assume the electrons to form a gas of extreme-relativistic free fermions, enclosed in a volume V. All available states up to the Fermi energy E_F are filled. This degenerate electron gas provides the pressure that balances the gravitational attraction. To compute the pressure, we first determine the total energy of n extremely relativistic electrons in a volume V, by following the steps from Eq. (14.13)–Eq. (14.18), but using the extreme relativistic relation $E = pc$. The total energy of the electrons then becomes

$$E = \left(\frac{\pi^2}{4}\right)^{1/3} \hbar c \, \frac{n^{4/3}}{V^{1/3}}. \qquad (18.15)$$

The pressure due to this Fermi gas is

$$p = -\frac{\partial E}{\partial V} = \frac{1}{3}\left(\frac{\pi^2}{4}\right)^{1/3} \hbar c \left(\frac{n}{V}\right)^{4/3}, \qquad (18.16)$$

and this pressure balances the gravitational attraction at this point in the development of the neutron star.

At densities above about 10^7 g/cm³, ^{56}Fe is no longer the lowest energy state; the electrons have acquired sufficient energy to transform ^{56}Fe nuclei by multiple electron capture processes into ^{62}Ni. As earlier stated, the neutrinos emitted in these electron capture processes remove energy and lower the total energy of the star. As the pressure increases, the electron capture processes continue and at about 4×10^{11} g/cm³, nuclei with 82 neutrons, such as ^{118}Kr, are most stable.[38] Ordinary krypton on earth has $A = 84$. The most stable nuclides at high pressure thus are very neutron-rich. Under ordinary circumstances, such nuclides would decay by electron emission. However, at the pressure under discussion here, all available energy levels are already occupied by electrons and the Pauli principle prevents electron decay.

The last neutron in ^{118}Kr is barely bound. As the density increases beyond 4×10^{11} g/cm³, the neutrons begin to leak out of the nuclei and form a degenerate liquid. As the pressure increases further, the nuclei in this *neutron drip regime* become more neutron rich and grow in size. At a density of about 2.5×10^{14} g/cm³, they essentially touch, merge together, and form a continuous fluid of neutrons, protons, and electrons.[39] Neutrons predominate and protons constitute only about 4% of the matter. Neutrons cannot decay to protons because the decay electron would have an energy below the electron Fermi energy; the decay is thus forbidden by the Pauli principle.

At still higher densities, it becomes energetically feasible to create more massive elementary particles through electron captures such as

$$e^- n \longrightarrow \nu \Sigma^-;$$

these particles can again be stable because of the exclusion principle.[40] The number of constituents of matter as a function of density is shown in Fig. 18.7. Of course, these curves are the result of a particular calculation and could be wrong!

38. G. Baym, C. Pethick, and P. Sutherland, *Astrophys. J.* **170**, 299 (1971).
39. G. Baym, H. A. Bethe, and C. J. Pethick, *Nucl. Phys.* **A175**, 225 (1971).
40. V. R. Pandharipande, *Nucl. Phys.* **A178**, 123 (1971).

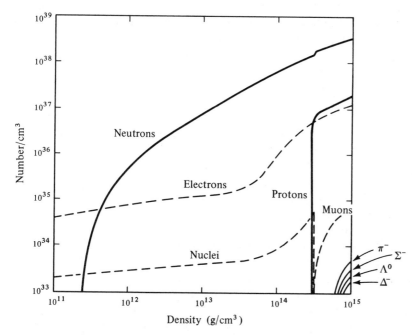

Fig. 18.7. Number of the constituents of matter versus density. The *neutron drip regime*, where neutrons leak out from the nuclei, starts at 4×10^{11} g/cm³. At a density of about 2.5×10^{14} g/cm³, nuclei begin dissolve. At even higher densities, muons and strange particles appear. (Courtesy M. A. Ruderman.)

We now turn again to the internal pressure in a neutron star. We have seen above that the degenerate electron gas provides pressure that prevents collapse at lower pressures. At higher pressure, complete collapse is prevented by a combination of two features, the repulsive core in the nucleon-nucleon force (Fig. 12.15), and the degeneracy energy of the neutrons. Fig. 18.7 indicates that neutrons predominate at the highest pressure. They form a degenerate Fermi gas and the arguments leading to Eq. (18.16) can be repeated nonrelativistically. Again, as in Eq. (18.16), the degeneracy pressure increases with decreasing volume until it, together with the hard core repulsion, balances the gravitational attraction.

Neutron stars were predicted long ago,[41] but hope for their observation was small and they remained mythical objects for a long time. Their discovery was unexpected: In 1967, a strange new class of celestial objects was observed at the University of Cambridge.[42] The objects were point-like, definitely outside the solar system, and emitted periodic radio signals.

41. W. Baade and F. Zwicky, *Proc. Nat. Acad. Sci. Amer.* **20**, 259 (1934); L. D. Landau, *Phys. Z. Sowiet* **1**, 285 (1932).

42. A. Hewish, S. J. Bell, J. D. S. Pilkington, P. F. Scott, and R. A. Collins, *Nature* **217**, 709 (1968). A. Hewish, *Sci. Amer.* **219**, 25 (Oct. 1968); J. P. Ostriker, *Sci. Amer.* **224**, 48 (Jan. 1971).

They were nicknamed *pulsars* and, despite the fact that the objects are not pulsating, but rotating, the name has been accepted. More than 100 pulsars are known; each has its own characteristic signature. The pulsar periods range from 33 msec to 3.75 sec, and they lengthen in a very regular fashion. The pulsar with the shortest period (33 msec) is in the Crab nebula in the constellation Taurus, where a supernova was observed in 1054 A.D. by the Chinese and Koreans; the shortest-period pulsar consequently appears to be associated with a young supernova.

As suggested by Gold, a pulsar is a neutron star.[43] The pulsar period is associated with the rotational period of the neutron star; its slowing down is caused by the loss of rotational energy. The rotational energy lost by the Crab pulsar, for instance, is of the same order as the total energy emitted by the nebula. The neutron star thus is the power source of the huge Crab nebula.

Pulsars have not only been observed as radio stars; in one instance, periodic light emission has been seen. The periods, the slow-down rates, and the sudden changes in the period are being studied very carefully. Step by step, pulsars reveal properties of neutron stars. In an indirect way, astrophysicists and nuclear physicists have obtained access to a laboratory in which densities beyond 10^{15} g/cm^3 are available; the properties of nuclear matter can thus be studied in a beautiful combination of various disciplines.

18.6 References

Nuclear astrophysics advances so fast that no text can be up-to-date for long. We suggest that a reader, after getting some information from one of the texts or reviews listed below, search in the three following periodicals for the latest news:

Comments on Astrophysics and Space Physics

Annual Review of Nuclear Science

Annual Review of Astronomy and Astrophysics

The entire field of nuclear astrophysics, up to about 1966, is treated in a lucid and easily readable fashion in W. A. Fowler, *Nuclear Astrophysics*, American Philosophical Society, Philadelphia, 1967.

Cosmic Rays. The literature on cosmic rays up to 1966 is reviewed in J. R. Winckler and D. J. Hofmann, "Resource Letter CR-1 on Cosmic Rays," *Am. J. Phys.* **35**, 1 (1967). This resource letter and a number of the important papers are collected in *Cosmic Rays, Selected Reprints*. American Institute of Physics, New York.

43. T. Gold, *Nature* **218**, 731 (1968).

Stellar Energy. An easily readable account of the energy production in stars is Oakes Ames, "Stars and Nuclei," *The Physics Teacher* (April and May 1972). A somewhat dated, but still worthwhile discussion of the energy production in stars is given in G. Gamow and C. L. Critchfield, *Theory of the Atomic Nucleus and Nuclear Energy Sources*, Clarendon Press, Oxford, 1949. For a modern account, see D. D. Clayton, *Principles of Stellar Evolution and Nucleosynthesis*. McGraw-Hill, New York, 1968. A masterpiece of popular writing is G. Gamow, *The Birth and Death of the Sun*, Viking, New York, 1953.

Neutrino Astronomy. In addition to the reviews listed in the text, consult L. M. Lederman, Resource Letter Neu-1, History of the Neutrino, *Am. J. Phys.* **38**, 129 (1970). See also J. Bahcall and R. L. Sears, "Solar Neutrinos," *Ann. Review Astronomy and Astrophys.* **10**, 25 (1972). M. A. Ruderman, "Astrophysical Neutrinos," *Rept. Progr. Phys.* **28**, 411 (1965); reprinted in *Astrophysics*, Benjamin, Reading, Mass., 1969.

Nucleosynthesis. The literature is reviewed in W. A. Fowler and W. E. Stephens, "Resource Letter OE-1, Origin of the Elements," *Am. J. Phys.* **36**, 289 (1958). This resource letter is contained, together with selected reprints, in two collections: *Origin of the Elements*. American Institute of Physics, New York, and *Synthesis and Abundances of the Elements*, American Institute of Physics, New York.

Neutron Stars. The following reviews cover neutron stars and pulsars:

A. G. W. Cameron, "Neutron Stars," *Ann. Rev. Astronomy Astrophys.* **8**, 179 (1970).

V. L. Ginzburg, "Pulsars," *Soviet Phys. Usp.* **14**, 83 (1971).

F. G. Smith, "Pulsars," *Rept. Progr. Phys.* **35**, 399 (1972).

D. Ter Haar, *Physics Reports* **3C**, 57 (1972).

M. Ruderman, "Pulsars," *Ann. Rev. Astronomy Astrophys.* **10**, 427 (1972).

19

Nuclei in Chemistry

Subatomic physics and chemistry display a mutual interaction: Nuclei and particles can be studied with chemical tools, and chemistry can be investigated with subatomic tools. We shall be concerned only with the second aspect here because we are trying to show how subatomic physics influences other fields.

19.1 Synthetic Elements

Before the construction of particle accelerators and nuclear reactors, the number of elements to be studied was fixed. If an element was missing, its properties remained hypothetical. With the advent of accelerators and reactors, new elements became accessible, and we shall sketch three ways in which nuclear physics enriched chemistry, namely by producing elements missing in the periodic table below uranium, by creating transuranium elements, and by making exotic atoms, such as positronium and muonium. It is possible that the superheavy elements, with $Z \approx 111$, will soon be added to this list.

Missing Elements. Before 1937, four elements below uranium were unknown; the missing nuclear charge numbers were 43, 61, 85, 87. Despite various claims and a plethora of proposed names, no firm evidence for any of these elements existed. In 1935, Perrier and Segrè, in Italy, obtained molybdenum that had been bombarded with deuterons in the Berkeley cyclotron. They established unambiguously that a radioactive substance contained in their sample was due to the element 43.[1] The new element was produced in the reaction

$$d + \text{Mo} \longrightarrow n + 43, \quad \text{or} \quad \text{Mo}(d, n)43,$$

and it was called *technetium* (Tc). To establish its elemental character, the irradiated molybdenum was dissolved in aqua regia. Perrier and Segrè then checked if the radioactivity would be carried down with zirconium ($Z = 40$), niobium ($Z = 41$), molybdenum ($Z = 42$), or manganese and rhenium. They found that it did *not* follow the first three but precipitated with the last two. Rhenium and manganese are homologs of 43, one above and one below in the periodic table, and the chemical nature of Tc was thus determined. By now, 16 isotopes of Tc, with A ranging from 92 to 107, are known. The element 87 was found in 1939 in the decay of the naturally radioactive nuclide ^{227}Ac, and it was named *francium* (Fr). The element 85, *astatine*, was produced in 1940 through the reaction $\text{Bi}(\alpha, 2n)\text{At}$. The last holdout, element 61, turned up as a fission product in 1945; it was baptized *promethium*, because Prometheus brought fire from heaven for the use of man, just as fission has made available the energy of the nucleus.

Transuranium Elements. We mentioned in Section 17.1 that Fermi bombarded heavy nuclei with neutrons as early as 1934. He expected to produce transuranium elements and thought that he had done so. In reality, however, he had induced fission. The first true transuranium element, *neptunium* ($Z = 93$), was discovered in 1940 by McMillan and Abelson[2] through the sequence

$$^{238}\text{U}(n, \gamma)^{239}\text{U}; \qquad ^{239}\text{U} \xrightarrow{2.3d} {}^{239}\text{Np} \, e^- \bar{\nu}.$$

The next element, *plutonium*, was found soon thereafter through the chain

$$^{238}\text{U}(d, 2n)^{238}\text{Np}; \qquad ^{238}\text{Np} \xrightarrow{2.1d} {}^{238}\text{Pu} \, e^- \bar{\nu}.$$

Plutonium is by far the most important transuranium element because of its almost unlimited use as a nuclear fuel (See Section 17.2).

One unexpected feature turned up when the chemical properties of the new transuranium elements were explored. It was generally anticipated that the new elements would be homologs of the period V elements, because Ac, Th, Pa, and U resemble Y, Hf, Ta, and W, respectively. Thus neptunium should be chemically homolog to rhenium. In fact, however, it behaves much more like a rare earth. This similarity between the rare earths (lanthanides) and the transuranium elements holds also for the ones

1. C. Perrier and E. Segrè, *J. Chem. Phys.* **5**, 715 (1937).
2. E. M. McMillan and P. H. Abelson, *Phys. Rev.* **57**, 1185 (1940).

Table 19.1 LANTHANIDES AND ACTINIDES. In the periodic table, all lanthanides have chemical properties that are similar to the ones of La, $Z = 57$. All actinides are similar to actinium, $Z = 89$. The key to the table is given below.

Lanthanides

Atomic No.	Symbol	Oxidation States	Atomic Weight	Electron Config.
58	Ce	+3 +4	140.12	-20-8-2
59	Pr	+3 +4	140.9077	-21-8-2
60	Nd	+3	144.24	-22-8-2
61	Pm	+3	(145)	-23-8-2
62	Sm	+2 +3	150.4	-24-8-2
63	Eu	+2 +3	151.96	-25-8-2
64	Gd	+3	157.25	-25-9-2
65	Tb	+3	158.9254	-27-8-2
66	Dy	+3	162.50	-28-8-2
67	Ho	+3	164.9303	-29-8-2
68	Er	+3	167.26	-30-8-2
69	Tm	+3	168.9342	-31-8-2
70	Yb	+2 +3	173.04	-32-8-2
71	Lu	+3	174.97	-32-9-2

Actinides

Atomic No.	Symbol	Oxidation States	Atomic Weight	Electron Config.
90	Th	+4	232.0381	-18-10-2
91	Pa	+5 +4	231.0359	-20-9-2
92	U	+3 +4 +5 +6	238.029	-21-9-2
93	Np	+3 +4 +5 +6	237.0482	-22-9-2
94	Pu	+3 +4 +5 +6	(244)	-24-8-2
95	Am	+3 +4 +5 +6	(243)	-25-8-2
96	Cm	+3 +4	(247)	-25-9-2
97	Bk	+3 +4	(247)	-27-8-2
98	Cf	+3	(251)	-28-8-2
99	Es	+3	(254)	-29-8-2
100	Fm		(257)	-30-8-2
101	Md		(256)	-31-8-2
102	No		(254)	-32-8-2
103	Lr		(254)	-32-9-2

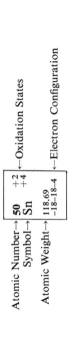

Key:

Atomic Number → **50**
Symbol → **Sn**   +2 +4 ← Oxidation States
Atomic Weight → 118.69
-18-18-4 ← Electron Configuration

Numbers in parentheses are mass numbers of most stable isotope of that element.

that were discovered later. The various transuranium elements fill inner shells and belong chemically in the same place, the one of actinium. Consequently, they are called *actinides*. In the course of time, actinides up to $Z = 105$ have been produced.[3] With increasing Z the difficulty of creating and identifying new elements increases rapidly. On the one hand, the targets from which the production is started may in itself be radioactive and therefore exist only in small quantities. On the other hand, the lifetimes of the new nuclides become increasingly shorter, and identification must be based on decay products. The actinides up to $Z = 103$ are listed in Table 19.1, together with the corresponding lanthanides. Claims to the discovery of elements 104 and 105 have been filed by an American and by a Russian group, and the elements have not yet been officially named.

Exotic (New) Atoms. The nuclei of ordinary atoms are built from protons and neutrons, and their shells from negative electrons. Exotic atoms result when one (or more) of these stable building blocks is replaced by another elementary particle. The first exotic atom, positronium, was discovered by Deutsch.[4] At the present time, many new atoms are known, and they are all of interest in various fields of physics.[5] We list here only the ones that are at present used as tools in chemistry,[6] namely positronium ($e^+ e^-$), muonium ($\mu^+ e^-$), and μ and π mesoatoms. In mesoatoms, a negative muon or pion plays the role of an electron. All exotic atoms are unstable, and the chemist's interest is based on this property: Lifetime and decay characteristics are affected by the chemical environment, and they can serve as probes for it.

Superheavy Elements.[7,8] Figure 17.3 shows that the half-life against spontaneous fission decreases as a function of Z^2/A. From the figure it would appear as if nuclides beyond about $Z = 105$ decay so rapidly that they cannot be observed. However, nuclear shell effects change the picture. In Chapter 15 it was shown that $N = 126$ is a shell closure for neutrons. It was therefore assumed for many years that $Z = 126$ would also be magic for protons. A nucleus with $Z = 126$ is so far removed from the last stable one that it is nearly impossible to create it. However, in 1959, Mottelson and Nilsson pointed out that the proton closed shell should already occur at $Z = 114$; in 1966, the conclusion was reached that a nucleus with $Z =$

3. The story of element 101 is told by one of the discoverers, A. Ghiorso, in *Adventures in Experimental Physics*, Beta, 1972. Chapter 11.

4. M. Deutsch, *Phys. Rev.* **82**, 455 (1951); **83**, 866 (1951).

5. See, for instance, V. W. Hughes, *Ann. Rev. Nucl. Sci.* **16**, 445 (1966) (muonium); G. Backenstoss, *Ann. Rev. Nucl. Sci.* **20**, 467 (1970) (pionic atoms); V. I. Goldanskii, *At. Energy Rev.* **6**, 3 (1968) (positronium); C. S. Wu and L. Wilets, *Ann. Rev. Nucl. Sci.* **19**, 527 (1969) (muonic atoms); and E. H. S. Burhop, *High Energy Phys.* **3** (1969) (mesoatoms).

6. V. I. Goldanskii and V. G. Firsov, *Ann. Rev. Phys. Chem.* **22**, 209 (1971).

7. S. G. Thompson and C. F. Tsang, *Science* **178**, 1047 (1972).

8. J. R. Nix, *Phys. Today* **25**, 30 (April 1972); *Ann. Rev. Nucl. Sci.* **22**, 65 (1972); G. N. Flerov, V. A. Druin, and A. A. Pleve, *Soviet Phys. Usp.* **13**, 24 (1970).

114 and $N = 184$ could be doubly magic[9] The prediction that $Z = 114$ rather than $Z = 126$ should be magic can be partially understood with Fig. 15.10: A nucleus with $Z \approx 120$ has a mass number A of about 300 whereas $N \approx 120$ corresponds to $A \approx 200$. A nucleus with $Z \approx 120$ consequently has a much larger radius than one with $N \approx 120$. In a potential with larger radius, higher angular momentum states move lower; the levels $1h_{9/2}$, $2f_{7/2}$, and $1i_{13/2}$ thus move below and separate from $2f_{5/2}$, $3p_{3/2}$, and $3p_{1/2}$, and the shell closure occurs at $Z = 114$. (However, it should be remembered that this prediction has not yet been verified experimentally!) Since 1966, calculations of fission barriers for nuclei in the neighborhood of the doubly magic nuclide ($Z = 114$, $N = 184$) have been performed with more and more refined techniques[10], and one result is shown in Fig. 19.1. The peninsula (mainland) of stable nuclei, which has been displayed in Fig. 5.20, is separated by a strait from the magic island of the superheavy elements. Some of the superheavy nuclides are expected to have lifetimes between 10^3 and 10^{15} yr. For the past few years, the search for such elements has been very active. None have been seen so far with certainty, either in nature (as remnants from the big bang or from supernovae) or at accelerators. However, new heavy-ion accelerators may permit the leap across the strait in reactions such as

$$^{76}\text{Ge} + {}^{232}\text{Th} \longrightarrow {}^{304}122 + 4n$$
$$\searrow {}^{301}120 + {}^4\text{He} + 3n.$$

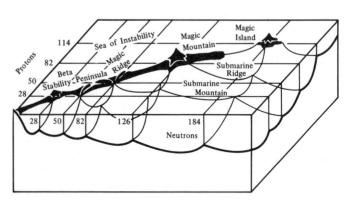

Fig. 19.1. Nuclear stability is illustrated in a scheme that shows a peninsula of known elements and an island of predicted stability (nuclei around $Z = 114$ and $N = 184$) in a "sea of instability." Grid lines show magic numbers of protons and neutrons giving rise to exceptional stability. Magic regions on the mainland peninsula are represented by mountains or ridges. [From S. G. Thompson and C. F. Tsang, *Science* **178**, 1047 (1972). Copyright © 1972 by the American Association for the Advancement of Science.]

9. B. R. Mottelson and S. G. Nilsson, *Kgl. Danske Videnskab Selskab, Mat.-fys. Medd.* **1**, No. 8 (1959); H. W. Meldner, private communication to W. D. Myers and W. J. Swiatecki, *Nucl. Phys.* **81**, 1 (1966); A. Sobiczewski, F. A. Gareev, and B. N. Kalinkin, *Phys. Letters* **22**, 500 (1966); S. G. Nilsson et al., *Nucl. Phys.* **A131**, 1 (1969).

10. V. M. Strutinsky, *Soviet J. Nucl. Phys.* **3**, 449 (1966); *Nucl. Phys.* **A95**, 420 (1967); **A122**, 1 (1968). M. Brack et al., *Rev. Mod. Phys.* **44**, 320 (1972).

One way to ascertain that superheavy elements have been produced is to determine the number of neutrons per fission. Figure 17.6 indicates that the fragment excitation energy increases rapidly with Z; superheavy nuclides are expected to emit as many as 10 neutrons/fission.

The production of nuclides in the magic island would be exciting not only for nuclear physicists but also for atomic physicists and chemists. Relativistic effects in the atomic shell will be large and the spin-orbit splitting of the atomic levels will become a dominant feature. Some of the new elements may exhibit radically new chemical properties. Element 115, for instance, is a homolog of arsenic, antimony, and bismuth. However, while these elements show oxidation states of $+3$, $+5$, and -3, 115 is expected to be monovalent.[11]

19.2 Chemical Analysis

> *Sherlock Holmes was good, but*
> *Could he find a thimbleful of poison in 10 tank cars of water?*
> *Could he trap a suspect through an invisible trace of antimony on his hand?*
>
> From the advertisement of a company providing
> an activation analysis service

Techniques from subatomic physics have become indispensable in the analysis of chemical systems. We shall sketch two powerful tools, mass spectrometry and activation analysis.

Mass spectrometry has already been discussed in Section 5.3, and Fig. 5.7 shows the basic principle of Aston's mass spectrometer. Modern instruments still work on the same principle but are far more sophisticated and have a much higher resolution.[12,13] They also have been perfected to the point where they can be used as routine analysis tools; it is estimated that more than 10,000 instruments are in service at present. The high resolution makes it possible to distinguish between two compounds that have the same mass number but differ slightly in actual mass. An example is shown in Fig. 19.2: Two hydrocarbons with the mass number 432 but different composition can be separated. Routine and efficient use of mass spectrometry is helped by a number of advances. On-line computers, direct coupling to a gas chromatograph, and the existence of a mass spectrometry data center are but some of the features that have made mass spectrometry a universal tool in chemistry laboratories.[14]

Activation analysis helps to determine which *elements* are present in a given sample. The method is simple:[15] A sample is irradiated with suitable particles (usually thermal or high-energy neutrons, but sometimes

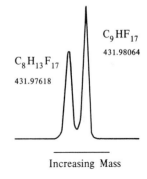

$C_9 HF_{17}$
431.98064

$C_8 H_{13} F_{17}$
431.97618

Increasing Mass

Fig. 19.2. Doublet of a perfluorokerosene at mass 432 showing separation for two hydrocarbons differing in mass by only four parts in 100,000. (Courtesy W. N. Howald, Mass Spectrographic Laboratory, Univ. of Washington.)

11. B. Fricke and J. T. Waber, *Actinides Rev.* **1**, 433 (1971).

12. See, for instance, *Mass Spectrometry; Techniques and Applications* (G. W. A. Milne, ed.), Wiley-Interscience, New York, 1971.

13. A. O. C. Nier, *Am. Sci.* **54**, 59 (1966).

14. A. Quayle, ed., *Advances in Mass Spectrometry*, Elsevier, Amsterdam, 1971.

15. E. V. Sayre, *Ann. Rev. Nucl. Sci.* **13**, 145 (1963).

photons or charged particles), nuclear reactions lead to radioactive nuclides, and these are identified through their radiations. Activation analysis
is selective, nondestructive, and samples can be studied from a distance.
More than 70 elements can be detected by activation analysis; quantities
can be as small as 10^{-12} g, and the typical accuracy lies between 2% and
5%.

Activation analysis has been applied to a wide variety of samples,[16]
from moon rocks and oil-well cores to archaeological artifacts; it is used
in biochemistry, medicine, criminology and the testing of highly purified
material. The surveillance of the environment against pollution is one field
in which neutron activation analysis has already played a considerable
role. Mercury, arsenic, and selenium are three toxic elements whose con-
centrations must be continuously monitored. To do so, samples of water,
soil, plants, fish, coal, oil, and filtered-air particulates are sealed in plastic
or quartz vials and placed in a nuclear reactor. Mercury, for instance, then
undergoes the reactions ^{196}Hg(n, γ) ^{197}Hg, ^{196}Hg(n, γ) ^{197m}Hg, and ^{202}Hg
(n, γ) ^{203}Hg. The three isotopes that are produced are radioactive and have
characteristic half-lives and gamma radiations:

$$^{197}\text{Hg:} \quad T_{1/2} = 2.7 \text{ days} \qquad E_\gamma = \quad 0.07734 \text{ MeV}$$

$$^{197m}\text{Hg:} \qquad \quad 24 \text{ hours} \qquad \qquad 0.1340, 0.1653 \text{ MeV}$$

$$^{203}\text{Hg:} \qquad \quad 47 \text{ days} \qquad \qquad 0.27918 \text{ MeV.}$$

The gamma rays and, if necessary, the half-lives can be studied with solid-
state detectors. Figure 4.8 makes it plausible that the high resolution of
such a counter allows separation of the Hg gamma rays from others; Hg
thus has a unique signature and can be detected in very small quantities.

19.3 Chemical Structure

The task of the chemist is not finished when the composition of a
molecule is known. The next step is the determination of its structure, the
spatial arrangement of the atoms in the molecule. Here also subatomic
tools contribute greatly. We shall discuss three very different aspects of the
study of chemical structure by nuclear means: synchrotron radiation,
hyperfine interactions, and inner-shell spectroscopy.

Synchrotron Radiation. The most powerful method for the determination of the three-dimensional structure of molecules is *X-ray diffraction*.
The basic idea is identical to that discussed in Section 6.9 for particles and
nuclei: X rays, with a wavelength comparable to, or smaller than, the
dimensions to be explored are diffracted by the sample. The scattering

16. P. Kruger, *Principles of Activation Analysis*, Wiley, New York, 1971; M. Rakovic,
Neutron Activation Analysis, CRC, Cleveland, 1970. *Advances in Activation Analysis*. (J. M. A.
Lenihan, S. J. Thomson, and V. P. Guinn, eds.), Academic Press, New York, 1972.

results in a diffraction pattern from which, with much labor and computer time, the structure of the scatterer is deduced. The X rays are produced by hitting a metal target with an intense electron beam; either characteristic X rays or bremsstrahlung are then used. Two facts restrict structure studies: The intensities of such X-ray beams are limited by the power dissipated in the target, and only a few useful X-ray lines are available. *Synchrotron radiation* promises to overcome both of these limitations. We stated at the end of Section 2.5 that electron synchrotrons lose a sizable fraction of the input energy as radiation. The radiation from any given point is pencil-like and tangential to the electron beam orbit. For a given beam energy and radius of curvature, the intensity spectrum and the polarization of the synchrotron radiation are well known.[17,18] As an example, we show in Fig. 19.3 the intensity of the radiation emitted by DESY, the electron synchrotron in Hamburg, as a function of the machine energy. A comparison of the spectrum with other light sources in the same region (between 0.1 Å and 10^3 Å, 1 Å = 0.1 nm = 10^{-10} m) gives the following results:

1. Between 1000 and 500 Å, synchrotron radiation is considerably stronger than any competing light source.

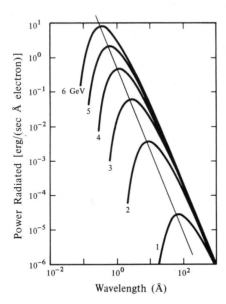

Fig. 19.3. Intensity of the synchrotron radiation emitted by electrons on a radius of 31.7 m (DESY) at various energies. (After R. P. Godwin, "Synchrotron Radiation as a Light Source," *Springer Tracts in Modern Physics*, **51**, Springer, Berlin, Heidelberg, New York, 1969.)

17. Feynman Lectures 1-34-3; Jackson, Chapter 14; A. A. Sokolov and J. M. Ternov, *Synchrotron Radiation*, Akademie, Berlin, 1968.

18. R. P. Godwin, *Springer Tracts Modern Phys.*, **51**, 1 (1969). R. Codling, *Rept. Progr. Phys.* **36**, 541 (1973).

2. Between 500 and 200 Å, it is the only intense source.

3. Below 200 Å, it is much more intense than all other sources. Even in a region where powerful X-ray sources exist, around 1 Å, synchrotron radiation is orders of magnitude more intense.[19]

4. Synchrotron radiation is linearly polarized.

These advantages are partially offset by the fact that a synchrotron is not easily available. However, it is likely that storage rings will be built that are cheaper than entire synchrotrons and will serve as excellent light sources.

While X-ray diffraction is the tool to establish the overall structure of a molecule, the nucleus can be used as a sensitive probe to study particular sites. Three properties make nuclei good probes: their extended charge distributions (finite radius), their magnetic dipole moments, and their electric quadrupole moments. The surrounding "chemical" produces electromagnetic fields at the nuclear site; the interaction of these fields with the nucleus shifts or splits the nuclear levels and thus provides information about the environment. We shall discuss the basic ideas underlying these hyperfine interactions and then treat some of the relevant techniques.

Hyperfine Interactions. The hyperfine energy is the interaction energy between the nucleus and the electric and magnetic fields produced by the surrounding atomic and molecular electrons. The mathematical treatment of this energy through the multipole expansion is well known.[20] Three types of hyperfine couplings dominate: the magnetic dipole and the electric monopole and quadrupole interactions. The *magnetic dipole interaction* has already been discussed in Chapter 5. If a nucleus with a magnetic moment μ is placed in a magnetic field B, its levels undergo Zeeman splitting, as shown in Fig. 5.5. A measurement of the energy difference between any two Zeeman levels provides a determination of the magnetic field, B, at the nucleus, if the nuclear magnetic moment is known. To determine the *electrostatic interaction*, we use Eq. (10.87) to write the interaction between a nucleus with charge density $Ze\rho(\mathbf{x})$ and a scalar potential A_0, as

$$H_{em} = Ze \int d^3x \, \rho(\mathbf{x}) A_0(\mathbf{x}), \qquad (19.1)$$

where the integration extends over the nuclear volume. Expansion of $A_0(\mathbf{x}) \equiv A_0(x_1, x_2, x_3)$ in a power series about the center of the nucleus gives, with $\int d^3x \, \rho(\mathbf{x}) = 1$ [Eq. (6.26)],

$$H_{em} = H_{em}^{(0)} + H_{em}^{(1)} + H_{em}^{(2)},$$
$$H_{em}^{(0)} = Ze \, A_0(0) \qquad (19.2)$$

19. G. Rosenbaum, K. C. Holmes, and J. Witz, *Nature* **230**, 434 (1971).
20. Jackson, Chapter 4.

$$H_{em}^{(1)} = Ze \sum_{i=1}^{3} \left(\frac{\partial A_0}{\partial x_i} \right)_0 \int d^3x \; \rho(\mathbf{x}) x_i$$

(19.2 cont.)

$$H_{em}^{(2)} = \frac{Ze}{2} \sum_{i=1}^{3} \left(\frac{\partial^2 A_0}{\partial x_i^2} \right)_0 \int d^3x \; \rho(\mathbf{x}) x_i^2.$$

The coordinate system has been so chosen that the mixed terms of the form $(\partial^2 A_0/\partial x_i \partial x_j) x_i x_j$ vanish. The term $H_{em}^{(0)}$ represents the interaction of a point nucleus with charge Ze with the potential A_0. The term $H_{em}^{(1)}$ vanishes because nuclei have a well defined parity. The third term, $H_{em}^{(2)}$, is the one of interest here. The second derivative $\partial^2 A_0/\partial x_i^2$ is the gradient of the electric field along the x_i axis at the point $\mathbf{x} = 0$. We use the customary notation

$$\left(\frac{\partial^2 A_0}{\partial x_i^2} \right)_0 \equiv V_{ii}, \qquad V_{33} \equiv V_{zz}, \qquad r^2 = x_1^2 + x_2^2 + x_3^2,$$

and split the term into two components:

$$H_{em}^{(2)} = H_M + H_Q$$

$$H_M = \frac{Ze}{6} \sum_i V_{ii} \int d^3x \; \rho(\mathbf{x}) r^2 \tag{19.3}$$

$$H_Q = \frac{Ze}{6} \sum_i V_{ii} \int d^3x \; \rho(\mathbf{x})(3x_i^2 - r^2).$$

The *monopole term*, H_M, can be rewritten by using Poisson's equation[21]

$$\nabla^2 V = \sum_i V_{ii} = -4\pi q \rho_e. \tag{19.4}$$

The charge density $q\rho_e$ at the nucleus is produced by the atomic and molecular electrons and is given by

$$q\rho_e(0) = -e |\psi(0)|^2, \tag{19.5}$$

where $|\psi(0)|^2$ is the electron probability density at the nucleus. The integral in H_M is the nuclear mean-square radius, Eq. (6.29), so that the monopole term becomes

$$H_M = \frac{2\pi}{3} Ze^2 \langle r^2 \rangle |\psi(0)|^2. \tag{19.6}$$

H_M produces a shift of the nuclear energy level when compared to that for a point nucleus, where $\langle r^2 \rangle = 0$. The shift is proportional to the mean-square radius and the electron probability density at the nucleus. In the discussion of the Mössbauer effect below, the term will turn out to be important for chemical studies.

The *quadrupole term*, H_Q, describes the interaction between the electric field gradient and the nuclear quadrupole moment. For a spherical nucleus, $\int d^3x \; \rho(\mathbf{x}) 3x_i^2 = \int d^3x \; \rho(\mathbf{x}) r^2$, and H_Q vanishes. To evaluate H_Q

---

21. Jackson, Section 1.7.

for a nonspherical nucleus, we first consider spherically symmetric electrons, for instance s electrons. For such electrons, $V_{11} = V_{22} = V_{33} \equiv V_{zz}$, and

$$H_Q = \tfrac{1}{6} Z e V_{zz} \int d^3x \, \rho(\mathbf{x}) \sum_{i=1}^{3} (3x_i^2 - r^2)$$

vanishes because $\sum_{i=1}^{3} (3x_i^2 - r^2) = 0$. The only contribution to H_Q comes from electrons with a nonspherical charge distribution. For such electrons, $\psi(0) = 0$, and Poisson's equation simplifies to that of Laplace,

$$V_{xx} + V_{yy} + V_{zz} = 0. \tag{19.7}$$

For an axially symmetric field gradient, where $V_{xx} = V_{yy}$, the quadrupole interaction becomes

$$H_Q = \tfrac{1}{4} e V_{zz} \int Z \, d^3x \, \rho(\mathbf{x})(3z^2 - r^2). \tag{19.8}$$

Comparison of Eq. (19.8) with Eq. (12.30) and particularly with Fig. (12.10) shows that the integral in Eq. (19.8) is the nuclear quadrupole moment *if* the nuclear spin points along the quantization axis, z. In general, however, the nuclear spin J will point in some other direction which will be described by the magnetic quantum number m defined in Fig. 19.4. With some work, the integral in Eq. (19.8) can then be expressed in terms of the nuclear quadrupole moment Q, defined in Eq. (12.30), and of J and m:

$$H_Q = \frac{1}{4} e V_{zz} Q \, \frac{3m^2 - J(J+1)}{J(2J-1)}. \tag{19.9}$$

The quadrupole interaction gives rise to a splitting of the nuclear energy levels with spins $J \geq 1$. The splitting is proportional to the nuclear quadrupole moment, Q, and the electric field gradient, V_{zz}. States m and $-m$ have the same energy.

Equations (5.21), (19.6), and (19.9) provide the background needed to describe the use of the hyperfine interaction to study chemical structure. We shall sketch a few applications which are useful not only in chemistry but in many other fields as well.

Nuclear Magnetic Resonance (NMR).[22] The salient features of a nuclear magnetic resonance (NMR) spectrometer are shown in Fig. 19.5. The sample is placed in a magnetic field, B_0, and surrounded by a coil that produces a small oscillating magnetic field of frequency $\omega = 2\pi f$. Consider, for simplicity, a proton in the sample. It has spin $\tfrac{1}{2}$ and a magnetic moment of $\mu = 2.79278 \, \mu_N$. The applied field B_0 splits the two levels $m = \pm\tfrac{1}{2}$ by an amount $2\mu B_0$. The lower level will be more populated,

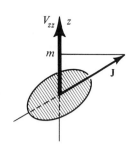

Fig. 19.4. The energy of a nucleus with quadrupole moment Q in an electric field gradient V_{zz} depends on the orientation of the spin, J, with respect to the direction of the field gradient V_{zz}.

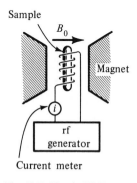

Fig. 19.5. Simple NMR spectrometer.

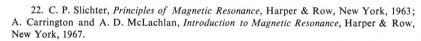
22. C. P. Slichter, *Principles of Magnetic Resonance*, Harper & Row, New York, 1963; A. Carrington and A. D. McLachlan, *Introduction to Magnetic Resonance*, Harper & Row, New York, 1967.

according to Eq. (9.31). If the external radio frequency satisfies the resonance condition

$$2 \mu B_o = \hbar \omega,$$

energy will be absorbed by the sample until both levels are equally populated, and this absorption can be seen as a change in the current i in Fig.
19.5. In practice, the rf is kept constant (for instance, at $f = 100$ MHz)
and the magnetic field B_o is changed. To see what can be learned from even
a simple NMR spectrum, consider the one of ethyl alcohol (CH_3–CH_2–
OH) in Fig. 19.6. Three features can be discerned: (1) The ratio of areas
under the three groups is $1:2:3$; it indicates the number of protons in each

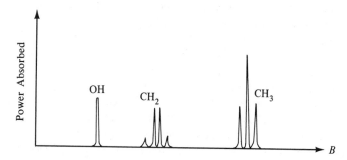

Fig. 19.6. Schematic appearance of the proton NMR spectrum of ethyl alcohol
(CH_3—CH_2—OH).

group. (2) The protons in the three groups are not equivalent; their energy
splittings are slightly different. The reason is the NMR chemical shift: The
external magnetic field induces electric currents in the molecule. These
currents in turn produce magnetic fields at the protons. In general, the
induced fields will be opposed to the external one and the resulting field
at the proton will be smaller than the applied one. The size of this effect
permits conclusions concerning the electronic surrounding. (3) Where two
or three protons are in the same group, the NMR line is split. This splitting is caused by the spin-spin interaction between the protons in the same
group; the magnetic field of one proton at the site of the other can either
add or subtract from the external one. In the situation shown in Fig. 19.6,
the magnetic dipole-interaction does not act directly between the protons
but is mediated by the electrons of the carbon atom that lies between the
hydrogen atoms.

In the spectrum shown in Fig. 19.6, the splitting of the various lines is
caused by a magnetic interaction and it is thus governed by Eq. (5.21). In
many molecules, the dominant splitting is due to electric field gradients:
p electrons can produce very large electric field gradients at a nuclear site.
The proton resonance is not sensitive to such field gradients, because the
proton has no quadrupole moment. Nuclides with quadrupole moments,
such as ^{35}Cl, ^{79}Br, and ^{127}I, experience a splitting of the nuclear ground

state due to the field gradients of the atomic and molecular electrons. The quadrupole-split spectrum can be explored in essentially the same way as the magnetically split one. The information so obtained, according to Eq. (19.9), provides a value for the product $V_{zz}Q$. If the quadrupole moment is known, V_{zz} is determined. From V_{zz} conclusions about the nature of the chemical bond and the symmetry of the site can be obtained.

So far we have discussed the splitting of NMR lines. *Relaxation phenomena*[22] are equally important for the application of NMR to chemistry. As we have described the principle of proton magnetic resonance, the power absorption by the sample would stop as soon as the two magnetic sublevels are equally populated. Such a situation corresponds to an infinitely high "spin temperature," as can be seen from Eq. (9.31). If the protons were isolated from their surrounding, they would remain in such a state for a long time. However, the temperature of the surrounding (the *lattice*) is finite, and the protons, therefore, try to reestablish thermal equilibrium by interaction with the surrounding. Said in another way, the interaction between the protons and the lattice will lower the *infinite* spin temperature and slightly raise the temperature of the lattice. The time characteristic of this spin-lattice relaxation can be measured in various ways, for instance by determining the time it takes the proton spins to return to thermal equilibrium after the external rf power has been turned off. The spin-lattice relaxation time depends, for instance, on the chemical binding, and relaxation techniques, therefore, have become indispensable in chemistry.

Mössbauer Effect.[23] Nuclear gamma transitions have been treated in Section 10.5; we return here again to this phenomenon to discuss an effect that was startling when discovered, namely recoilless gamma emission. Consider a specific example, the emission of the 14.4-keV gamma ray from the first excited state of ^{57}Fe. The main features of the decay scheme are shown in Fig. 19.7. The half-life of the first excited state has been de-

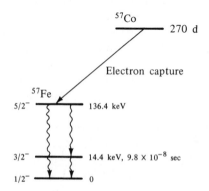

Fig. 19.7. Main features of the decay ^{57}Co $\longrightarrow$ ^{57}Fe.

23. R. H. Herber, *Sci. Amer.* **225**, 86 (Oct. 1971).

termined to be 9.8×10^{-8} sec. In Section 5.7, it was shown that the radiation from a level with finite half-life is not monenergetic but does have a natural line width, as sketched in Fig. 5.17. The width, Γ, is given by Eq. (5.45), and for the $(\frac{3}{2})^-$ level in ^{57}Fe, it is

$$\Gamma\left[\left(\frac{3}{2}\right)^-\right] = \frac{\hbar}{\tau} = \frac{\hbar \ln 2}{t_{1/2}} = 4.67 \times 10^{-9} \text{ eV}. \qquad (19.10)$$

The ratio of decay energy to width is enormous,

$$\frac{E}{\Gamma} = 3.2 \times 10^{12}.$$

Two questions arise: Do gamma rays really possess such a narrow natural line width? Can it be observed? Certainly, the energy-sensitive detectors of Chapter 4 are too crude to study line widths as small as they are here, and the first question cannot thus be answered by examining the gamma rays with conventional detectors. However, theory provides part of the answer to the first question: In most situations, nuclear gamma rays are emitted by nuclei that form part of an atomic system. The thermal motion of the system broadens the emitted gamma ray. Moreover, the gamma ray imparts a recoil to the emitting nucleus, and this recoil energy is lost from the gamma-ray energy. Consequently, the emitted gamma ray is very much broadened and is shifted downward in energy. However, in 1957, Mössbauer found gamma transitions with the full transition energy and with the natural line width.[24] To understand the occurrence of such recoilless transitions, we return for a moment to elastic electron scattering in Chapter 6.[25] In elastic scattering the nucleus does not take up internal excitation energy; the initial and the final nuclear state are identical. In analogy, we consider the gamma-ray emitting nucleus to be part of a solid. The solid is a quantum mechanical system and a recoilless transition occurs if the initial and the final state of the solid are identical. The solid, then, has not taken up internal energy, and consequently the gamma ray must have the full energy and also show the natural line width. The probability, f, for the emission of a gamma ray without recoil energy loss can be computed; at low temperatures it is given by

$$f = e^{-3R/2k\theta}. \qquad (19.11)$$

Here, θ is the Debye temperature of the solid, $k = 8.62 \times 10^{-5}$ $eV/$K is the Boltzmann constant, and R is the recoil energy of the free nucleus, given by

$$R = \frac{E_\gamma^2}{2Mc^2}. \qquad (19.12)$$

For the 14.1-keV transition in ^{57}Fe, R is only 1.9×10^{-3} eV. The Debye temperature of many solids is of the order of 200 K and $k\theta$ is about $1.7 \times$

24. R. L. Mössbauer, *Z. Physik* **151**, 124 (1958); *Naturwissenschaften* **45**, 538 (1958).
    25. A very simple discussion of the theory of Mössbauer effect is given by P. G. Debrunner and Hans Frauenfelder, *An Introduction to Mössbauer Spectroscopy* (L. May, ed.), Plenum, New York, 1971.

10^{-2} eV. The exponent in Eq. (19.11) then is small, f is close to one, and the 14.4-keV gamma rays from ^{57}Fe will normally be emitted without energy loss and with the natural line width. Why was this fact not observed before 1957? A partial answer has already been given above: No conventional detector has an energy resolution comparable to the natural line width. Mössbauer overcame this limitation by an ingenious approach, *resonance absorption*. Equation (12.7) expresses the fact that scattering at a resonance assumes a maximum value. Similarly, if a photon with energy E_0 impinges on an absorber that has an excited state at just this energy and if no energy is lost in absorption, the absorption cross section will assume its largest value. The elements of a Mössbauer spectrometer are shown in Fig. 19.8.

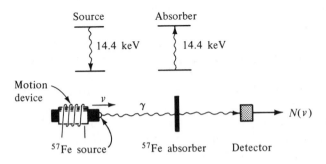

Fig. 19.8. Elements of a Mössbauer spectrometer.

Source and absorber contain the same nuclide, say ^{57}Fe. Assume for a moment that both nuclides are embedded in the same chemical surrounding; the nuclear energy levels are then identical in the source and the absorber. The emitted gamma ray will, if recoilless, show a natural line centered around the energy E_0 [see Eq. (5.44)]. The profile for recoilless absorption will be the same, and the gamma-ray beam, incident on the absorber, will be attenuated maximally when source and absorber are both at rest ($v = 0$). To study the shape of the lines, the source (or the absorber) is placed on a motion device, for instance an electromagnetic transducer, that can move with a constant velocity v. The energy of the emitted gamma-ray then is Doppler-shifted by an amount

$$\Delta E = \frac{v}{c} E_0, \tag{19.13}$$

and the gamma-ray intensity at the energy $E_0 + \Delta E$ can be measured. The intensity is then observed as a function of v. The result is shown in Fig. 19.9: The overlap of two Lorentzians with width Γ is again a Lorentzian, but with width 2Γ. A Mössbauer spectrometer thus permits the exploration of the line shape of the absorber if the line shape of the source is known.

The application of Mössbauer spectrometry to chemical problems can be understood with Fig. 19.10. Here, the energy levels of ^{57}Fe are shown for three different situations, for point nuclei, for actual nuclei in a sur-

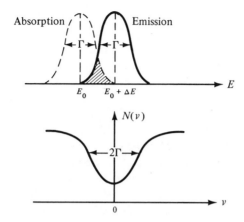

Fig. 19.9. Overlap of an emission and absorption line. $N(v)$ is the transmitted intensity as a function of the source velocity v.

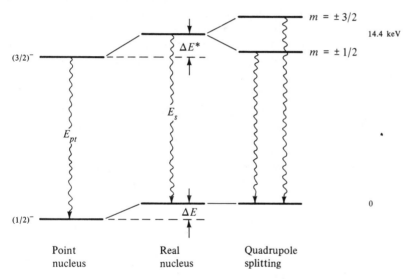

Fig. 19.10. Energy levels of ^{57}Fe in a point nucleus, a finite nucleus, and a finite nucleus in an external electric field gradient.

rounding without external electromagnetic field, and in one with electric field gradient. The shift of the energy levels from point to real nuclei is given by Eq. (19.6); the splitting of the excited state in the electric field gradient is determined by Eq. (19.9).

Consider, first, the quadrupole interaction. As Fig. 19.10 shows, the external electric field gradient splits the transition $\frac{3}{2}^{-} \longrightarrow \frac{1}{2}^{-}$ into a *quadrupole doublet*. This quadrupole doublet can be seen by using the quadrupole-split transitions in the absorber and determining the transmission of a single-line source. The energy difference between the two lines follows from

Eq. (19.9) as

$$\Delta E_Q = \tfrac{1}{2} e V_{zz} Q, \tag{19.14}$$

where Q is the quadrupole moment of the first excited state of ^{57}Fe. This quadrupole moment has been determined by some Mössbauer experiments; it is $Q(\tfrac{3}{2}) \approx +0.28 \times 10^{-24}$ cm^2. From the measured splitting ΔE_Q, the field gradient V_{zz} can be determined, and conclusions about the structure and electron configuration giving rise to V_{zz} can be drawn. This technique has found applications in solid-state physics, metallurgy, and many branches of chemistry. Here we select an example from *biochemistry* and show in Figs. 19.11 and 19.12 the Mössbauer spectra of oxidized and reduced putidaredoxin.[26] We first note the "energy" scale; it is given in mm/sec. These units are customary in Mössbauer spectrometry, and they can be understood with Eq. (19.13): The energy shift is proportional to the velocity v with which the source moves toward or away from the absorber. For the 14.4-keV transition in ^{57}Fe, the conversion is 1 mm/sec = 4.8×10^{-8} eV. Putidaredoxin is an iron-sulfur protein that acts as a one-electron transfer enzyme. Each molecule, with a molecular weight of 12,500 Daltons, contains two iron atoms. The Mössbauer spectrum in the oxidized form shows a quadrupole doublet. Since putidaredoxin has two iron atoms per molecule and only one quadrupole doublet is observed, it is concluded that the two iron atoms sit in equivalent sites. Upon reduction (adding one electron per molecule), the spectrum becomes rather complicated, as is evident from Fig. 19.12. A detailed analysis[26] gives the following results: The two iron atoms are now inequivalent. The spin of

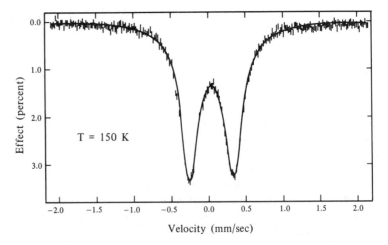

Fig. 19.11. Mössbauer spectrum of oxidized putidaredoxin. [From E. Münck et al., *Biochem.* **11**, 855 (1972). Copyright © 1972 by the American Chemical Society. Reprinted by permission of the copyright holder.]

26. E. Münck, P. G. Debrunner, J. C. M. Tsibris, and I. C. Gunsalus, *Biochem.* **11**, 855 (1972).

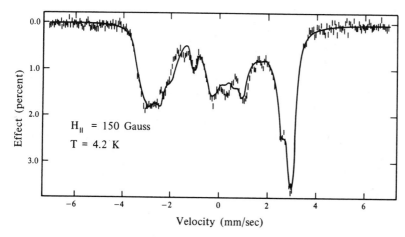

$H_{\parallel} = 150$ Gauss

$T = 4.2$ K

Fig. 19.12. Mössbauer spectrum of reduced putidaredoxin, measured in a magnetic field of 150 G, applied parallel to the γ rays. [From E. Münck et al., *Biochem.* **11**, 855 (1972). Copyright © 1972 by the American Chemical Society. Reprinted by permission of the copyright holder.]

one atom is $S = 2(Fe^{2+})$; the other is $S = 5/2(Fe^{3+})$. Both spins couple antiferromagnetically to produce an electronic ground state with $S = 1/2$. The spectrum of the reduced putidaredoxin thus consists of a superposition of two spectra, one from each iron. Each of the partial spectra displays magnetic and electric hyperfine structure, and the complete spectrum requires 21 parameters for a full description. These parameters can be obtained by combining the Mössbauer data, taken with various applied magnetic fields and at various temperatures, with results from other experiments.[26]

Next we turn to the shift caused by the finite nuclear size. From Fig. 19.10 and Eq. (19.16), the energy of the emitted gamma ray becomes

$$E_s = E_{pt} + \Delta E^* - \Delta E = E_{pt} + \frac{2\pi}{3} Ze^2 |\psi(0)|_s^2 \{\langle r^2 \rangle_e - \langle r^2 \rangle_g\},$$

(19.15)

where the subscript s denotes *source* and e and g refer to the excited and the ground state, respectively. A similar expression holds for the energy of the absorbed photon,

$$E_a = E_{pt} + \frac{2\pi}{3} Ze^2 |\psi(0)|_a^2 \{\langle r^2 \rangle_e - \langle r^2 \rangle_g\}.$$

(19.16)

The difference between source and absorber energy is

$$\delta = E_a - E_s = \frac{2\pi}{3} Ze^2 \{\langle r^2 \rangle_e - \langle r^2 \rangle_g\}[|\psi(0)|_a^2 - |\psi(0)|_s^2].$$

(19.17)

The quantity δ is called the *isomer* or *chemical shift*, relative to a given source of radiation. (This chemical shift is different from the one introduced in NMR; δ is an electric effect, whereas the NMR chemical shift is magnetic.) The isomer shift is a good tool in studying chemical compounds: Assume that we use a $^{57}\text{Co} \longrightarrow {}^{57}\text{Fe}$ source embedded in a "standard" material, and assume that we have already determined $\{\langle r^2 \rangle_e - \langle r^2 \rangle_g\}$. The quantity δ can then be measured for the compound to be investigated, and it yields a value for $|\psi(0)|_a^2$ for this compound with respect to the standard source. Since the value of $|\psi(0)|^2$ is determined by the s electrons, it is thus possible to get information about the s-electron density for the compound. ^{57}Co embedded in stainless steel is sometimes used as the standard source. More often, a sodium nitroprusside absorber serves as the standard. The information provided by the isomer shift is evident from Fig. 19.13.[27]

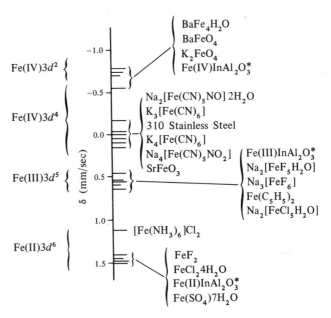

Fig. 19.13. Isomer shifts for iron compounds. [From J. Danon, *Mössbauer Spectroscopy and Its Application*, International Atomic Energy Agency, Vienna, 1972.]

Inner Shell Spectroscopy.[28] The inner electrons in an atom feel only a small influence from the valence electrons and hence from chemical binding and structure effects. For many years, they have therefore been largely ignored by the chemists. Improved spectroscopic techniques, borrowed from nuclear physics,[29] have changed the situation, and inner shell spectroscopy is rapidly becoming a powerful tool in chemistry. An inner shell spectrometer (often called an ESCA spectrometer, for electron

27. J. Danon, *Mössbauer Spectroscopy and Its Applications*, International Atomic Energy Agency, Vienna, 1972.
28. J. M. Hollander and D. A. Shirley, *Ann. Rev. Nucl. Sci.* **20**, 435 (1970).
29. S. B. M. Hagström, C. Nordling, and K. Siegbahn, *Z. Phys.* **178**, 439 (1964).

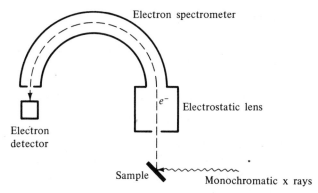

Fig. 19.14. Inner-shell spectrometer.

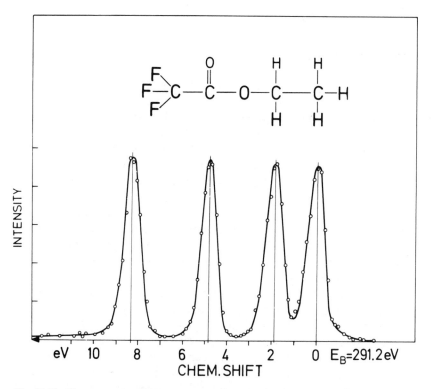

Fig. 19.15. Electron spectrum from carbon in ethyl trifluoroacetate. All four atoms in this molecule are distinguished in the spectrum. The lines appear in the same order from left to right as do the corresponding carbon atoms in the structure that has been drawn in the figure. [From K. Siegbahn, *Endeavour* **32**, No. 116, p. 51 (1973).]

Spectroscopy for Chemical Analysis) is shown schematically in Fig. 19.14.[30] Monochromatic X rays fall on the sample and knock out electrons. These are focused by an electrostatic lens system onto an electron

30. K. Siegbahn, D. Hammond, H. Fellner-Feldegg, and E. F. Barnett, *Science* **176**, 245 (1972).

spectrometer. The electron spectrometer is the crucial contribution from nuclear physics. It was developed to near perfection in the search for better beta spectroscopy.[31]

The energy of the ejected inner shell electron is given by

$$E_i = E_x - B,$$

where E_x is the energy of the incident X rays and B is the binding energy of the ejected electron. B depends on the chemical binding through shielding by the valence electrons. This shielding is proportional to the total electron population of the valence shell.

An example of the information that is provided by inner shell photo-electron spectroscopy is shown in Fig. 19.15.[32] The four carbon atoms in ethyl trifluoroacetate can be clearly distinguished.

19.4 Radiotracers

The radiations emitted by radionuclides make them excellent beacons, and the whereabouts of a certain element can be followed even in minute quantities by tracing the proper radiation. Radiotracers were first used by G. Hevesy in 1911, and the story is told by Glasstone:[33] "In order to determine whether his landlady, in spite of her denials, was incorporating the remains of the Sunday pie in meals served later in the week G. Hevesy made an experiment in 1911 which he later described as follows: The coming Sunday, in an unguarded moment, I added some active deposit of thorium to the freshly prepared pie, and on the following Wednesday, with the aid of an electroscope, I demonstrated to the landlady the presence of the active material in the souffle." This short description contains all the essential features of tracer techniques. Only three things have changed since 1911: The number of radioactive nuclides and their availability have increased enormously; the detection methods have improved in sensitivity, speed, and energy selectivity; and nobody can do such an experiment now without notifying the health physicist and filling out forms.

Radiotracers offer two advantages over other methods, *sensitivity* and *specificity*. The enormous sensitivity can be seen from the fact that element 101, Mv, was identified with about 20 atoms.[34] The specificity comes from the unique assignment possible through measurements of the emitted

31. K. Siegbahn, in *Alpha-, Beta- and Gamma-Ray Spectroscopy* (K. Siegbahn, ed.), North-Holland, Amsterdam, 1965.

32. K. Siegbahn et al., *ESCA—Atomic, Molecular and Solid State Structure Studied by Means of Electron Spectroscopy*. Nova Acta Regiae Societatis Scientiarum Upsaliensis Ser. IV, Vol. 20. Almqvist & Wiksell Uppsala, 1967. K. Siegbahn, *Endeavour*, **32**, No. 116, p. 51 (1973).

33. S. Glasstone, *Sourcebook on Atomic Energy*, Van Nostrand Rinehold, New York, 1967, p. 666.

34. A. Ghiorso, B. G. Harvey, G. R. Choppin, S. G. Thompson, and G. T. Seaborg, *Phys. Rev.* **98**, 1518 (1955).

radiations. Figure 4.8 shows that semiconductor detectors, with their high resolution and their accurate energy calibration, permit unambiguous assignment of most gamma rays.

Two assumptions are implicit in the use of radionuclides, the chemical identity of a radioisotope with the corresponding stable isotope and non-interference of the radioactive decay with the properties under investigation. The first assumption is incorrect only if mass effects are important; in practice such effects must be considered only for hydrogen-deuterium-tritium substitutions. The second assumption is correct as long as the tracer concentration is very small.

Many radionuclides are available, and they have been used in all branches of physics, in biology, in medicine, and in many industrial applications.[35] Since the idea underlying tracer studies is so simple, we shall not pursue it further.

19.5 Radiation Chemistry

In radiotracer studies, the nuclear radiation reveals the presence and location of the radionuclide. In radiation chemistry, the radiation induces changes in chemical properties. Among the enormous number of processes that have been studied, we pick two: radiation-induced chemical processing and pulse radiolysis. The first example is typical for the "applied" side of radiation chemistry, whereas the second one is "fundamental."

Radiation-induced chemical processing[36] implies the use of ionizing radiation for the synthesis or modification of a chemical product. This technique is widely used, from the routine sterilization of disposable medical supplies by gamma radiation to food sterilization.[37] We shall sketch here, as an example, the radiation cross linking of polyethylene.[38] Polyethylene, C_nH_{2n}, is a linear chain polymer with up to 2000 C atoms. Upon irradiation by neutrons, electrons, or gamma rays, molecules link by forming new C–C bonds at the expense of C–H bonds. The result is a polymer with a stronger lattice. Very likely, cross linking proceeds as follows (Fig. 19.16): The radiation produces a hydrogen atom and a polymer radical. The hydrogen atom removes another hydrogen from a nearby polymer chain and thus forms hydrogen gas and a second radical. Two radicals then combine to form a C–C bond. Cross-linked polyethylene finds a number of applications; it is, for instance, used for wire and cable coverings. The same technique is also applied to textiles to obtain fabrics with improved shrink resistance and water repellency. The most important

35. Y. Wang, ed., *Handbook of Radioactive Nuclides*, Chemical Rubber Co., Cleveland, 1969.

36. V. T. Stannett and E. P. Stahel, *Ann. Rev. Nucl. Sci.* **21**, 397 (1971).

37. F. E. Fowler, *Isotopes Radiation Technol.* **9**, 253 (1972).

38. A. Chapiro, *Radiation Chemistry of Polymeric Systems*, Wiley-Interscience, New York, 1962.

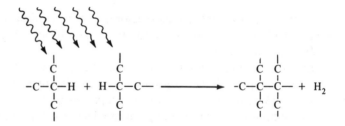

Fig. 19.16. Cross linking of polymer chains through radiation.

application may well be to waste management, based on the following consideration: Irradiated concrete-polymer combinations far surpass ordinary concrete in strength and durability. Similar combinations involving waste (crushed discarded bottles or sewage solids, for instance) may alleviate pollution and at the same time provide superior construction materials. In all irradiation technologies, powerful electron accelerators are the best radiation sources. It is interesting to note that another cross linking between basic and applied research has occurred here: The design of the Los Alamos linear proton accelerator has led to much improved low-energy electron linacs that are increasingly used in clinical and industrial work.

Pulse radiolysis[39] is one of the relatively new techniques that permit refined investigations of the kinetics and the reaction mechanisms in chemical and biochemical processes. A short (nsec to μsec) and very intense pulse of ionizing radiation, usually electrons, is directed into the sample under investigation. The detailed processes that follow such a pulse are complex, but the main features are reasonably well understood. In water, after about 10^{-11} sec, the main products are

$$H, \qquad OH, \qquad H_3O_{aq}^+, \qquad and \qquad e_{aq}^-.$$

All of these species can undergo further reactions, but of particular interest is the *hydrated electron*,[40] e_{aq}^-, because it is the simplest negative aqueous ion and is highly reactive. It has a standard potential of 2.7 V and thus is the most reactive species known. But what is a hydrated electron? Its constitution can be explained with reference to Fig. 19.17. The free electron, moving much faster than the heavier water molecules, polarizes some of these and digs itself a potential well. It then moves around, dragging water molecules along. Its *Bohr radius* is about 3 Å, and it has an absorption spectrum with a maximum at 700 nm. Since e_{aq}^- is so highly reactive, it can be used to initiate a large variety of reactions. Pulsed production of hydrated electrons thus permits kinetic studies down into the nsec range. A

39. M. S. Matheson and L. M. Dorfman, *Pulse Radiolysis*, M.I.T. Press, Cambridge, Mass., 1969.

40. E. J. Hart, *Accounts Chem. Res.* **2**, 161 (1969); E. J. Hart and M. Anbar, *The Hydrated Electron*, Wiley-Interscience, New York, 1970.

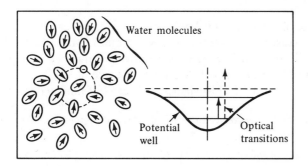

Fig. 19.17. The hydrated electron polarizes water molecules and moves around with its entourage. At right are the energy levels.

nuclear tool, namely a pulsed electron accelerator, thus allows rapid chemical studies with a remarkable entity, the hydrated electron.

19.6 References

Books and reviews on nuclear chemistry, with emphasis on radiochemistry, are listed and briefly described in "Source Material for Radiochemistry", *Nuclear Science Series Report Number 42*, National Academy of Sciences, Washington, D. C. 1970. The following two texts cover a large amount of material on nuclear and radiochemistry:

G. Friedlander, J. W. Kennedy, and J. M. Miller, *Nuclear and Radiochemistry*, 2nd ed., Wiley, New York, 1964.

M. Haissinsky, *Nuclear Chemistry and Its Applications*, Addison-Wesley, Reading, Mass., 1964.

In addition to the references already listed in the text, the following books and reviews give further information:

Synthetic Elements

G. T. Seaborg, *Man-Made Transuranium Elements*, Prentice-Hall, Englewood Cliffs, N.J., 1963.

G. T. Seaborg, *Ann. Rev. Nucl. Sci.* **18**, 53 (1968).

Chemical Structure. The entire field of hyperfine interactions is reviewed in A. J. Freeman and R. B. Frankel, eds., *Hyperfine Interactions*, Academic Press, New York, 1967. The number of publications on NMR exceeds 500 per month and new books appear every year. To keep up-to-date, consult the following serial publications:

Advances in Magnetic Resonance
Progress in Nuclear Magnetic Resonance Spectroscopy
Annual Review of NMR Spectroscopy

The earlier papers and books on the Mössbauer effect are listed in
G. K. Wertheim, "Resource Letter ME-1 on the Mössbauer Effect," *Am.
J. Phys.* **31**, 1(1963). The resource letter and a number of papers are re-
printed in *Mössbauer Effect*, American Institute of Physics, New York.

Introductions to the Mössbauer effect can be found in two small books:

G. K. Wertheim, *Mössbauer Effect, Principles and Applications*, Academic Press,
New York, 1964.

L. May, ed., *An Introduction to the Mössbauer Effect*, Plenum, N.Y., 1971.

Comprehensive treatments of the applications, particularly to chemistry, are given in

V. I. Goldanskii and R. H. Herber, eds., *Chemical Applications of Mössbauer
Spectroscopy*. Academic Press, New York, 1968.

N. N. Greenwood and T. C. Gibb, *Mössbauer Spectroscopy*, Chapman & Hall,
London, 1971.

The series *Mössbauer Effect Methodology*, (I. J. Gruverman, ed.),
Plenum, New York, provides many authoritative articles.

The Mössbauer literature is reviewed yearly in *Mössbauer Effect Data
Index* (J. G. Stevens and V. E. Stevens, eds.), Plenum, New York, 1972.

PART VII

Appendix Tables

Table A1 MOST FREQUENTLY USED CONSTANTS. The constants here are given only about to sliderule accuracy. For better values, use the following Tables.

Velocity of light	c	2.998×10^{23} fm/sec
Dirac's $\hbar$	$\hbar$	6.58×10^{-22} MeV sec
	$\hbar c$	197.3 MeV fm
Boltzmann constant	k	8.62×10^{-11} MeV/K
Fine structure constant	$e^2/\hbar c$	1/137.0
"Electron radius"	$e^2/m_e c^2$	2.818 fm
Compton wave length		
electron	$\hbar/m_e c$	386 fm
pion	$\hbar/m_\pi c$	1.414 fm ($\approx \sqrt{2}$ fm)
proton	$\hbar/m_p c$	0.210 fm
Nucleon magneton	$e\hbar/2m_p c$	3.153×10^{-18} MeV/Gauss
Masses: electron	m_e	0.511 MeV/c^2
pion, neutral	m_π^0	135.0 MeV/c^2
pion, charged	$m_\pi^\pm$	139.6 MeV/c^2
proton	m_p	938.3 MeV/c^2

Table A2 A More Complete Collection of Constants.
Physical and numerical constants.[1]

PHYSICAL CONSTANTS

N = $6.022169(40) \times 10^{23}$ mole^{-1} (based on $A_{C12} = 12$)

c = $2.9979250(10) \times 10^{10}$ cm sec^{-1}

e = $4.803250(21) \times 10^{-10}$ esu = $1.6021917(70) \times 10^{-19}$ coulomb

1 MeV = $1.6021917(70) \times 10^{-6}$ erg

$\hbar$ = $6.582183(22) \times 10^{-22}$ MeV sec = $1.0545919(80) \times 10^{-27}$ erg sec

$\hbar c$ = $1.9732891(66) \times 10^{-11}$ MeV cm = $197.32891(66)$ MeV fermi

   = $0.6240088(21)$ GeV mb$^{1/2}$

α = $e^2/\hbar c = 1/137.03602(21)$

$k_{Boltzmann}$ = $1.380622(59) \times 10^{-16}$ erg K^{-1}

   = $8.61708(37) \times 10^{-11}$ MeV K^{-1} = 1 eV/11604.85(49)K

m_e = $0.5110041(16)$ MeV = $9.109558(54) \times 10^{-31}$ kg

m_p = $938.2592(52)$ MeV = $1836.109(11) m_e = 6.72211(63) m_{\pi\pm}$

   = $1.00727661(8) m_1$ (where $m_1 = 1$ amu $= \frac{1}{12} m_{C12} = 931.4812(52)$ MeV)

m_d = $1875.587(10)$ MeV

r_e = $e^2/m_e c^2 = 2.817939(13)$ fermi (1 fermi = 10^{-13} cm)

λ_e = $\hbar/m_e c = r_e \alpha^{-1} = 3.861592(12) \times 10^{-11}$ cm

$a_{\infty Bohr}$ = $\hbar^2/m_e e^2 = r_e \alpha^{-2} = 0.52917715(81)$ A (1A = 10^{-8} cm)

$\sigma_{Thomson}$ = $\frac{8}{3} \pi r_e^2 = 0.6652453(61) \times 10^{-24}$ cm^2 = $0.6652453(61)$ barns

μ_{Bohr} = $e\hbar/2m_e c = 0.5788381(18) \times 10^{-14}$ MeV gauss^{-1}

$\mu_{nucleon}$ = $e\hbar/2m_p c = 3.152526(21) \times 10^{-18}$ MeV gauss^{-1}

$\frac{1}{2}\omega^e_{cyclotron}$ = $e/2m_e c = 8.794014(27) \times 10^6$ rad sec^{-1} gauss^{-1}

$\frac{1}{2}\omega^p_{cyclotron}$ = $e/2m_p c = 4.789484(27) \times 10^3$ rad sec^{-1} gauss^{-1}

Hydrogen-like atom (nonrelativistic, μ = reduced mass):

$$\frac{v}{c}\big)_{rms} = \frac{ze^2}{n\hbar c} ; \quad E_n = \frac{\mu}{2} v^2 = \frac{\mu z^2 e^4}{2(n\hbar)^2} ; \quad a_n = \frac{n^2 \hbar^2}{\mu z e^2}$$

$R_\infty = m_e e^4/2\hbar^2 = m_e c^2 \alpha^2/2 = 13.605826(45)$ eV (Rydberg)

$pc = 0.3 \, H\rho$(MeV, kilogauss, cm); 0.3 (which is $10^{-11}c$) enters because there are ≈ 300 "volts"/esu volt.

1 year (sidereal)   = 365.256 days = 3.1558×10^7 sec ($\approx \pi \times 10^7$ sec)

density of dry air   = 1.205 mg cm^{-3} (at 20°C, 760 mm)

acceleration by gravity   = 980.62 cm sec^{-2} (sea level, 45°)

gravitational constant   = $6.6732(31) \times 10^{-8}$ cm^3 g^{-1} sec^{-2}

1 calorie (thermochemical)   = 4.184 joules

1 atmosphere   = 1033.2275 g cm^{-2}

1 eV per particle   = 11604.85(49) °K (from E = kT)

NUMERICAL CONSTANTS

π	= 3.1415927	1 rad	= 57.2957795 deg	$\sqrt{\pi}$	=	1.7724539
e	= 2.7182818	1/e	= 0.3678794	$\sqrt{2}$	=	1.4142136
ln 2	= 0.6931472	ln 10	= 2.3025851	$\sqrt{3}$	=	1.7320508
$\log_{10} 2$	= 0.3010300	$\log_{10} e$	= 0.4342945	$\sqrt{10}$	=	3.1622777

1. From "Review of Particle Properties," *Rev. Mod. Phys.* **45**, (Supplement) S29 (1973). Compiled by S. J. Brodsky, based mainly on the adjustment of the fundamental physical constants by B. N. Taylor, W. H. Parker, and D. N. Langenberg, *Rev. Mod. Phys.* **41**, 375 (1969). The figures in parentheses correspond to the 1 standard deviation uncertainty in the last digits of the main number.

Table A3 PROPERTIES OF STABLE PARTICLES[2]

April 1973

N. Barash-Schmidt, A. Barbaro-Galtieri, C. Bricman, V. Chaloupka
R. L. Kelly, T. A. Lasinski, A. Rittenberg, M. Roos, A. H. Rosenfeld,
P. Söding, and T. G. Trippe

(Closing date for data: Feb. 1, 1973)

Stable Particle Table

For additional parameters, see Addendum to this table.

Quantities in italics have changed by more than one (old) standard deviation since April 1972.

Particle	$I^G(J^P)C_n$	Mass (MeV) Mass² (GeV)²	Mean life (sec) cτ (cm)	Partial decay mode		
				Mode	Fraction[a]	p or p_{max}[b] (MeV/c)
γ	$0,1(1^-)^-$	$0(<2)10^{-21}$	stable	stable		
ν	ν_e ν_μ $\;J=\tfrac{1}{2}$	$0(<60\text{ eV})$ $0(<1.2)$	stable	stable		
e	$J=\tfrac{1}{2}$	0.5110041 $\pm.0000016$	stable $(>2\times10^{21}\text{y})$	stable		
μ	$J=\tfrac{1}{2}$	105.6595 $\pm.0003$ $m^2=0.0112$ $m_\mu - m_{\pi^\pm} = -33.909$ $\pm.006$	2.1994×10^{-6} $\pm.0006\;S=1.1^*$ $c\tau=6.593\times10^4$	$e\nu\bar{\nu}$ $e\gamma\gamma$ $3e$ $e\gamma$	100 $(<1.6\;)10^{-5}$ $(<6\;)10^{-9}$ $(<2.2\;)10^{-8}$	53 53 53 53
$\pi^\pm$	$1^-(0^-)$	139.5688 $\pm.0064$ $m^2=0.0195$	2.6024×10^{-8} $\pm.0024$ $c\tau=780.2$ $(\tau^+-\tau^-)/\bar{\tau}=$ $(0.05\pm0.07)\%$ (test of CPT)	$\mu\nu$ $e\nu$ $\mu\nu\gamma$ $\pi^0 e\nu$ $e\nu\gamma$ $e\nu e^+e^-$	$100\;\%$ $^c(\;1.24\pm0.03)10^{-4}$ $(\;1.24\pm0.25)10^{-4}$ $(\;1.02\pm0.07)10^{-8}$ $^c(\;3.0\pm0.5\;)10^{-8}$ $(<3.4\;)10^{-8}$	30 70 30 5 70 70
π^0	$1^-(0^-)^+$	134.9645 $\pm.0074$ $m^2=0.0182$ $m_{\pi^\pm}-m_{\pi^0} = 4.6043$ $\pm.0037$	0.84×10^{-16} $\pm.10\;S=2.1^*$ $c\tau=2.5\times10^{-6}$	$\gamma\gamma$ γe^+e^- $\gamma\gamma\gamma$ $e^+e^-e^+e^-$	$(\;98.83\pm0.05)\%$ $(\;1.17\pm0.05)\%$ $(<5\;)10^{-6}$ $^d(\;3.47\;)10^{-5}$	67 67 67 67

2. From "Review of Particle Properties," *Rev. Mod. Phys.* **45**, (Supplement) S15 (1973).

Particle	$I^G(J^P)C_n$	Mass (MeV) / Mass^2 (GeV)^2	Mean life (sec) / cτ (cm)	Mode	Fraction^a	p or p_{max}^b (MeV/c)
$K^\pm$	$\frac{1}{2}(0^-)$	493.715 ±0.037 $m^2=0.244$ $m_{K^\pm}-m_{K^0}=-3.99$ ±0.13 S=1.1*	1.2371×10^{-8} $\pm.0026$ S=1.9* $c\tau=370.8$ $(\tau^+-\tau^-)/\bar\tau=$ $(.11\pm.09)\%$ (test of CPT) S=1.2*	$\mu\nu$	$(63.52\pm0.19)\%$	236
				$\pi\pi^0$	$(21.06\pm0.18)\%$ S=1.1*	205
				$\pi\pi^-\pi^+$	$(5.59\pm0.03)\%$ S=1.1	125
				$\pi\pi^0\pi^0$	$(1.73\pm0.05)\%$ S=1.4*	133
				$\mu\pi^0\nu$	$(3.24\pm0.10)\%$ S=1.9*	215
				$e\pi^0\nu$	$(4.85\pm0.06)\%$ S=1.1*	228
				$e\pi^0\pi^0\nu$	$(1.8^{+0.4}_{-0.6})\,10^{-5}$	207
				$\pi\pi^\mp e^\pm\nu$	$(3.7\pm0.2)\,10^{-5}$	203
				$\pi\pi^\pm e^\mp\nu$	$(<5)\,10^{-7}$	203
				$\pi\pi^\pm\mu^\pm\nu$	$(0.9\pm0.4)\,10^{-5}$	151
				$\pi\pi^\pm\mu^\mp\nu$	$(<3)\,10^{-6}$	151
				$e\nu$	$(1.38\pm0.20)\,10^{-5}$	247
				$e\nu\gamma$	$^c(<7)\,10^{-5}$	247
				$\pi\pi^0\gamma$	$^{h,c}(2.66\pm0.18)\,10^{-4}$	205
				$\pi\pi^+\pi^-\gamma$	$^c(10\pm4)\,10^{-5}$	125
				$\pi e\nu\gamma$	$^c(3.7\pm1.4)\,10^{-4}$	227
				πe^+e^-	$(<0.4)\,10^{-6}$	227
				$\pi^\mp e^\pm e^\pm$	$(<1.5)\,10^{-5}$	227
				$\pi\mu^+\mu^-$	$(<2.4)\,10^{-6}$	172
				$\pi\gamma\gamma$	$^c(<3.5)\,10^{-5}$	227
				$\pi\gamma\gamma\gamma$	$^c(<3)\,10^{-4}$	227
				$\pi\nu\bar\nu$	$(<1.4)\,10^{-6}$	227
				$\pi\gamma$	$(<4)\,10^{-6}$	227
				$\pi^\mp e^\pm\mu^\pm$	$(<3)\,10^{-8}$	214
				$\pi^\pm e^\mp\mu^\mp$	$(<1.4)\,10^{-8}$	214
				$\mu\nu\nu\nu$	$(<7)\,10^{-6}$	236
K^0	$\frac{1}{2}(0^-)$	497.71 ±0.13 S=1.1* $m^2=0.248$	50% K_{Short}, 50% K_{Long}			
K_S^0	$\frac{1}{2}(0^-)$		$^e0.882\times10^{-10}$ $\pm.008$ S=2.5* $c\tau=2.65$	$\pi^+\pi^-$	$(68.81\pm0.29)\%$ S=1.1*	206
				$\pi^0\pi^0$	$(31.19\pm0.29)\%$ S=1.1*	209
				$\mu^+\mu^-$	$(<0.7)\,10^{-5}$	225
				e^+e^-	$(<35)\,10^{-5}$	249
				$\pi^+\pi^-\gamma$	$^c(2.3\pm0.8)\,10^{-3}$	206
				$\gamma\gamma$	$(<0.7)\,10^{-3}$	249
K_L^0	$\frac{1}{2}(0^-)$	$m_{K_L}-m_{K_S}=0.5402\times10^{10}\,\hbar\,sec^{-1}$ ±0.0035	5.181×10^{-8} ±0.041 $c\tau=1553$	$\pi^0\pi^0\pi^0$	$(21.5\pm0.8)\%$ S=1.4*	139
				$\pi^+\pi^-\pi^0$	$(12.6\pm0.3)\%$	133
				$\pi\mu\nu$	$(26.9\pm0.6)\%$ S=1.1*	216
				$\pi e\nu$	$(38.8\pm0.6)\%$ S=1.1*	229
				$\pi e\nu\gamma$	$^c(1.3\pm0.8)\%$	229
				$\pi^+\pi^-$	$(0.157\pm0.005)\%$	206
				$\pi^0\pi^0$	$(0.094\pm0.019)\%$ S=1.5*	206
				$\pi^+\pi^-\gamma$	$^c(<0.4)\,10^{-3}$	206
				$\pi^0\gamma\gamma$	$(<2.4)\,10^{-4}$	231
				$\gamma\gamma$	$(4.9\pm0.4)\,10^{-4}$	249
				$e\mu$	$(<1.6)\,10^{-9}$	238
				$\mu^+\mu^-$	$^i(<1.9)\,10^{-9}$	225
				e^+e^-	$(<1.6)\,10^{-9}$	249
η	$0^+(0^-)^+$	548.8 ±0.6 S=1.4* $m^2=0.301$	$^e\Gamma=(2.63\pm0.58)keV$ Neutral decays 71.1% Charged decays 28.9%	$\gamma\gamma$	$(38.0\pm1.0)\%$ S=1.2*	274
				$\pi^0\gamma\gamma$	$^e(3.1\pm1.1)\%$ S=1.2*	258
				$3\pi^0$	$(30.0\pm1.1)\%$ S=1.1*	180
				$\pi^+\pi^-\pi^0$	$(23.9\pm0.6)\%$ S=1.1*	175
				$\pi^+\pi^-\gamma$	$(5.0\pm0.1)\%$	236
				$\pi^0e^+e^-$	$(<0.04)\%$	258
				$\pi^+\pi^-e^+e^-$	$(0.1\pm0.1)\%$	236
				$\pi^+\pi^-\pi^0\gamma$	$(<0.2)\%$	175
				$\pi^+\pi^-\gamma\gamma$	$(<0.2)\%$	236
				$\mu^+\mu^-$	$(2.2\pm0.8)\,10^{-5}$	253
				$\mu^+\mu^-\pi^0$	$(<5)\,10^{-4}$	211
p	$\frac{1}{2}(\frac{1}{2}^+)$	938.2592 ±0.0052 $m^2=0.8803$	stable $(>2\times10^{28}y)$			
n	$\frac{1}{2}(\frac{1}{2}^+)$	939.5527 ±0.0052 $m^2=0.8828$ $m_p-m_n=-1.29344$ ±0.00007	$(0.918\pm0.014)10^3$ $c\tau=2.75\times10^{13}$	$pe^-\nu$	100%	1

Particle	$I^G(J^P)C_n$	Mass (MeV) / Mass² (GeV)²	Mean life (sec) / $c\tau$ (cm)	Partial decay mode — Mode	Fraction[a]	p or p_{max}[b] (MeV/c)
Λ	$0(\frac{1}{2}^+)$	1115.59 ±0.05 S=1.1* m²=1.245	2.521×10⁻¹⁰ ±.021 S=1.2* $c\tau$=7.56	$p\pi^-$	(64.2±0.5)%	100
				$n\pi^0$	(35.8±0.5)%	104
				$pe\nu$	(8.13±0.29)10⁻⁴	163
				$p\mu\nu$	(1.57±0.35)10⁻⁴	131
				$p\pi^-\gamma$	c(0.85±0.14)10⁻³	100
Σ^+	$1(\frac{1}{2}^+)$	1189.41 ±0.07 S=1.6* m²=1.415 $m_{\Sigma^+}-m_{\Sigma^-}$=-7.94 ±.09 S=1.2	0.800×10⁻¹⁰ ±.006 $c\tau$=2.40 $\frac{\Gamma(\Sigma^+\to\ell^+n\nu)}{\Gamma(\Sigma^-\to\ell^-n\nu)}$<.035	$p\pi^0$	(51.6±0.7)%	189
				$n\pi^+$	(48.4±0.7)%	185
				$p\gamma$	c(1.24±0.18)10⁻³ S=1.4*	225
				$n\pi^+\gamma$	(1.31±0.24)10⁻⁴	185
				$\Lambda e^+\nu$	(2.02±0.47)10⁻⁵	72
				$n\mu^+\nu$	(<2.4)10⁻⁵	202
				$ne^+\nu$	(<1.0)10⁻⁵	224
				pe^+e^-	(<7)10⁻⁶	225
Σ^0	$1(\frac{1}{2}^+)$	1192.48 ±0.10 S=1.1* m²=1.422	<1.0×10⁻¹⁴ $c\tau$<3×10⁻⁴	$\Lambda\gamma$	100 %	74
				Λe^+e^-	d(5.45)10⁻³	74
Σ^-	$1(\frac{1}{2}^+)$	1197.34 ±0.07 S=1.2* m²=1.434 $m_{\Sigma^0}-m_{\Sigma^-}$=-4.86 ±.06	1.484×10⁻¹⁰ ±.019 S=1.6* $c\tau$=4.45	$n\pi^-$	100 %	193
				$ne^-\nu$	(1.10±0.05)10⁻³	230
				$n\mu^-\nu$	(0.45±0.04)10⁻³	210
				$\Lambda e^-\nu$	(0.60±0.06)10⁻⁴	79
				$n\pi^-\gamma$	c(1.0±0.2)10⁻⁴	193
Ξ^0	$\frac{1}{2}(\frac{1}{2}^+)^f$	1314.9 ±0.6 m²=1.729 $m_{\Xi^0}-m_{\Xi^-}$=-6.4 ±.6	2.98×10⁻¹⁰ ±.12 $c\tau$=8.93	$\Lambda\pi^0$	100 %	135
				$p\pi^-$	(<0.9)10⁻³	299
				$pe^-\nu$	(<1.3)10⁻³	323
				$\Sigma^+e^-\nu$	(<1.5)10⁻³	119
				$\Sigma^-e^+\nu$	(<1.5)10⁻³	112
				$\Sigma^+\mu^-\nu$	(<1.5)10⁻³	64
				$\Sigma^-\mu^+\nu$	(<1.5)10⁻³	49
				$p\mu^-\nu$	(<1.3)10⁻³	309
Ξ^-	$\frac{1}{2}(\frac{1}{2}^+)^f$	1321.29 ±0.14 m²=1.746	1.672×10⁻¹⁰ ±.032 S=1.1* $c\tau$=5.01	$\Lambda\pi^-$	100 %	139
				$\Lambda e^-\nu$	g(0.70±0.21)10⁻³	190
				$\Sigma^0e^-\nu$	(<0.5)10⁻³	123
				$\Lambda\mu^-\nu$	(<1.3)10⁻³	163
				$\Sigma^0\mu^-\nu$	(<0.5)%	70
				$n\pi^-$	(<1.1)10⁻³	303
				$ne^-\nu$	(<1.0)%	327
Ω^-	$0(\frac{3}{2}^+)^f$	1672.5±.5 m²=2.797	$1.3^{+0.4}_{-0.3}\times10^{-10}$ $c\tau$=3.9	$\Xi^0\pi^-$ $\Xi^-\pi^0$ ΛK^-	Total of 28 events seen	294 290 211

*S = Scale factor = $\sqrt{\chi^2/(N-1)}$, where N ≈ number of experiments. S should be ≈ 1. If S > 1, we have enlarged the error of the mean, δx, i.e., δx→Sδx. This convention is still inadequate, since if S >> 1, the experiments are probably inconsistent, and therefore the real uncertainty is probably even greater than Sδx. See text and ideogram in Stable Particle Data Card Listings.

a. Quoted upper limits correspond to a 90% confidence level.
b. In decays with more than two bodies, P_{max} is the maximum momentum that any particle can have.
c. See Stable Particle Data Card Listings for energy limits used in this measurement.
d. Theoretical value; see also Stable Particle Data Card Listings.
e. See note in Stable Particle Data Card Listings.
f. P for Ξ and J^P for Ω^- not yet measured. Values reported are SU(3) predictions.
g. Assumes rate for $\Xi^-\to\Sigma^0e^-\nu$ small compared with $\Xi^-\to\Lambda e^-\nu$.
h. The direct emission branching ratio is $(1.56\pm.35)\times10^{-5}$.
i. A contradictory unpublished result of ~9×10⁻⁹ (with 6 events seen) has been reported by Carithers et al. See note in Stable Particle Data Card Listings.

Table A3 ADDENDUM

	Magnetic moment				
e	$1.001\ 159\ 6577$ $\pm.000\ 000\ 0035$ $\dfrac{e\hbar}{2m_e c}$		μ **Decay parameters** [a]		
μ	$1.001\ 166\ 16$ $\pm.000\ 000\ 31$ $\dfrac{e\hbar}{2m_\mu c}$	$\rho = 0.752\pm0.003$ $\quad \eta = -0.12\ \pm0.21$ $\xi = 0.972\pm0.013$ $\quad \delta = 0.755\pm0.009$ $\quad h = 1.00\pm0.13$ $\|g_A/g_V\| = 0.86^{+0.33}_{-0.11}$ $\qquad \phi = 180°\ \pm15°$			

	Mode	Partial rate (sec^{-1})	$\Delta I = \tfrac{1}{2}$ rule for $K^\pm \to 3\pi$	Form factors for leptonic decays
$K^\pm$	$\mu\nu$ $\pi\pi^0$ $\pi^+\pi^-$ $\pi^0\pi^0$ $\mu\pi^0\nu$ $e\pi^0\nu$	$(51.35\pm0.19)10^6$ $\quad$ S=1.2* $(17.02\pm0.15)10^6$ $\quad$ S=1.1* $(4.52\pm0.02)10^6$ $\quad$ S=1.1* $(1.40\pm0.04)10^6$ $\quad$ S=1.4* $(2.62\pm0.08)10^6$ $\quad$ S=1.9* $(3.92\pm0.05)10^6$ $\quad$ S=1.1*	$\pi^+\pi^+\pi^-$ $^c g = -.214\pm.005$ $\quad$ S=1.7* $\pi^-\pi^+\pi^-$ $^c g = -.214\pm.007$ $\quad$ S=2.7* $\pi^+\pi^0\pi^0$ $^c g = .523\pm.023$ $\quad$ S=1.4* See also Stable Particle Data Card Listings and Appendix I	$\lambda^e_+ = 0.028\pm0.005$ See Stable Particle Data Card Listings for ξ and λ^μ_+.
K^0_S	$\pi^+\pi^-$ $\pi^0\pi^0$ '	$(0.780\pm.008)10^{10}$ $\quad$ S=1.9* $(0.353\pm.005)10^{10}$ $\quad$ S=1.4*	**CP violation parameters** $\|\eta_{+-}\| = (1.98\pm0.04)10^{-3}, \phi_{+-} = (42\pm3)°$ $\quad$ S=1.1* $\|\eta_{00}\| = (2.09\pm0.10)10^{-3}$ $\phi_{00} = (43\pm19)°$ $\quad$ S=1.2* $^d \delta = (.33\pm.04)10^{-2}$ $\quad$ S=1.5* $^f y^2 < 0.27$	$\Delta I = \tfrac{1}{2}$ rule for $K^0_L \to 3\pi$ $\pi^+\pi^-\pi^0$ $^c g = .60\pm.02$ S=2.7* See Data Cards & App. I $\Delta S = -\Delta Q$ Re x = -.003±.027 S=1.6* Im x = -.005±.038 S=1.2* Form Factors for leptonic decays $\lambda^e_+ = 0.025\pm0.005$ S=1.3* See Stable Particle Data Card Listings for λ^μ_+ and ξ
K^0_L	$\pi^0\pi^0\pi^0$ $\pi^+\pi^-\pi^0$ $\pi\mu\nu$ $\pi e\nu$ $\pi^+\pi^-$ $\pi^0\pi^0$	$(4.15\pm0.16)10^6$ $\quad$ S=1.3* $(2.43\pm0.05)10^6$ $(5.19\pm0.12)10^6$ $\quad$ S=1.1* $(7.48\pm0.13)10^6$ $\quad$ S=1.1* $(3.02\pm0.10)10^4$ $(1.82\pm0.38)10^4$ $\quad$ S=1.5*		

	Mode	Asymmetry parameter					
η	$\pi^+\pi^-\pi^0$ $\pi^+\pi^-\gamma$	$^e(0.24\pm.40)\%$ $\quad$ S=2.0* $(0.61\pm.54)\%$					

	Magnetic moment $(e\hbar/2m_p c)$	Decay parameters [b]				g_A/g_V [b]	g_V/g_A [b]
		Measured		Derived			
		α	ϕ(degree)	γ	Δ(degree)		
p	2.792782 $\pm.000017$						
n	-1.913148 $\pm.000066$	$pe^-\nu$				$-1.248\pm.010$ $\delta = (181.1\pm1.3)°$	
Λ	-0.67 $\pm.06$	$p\pi^-$ $n\pi^0$ $pe\nu$	0.647 ± 0.013 0.651 ± 0.045	$(-6.5\pm3.5)°$	0.76	$\left(7.6^{+4.0}_{-4.1}\right)°$ -0.66 ± 0.06 $\quad$ S=1.2*	
Σ^+	2.59 $\pm.46$	$p\pi^0$ $n\pi^+$ $p\gamma$	-0.984 ± 0.017 $+0.066\pm0.016$ $-1.03^{+.52}_{-.42}$	$(22\pm90)°$ $(167\pm20)°$ S=1.1*	0.17 -0.97	$\left(184\pm15\right)°$ $\left(-73^{+136}_{-10}\right)°$	
Σ^-		$n\pi^-$ $ne^-\nu$ $\Lambda e^-\nu$	-0.069 ± 0.008	$(10\pm15)°$	0.98	$\left(249^{+12}_{-115}\right)°$ See Data Cds. 0.37 ± 0.20	
Ξ^0	$-$	$\Lambda\pi^0$	-0.39 ± 0.09 S=1.2*	$(25\pm21)°$ S=1.3*	0.84	$\left(225^{+16}_{-35}\right)°$	
Ξ^-	-1.93 $\pm.75$	$\Lambda\pi^-$	-0.40 ± 0.03	$(-4\pm8)°$ S=1.1*	0.91	$\left(170^{+18}_{-17}\right)°$	

Table A3 ADDENDUM

a. $|g_A/g_V|$ defined by

$$g_V^2 = |C_V|^2 + |C'_V|^2,$$

$$g_A^2 = |C_A|^2 + |C'_A|^2,$$

$$\Sigma \langle \bar{e}|\Gamma_i|\mu \rangle \langle \bar{\nu}|\Gamma_i(C_i + C'_i \gamma_5)|\nu \rangle ;$$

ϕ defined by $\cos \phi = -\,R_e(C_A^* C'_V + C'_A C_V^*)/g_A g_V$ [for more details, see text Section IV E]

b. The definition of these quantities is as follows [for more details on sign convention, see text Section IV H]:

$$\alpha = \frac{2|s||p|\cos \Delta}{|s|^2 + |p|^2} ;$$

$$\beta = \frac{-2|s||p|\sin \Delta}{|s|^2 + |p|^2} .$$

$$\beta = \sqrt{1-\alpha^2} \sin \phi;$$

$$\gamma = \sqrt{1-\alpha^2} \cos \phi .$$

g_A/g_V defined by $\langle B_f |\gamma_\lambda(g_V - g_A \gamma_5)|B_i \rangle$;

δ defined by $g_A/g_V = |g_A/g_V|e^{i\delta}$.

c. The definition of the slope parameter of the Dalitz plot is as follows:

$$|M|^2 = 1 + g \left(\frac{s_3 - s_0}{m_{\pi^+}^2} \right).$$

d. The definition for the charge asymmetry is as follows:

$$\delta = \frac{\Gamma(K_L^0 \to \ell^+) - \Gamma(K_L^0 \to \ell^-)}{\Gamma(K_L^0 \to \ell^+) + \Gamma(K_L^0 \to \ell^-)}$$

e. See note in Stable Particle Data Card Listings.

f. The quantity y^2 is defined as follows:

$$y^2 = \frac{\Gamma(K_S^0 \to \pi^+\pi^-\pi^0)}{\Gamma(K_L^0 \to \pi^+\pi^-\pi^0)}$$

where CPT is assumed valid.

April 1973

In addition to the entries in the Meson Table, the Meson Data Card Listings contain all substantial claims for meson resonances. See Contents of Meson Data Card Listings[1].

Quantities in italics have changed by more than one (old) standard deviation since April 1972.

Name $\frac{G \ \frac{I \ 0 \ 1}{-\ \omega/\phi \ \pi}}{+\ \eta \ \rho}$ $I^G(J^P)C_n$ ⊢estab.	Mass M (MeV)	Full Width Γ (MeV)	$M^2 \pm \Gamma M$[a] $(GeV)^2$	Mode	Partial decay mode Fraction (%) [Upper limits are 1σ (%)]	p or p_{max}[b] (MeV/c)
$\pi^\pm(140)$ $\quad 1^-(0^-)+$	139.57	0.0	0.019483		See Stable Particle Table	
$\pi^0(135)$	134.96	7.8 eV ±.9 eV	0.018217			
$\eta(549)$ $\quad 0^+(0^-)+$	548.8 ±0.6	2.63 keV ±.58 keV	0.301 ±.000	All neutral $\pi^+\pi^-\pi^0 + \pi^+\pi^-\gamma$	71 See Stable 29 Particle Table	
ϵ $\quad 0^+(0^+)+$	≲ 700[c]	≳ 600[c]		$\pi\pi$		
Existence of pole not established. See note on $\pi\pi$ S wave¶.						
$\rho(770)$ $\quad 1^+(1^-)-$	770§ ±5§	146§ ±10§	0.593 ±.112	$\pi\pi$ e^+e^- $\mu^+\mu^-$ For upper limits, see footnote (e)	≈100 0.0043±.0005 (d) 0.0067±.0012 (d)	359 385 370
$\omega(784)$ $\quad 0^-(1^-)-$	783.8[f] ±0.3 S=1.3*	9.8 ±.5 S=1.1*	0.614 ±.008	$\pi^+\pi^-\pi^0$ $\pi^+\pi^-$ $\pi^0\gamma$ e^+e^- For upper limits, see footnote (g)	89.6±0.6 S=1.1* 1.3±0.3 S=1.5* 9.1±0.5 0.0076±.0017 S=1.9*	328 366 380 392
$\eta'(958)$ or X^0 $\quad 0^+(0^-)+$	958.1 ±0.4 S=1.4*	< 2	0.918 <.002	$\eta\pi\pi$ $\pi^+\pi^-\gamma$ (mainly $\rho^0\gamma$) $\gamma\gamma$ For upper limits, see footnote (h)	*71.8±3.9* S=2.0* *28.2±3.5* S=2.2* 1.9±0.3	234 458 479
$\delta(970)$ $\quad 1^-(0^+)+$ formerly called $\pi_N(975)$	∿ 970	50§ ±30§	0.941 ±.049	$\eta\pi$		311
Possibly a virtual bound state of the I = 1 $K\bar{K}$ system¶.						
S^* $\quad 0^+(0^+)+$	∿ 997[c]	50-150[c]	0.993	$\pi\pi$ $K\bar{K}$	near threshold	479
See notes on $\pi\pi$ and $K\bar{K}$ S wave¶.						
$\Phi(1019)$ $\quad 0^-(1^-)-$	1019.6 ±0.3 S=1.9	4.2 ±.2	1.040 ±.004	K^+K^- K_LK_S $\pi^+\pi^-\pi^0$ (incl. $\rho\pi$) $\eta\gamma$ e^+e^- $\mu^+\mu^-$ For upper limits, see footnote (i)	*46.8±2.7* S=1.6 *35.0±2.8* S=1.6* *15.2±3.6* S=1.8* 3.0±1.1 S=1.6* .032±.003 S=1.9* .025±.003	127 110 462 362 510 499
$A_1(1100)$ $\quad 1^-(1^+)+$	∿ 1100	200-400	1.21	$\rho\pi$	∿ 100	253
Broad enhancement in the $J^P=1^+$ $\rho\pi$ partial wave; not a Breit-Wigner resonance¶.						
$B(1235)$ $\quad 1^+(1^+)-$	1237§ ±10§	120§ ±20§	1.53 ±.12	$\omega\pi$ For upper limits, see footnote (j)	only mode seen	351

3. From "Review of Particle Properties," *Rev. Mod. Phys.* **45**, (Supplement) S20 (1973).

Table A4 (cont.)

Name $I^G(J^P)C_n$ estab.	Mass M (MeV)	Full Width Γ (MeV)	M^2 $\pm\Gamma M$ [a] $(GeV)^2$	Mode	Fraction (%) [Upper limits are 1σ (%)]	p or P_{max} [b] (MeV/c)
f(1270) $\underline{0^+(2^+)}+$	$1270_\S$ $\pm 5^\S$	$163_\S$ $\pm 15^\S$	1.61 $\pm.21$	$\pi\pi$ $2\pi^+2\pi^-$ $K\bar{K}$	~ 80 $5\pm2\S$ $5\pm3\S$	619 556 394
D(1285) $\underline{0^+(A)}\,_\mp^\pm$	$1286_\S$ $\pm 10^\S$	$30_\S$ $\pm 20^\S$	1.65 $\pm.03$	$K\bar{K}\pi$ $\eta\pi\pi$ $+[\delta(970)\pi$ $2\pi^+2\pi^-$ (prob. $\rho^0\pi^+\pi^-$)	seen seen seen] seen	305 484 250 565
$J^P = 0^-, 1^+, 2^-$, with 1^+ favoured						
A_2(1310) $\underline{1^-(2^+)}+$	$1310_\S$ $\pm 10^\S$	$100_\S$ $\pm 10^\S$	1.72 $\pm.13$	$\rho\pi$ $\eta\pi$ $\omega\pi\pi$ $K\bar{K}$ $\eta'(958)\pi$	72.4 ± 2.1 15.3 ± 1.3 7.6 ± 2.2 4.7 ± 0.6 <1	413 529 353 428 279
E(1420) $\underline{0^+(A)}\,_\mp^\pm$	$1416_\S$ $\pm 10^\S$	$60_\S$ $\pm 20^\S$	2.01 $\pm.08$	$K\bar{K}\pi$ $+[K^*\bar{K} + \bar{K}^*K$ $\eta\pi\pi$ $+[\delta(970)\pi$	~ 40 $\sim 20]$ ~ 60 possibly seen]	421 131 564 356
f'(1514) $\underline{0^+(2^+)}+$	1516 ± 3	40 ± 10	2.29 $\pm.06$	$K\bar{K}$	only mode seen	572
				For upper limits, see footnote (k)		
F_1(1540) $\underline{1}\,(A)$	1540 ± 5	40 ± 15	2.37 $\pm.06$	$K^*\bar{K} + \bar{K}^*K$	only mode seen	321
Evidence based on only one experiment						
ρ'(1600) $\underline{1^+(1^-)}-$	~ 1600	~ 500	2.56	4π $+[\rho\pi\pi$ $\pi\pi$	only mode seen ~ 80 < 1 (p)¶	575 788 629
Resonance interpretation uncertain.				For upper limits, see footnote (p)		
A_3(1640) $\underline{1^-(2^-)}+$	~ 1645	100-400	2.71	$f\pi$	~ 100	310
Broad enhancement in the $J^P = 2^-$ $f\pi$ partial wave; not a Breit-Wigner resonance.¶						
ω(1675) $\underline{0^-(N)}\,_-$ formerly called ϕ(1675)	1664 ± 13 S=1.2*	141 ± 17	2.77 $\pm.23$	$\rho\pi$ 3π 5π	dominant possibly observed 10 ± 10	645 804 777
g(1680) $\underline{1^+(3^-)}-$	$1680_\S$ $\pm 20^\S$	$160_\S$ $\pm 30^\S$	2.82 $\pm.27$	2π 4π (incl. $\pi\pi\rho,\rho\rho,A_2\pi,\omega\pi$) $K\bar{K}$ $K\bar{K}\pi$ (incl. $K^*\bar{K}$)	~ 40 ~ 50 (ℓ) ~ 3 ~ 3	828 781 677 617
J^P, M and Γ from the 2π mode[ℓ].						
See note (1) for possible heavier states.						
K^+(494) $1/2(0^-)$ K^0(498)	493.71 497.71		0.244 0.248	See Stable Particle Table		
K^*(892) $1/2(1^-)$	891.7 ± 0.5	50.1 ± 1.1	0.795 $\pm.045$	$K\pi$ $K\pi\pi$	≈ 100 < 0.2 < 0.16	288 216 309
	(Charged mode; $m^0 - m^\pm = 6.1\pm1.5$ MeV)					

Name $\frac{G \mid \frac{I}{0} \mid 1}{- \mid \omega/\phi \mid \pi}$ $\frac{}{+ \mid \eta \mid \rho}$	$I^G(J^P)C_n$ estab.	Mass M (MeV)	Full Width Γ (MeV)	M^2 $\pm\Gamma M^{(a)}$ $(GeV)^2$	Mode	Fraction (%) [Upper limits are 1σ (%)]	p or $P_{max}^{(b)}$ (MeV/c)
κ See note on Kπ S wave¶.	$1/2(0^+)$				δ_0^1 is near 90°, with slow variation, in mass region 1200-1400 MeV.		
Q $\left\{ \begin{array}{l} K_A(1240)1/2(1^+) \text{ or C} \\ \text{seen in } \bar{p}p \text{ at rest} \\ \hline K_A(1280\ 1/2(1^+) \text{ to 1400)} \\ \text{See note (m).} \end{array} \right.$		1242 ±10 1280 to 1400	127 ±25	1.54 ±.16	Kππ †[K*π †[Kρ †[K(ππ)$_{\ell=0}$	only mode seen large] seen] possibly seen]	
$K_N(1420)$	$1/2(2^+)$	1421_6 ±5§	100_6 ±10§ See note (n).	2.02 ±.14	Kπ K*π Kρ Kω Kη	55.0±3.3 S=1.2* 29.5±2.7 9.2±2.9 S=1.2* 4.4±1.7 2.0±1.8	616 415 319 304 482
L(1770)	$1/2(A)$ $J^P=2^-$ favoured, 1^+ and 3^+ not excluded.	1765_6 ±10§	140_6 ±50§	3.11 ±.25	Kππ Kπππ †[K_N(1420)π and other subreactions¶]	dominant seen	788 757

See note (1) for possible heavier states.

(1) Contents of Meson Data Card Listings

	Non-strange (Y = 0)					Strange (|Y| = 1)	
entry	$I^G(J^P)C_n$	entry	$I^G(J^P)C_n$	entry	$I^G(J^P)C_n$	entry	$I\ (J^P)$
π (140)	$1^-(0^-)+$	→ η_N (1080)	$0^+(N\ )+$	A₃ (1640)	$1^-(2^-)+$	K (494)	$1/2(0^-)$
η (549)	$0^+(0^-)+$	A₁ (1100)	$1^-(1^+)+$	ω (1675)	$0^-(N\ )-$	K* (892)	$1/2(1^-)$
ε (600)	$0^+(0^+)+$	→ M (1150)		g (1680)	$1^+(3^-)-$	κ	$1/2(0^+)$
ρ (770)	$1^+(1^-)-$	→ $A_{1.s}$(1170)	1^-	↳ X (1690)	$-$	→ K_A(1175)	$3/2$
ω (784)	$0^-(1^-)-$	B (1235)	$1^+(1^+)-$	→ X (1795)	1	→ K_A(1265)	$3/2$
→ M (940)	$+$	F (1270)	$0^+(2^+)+$	→ η/ρ(1830)	$+$	Q	$1/2(1^+)$
→ M (953)		D (1285)	$0^+(A\ )+$	→ ω/π(1830)	$-$	K_N(1420)	$1/2(2^+)$
η' (958)	$0^+(0^-)+$	A₂ (1310)	$1^-(2^+)+$	→ S (1930)		→ K_N(1660)	$1/2$
δ (970)	$1^-(0^+)+$	E (1420)	$0^+(A\ )+$	→ ρ (2100)	1^+	→ K_N(1760)	$1/2$
→ H (990)	$0^-(A\ )-$	→ X (1430)	0	→ T (2200)	1	L (1770)	$1/2(A\ )$
S*(1000)	$0^+(0^+)+$	→ X (1440)	1	→ ρ (2275)	1^+	→ K_N(1850)	
φ (1019)	$0^-(1^-)-$	f' (1514)	$0^+(2^+)+$	→ U (2360)	1	→ K*(2200)	
→ M (1033)		F₁ (1540)	1 (A)	→ N$\bar{\text{N}}$ (2375)	0	→ K*(2800)	
→ B₁(1040)	1^+	ρ' (1600)	$1^+(1^-)-$	→ X(2500-3600)			

Table A4 (cont.)

→   indicates an entry in Meson Data Card Listings not entered in the Meson Table. We do not regard these as established resonances.

¶   See Meson Data Card Listings.

*   Quoted error includes scale factor $S = \sqrt{\chi^2/(N-1)}$. See footnote to Stable Particle Table.

†   Square brackets indicate a subreaction of the previous (unbracketed) decay mode(s).

§   This is only an educated guess; the error given is larger than the error of the average of the published values. (See Meson Data Card Listings for the latter.)

(a) ΓM is approximately the half-width of the resonance when plotted against M^2.

(b)   For decay modes into ≥ 3 particles, p_{max} is the maximum momentum that any of the particles in the final state can have. The momenta have been calculated by using the averaged central mass values, without taking into account the widths of the resonances.

(c)   From pole position $(M - i\Gamma/2)$. For both ϵ and S^* the pole is on Riemann Sheet 2.

(d)   The e^+e^- branching ratio is from $e^+e^- \to \pi^+\pi^-$ experiments only. The $\omega\rho$ interference is then due to $\omega\rho$ mixing only, and is expected to be small. See note in Meson Data Card Listings. The $\mu^+\mu^-$ branching ratio is compiled from 3 experiments; each possibly with substantial $\omega\rho$ interference. The error reflects this uncertainty; see notes in Meson Data Card Listings. If $e\mu$ universality holds, $\Gamma(\rho^0 \to \mu^+\mu^-) = \Gamma(\rho^0 \to e^+e^-)$ × phase space correction.

(e)   Empirical limits on fractions for other decay modes of $\rho(765)$ are $\pi^\pm\gamma < 0.5\%$, $\pi^\pm\eta < 0.8\%$, $\pi^+\pi^+\pi^-\pi^- < 0.15\%$, $\pi^\pm\pi^+\pi^-\pi^0 < 0.2\%$.

(f)   Note that experiments with final state $K_S K_S \omega$ ($\bar{p}p$ at rest) give $M_\omega = 780.6 \pm 0.5$¶.

(g)   Empirical limits on fractions for other decay modes of $\omega(784)$ are $\pi^+\pi^-\gamma < 5\%$, $\pi^0\pi^0\gamma < 1\%$, η + neutral(s) < 1.5%, $\mu^+\mu^- < 0.02\%$, $\pi^0\mu^+\mu^- < 0.2\%$, $\eta\gamma < 0.5\%$.

(h)   Empirical limits on fractions for other decay modes of $\eta'(958)$: $\pi^+\pi^- < 2\%$, $\pi^+\pi^-\pi^0 < 5\%$, $\pi^+\pi^+\pi^-\pi^- < 1\%$, $\pi^+\pi^+\pi^-\pi^-\pi^0 < 1\%$, $6\pi < 1\%$, $\pi^+\pi^-e^+e^- < 0.6\%$, $\pi^0e^+e^- < 1.3\%$, $\eta e^+e^- < 1.1\%$, $\pi^0\rho^0 < 4\%$, $\pi^0\omega < 8\%$.

(i)   Empirical limits on fractions for other decay modes of $\phi(1019)$ are $\pi^+\pi^- < 0.03\%$, $\pi^+\pi^-\gamma < 4\%$, $\omega\gamma < 5\%$, $\rho\gamma < 2\%$, $\pi^0\gamma < 0.35\%$, $2\pi^+2\pi^-\pi^0 < 9\%$.

(j)   Empirical limits on fractions for other decay modes of $B(1235)$: $\pi\pi < 15\%$, $K\bar{K} < 2\%$, $4\pi < 50\%$, $\phi\pi < 1.5\%$, $\eta\pi < 25\%$, $(\bar{K}K)^\pm\pi^0 < 8\%$, $K_S K_S \pi^\pm < 2\%$, $K_S K_L \pi^\pm < 6\%$.

(k)   Empirical limits on fractions for other decay modes of $f'(1514)$ are $\pi^+\pi^- < 20\%$, $\eta\eta < 50\%$, $\eta\pi\pi < 30\%$, $K\bar{K}\pi + K^*\bar{K} < 35\%$, $2\pi^+2\pi^- < 32\%$.

(ℓ)   We assume as a working hypothesis that peaks with $I^G = 1^+$ observed around 1.7 GeV all come from g(1680). For indications to the contrary see Meson Data Card Listings.

(m)   See Q-region note in Meson Data Card Listings. Some investigators see a broad enhancement in mass (K$\pi\pi$) from 1250-1400 MeV (the Q region), and others see structure. The Kη, Kω, and Kπ are less than a few percent.

(n)   The tabulated mass of 1421 MeV comes only from charged $K_N(1420) \to K\pi$ measurements; the average of the neutral $K_N(1420)$ mass is 1423 MeV. K$\pi\pi$ mode can be contaminated with diffractively produced $Q^\pm$.

(o)   Empirical limits on fractions for other decay modes of $f(1270)$ are $\eta\eta < 15\%$; $K^0\bar{K}^-\pi^+$ + c.c. < 6%.

(p)   The tiny partial width for $\rho' \to \pi\pi$ ($\Gamma < 2$ MeV) is based on an OPE model.¶ Empirical limits are $\pi\pi < 20\%$, $K\bar{K} < 8\%$.

Established Nonets, and octet-singlet mixing angles from Appendix IIB, Eq. (2'). Of the two isosinglets, the "mainly octet" one is written first, followed by a semi-colon.

$(J^P)C_n$	Nonet members	$\theta_{lin.}$	$\theta_{quadr.}$
$(0^-)+$	π, K, η; η'	$24 \pm 1°$	$10 \pm 1°$
$(1^-)-$	ρ, K^*, ϕ; ω	$36 \pm 1°$	$39 \pm 1°$
$(2^+)+$	A_2, $K_N(1420)$, f'; f	$29 \pm 2°$	$31 \pm 2°$

Table A5[4] STABLE AND UNSTABLE BARYONS

April 1973

Baryon States for which information can be found in the Data Card Listings. The name, the mass, the quantum numbers, and the status are shown. Those states with four or three stars can be found in the following Table, the others have been omitted because the evidence for the existence of the effect and/or for its interpretation as a resonance is open to considerable question.

N(940)	P11	****	Δ(1236)	P33	****	Λ(1115)	P01	****	Σ(1190)	P11	****	Ξ(1320)	P11	****
N(1470)	P11	****	Δ(1650)	S31	****	Λ(1330)		Dead	Σ(1385)	P13	****	Ξ(1530)	P13	****
N(1520)	D13	****	Δ(1670)	D33	***	Λ(1405)	S01	****	Σ(1440)	PE	Dead	Ξ(1630)		**
N(1535)	S11	****	Δ(1690)	P33	*	Λ(1520)	D03	****	Σ(1480)	PE	*	Ξ(1820)		***
N(1670)	D15	****	Δ(1890)	F35	***	Λ(1670)	S01	****	Σ(1620)	S11	**	Ξ(1940)		***
N(1688)	F15	****	Δ(1910)	P31	***	Λ(1690)	D03	****	Σ(1620)	P11	**	Ξ(2030)		**
N(1700)	S11	****	Δ(1950)	F37	****	Λ(1750)	P01	**	Σ(1620)	PE	**	Ξ(2250)		*
N(1700)	D13	**	Δ(1960)	D35	*	Λ(1815)	F05	****	Σ(1670)	D13	****	Ξ(2500)		**
N(1780)	P11	***	Δ(2160)	P33	*	Λ(1830)	D05	****	Σ(1670)	PE	**			
N(1860)	P13	***	Δ(2420)	H311	***	Λ(1860)	P03	**	Σ(1690)	PE	**			
N(1990)	F17	**	Δ(2850)		***	Λ(1870)	S01	**	Σ(1750)	S11	***			
N(2040)	D13	**	Δ(3230)		***	Λ(2010)	D03	**	Σ(1765)	D15	****	Ω(1670)	P03	****
N(2100)	S11	*				Λ(2020)	F07	**	Σ(1840)	P13	*			
N(2100)	D15	*				Λ(2100)	G07	****	Σ(1880)	P11	**			
N(2175)	F15	*				Λ(2110)		*	Σ(1915)	F15	****			
N(2190)	G17	***	Z0(1780)	P01	*	Λ(2350)		****	Σ(1940)	D13	***			
N(2220)	H19	***	Z0(1865)		*	Λ(2585)		***	Σ(2000)	S11	*			
N(2650)		***	Z1(1900)	P13	*				Σ(2030)	F17	****			
N(3030)		***	Z1(2150)		*				Σ(2070)	F15	*			
N(3245)		*	Z1(2500)		*				Σ(2080)	P13	**			
N(3690)		*							Σ(2100)	G17	*			
N(3755)		*							Σ(2250)		****			
									Σ(2455)		***			
									Σ(2620)		***			
									Σ(3000)		**			

**** Good, clear, and unmistakable.   *** Good, but in need of clarification or not absolutely certain.
** Needs confirmation.   * Weak.

[See notes on N's and Δ's, on possible Z^*'s, and on Y^*'s at the beginning of those sections in the Baryon Data Card Listings; also see notes on individual resonances in the Baryon Data Card Listings.]

Particle[a]	I (J^P) ⊢ estab.	π or K Beam T(GeV) p(GeV/c) $\sigma = 4\pi \lambdabar^2$ (mb)	Mass M^b (MeV)	Full Width Γ^b (MeV)	M^2 ± ΓM^c (GeV^2)	Partial decay mode		
						Mode	Fraction %	p or p$_{max}$^{d} (MeV/c)
p	1/2(1/2$^+$)		938.3		0.880	See Stable Particle Table		
n			939.6		0.883			
N'(1470)	1/2(1/2$^+$) P'$_{11}$	T=0.53πp p=0.66 σ=27.8	~1470	165 to 300	2.16 ±0.41	Nπ	60	420
						Nππ	40	368
						[Nε	5-30]^e	
						[Δπ	20-30]^e	173
						[Nρ	~7]^e	
						pγ^g	0.05	435
						nγ^g	0.0	435
N'(1520)	1/2(3/2$^-$) D'$_{13}$	T=0.61 p=0.74 σ=23.5	1510 to 1540	105 to 150	2.31 ±0.18	Nπ	50	456
						Nππ	~50	410
						[Nε	0-2]^e	
						[Nρ	7-25]^e	
						[Δπ	15-40]^e	224
						Nη	0.2-1.4	
						pγ^g	0.55	471
						nγ^g	0.30	471
N'(1535)	1/2(1/2$^-$) S'$_{11}$	T=0.64 p=0.76 σ=22.5	1500 to 1600	50 to 160	2.36 ±0.18	Nπ	35	467
						Nη	55	182
						Nππ	~10	422
						[Nρ	1-2]^e	
						pγ^g	0.2-0.4	481
						nγ^g	0.12	481

4. From "Review of Particle Properties," *Rev. Mod. Phys.* **45**, (Supplement) S24 (1973).

Table A5 (cont.)

Particle[a]	I (J^P) ⊢ estab.	π or K Beam T(GeV) p(GeV/c) $\sigma = 4\pi\lambdabar^2$ (mb)	Mass M[b] (MeV)	Full Width Γ[b] (MeV)	M² ± ΓM[c] (GeV²)	Partial decay mode Mode	Fraction %	p or p_max[d] (MeV/c)
N'(1670)[i]	1/2(5/2⁻) D'₁₅	T=0.87 p=1.00 σ=15.6	1670 to 1685	115 to 175	2.79 ±0.24	Nπ N$\pi\pi$ [Δπ ΛK Nη pγ[g] nγ[g]	40 60 50-60][e] <1 <1[j] 0.01 0.02	560 525 357 200 368 572 572
N'(1688)[i]	1/2(5/2⁺) F'₁₅	T=0.90 p=1.03 σ=14.9	1680 to 1690	105 to 180	2.85 ±0.21	Nπ N$\pi\pi$ [Nϵ [Nρ [Δπ ΛK Nη pγ[g] nγ[g]	60 40 12][e] 15][e] 13-40][e] <0.1[j] <0.3[j] 0.20 0.01	572 538 340 372 231 388 583 583
N''(1700)[i]	1/2(1/2⁻) S''₁₁	T=0.92 p=1.05 σ=14.3	1665 to 1765	100 to 300	2.89 ±0.42	Nπ N$\pi\pi$ [Nϵ [Nρ ΛK Nη pγ[g] nγ[g]	60 25-30][e] 10-20][e] 5. ~3[j] 0.05-0.1 0.05	580 547 355 250 340 591 591
N''(1780)[i]	1/2(1/2⁺) P''₁₁	T=1.07 p=1.20 σ=12.2	1650 to 1860	50 to 350	3.17 ±0.51	Nπ N$\pi\pi$ [Nϵ [Δπ ΛK Nη pγ[g] nγ[g]	~20 30-40][e,h] 25-35][e,h] <7 10-20[j] 0.01 0.01	633 603 440 445 353 476 643 643
N(1860)	1/2(3/2⁺) P₁₃	T=1.22 p=1.36 σ=10.4	1770 to 1860	180 to 330	3.46 ±0.57	Nπ N$\pi\pi$ [Nρ ΛK Nη	25 55-65][e,h] ~5 ~4[j]	685 657 366 437 545
N(2190)	1/2(7/2⁻) G₁₇	T=1.94 p=2.07 σ=6.21	2000 to 2260	270 to 325	4.80 ±0.67	Nπ N$\pi\pi$	25	888 868
N(2220)	1/2(9/2⁺) H₁₉	T=2.00 p=2.14 σ=5.97	2200 to 2245	260 to 330	4.93 ±0.65	Nπ N$\pi\pi$	15	905 887
N(2650)	1/2(?⁻)	T=3.12 p=3.26 σ=3.67	~2650	~360	7.02 ±0.95	Nπ N$\pi\pi$	(J+1/2)x =0.45[f]	1154 1140
N(3030)	1/2(?)	T=4.27 p=4.41 σ=2.62	~3030	~ 400	9.18 ±1.21	Nπ N$\pi\pi$	(J+1/2)x =0.05[f]	1366 1354
Δ'(1236)[m]	3/2(3/2⁺) P'₃₃	T=0.195 (++) p=0.304 σ=91.8	1230 to 1236	110 to 122	1.53 ±0.14	Nπ N$\pi^+\pi^-$ Nγ[g]	99.4 0 ~0.6	231 90 262
		Pole position[m]: M-iΓ/2 = (1211.6±0.7) -i(49.5±1.8)						
Δ(1650)	3/2(1/2⁻) S₃₁	T=0.83 p=0.96 σ=16.4	1615 to 1695	130 to 200	2.72 ±0.28	Nπ N$\pi\pi$ [Nρ [Δπ Nγ[g]	28 72 8-16][e] 26-32][e] 0.30	547 511 558 340 558

Table A5 (cont.)

Particle[a]	I (J^P) ⊢―⊣ estab.	π or K Beam: T(GeV), p(GeV/c), $\sigma = 4\pi\lambdabar^2$ (mb)	Mass M[b] (MeV)	Full Width Γ[b] (MeV)	$M^2 \pm \Gamma M$[c] (GeV^2)	Mode	Fraction %	p or p_{max}[d] (MeV/c)
Δ (1670)	$3/2(3/2^-)$ D_{33}	T=0.87 p=1.00 σ=15.6	1650 to 1720	175 to 300	2.79 ±0.40	$N\pi$ $N\pi\pi$ $[\Delta\pi$ $N\gamma$[g]	15 22-30][e] 0.05	560 525 357 572
Δ (1890)	$3/2(5/2^+)$ F_{35}	T=1.28 p=1.42 σ=9.88	1840 to 1920	200 to 350	3.57 ±0.49	$N\pi$ $N\pi\pi$ $[N\rho$ $N\gamma$[g]	17 55-70][e] 0.03	704 677 403 712
Δ (1910)	$3/2(1/2^+)$ P_{31}	T=1.33 p=1.46 σ=9.54	1780 to 1935	200 to 340	3.65 ±0.52	$N\pi$ $N\pi\pi$ $[N\rho$ $[\Delta\pi$ $N\gamma$[g]	25 3-16][e] 4-16][e] 0.03	716 691 429 543 725
Δ (1950)	$3/2(7/2^+)$ F_{37}	T=1.41 p=1.54 σ=8.90	1930 to 1980	170 to 270	3.80 ±0.44	$N\pi$ $N\pi\pi$ $[N\rho$ $[\Delta\pi$ $N\gamma$[g] ΣK $\Sigma(1385)K$	45 8-12][e] 14-19][e] 0.15 ~2 1.4	741 716 471 571 749 460 232
Δ (2420)	$3/2(11/2^+)$ ⊢―⊣	T=2.50 p=2.64 σ=4.68	2320 to 2450	270 to 350	5.86 ±0.75	$N\pi$ $N\pi\pi$	11 >20	1023 1006
Δ (2850)	$3/2($?$^+$ $)$	T=3.71 p=3.85 σ=3.05	~2850	~400	8.12 ±1.14	$N\pi$ $N\pi\pi$	(J+1/2)x =0.25[f]	1266 1254
Δ (3230)	$3/2($? $)$	T=4.94 p=5.08 σ=2.25	~3230	~440	10.4 ±1.4	$N\pi$ $N\pi\pi$	(J+1/2)x =0.05[f]	1475 1464
Z*		Evidence for states with hypercharge 2 is controversial. See the Baryon Data Card Listings for discussion and display of data.						
Λ	$0(1/2^+)$		1115.6		1.24	See Stable Particle Table		
Λ'(1405)	$0(1/2^-)$ S'_{01}	p<0 K^-p	1405 ±5[n]	40 ±10[n]	1.97 ±0.06	$\Sigma\pi$	100	142
Λ'(1520)	$0(3/2^-)$ D'_{03}	p=0.389 σ=84.5	1518 ±2[n]	16 ±2[n]	2.30 ±0.02	$N\overline{K}$ $\Sigma\pi$ $\Lambda\pi\pi$ $\Sigma\pi\pi$	45±1 41±1 10±.5 1.0±.1	234 258 250 140
Λ''(1670)	$0(1/2^-)$ S''_{01}	p=0.74 σ=28.5	~1670	15 to 38	2.79 ±0.04	$N\overline{K}$ $\Lambda\eta$ $\Sigma\pi$	15-35 15-25 30-50	410 64 393
Λ''(1690)	$0(3/2^-)$ D''_{03}	p=0.78 σ=26.1	~1690	27 to 85	2.86 0.09	$N\overline{K}$ $\Sigma\pi$ $\Lambda\pi\pi$ $\Sigma\pi\pi$	20-30 40-70 <25 <25	429 409 415 352
Λ'(1815)	$0(5/2^+)$ F'_{05}	p=1.05 σ=16.7	1820 ±5[n]	64 to 104	3.30 ±0.15	$N\overline{K}$ $\Sigma\pi$ $\Sigma(1385)\pi$	61 11 15-20	542 508 362
Λ'(1830)	$0(5/2^-)$ D'_{05}	p=1.09 σ=15.8	1810 to 1840	60 to 150	3.33 ±0.19	$N\overline{K}$ $\Sigma\pi$ $\Lambda\pi\pi$	~10 20-60	554 519 536
Λ (2100)	$0(7/2^-)$ G_{07}	p=1.68 σ=8.68	~2100	60 to 140	4.41 ±0.22	$N\overline{K}$ $\Sigma\pi$ $\Lambda\eta$ ΞK $\Lambda\omega$	25 ~5 <3 ~2 ~1	748 699 617 483 443

524

Table A5 (cont.)

Particle[a]	I (J^P) ⊢⊣ estab.	π or K Beam T(GeV) p(GeV/c) $\sigma = 4\pi \lambdabar^2$ (mb)	Mass M[b] (MeV)	Full Width Γ[b] (MeV)	M^2 $\pm \Gamma M$[c] (GeV²)	Mode	Fraction %	p or p_{max}[d] (MeV/c)
Λ (2350)	0(?)	p=2.29 σ=5.85	~ 2350	140 to 324	5.52 ±0.55	N$\overline{K}$	(J+1/2)x =0.7[f]	913
Λ(2585)	0(?)	p=2.91 σ=4.37	~ 2585	~ 300	6.66 ±0.77	N$\overline{K}$	(J+1/2)x =1.0[f]	1058
Σ	1(1/2^+)		(+)1189.4 (0)1192.5 (-)1197.3		1.41 1.42 1.43	See Stable Particle Table		
Σ'(1385)	1(3/2^+)P'13	p<0 K⁻p	(+)1383±1 S=1.3* (-)1386±2 S=2.2*	(+) 34±2 S=2.0* (-)36±6 S=3.5*!	1.92 ±0.05	$\Lambda \pi$ $\Sigma \pi$	89±2 11±2	208 117
Σ'(1670)[k]	1(3/2^-)D'13	p=0.74 σ=28.5	~ 1670	35-65	2.79 ±0.08	N$\overline{K}$ $\Sigma \pi$ $\Lambda \pi$ $\Sigma \pi \pi$ [Λ(1405)π][e] $\Lambda \pi \pi$	~8 ~40 ~12 5-15	410 387 447 326 207 397

Parameters here are obtained from partial wave analyses for a D13 resonance. Production experiments suggest two such states; see footnote k and the Baryon Data Card Listings.

Particle[a]	I (J^P)	π or K Beam	Mass	Full Width	$M^2 \pm \Gamma M$	Mode	Fraction %	p or p_{max}
Σ'(1750)	1(1/2^-)S''11	p=0.91 σ=20.7	1700 to 1790	50 to 100	3.05 ±0.13	N$\overline{K}$ $\Lambda \pi$ $\Sigma \eta$	seen seen seen	483 507 54
Σ (1765)	1(5/2^-) D15	p=0.94 σ=19.6	1765 ±5[n]	~120	3.12 ±0.21	N$\overline{K}$ $\Lambda \pi$ Λ(1520)π Σ(1385)π $\Sigma \pi$	~ 41 ~ 13 ~ 15 ~ 10 ~1	496 518 187 315 461
Σ (1915)[i]	1(5/2^+) F'15	p=1.25 σ=13.0	1900-1930	50-100	3.67 ±0.14	N$\overline{K}$ $\Lambda \pi$ $\Sigma \pi$	~14 ~ 6 ~ 6	612 619 568

Formation and production experiments do not agree on $\Sigma \pi / \Lambda \pi$ ratio.

Particle[a]	I (J^P)	π or K Beam	Mass	Full Width	$M^2 \pm \Gamma M$	Mode	Fraction %	p or p_{max}
Σ''(1940)	1(3/2^-)D''13	p=1.32 σ=12.0	~1940	~220	3.77 ±0.43	N$\overline{K}$ $\Lambda \pi$ $\Sigma \pi$	seen seen	678 680 589
Σ (2030)	1(7/2^+) F17	p=1.52 σ=9.93	~2030	100 to 170	4.12 ±0.27	N$\overline{K}$ $\Lambda \pi$ $\Sigma \pi$ Ξ K	~ 20 ~ 20 ~ 4 < 2	700 700 652 412
Σ (2250)	1(?)	p=2.04 σ=6.76	~2250	100 to 230	5.06 ±0.37	N$\overline{K}$ $\Sigma \pi$ $\Lambda \pi$	(J+1/2)x =0.3[f]	849 842 799
Σ (2455)	1(?)	p=2.57 σ=5.09	~2455	~120	6.03 ±0.29	N$\overline{K}$	(J+1/2)x =0.2[f]	979
Σ (2620)	1(?)	p=2.95 σ=4.30	~ 2620	~175	6.86 ±0.46	N$\overline{K}$	(J+1/2)x =0.3[f]	1064
Ξ[l]	1/2(1/2^+)		(0)1314.9 (-)1321.3		1.73 1.75	See Stable Particle Table		
Ξ (1530)[l]	1/2(3/2^+) P13		(0) 1531.6±0.4 S=1.3* (-) 1535.0±0.6	(0) 9.1±0.5 (-) 12.9±4.1	2.34 ±0.01	$\Xi \pi$	100	144
Ξ (1820)[l]	1/2(?)		1795 to 1870	12 to 99	3.31 ±0.10	$\Lambda \overline{K}$ $\Xi \pi$ Ξ (1530)π $\Sigma \overline{K}$		396 413 234 306

All four decay modes have been seen. Branching ratios not quoted because there may be more than one state here.

Particle[a]	I (J^P)	π or K Beam	Mass	Full Width	$M^2 \pm \Gamma M$	Mode	Fraction %	p or p_{max}
Ξ (1940)[l]	1/2(?)		1894 to 1961	42 to 140	3.72 ±0.18	$\Xi \pi$ Ξ(1530)π		499 336

Seen in both final states; not clear if one, or more, states present.

Particle[a]	I (J^P)		Mass		$M^2 \pm \Gamma M$			
Ω^-	0(3/2^+)		1672.5		2.80	See Stable Particle Table		

Table A5 (cont.)

Table A6 Nuclear Data[5]

It should be emphasized that the information presented here was not reviewed independently but is simply a summary of the results compiled in the papers listed below.

Mass excesses (*M-A, in keV*) are from Mattauch, Thiele and Wapstra, *Nuclear Physics* **67**, 1 (1965), except for a few values discussed in Lauritsen and Ajzenberg-Selove [*Nuclear Physics* **78**, 1 (1966); ibid. **A114**, 1 (1968)], and Endt and Van der Leun [*Nuclear Physics* **A105**, 1 (1967)].

J^{π} [listed as *J, P* and given in units of $\hbar$], *T* [*half life* in seconds (*S*), minutes (*M*), hours (*H*), days (*D*) and years (*Y*)], *A* [*abundance*, in percent] or *W* [*width* of the state in eV (*E*), keV (*K*) or MeV (*M*)] are derived principally from the GE "Chart of the Nuclides" (7/69); the sixth edition of the "Table of Isotopes" (Lederer, Hollander and Perlman, J. Wiley Inc. 1967); the "Nuclear Data Sheets" and the "Recent References" compilations of the O.R.N.L. Nuclear Data Group; and Fuller and Cohen, *Nuclear Data Tables* **A5**, 433 (1969). For the light nuclei, the compilations by Lauritsen and Ajzenberg-Selove and by Endt and Van der Leun have also been used.

5. Prepared by S. Morse and F. Ajzenberg-Selove, University of Pennsylvania, Philadelphia, Dec. 1971, with minor corrections. Courtesy F. Ajzenberg-Selove.

Table A6 (cont.)

Z	A	J,P	M-A(KEV)	T,A,OR W	Z	A	J,P	M-A(KEV)	T,A,OR W
					9	F 16	0-	10 693.	W=50K
0	N 1	1/2+	8 071.69	T=10.6M	7	N 17	1/2-	7 871.	T=4.16S
1	H 1	1/2+	7 289.22	A=99.9855	8	O 17	5/2+	- 807.4	A=0.039
					9	F 17	5/2+	1 951.9	T=66.0S
1	H 2	1+	13 136.27	A=0.015	10	NE 17	1/2-	16 480.	T=0.108S
1	H 3	1/2+	14 950.38	T=12.33Y	8	O 18	0+	- 782.50	A=0.205
2	HE 3	1/2+	14 931.73	A=E-4	9	F 18	1+	872.8	T=109.8M
					10	NE 18	0+	5 319.0	T=1.67S
2	HE 4	0+	2 424.94	A=100.					
					8	O 19	5/2+	3 332.3	T=26.91S
2	HE 5	3/2-	11 390.	W=0.58M	9	F 19	1/2+	-1 486.1	A=100.0
3	LI 5	3/2-	11 680.	W=1.5M	10	NE 19	1/2+	1 752.1	T=17.4S
2	HE 6	0+	17 597.3	T=0.802S	8	O 20	0+	3 800.	T=13.57S
3	LI 6	1+	14 087.5	A=7.5	9	F 20	2+	-15.7	T=11.03S
4	BE 6	0+	18 375.	W=92K	10	NE 20	0+	-7 041.7	A=90.5
					11	NA 20	2+	6 840.	T=0.450S
3	LI 7	3/2-	14 908.6	A=92.5					
4	BE 7	3/2-	15 770.3	T=53.4D	9	F 21	5/2+	-46.	T=4.4S
					10	NE 21	3/2+	-5 731.2	A=0.27
2	HE 8	0+	31 650.	T=0.122S	11	NA 21	3/2+	-2 183.	T=22.8S
3	LI 8	2+	20 947.5	T=0.848S	12	MG 21	5/2+	10 911.	T=0.121S
4	BE 8	0+	4 941.8	W=6.8EV					
5	B 8	2+	22 922.3	T=0.774S	9	F 22		2 828.	
					10	NE 22	0+	-8 025.1	A=9.2
3	LI 9	3/2-	24 966.	T=0.172S	11	NA 22	3+	-5 182.2	T=2.601Y
4	BE 9	3/2-	11 348.4	A=100.0	12	MG 22	0+	-384.	T=3.99S
5	B 9	3/2-	12 415.7	W=0.54K					
6	C 9	3/2-	28 912.	T=0.127S	10	NE 23	5/2+	-5 150.0	T=37.6S
					11	NA 23	3/2+	-9 529.0	A=100.0
4	BE 10	0+	12 608.1	T=2.7E+6Y	12	MG 23	3/2+	-5 472.4	T=12.S
5	B 10	3+	12 052.3	A=19.8	13	AL 23	5/2+	6 770.	T 0.48S
6	C 10	0+	15 702.7	T=19.41S					
					10	NE 24	0+	-5 948.	T=3.38M
4	BE 11	1/2+	20 177.	T=13.68S	11	NA 24	4+	-8 416.7	T=15.00H
5	B 11	3/2-	8 667.95	A=80.2	12	MG 24	0+	-13 931.3	A=78.99
6	C 11	3/2-	10 650.2	T=20.39M	13	AL 24	4+	-49.	T=2.09S
5	B 12	1+	13 370.4	T=0.0204S	11	NA 25	5/2+	-9 356.	T=59.6S
6	C 12	0+	0.	A=98.892	12	MG 25	5/2+	-13 191.5	A=10.00
7	N 12	1+	17 344.	T=0.0110S	13	AL 25	5/2+	-8 912.3	T=7.23S
					14	SI 25	5/2+	3 820.	T=0.22S
5	B 13	3/2-	16 562.0	T=0.0186S					
6	C 13	1/2-	3 125.27	A=1.108	11	NA 26	2,3+	-6 853.	T=1.00S
7	N 13	1/2-	5 345.7	T=9.961M	12	MG 26	0+	-16 213.4	A=11.01
					13	AL 26	5+	-12 208.8	T=7.4E+5Y
6	C 14	0+	3 019.95	T=5692.Y	14	SI 26	0+	-7 147.	T=2.1S
7	N 14	1+	2 863.82	A=99.64					
8	O 14	0+	8 008.59	T=70.98S	12	MG 27	1/2+	-14 584.7	T=9.45M
					13	AL 27	5/2+	-17 195.0	A=100.0
6	C 15	1/2+	9 873.5	T=2.33S	14	SI 27	5/2+	-12 385.4	T=4.16S
7	N 15	1/2-	101.8	A=0.36					
8	O 15	1/2-	2 861.1	T=122.24S	12	MG 28	0+	-15 017.	T=21.1H
					13	AL 28	3+	-16 848.8	T=2.246M
6	C 16	0+	13 693.	T=0.74S	14	SI 28	0+	-21 491.1	A=92.2
7	N 16	2-	5 683.5	T=7.13S	15	P 28	3+	-7 154.	T=0.270S
8	O 16	0+	-4 736.68	A=99.756					

Table A6 (cont.)

13	AL	29	5/2+	-18 213.	T=6.52M
14	SI	29	1/2+	-21 893.3	A=4.7
15	P	29	1/2+	-16 950.	T=4.18S
13	AL	30	2,3+	-15 890.	T=3.27S
14	SI	30	0+	-24 431.3	A=3.1
15	P	30	1+	-20 203.9	T=2.50M
16	S	30	0+	-14 065.	T=1.23S
14	SI	31	3/2+	-22 947.9	T=157.3M
15	P	31	1/2+	-24 439.6	A=100.0
16	S	31	1/2+	-18 998.	T=2.62S
14	SI	32	0+	-24 091.	T=2.8E+2Y
15	P	32	1+	-24 304.2	T=14.28D
16	S	32	0+	-26 014.3	A=95.0
17	CL	32	2+	-13 263.	T=0.297S
15	P	33	1/2+	-26 337.0	T=25.3D
16	S	33	3/2+	-26 586.0	A=0.75
17	CL	33	3/2+	-21 002.4	T=2.50S
15	P	34	1+	-24 830.	T=12.4S
16	S	34	0+	-29 929.2	A=4.2
17	CL	34	0+	-24 438.4	T=1.58S
18	AR	34	0+	-18 395.	T=0.9S
16	S	35	3/2+	-28 845.6	T=87.2D
17	CL	35	3/2+	-29 013.0	A=75.77
18	AR	35	3/2+	-23 049.4	T=1.77S
16	S	36	0+	-30 665.9	A=0.015
17	CL	36	2+	-29 521.8	T=3.0E+5Y
18	AR	36	0+	-30 230.5	A=0.34
16	S	37	7/2-	-26 907.	T=5.06M
17	CL	37	3/2+	-31 761.5	A=24.23
18	AR	37	3/2+	-30 947.4	T=34.8D
19	K	37	3/2+	-24 798.4	T=1.23S
20	CA	37	3/2+	-13 230.	T=0.173S
16	S	38	0+	-26 863.	T=170.M
17	CL	38	2-	-29 800.	T=37.2M
18	AR	38	0+	-34 714.4	A=0.07
19	K	38	3+	-28 792.	T=7.63M
20	CA	38	0+	-22 023.	T=0.46S
17	CL	39	3/2+	-29 802.	T=55.5M
18	AR	39	7/2-	-33 240.	T=269.Y
19	K	39	3/2+	-33 805.3	A=93.3
20	CA	39	3/2+	-27 283.	T=0.87S
17	CL	40	2-	-27 540.	T=1.42M
18	AR	40	0+	-35 039.2	A=99.59
19	K	40	4-	-33 534.1	A=0.012
					T=1.3E+9Y
20	CA	40	0+	-34 845.7	A=96.94
21	SC	40	4-	-20 521.	T=0.182S

18	AR	41	7/2-	-33 066.1	T=1.83H
19	K	41	3/2+	-35 558.3	A=6.7
20	CA	41	7/2-	-35 137.1	T=8.E+4Y
21	SC	41	7/2-	-28 641.	T=0.60S
18	AR	42	0+	-34 420.	T=33.Y
19	K	42	2-	-35 021.4	T=12.36H
20	CA	42	0+	-38 538.1	A=0.65
21	SC	42	0+	-32 107.0	T=0.683S
22	TI	42	0+	-25 121.	T=0.20S
19	K	43	3/2+	-36 582.	T=21.8H
20	CA	43	7/2-	-38 399.0	A=0.14
21	SC	43	7/2-	-36 179.0	T=3.89H
22	TI	43	7/2-	-29 320.	T=0.49S
19	K	44	2-	-35 805.	T=22.M
20	CA	44	0+	-41 463.6	A=2.08
21	SC	44	2+	-37 814.	T=3.92H
22	TI	44	0+	-37 548.	T=47.Y
19	K	45	3/2+	-36 611.	T=17.M
20	CA	45	7/2-	-40 806.3	T=162.7D
21	SC	45	7/2-	-41 063.1	A=100.0
22	TI	45	7/2-	-39 000.7	T=3.08H
19	K	46	2-	-35 426.	T=110.S
20	CA	46	0+	-43 138.	A=0.003
21	SC	46	4+	-41 758.4	T=83.80D
22	TI	46	0+	-44 125.8	A=8.00
23	V	46	0+	-37 071.4	T=0.426S
19	K	47	1/2+	-35 704.	T=18.S
20	CA	47	7/2-	-42 343.	T=4.54D
21	SC	47	7/2-	-44 328.9	T=3.41D
22	TI	47	5/2-	-44 929.2	A=7.5
23	V	47	3/2-	-42 004.8	T=31.2M
20	CA	48	0+	-44 222.	A=0.19
21	SC	48	6+	-44 495.	T=43.7H
22	TI	48	0+	-48 485.6	A=73.7
23	V	48	4+	-44 470.2	T=16.18D
24	CR	48	0+	-42 816.	T=22.96H
20	CA	49	3/2-	-41 292.	T=8.72M
21	SC	49	7/2-	-46 552.	T=57.3M
22	TI	49	7/2-	-48 557.3	A=5.5
23	V	49	7/2-	-47 956.1	T=331.D
24	CR	49	5/2-	-45 388.	T=42.0M
20	CA	50	0+	-39 578.	T=14.S
21	SC	50	5+	-44 545.	T=1.71M
22	TI	50	0+	-51 433.6	A=5.3
23	V	50	6+	-49 216.7	T 4.E+16Y
					A=0.25
24	CR	50	0+	-50 255.7	A=4.35
25	MN	50	0+	-42 624.6	T=0.29S
21	SC	51	7/2-	-43 227.	T=13.S

22 TI 51	3/2-	-49 739.	T=5.76M	
23 V 51	7/2-	-52 197.4	A=99.75	
24 CR 51	7/2-	-51 446.0	T=27.71D	
25 MN 51	5/2-	-48 240.	T=45.9M	
22 TI 52	0+	-49 470.	T=1.7M	
23 V 52	3+	-51 436.9	T=3.75M	
24 CR 52	0+	-55 415.0	A=83.79	
25 MN 52	6+	-50 705.	T=5.63D	
26 FE 52	0+	-48 333.	T=8.3H	
23 V 53	7/2-	-51 861.	T=1.55M	
24 CR 53	3/2-	-55 283.8	A=9.50	
25 MN 53	7/2-	-54 686.5	T=1.1E+7Y	
26 FE 53	7/2-	-50 942.	T=8.53M	
23 V 54	5+	-49 930.	T=43.S	
24 CR 54	0+	-56 932.3	A=2.36	
25 MN 54	3+	-55 557.	T=313.D	
26 FE 54	0+	-56 251.7	A=5.8	
27 CO 54	0+	-48 002.	T=0.194S	
24 CR 55	3/2-	-55 121.	T=3.56M	
25 MN 55	5/2-	-57 710.0	A=100.0	
26 FE 55	3/2-	-57 478.4	T=2.7Y	
27 CC 55	7/2-	-54 012.4	T=18.H	
24 CR 56	0+	-55 266.	T=5.9M	
25 MN 56	3+	-56 908.7	T=2.582H	
26 FE 56	0+	-60 609.4	A=91.7	
27 CC 56	4+	-56 041.2	T=77.3D	
28 NI 56	0+	-53 908.	T=6.10D	
25 MN 57	5/2-	-57 620.	T=1.59M	
26 FE 57	1/2-	-60 183.8	A=2.14	
27 CO 57	7/2-	-59 347.0	T=271.D	
28 NI 57	3/2-	-56 104.	T=36.1H	
25 MN 58		-56 060.	T=1.1M	
26 FE 58	0+	-62 155.1	A=0.31	
27 CC 58	2+	-59 847.2	T=71.4D	
28 NI 58	0+	-60 235.	A=68.	
29 CU 58	1+	-51 668.	T=3.21S	
26 FE 59	3/2-	-60 670.0	T=45.D	
27 CO 59	7/2-	-62 235.7	A=100.0	
28 NI 59	3/2-	-61 162.6	T=8E+4Y	
29 CU 59	3/2-	-56 363.	T=81.8S	
26 FE 60	0+	-61 435.	T=E+5Y	
27 CO 60	5+	-61 655.6	T=5.269Y	
28 NI 60	0+	-64 479.2	A=26.1	
29 CU 60	2+	-58 352.	T=23.M	
26 FE 61	3/2-	-59 030.	T=6.0M	
27 CC 61	7/2-	-62 920.	T=1.65H	
28 NI 61	3/2-	-64 227.0	A=1.1	
29 CU 61	3/2-	-61 981.8	T=3.37H	
30 ZN 61	3/2-	-56 580.	T=87.S	

27 CO 62	5+	-61 530.	T=13.9M	
28 NI 62	0+	-66 751.9	A=3.6	
29 CU 62	1+	-62 805.	T=9.78M	
30 ZN 62	0+	-61 115.	T=9.15H	
27 CO 63		-61 863.	T=27.S	
28 NI 63	1/2-	-65 521.5	T=92.Y	
29 CU 63	3/2-	-65 587.4	A=69.1	
30 ZN 63	3/2-	-62 222.	T=38.5M	
31 GA 63	3/2-	-56 720.	T=31.S	
28 NI 64	0+	-67 109.3	A=0.9	
29 CU 64	1+	-65 431.8	T=12.74H	
30 ZN 64	0+	-66 006.4	A=48.9	
31 GA 64	0+	-58 934.	T=2.6M	
28 NI 65	5/2-	-65 133.	T=2.54H	
29 CU 65	3/2-	-67 264.8	A=30.9	
30 ZN 65	5/2-	-65 914.1	T=243.7D	
31 GA 65	3/2-	-62 655.	T=15.2M	
32 GE 65		-56 360.	T=1.5M	
28 NI 66	0+	-66 060.	T=54.6H	
29 CU 66	1+	-66 259.8	T=5.1M	
30 ZN 66	0+	-68 894.5	A=27.8	
31 GA 66	0+	-63 719.	T=9.5H	
32 GE 66	0+	-61 617.	T=2.27H	
28 NI 67		-63 200.	T=50.S	
29 CU 67	3/2-	-67 302.	T=61.6H	
30 ZN 67	5/2-	-67 876.7	A=4.1	
31 GA 67	3/2-	-66 876.	T=78.2H	
32 GE 67		-62 450.	T=19.0M	
29 CU 68	1+	-65 420.	T=31.S	
30 ZN 68	0+	-70 004.3	A=18.6	
31 GA 68	1+	-67 085.	T=68.2M	
32 GE 68	0+	-66 698.	T=287.D	
30 ZN 69	1/2-	-68 416.2	T=57.M	
31 GA 69	3/2-	-69 323.0	A=60.2	
32 GE 69	5/2-	-67 097.5	T=39.1H	
33 AS 69		-63 130.	T=15.M	
30 ZN 70	0+	-69 559.7	A=0.62	
31 GA 70	1+	-68 906.	T=21.1M	
32 GE 70	0+	-70 559.5	A=20.7	
33 AS 70	4	-64 338.	T=52.5M	
34 SE 70	0+		T=39.M	
30 ZN 71	1/2-	-67 332.	T=2.4M	
31 GA 71	3/2-	-70 138.1	A=40.	
32 GE 71	1/2-	-69 903.0	T=11.D	
33 AS 71	5/2-	-67 894.	T=64.H	
34 SE 71	5/2-	-62 890.	T=4.9M	
30 ZN 72	0+	-68 131.	T=46.5H	
31 GA 72	3-	-68 587.6	T=14.1H	

Table A6 (cont.)

32 GE 72	0+	-72	580.7	A=27.5
33 AS 72	2-	-68	230.	T=26.H
34 SE 72	0+	-67	630.	T=8.5D
31 GA 73	3/2-	-69	740.	T=4.88H
32 GE 73	9/2+	-71	293.2	A=7.7
33 AS 73	3/2-	-70	954.	T=76.D
34 SE 73	9/2+	-68	214.	T=7.1H
31 GA 74	3-	-67	920.	T=8.2M
32 GE 74	0+	-73	422.4	A=36.4
33 AS 74	2-	-70	858.7	T=17.7D
34 SE 74	0+	-72	213.0	A=0.9
35 BR 74	1+	-65	210.	T=37.M
36 KR 74	0+	-62	110.	T=16.M
32 GE 75	1/2-	-71	841.	T=82.8M
33 AS 75	3/2-	-73	029.7	A=100.
34 SE 75	5/2+	-72	164.9	T=120.D
35 BR 75	3/2-	-69	155.	T=96.M
36 KR 75		-64	050.	T=5.5M
32 GE 76	0+	-73	212.3	A=7.7
33 AS 76	2-	-72	286.2	T=26.3H
34 SE 76	0+	-75	254.6	A=9.0
35 BR 76	1	-70	150.	T=16.H
36 KR 76	0+	-69	150.	T=14.8H
32 GE 77	7/2+	-71	160.	T=11.3H
33 AS 77	3/2-	-73	917.	T=38.8H
34 SE 77	1/2-	-74	601.4	A=7.5
35 BR 77	3/2-	-73	236.9	T=56.H
36 KR 77	7/2+	-70	237.	T=1.19H
32 GE 78	0+	-71	780.	T=1.45H
33 AS 78	2-	-72	760.	T=1.515H
34 SE 78	0+	-77	026.8	A=23.5
35 BR 78	1+	-73	453.	T=6.5M
36 KR 78	0+	-74	147.	A=0.35
37 RB 78				T=6.5M
33 AS 79	3/2-	-73	690.	T=9.M
34 SE 79	7/2+	-75	933.0	T 6.5E+4Y
35 BR 79	3/2-	-76	074.1	A=50.69
36 KR 79	1/2-	-74	443.	T=34.9H
37 RB 79	3/2-	-70	920.	T=23.M
33 AS 80	1+	-71	760.	T=15.3S
34 SE 80	0+	-77	757.0	A=50.
35 BR 80	1+	-75	885.3	T=17.4M
36 KR 80	0+	-77	896.	A=2.25
37 RB 80	1+	-72	100.	T=30.S
33 AS 81	3/2-	-72	590.	T=32.S
34 SE 81	1/2-	-76	387.	T=18.6M
35 BR 81	3/2-	-77	974.	A=49.31
36 KR 81	7/2+	-77	680.	T=2.1E+5Y
37 RB 81	3/2-	-75	420.	T=4.58H
33 AS 82	5-			T=13.3S
34 SE 82	0+	-77	587.	A=9.0
35 BR 82	5-	-77	503.	T=35.4H
36 KR 82	0+	-80	591.0	A=11.6
37 RB 82	1+	-76	194.	T=1.25M
38 SR 82	0+	-75	590.	T=25.D
34 SE 83	9/2+	-75	440.	T=22.6M
35 BR 83	3/2-	-79	018.	T=2.41H
36 KR 83	9/2+	-79	987.0	A=11.5
37 RB 83	5/2-	-78	949.	T=83.D
38 SR 83	7/2+	-76	699.	T=32.4H
34 SE 84	0+	-75	920.	T=3.3M
35 BR 84	2-	-77	730.	T=31.8M
36 KR 84	0+	-82	433.2	A=57.0
37 RB 84	2-	-79	753.0	T=33.D
38 SR 84	0+	-80	639.8	A=0.56
39 Y 84	4-	-73	690.	T=40.M
35 BR 85	3/2-	-78	670.	T=2.87M
36 KR 85	9/2+	-81	472.6	T=10.73Y
37 RB 85	5/2-	-82	159.6	A=72.17
38 SR 85	9/2+	-81	096.	T=65.2D
39 Y 85	9/2+	-77	836.	T=4.8H
35 BR 86	1,2	-75	960.	T=54.S
36 KR 86	0+	-83	261.3	A=17.3
37 RB 86	2-	-82	738.3	T=18.66D
38 SR 86	0+	-84	509.4	A=9.9
39 Y 86	4-	-79	236.	T=14.6H
40 ZR 86	0+	-77	940.	T=16.5H
35 BR 87		-74	200.	T=55.7S
36 KR 87	5/2+	-80	700.	T=76.4M
37 RB 87	3/2-	-84	592.6	T=5.E+10Y
				A=27.83
38 SR 87	9/2+	-84	866.1	A=7.0
39 Y 87	1/2-	-82	984.	T=80.3H
40 ZR 87		-79	484.	T=1.78H
35 BR 88				T=16.S
36 KR 88	0+	-79	700.	T=2.8H
37 RB 88	2-	-82	604.	T=17.8M
38 SR 88	0+	-87	907.6	A=82.6
39 Y 88	4-	-84	289.	T=107.D
40 ZR 88	0+	-83	610.	T=85.D
41 NB 88	8+	-76	410.	T=14.M
35 BR 89				T=4.4S
36 KR 89	5/2+	-76	560.	T=3.16M
37 RB 89	3/2-	-81	710.	T=15.2M
38 SR 89	5/2+	-86	196.	T=50.5D
39 Y 89	1/2-	-87	685.6	A=100.
40 ZR 89	9/2+	-84	851.	T=78.4H
41 NB 89	9/2+	-80	980.	T=1.9H
36 KR 90	0+	-74	890.	T=32.3S
37 RB 90	1-	-79	300.	T=2.7M

38 SR 90	0+	-85 927.9	T=28.9Y		
39 Y 90	2-	-86 473.9	T=64.H		
40 ZR 90	0+	-88 762.6	A=51.4		
41 NB 90	8+	-82 652.	T=14.6H		
42 MC 90	0+	-80 165.	T=5.7H		

38 SR 90   0+   -85 927.9   T=28.9Y
39 Y 90   2-   -86 473.9   T=64.H
40 ZR 90   0+   -88 762.6   A=51.4
41 NB 90   8+   -82 652.   T=14.6H
42 MC 90   0+   -80 165.   T=5.7H

36 KR 91   5/2+ -71 500.   T=9.0S
37 RB 91   -   -78 000.   T=58.5S
38 SR 91   5/2+ -83 684.   T=9.48H
39 Y 91   1/2- -86 349.   T=58.6D
40 ZR 91   5/2+ -87 893.5 A=11.2
41 NB 91   9/2+ -86 632.   T=LONG
42 MC 91   9/2+ -82 188.   T=15.49M

36 KR 92   0+   T=1.9S
37 RB 92   -75 020.   T=4.48S
38 SR 92   0+ -82 920.   T=2.7H
39 Y 92   2- -84 834.   T=3.53H
40 ZR 92   0+ -88 456.9 A=17.1
41 NB 92   7+ -86 453.   T=2.E+7Y
42 MC 92   0+ -86 808.4 A=14.8
43 TC 92   8,9+ -78 860. T=4.4M

36 KR 93   T=1.29S
37 RB 93   -73 050.   T=5.87S
38 SR 93   -79 950.   T=7.5M
39 Y 93   1/2- -84 254. T=10.2H
40 ZR 93   5/2+ -87 143.7 T=9.5E+5Y
41 NB 93   9/2+ -87 207.1 A=100.
42 MC 93   5/2+ -86 809. T=3.E+3Y
43 TC 93   9/2+ -83 623. T=2.73H

36 KR 94   0+   T=0.20S
37 RB 94   T=2.71S
38 SR 94   0+ -78 740.   T=78S
39 Y 94   2- -82 260.   T=20.3M
40 ZR 94   0+ -87 263.1 A=17.5
41 NB 94   6+ -86 364.3 T=2.E+4Y
42 MC 94   0+ -88 409.9 A=9.1
43 TC 94   6,7+ -84 150. T=4.9H
44 RU 94   0+ -82 569.   T=52.M

37 RB 95   T=0.36S
38 SR 95   -75 540.   T=26.S
39 Y 95   -   -81 236.   T=10.5M
40 ZR 95   5/2+ -85 666.0 T=65.5D
41 NB 95   9/2+ -86 788.5 T=35.D
42 MC 95   5/2+ -87 713.3 A=15.9
43 TC 95   9/2+ -86 012.   T=20.H
44 RU 95   5/2+ -83 450. T=1.65H

38 SR 96   0+   T=4.S
39 Y 96   -78 630.   T=2.3M
40 ZR 96   0+ -85 426.0 A=2.8
41 NB 96   5+ -85 609.   T=23.5H
42 MO 96   0+ -88 795.9 A=16.7
43 TC 96   7+ -85 860.   T=4.3D
44 RU 96   0+ -86 073.   A=5.5

39 Y 97   1/2- -76 830. T=1.11S
40 ZR 97   1/2+ -82 933. T=16.8H
41 NB 97   9/2+ -85 605. T=73.6M
42 MC 97   5/2+ -87 540.2 A=9.5
43 TC 97   9/2+ -87 195.   T=2.6E+6Y
44 RU 97   5/2+ -86 040. T=2.89D
45 RH 97   9/2+ -82 550. T=33.M

40 ZR 98   0+ -81 273.   T=31.S
41 NB 98   1+ -83 510.   T=51.M
42 MC 98   0+ -88 110.9 A=24.4
43 TC 98   7,6+ -86 520. T=1.5E+6Y
44 RU 98   0+ -88 223.0 A=1.9
45 RH 98   2+ -83 166.   T=8.7M

40 ZR 99   -78 360.   T=2.4S
41 NB 99   9/2+ -82 860. T=2.4M
42 MC 99   1/2+ -85 956. T=66.3H
43 TC 99   9/2+ -87 328. T=2.1E+5Y
44 RU 99   5/2+ -87 620.2 A=12.7
45 RH 99   1/2- -85 568. T=15.0D
46 PD 99   5/2+ -82 163. T=21.4M
47 AG 99   -76 130.   T=1.8M

41 NB 100   -80 190.   T=2.9M
42 MO 100   0+ -86 185.1 A=9.6
43 TC 100   1+ -85 850. T=16.S
44 RU 100   0+ -89 221.9 A=12.6
45 RH 100   1- -85 592. T=21.H
46 PD 100   0+ -85 190. T=3.7D
47 AG 100   5+ -77 890. T=8.M

41 NB 101   -79 400.   T=7.S
42 MO 101   1/2+ -83 504. T=14.6M
43 TC 101   9/2+ -86 325. T=14.2M
44 RU 101   5/2+ -87 955.7 A=17.1
45 RH 101   1/2- -87 402. T=3.Y
46 PD 101   5/2+ -85 412. T=8.3H
47 AG 101   9/2+ -81 010. T=10.8M

42 MC 102   0+ -83 600.   T=11.M
43 TC 102   1+ -84 600.   T=5.3S
44 RU 102   0+ -89 100.2 A=31.6
45 RH 102   1- -86 778.   T=207.D
46 PD 102   0+ -87 927.   A=1.0
47 AG 102   5+ -82 367. T=13.M
48 CD 102   0+ -79 470. T=5.5M

42 MO 103   -80 500.   T=60.S
43 TC 103   -84 900.   T=50.S
44 RU 103   5/2+ -87 253. T=39.6D
45 RH 103   1/2- -88 016.0 A=100.
46 PD 103   5/2+ -87 463. T=17.5D
47 AG 103   7/2+ -84 780. T=1.1H
48 CD 103   -80 380.   T=7.3M

42 MC 104   0+ -80 190.   T=1.6M
43 TC 104   -82 790.   T=18.M
44 RU 104   0+ -88 094.0 A=18.6

45	RH	104	1+	-86 944.	T=42.S
46	PD	104	0+	-89 411.	A=11.0
47	AG	104	5+	-85 311.	T=67.M
48	CD	104	0+	-84 010.	T=56.M
42	MC	105			T=0.9M
43	TC	105		-82 530.	T=8.0M
44	RU	105	5/2+	-85 930.	T=4.44H
45	RH	105	7/2+	-87 847.	T=35.5H
46	PD	105	5/2+	-88 413.	A=22.2
47	AG	105	1/2-	-87 078.	T=41.C
48	CD	105	5/2+	-84 280.	T=56.D
43	TC	106		-79 820.	T=37.S
44	RU	106	0+	-86 323.	T=1.01Y
45	RH	106	1+	-86 362.	T=30.S
46	PD	106	0+	-89 902.	A=27.3
47	AG	106	1+	-86 928.	T=24.0M
48	CD	106	0+	-87 130.2	A=1.2
49	IN	106		-80 390.	T=5.32M
44	RU	107		-83 710.	T=4.2M
45	RH	107	5/2+	-86 860.	T=21.7M
46	PD	107	5/2+	-88 373.0	T=6.5E+6Y
47	AG	107	1/2-	-88 408.0	A=51.83
48	CD	107	5/2+	-86 991.	T=6.49H
49	IN	107	9/2+	-83 500.	T=32.7M
44	RU	108	0+	-83 710.	T=4.5M
45	RH	108	1+	-85 030.	T=16.8S
46	PD	108	0+	-89 526.	A=26.7
47	AG	108	1+	-87 605.	T=2.41M
48	CD	108	0+	-89 248.0	A=0.9
49	IN	108	2+	-84 100.	T=40.M
50	SN	108	0+	-82 000.	T=10.5M
45	RH	109		-85 110.	T 90.S
46	PD	109	5/2+	-87 606.	T=13.46H
47	AG	109	1/2-	-88 721.5	A=48.17
48	CD	109	5/2+	-88 539.	T=453.D
49	IN	109	9/2+	-86 520.	T=4.2H
50	SN	109		-82 720.	T=18.0M
45	RH	110	1+	-82 940.	T=3.0S
46	PD	110	0+	-88 340.	A=11.8
47	AG	110	1+	-87 455.5	T=24.57S
48	CD	110	0+	-90 346.4	A=12.4
49	IN	110	2+	-86 420.	T=69.1M
50	SN	110	0+	-85 824.	T=4.0H
46	PD	111	5/2+	-86 020.	T=22.M
47	AG	111	1/2-	-88 224.	T=7.47D
48	CD	111	1/2+	-89 251.6	A=12.8
49	IN	111	9/2+	-88 426.	T=2.83D
50	SN	111	7/2+	-85 918.	T=35.3M
46	PD	112	0+	-86 280.	T=20.12H
47	AG	112	2-	-86 580.	T=3.12H
48	CD	112	0+	-90 576.9	A=24.0
49	IN	112	1+	-87 989.	T=14.4M
50	SN	112	0+	-88 648.	A=1.0
51	SB	112	3+	-81 850.	T=53.S
46	PD	113			T=1.5M
47	AG	113	1/2-	-87 035.	T=5.37H
48	CD	113	1/2+	-89 044.9	A=12.3
49	IN	113	9/2+	-89 342.	A=4.3
50	SN	113	1/2+	-88 317.	T=115.2D
51	SB	113	5/2+	-84 419.	T=6.7M
46	PD	114	0+		T=2.4M
47	AG	114		-85 010.	T=5.1S
48	CD	114	0+	-90 014.2	A=28.8
49	IN	114	1+	-88 584.	T=71.9S
50	SN	114	0+	-90 565.	A=0.66
51	SB	114	3+	-84 870.	T=3.6M
52	TE	114	0+	-82 170.	T=17.M
47	AG	115	1/2-	-84 910.	T=21.M
48	CD	115	1/2+	-88 090.	T=53.5H
49	IN	115	9/2+	-89 541.	A=95.7
					T=5.E+14Y
50	SN	115	1/2+	-90 027.	A=0.35
51	SB	115	5/2+	-86 997.	T=31.9M
52	TE	115	1/2+	-82 460.	T=6.M
47	AG	116		-82 420.	T=2.68M
48	CD	116	0+	-88 715.0	A=7.6
49	IN	116	1+	-88 248.	T=14.2S
50	SN	116	0+	-91 521.8	A=14.4
51	SB	116	3+	-87 020.	T=16.M
52	TE	116	0+	-85 460.	T=2.5H
48	CD	117	1/2+	-86 408.	T=2.6H
49	IN	117	9/2+	-88 929.	T=44.M
50	SN	117	1/2+	-90 392.6	A=7.6
51	SB	117	5/2+	-88 640.	T=2.8H
52	TE	117	1/2+	-85 150.	T=64.M
53	I	117	5/2+	-80 840.	T=2.5M
48	CD	118	0+	-86 704.	T=50.3M
49	IN	118	1+	-87 450.	T=5.S
50	SN	118	0+	-91 648.3	A=24.1
51	SB	118	1+	-87 953.	T=3.5M
52	TE	118	0+	-87 650.	T=6.0D
53	I	118		-81 550.	T=14.2M
54	XE	118	0+	-78 250.	T=6.M
48	CD	119		-84 210.	T=9.45M
49	IN	119	9/2+	-87 714.	T=2.83M
50	SN	119	1/2+	-90 061.6	A=8.6
51	SB	119	5/2+	-89 483.	T=38.1H
52	TE	119	1/2+	-87 189.	T=15.9H
53	I	119	5/2+	-83 990.	T=18.23M
54	XE	119		-79 000.	T=6.M
49	IN	120	1+	-85 490.	T=3.2S
50	SN	120	0+	-91 094.3	A=32.8

Z	El	A	Jπ	Mass		Value
51	SB	120	1+	-88	414.	T=15.9M
52	TE	120	0+	-89	402.	A=0.09
53	I	120		-84	100.	T=53.4M
54	XE	120	0+	-81	900.	T=40.1M
49	IN	121	9/2+	-85	820.	T=28.S
50	SN	121	3/2+	-89	202.7	T=27.06H
51	SB	121	5/2+	-89	589.9	A=57.3
52	TE	121	1/2+	-88	590.	T=17.1D
53	I	121	5/2+	-86	220.	T=2.12H
54	XE	121		-82	430.	T=38.8M
49	IN	122		-83	240.	T=10.0S
50	SN	122	0+	-89	935.6	A=4.7
51	SB	122	2-	-88	325.6	T=64.34H
52	TE	122	0+	-90	303.8	A=2.4
53	I	122	1+	-86	160.	T=3.62M
54	XE	122	0+	-85	060.	T=20.1H
49	IN	123	9/2+	-83	420.	T=5.97S
50	SN	123	11/2-	-87	809.	T=129.2D
51	SB	123	7/2+	-89	219.1	A=42.7
52	TE	123	1/2+	-89	162.	A=0.87
						T 1.2E+13Y
53	I	123	5/2+	-87	960.	T=13.2H
54	XE	123	1/2+	-85	290.	T=2.08H
49	IN	124		-80	830.	T=4.S
50	SN	124	0+	-88	229.0	A=5.8
51	SB	124	3-	-87	614.2	T=60.2D
52	TE	124	0+	-90	514.1	A=4.6
53	I	124	2-	-87	354.	T=4.17D
54	XE	124	0+	-87	450.	A=0.10
50	SN	125	11/2-	-85	890.	T=9.63D
51	SB	125	7/2+	-88	262.	T=2.77Y
52	TE	125	1/2+	-89	027.3	A=7.0
53	I	125	5/2+	-88	879.3	T=59.9D
54	XE	125	1/2+	-87	140.	T=17.H
55	CS	125	1/2+	-84	070.	T=45.M
56	BA	125		-79	570.	T=3.0M
50	SN	126	0+	-86	013.	T=E+5Y
51	SB	126	8-	-86	330.	T=12.4D
52	TE	126	0+	-90	064.9	A=18.7
53	I	126	2-	-87	914.	T=13.02D
54	XE	126	0+	-89	165.	A=0.09
55	CS	126	1+	-84	160.	T=98.6S
56	BA	126	0+	-82	360.	T=97.M
50	SN	127	11/2-	-83	510.	T=2.12H
51	SB	127	7/2+	-86	708.	T=91.2H
52	TE	127	3/2+	-88	289.	T=9.23H
53	I	127	5/2+	-88	981.4	A=100.
54	XE	127	1/2+	-88	317.	T=36.4D
55	CS	127	1/2+	-86	227.	T=6.3H
56	BA	127		-82	730.	T=18.M
50	SN	128	0+	-83	400.	T=59.3M
51	SB	128	8-	-84	700.	T=9.01H
52	TE	128	0+	-88	988.9	A=31.8
53	I	128	1+	-87	735.1	T=25.1M
54	XE	128	0+	-89	860.1	A=1.9
55	CS	128	1+	-85	953.	T=3.8M
56	BA	128	0+	-85	250.	T=2.43D
57	LA	128		-78	450.	T=4.2M
50	SN	129				T=7.5M
51	SB	129	7/2+	-84	591.	T=4.31H
52	TE	129	3/2+	-87	004.	T=68.7M
53	I	129	7/2+	-88	503.	T=1.6E+7Y
54	XE	129	1/2+	-88	694.0	A=26.4
55	CS	129	1/2+	-87	590.	T=33.1H
56	BA	129		-85	150.	T=2.13H
57	LA	129	3/2+	-81	150.	T=10.0M
51	SB	130	5+	-82	350.	T=5.7M
52	TE	130	0+	-87	345.4	A=34.5
53	I	130	5-	-86	888.	T=12.3H
54	XE	130	0+	-89	880.1	A=3.9
55	CS	130	1+	-86	857.	T=29.1M
56	BA	130	0+	-87	297.	A=0.10
57	LA	130	3+	-81	600.	T=8.7M
58	CE	130	0+			T=30.M
51	SB	131	7/2+	-82	090.	T=23.0M
52	TE	131	3/2+	-85	191.	T=25.0M
53	I	131	7/2+	-87	443.2	T=8.065D
54	XE	131	3/2+	-88	414.0	A=21.2
55	CS	131	5/2+	-88	059.	T=9.7D
56	BA	131	1/2+	-86	719.	T=11.7D
57	LA	131	1/2+	-83	760.	T=1.0H
58	CE	131		-79	460.	T=10.M
51	SB	132	7,8-	-79	590.	T=4.1M
52	TE	132	0+	-85	193.	T=78.H
53	I	132	4+	-85	698.	T=2.3H
54	XE	132	0+	-89	278.4	A=27.0
55	CS	132	2-	-87	179.	T=6.58D
56	BA	132	0+	-88	451.	A=0.095
57	LA	132	2-	-83	740.	T=4.8H
58	CE	132	0+	-82	340.	T=4.8H
51	SB	133	7/2+	-79	000.	T=2.3M
52	TE	133	3/2+	-82	900.	T=12.5M
53	I	133	7/2+	-85	860.	T=20.9H
54	XE	133	3/2+	-87	660.	T=5.31D
55	CS	133	7/2+	-88	087.	A=100.
56	BA	133	1/2+	-87	572.	T=10.35Y
57	LA	133	5/2+	-85	670.	T=4.H
58	CE	133	5/2+	-82	370.	T=5.40H
52	TE	134	0+	-82	570.	T=41.8
53	I	134	4+	-83	970.	T=52.5M
54	XE	134	0+	-88	123.0	A=10.5
55	CS	134	4+	-86	906.	T=2.06Y
56	BA	134	0+	-88	965.	A=2.4
57	LA	134	1+	-85	255.	T=6.7M

Table A6 (cont.)

58 CE 134	0+	-84 750.	T=72.H		
59 PR 134		-78 550.	T=16.4M		
53 I 135	7/2+	-83 776.	T=6.7H		
54 XE 135	3/2+	-86 502.	T=9.14H		
55 CS 135	7/2+	-87 659.	T=2.3E+6Y		
56 BA 135	3/2+	-87 868.	A=6.5		
57 LA 135	5/2+	-86 830.	T=19.5H		
58 CE 135	1/2+	-84 530.	T=17.7H		
59 PR 135	5/2+	-80 950.	T=25.4M		
60 ND 135			T=5.5M		
53 I 136	2-	-79 420.	T=82.8S		
54 XE 136	0+	-86 423.	A=8.9		
55 CS 136	5+	-86 356.	T=13.5D		
56 BA 136	0+	-88 904.	A=7.8		
57 LA 136	1+	-86 030.	T=9.87M		
58 CE 136	0+	-86 462.	A=0.19		
59 PR 136	2+	-81 260.	T=13.1M		
60 ND 136	0+	-78 800.	T=55.M		
53 I 137		-76 810.	T=24.6S		
54 XE 137	7/2-	-82 213.	T=3.82M		
55 CS 137	7/2+	-86 561.	T=30.13Y		
56 BA 137	3/2+	-87 734.	A=11.2		
57 LA 137	7/2+	-87 230.	T=6.E+4Y		
58 CE 137	3/2+	-86 030.	T=9.0H		
59 PR 137	5/2+	-83 280.	T=76.6M		
60 ND 137	1/2+	-79 280.	T=40.M		
54 XE 138	0+	-80 070.	T=14.2M		
55 CS 138	3-	-82 870.	T=32.2M		
56 BA 138	0+	-88 274.	A=71.9		
57 LA 138	5-	-86 480.	A=0.09 T=1.1E+11Y		
58 CE 138	0+	-87 536.	A=0.26		
59 PR 138	1+	-83 099.	T=1.44M		
54 XE 139		-75 980.	T=39.7S		
55 CS 139		-80 780.	T=9.53M		
56 BA 139	7/2-	-84 926.	T=85.2M		
57 LA 139	7/2+	-87 186.	A=99.91		
58 CE 139	3/2+	-86 911.	T=137.5D		
59 PR 139	5/2+	-84 799.	T=4.41H		
60 ND 139	3/2+	-82 000.	T=30.M		
54 XE 140	0+	-73 240.	T=13.6S		
55 CS 140		-77 540.	T=63.8S		
56 BA 140	0+	-83 241.	T=12.8D		
57 LA 140	3-	-84 276.	T=40.23H		
58 CE 140	0+	-88 042.	A=88.5		
59 PR 140	1+	-84 654.	T=3.39M		
60 ND 140	0+	-84 180.	T=3.37D		
54 XE 141			T=1.73S		
55 CS 141		-74 870.	T=24.9S		
56 BA 141		-79 970.	T=18.3M		
57 LA 141		-82 969.	T=3.87H		
58 CE 141	7/2-	-85 399.	T=32.53D		
59 PR 141	5/2+	-85 980.	A=100.		
60 ND 141	3/2+	-84 175.	T=2.46H		
61 PM 141	5/2+	-80 450.	T=20.9M		
54 XE 142	0+		T=1.22S		
55 CS 142		-71 070.	T=1.7S		
56 BA 142	0+	-77 770.	T=10.7M		
57 LA 142	2-	-79 970.	T=92.4M		
58 CE 142	0+	-84 487.	A=11.1 T 5.E+16Y		
59 PR 142	2-	-83 752.	T=19.2H		
60 ND 142	0+	-85 916.	A=27.1		
61 PM 142	1+	-81 100.	T=40.S		
62 SM 142	0+	-79 050.	T=72.5M		
63 EU 142	1+		T=1.2M		
54 XE 143			T=1.CS		
55 CS 143			T=1.7S		
56 BA 143		-74 010.	T=13.6S		
57 LA 143		-78 210.	T=14.M		
58 CE 143	3/2-	-81 593.	T=33.0H		
59 PR 143	7/2+	-83 038.	T=13.6D		
60 ND 143	7/2-	-83 970.	A=12.2		
61 PM 143	5/2+	-82 901.	T=265.D		
62 SM 143	3/2+	-79 422.	T=8.83M		
63 EU 143	5/2+	-74 420.	T=2.6M		
58 CE 144	0+	-80 403.	T=284.4D		
59 PR 144	0-	-80 719.	T=17.3M		
60 ND 144	0+	-83 716.	A=23.9 T=2.1E+15Y		
61 PM 144	6-	-81 340.	T=1.0Y		
62 SM 144	0+	-81 904.	A=3.1		
63 EU 144	1+	-75 577.	T=10.5S		
58 CE 145		-77 110.	T=3.3M		
59 PR 145	7/2+	-79 599.	T=5.98H		
60 ND 145	7/2-	-81 404.	A=8.3		
61 PM 145	5/2+	-81 234.	T=17.7Y		
62 SM 145	7/2-	-80 596.	T=340.D		
63 EU 145	5/2+	-77 876.	T=5.96D		
64 GD 145	1/2+	-72 880.	T=22.9M		
58 CE 146	0+	-75 740.	T=14.2M		
59 PR 146	3-	-76 820.	T=24.2M		
60 ND 146	0+	-80 898.	A=17.2		
61 PM 146	3,4-	-79 421.	T=5.5Y		
62 SM 146	0+	-80 947.	T=1.E+8Y		
63 EU 146	4-	-77 075.	T=4.6D		
64 GD 146	0+	-75 880.	T=48.3D		
59 PR 147		-75 430.	T=12.M		
60 ND 147	5/2-	-78 129.	T=10.99D		
61 PM 147	7/2+	-79 023.	T=2.623Y		
62 SM 147	7/2-	-79 248.	A=15.0 T=1.1E+11Y		
63 EU 147	5/2+	-77 486.	T=24.3D		
64 GD 147	7/2-	-75 158.	T=38.H		
65 TB 147	5/2-	-70 560.	T=1.6H		

59 PR 148		-72 480.	T=2.M		
60 ND 148	0+	-77 381.	A=5.7		
61 PM 148	1-	-76 852.	T=5.4D		
62 SM 148	0+	-79 317.	A=11.2		
			T=8.E+15Y		
63 EU 148	5-	-76 217.	T=54.5D		
64 GD 148	0+	-76 207.	T=93.Y		
65 TB 148		-70 590.	T=70.M		
59 PR 149		-71 380.	T=2.3M		
60 ND 149	5/2-	-74 377.	T=1.73H		
61 PM 149	7/2+	-76 046.	T=53.1H		
62 SM 149	7/2-	-77 118.	A=13.8		
			T 1.E+16Y		
63 EU 149	5/2+	-76 360.	T=93.D		
64 GD 149	7/2-	-75 072.	T=9.3D		
65 TB 149		-71 375.	T=4.13H		
60 ND 150	0+	-73 662.	A=5.6		
61 PM 150	1	-73 530.	T=2.68H		
62 SM 150	0+	-77 033.	A=7.4		
63 EU 150	0,1-	-74 719.	T=12.6H		
64 GD 150	0+	-75 728.	T=1.8E+6Y		
65 TB 150		-71 060.	T=3.1H		
66 DY 150	0+	-69 100.	T=7.2M		
60 ND 151		-70 899.	T=12.4M		
61 PM 151	5/2+	-73 365.	T=28.4H		
62 SM 151	3/2-	-74 553.	T=93.Y		
63 EU 151	5/2+	-74 629.	A=47.8		
64 GD 151	7/2-	-74 165.	T=120.D		
65 TB 151	1/2	-71 557.	T=17.6H		
66 DY 151	7/2	-68 552.	T=18.M		
67 HO 151		-63 500.	T=35.6S		
61 PM 152		-71 350.	T=4.2M		
62 SM 152	0+	-74 749.	A=26.7		
63 EU 152	3-	-72 863.	A=14.Y		
64 GD 152	0+	-74 691.	A=0.20		
			T=1.1E+14Y		
65 TB 152	2-	-70 871.	T=17.6H		
66 DY 152	0+	-70 057.	T=2.38H		
67 HO 152		-63 670.	T=2.5M		
61 PM 153		-70 740.	T=5.5M		
62 SM 153	3/2+	-72 544.	T=46.6H		
63 EU 153	5/2+	-73 347.	A=52.2		
64 GD 153	3/2+	-73 106.	T=241.D		
65 TB 153	5/2-	-71 310.	T=2.34D		
66 DY 153	7/2	-69 090.	T=6.3H		
67 HO 153		-64 832.	T=9.3M		
68 ER 153		-60 250.	T=36.S		
61 PM 154		-68 450.	T=2.5M		
62 SM 154	0+	-72 451.	A=22.8		
63 EU 154	3-	-71 713.	T=8.Y		
64 GD 154	0+	-73 691.	A=2.2		
65 TB 154		-70 290.	T=20.H		

66 DY 154	0+	-70 356.	T=E+6Y		
67 HO 154	1	-64 598.	T=11.8M		
68 ER 154	0+	-62 400.	T=4.5M		
62 SM 155	3/2-	-70 193.	T=22.3M		
63 EU 155	5/2+	-71 818.	T=4.8Y		
64 GD 155	3/2-	-72 065.	A=14.9		
65 TB 155	3/2+	-71 220.	T=5.3D		
66 DY 155	3/2-	-69 121.	T=9.9H		
67 HO 155	5/2	-65 820.	T=47.M		
68 ER 155		-62 010.	T=5.5M		
62 SM 156	0+	-69 359.	T=9.4H		
63 EU 156	0+	-70 072.	T=15.17D		
64 GD 156	0+	-72 524.	A=20.6		
65 TB 156	3-	-70 220.	T=5.1D		
66 DY 156	0+	-70 491.	A=0.06		
			T=2.E+14Y		
67 HO 156	1	-65 390.	T=55.M		
62 SM 157			T=0.5M		
63 EU 157		-69 461.	T=15.2H		
64 GD 157	3/2-	-70 821.	A=15.7		
65 TB 157	3/2+	-70 757.	T=160.Y		
66 DY 157	3/2+	-69 394.	T=8.1H		
67 HO 157	7/2	-66 890.	T=15.M		
68 ER 157	3/2	-62 990.	T=24.M		
63 EU 158		-67 250.	T=46.M		
64 GD 158	0+	-70 680.	A=24.7		
65 TB 158	3-	-69 440.	T=150.Y		
66 DY 158	0+	-70 384.	A=0.10		
67 HO 158	5+	-66 407.	T=11.3M		
68 ER 158	0+	-64 910.	T=2.3H		
63 EU 159	5/2+	-65 920.	T=18.1M		
64 GD 159	3/2-	-68 553.	T=18.6H		
65 TB 159	3/2+	-69 503.	A=100.		
66 DY 159	3/2-	-69 138.	T=144.D		
67 HO 159	7/2-	-67 440.	T=33.M		
68 ER 159	3/2-	-64 340.	T=36.M		
64 GD 160	0+	-67 934.	A=21.7		
65 TB 160	3-	-67 813.	T=72.4D		
66 DY 160	0+	-69 648.	A=2.3		
67 HO 160	5+	-66 728.	T=25.M		
68 ER 160	0+	-65 930.	T=28.6H		
64 GD 161	5/2-	-65 494.	T=3.7M		
65 TB 161	3/2+	-67 445.	T=6.92D		
66 DY 161	5/2+	-68 027.	A=18.9		
67 HO 161	7/2-	-67 210.	T=2.5H		
68 ER 161	3/2-	-65 161.	T=3.1H		
69 TM 161	7/2	-61 640.	T=39.M		
65 TB 162	1-	-65 690.	T=7.47M		
66 DY 162	0+	-68 151.	A=25.5		
67 HO 162	1+	-65 981.	T=13.M		
68 ER 162	0+	-66 299.	A=0.14		

Table A6 (cont.)

69	TM	162	1-	-61 600.	T=21.8M
65	TB	163	3/2+	-64 670.	T=19.5M
66	DY	163	5/2-	-66 351.	A=24.9
67	HO	163	7/2-	-66 342.	T=33.Y
68	ER	163	5/2-	-65 134.	T=75.M
69	TM	163	1/2+	-62 717.	T=1.8H
65	TB	164		-62 590.	T=3.04M
66	DY	164	0+	-65 934.	A=28.2
67	HO	164	1+	-64 955.	T=24.M
68	ER	164	0+	-65 918.	A=1.6
69	TM	164	1+	-61 956.	T=1.9M
70	YB	164	0+	-60 860.	T=77.M
66	DY	165	7/2+	-63 577.	T=2.35H
67	HO	165	7/2-	-64 873.	A=100.
68	ER	165	5/2-	-64 501.	T=10.36H
69	TM	165	1/2+	-62 936.	T=30.1H
70	YB	165	5/2-	-60 184.	T=10.M
66	DY	166	0+	-62 563.	T=81.5H
67	HO	166	0-	-63 044.	T=26.8H
68	ER	166	0+	-64 904.	A=33.4
69	TM	166	2+	-61 869.	T=7.7H
70	YB	166	0+	-61 609.	T=56.7H
67	HO	167	7/2-	-62 298.	T=3.1H
68	ER	167	7/2+	-63 268.	A=22.9
69	TM	167	1/2+	-62 521.	T=9.3D
70	YB	167	5/2-	-60 566.	T=17.5M
71	LU	167		-57 500.	T=55.M
67	HO	168		-60 200.	T=3.0M
68	ER	168	0+	-62 968.	A=27.0
69	TM	168	3+	-61 270.	T=93.D
70	YB	168	0+	-61 549.	A=0.14
71	LU	168	1-	-57 190.	T=6.1M
72	HF	168	0+	-55 190.	T=26.M
67	HO	169		-58 750.	T=4.7M
68	ER	169	1/2-	-60 899.	T=9.3D
69	TM	169	1/2+	-61 251.	A=100.
70	YB	169	7/2+	-60 344.	T=31.D
71	LU	169	7/2+	-58 074.	T=1.42D
72	HF	169	5/2-	-54 700.	T=3.25M
68	ER	170	0+	-60 091.	A=15.0
69	TM	170	1-	-59 773.	T=129.D
70	YB	170	0+	-60 741.	A=3.0
71	LU	170	0+	-57 301.	T=2.0D
72	HF	170	0+	-56 100.	T=16.0H
68	ER	171	5/2-	-57 700.	T=7.5H
69	TM	171	1/2+	-59 190.	T=1.92Y
70	YB	171	1/2-	-59 287.	A=14.3
71	LU	171	7/2+	-57 890.	T=8.2D
72	HF	171		-55 290.	T=12.2H
68	ER	172	0+	-56 480.	T=49.H
69	TM	172	2-	-57 369.	T=63.6H
70	YB	172	0+	-59 239.	A=21.9
71	LU	172	4-	-56 740.	T=6.70D
72	HF	172	0+	-56 340.	T=5.Y
73	TA	172		-51 340.	T=44.M
68	ER	173		-53 420.	T=12.M
69	TM	173	1/2+	-56 215.	T=8.2H
70	YB	173	5/2-	-57 535.	A=16.2
71	LU	173	7/2+	-56 845.	T=500.D
72	HF	173	1/2-	-55 250.	T=23.6H
73	TA	173	7/2+	-52 350.	T=3.6H
69	TM	174	4-	-53 870.	T=5.5M
70	YB	174	0+	-56 933.	A=31.8
71	LU	174	1-	-55 562.	T=3.6Y
72	HF	174	0+	-55 760.	A=0.18
					T=2.E+15Y
73	TA	174	4-	-51 760.	T=1.2H
74	W	174	0+	-49 860.	T=29.M
69	TM	175	1/2+	-52 280.	T=16.M
70	YB	175	7/2-	-54 681.	T=4.19D
71	LU	175	7/2+	-55 149.	A=97.4
72	HF	175	5/2-	-54 542.	T=70.D
73	TA	175	7/2+	-52 340.	T=10.5H
74	W	175	1/2-	-49 340.	T=34.M
69	TM	176		-49 340.	T=1.4M
70	YB	176	0+	-53 485.	A=12.7
71	LU	176	7-	-53 370.	A=2.6
72	HF	176	0+	-54 559.	A=5.2
73	TA	176	1-	-51 460.	T=8.1H
74	W	176	0+	-50 460.	T=2.5H
78	PT	176	0+	-27 850.	T=6.6S
70	YB	177	9/2+	-50 975.	T=1.9H
71	LU	177	7/2+	-52 371.	T=6.7D
72	HF	177	7/2-	-52 868.	A=18.5
73	TA	177	7/2+	-51 710.	T=56.6H
74	W	177		-49 710.	T=2.2H
75	RE	177		-46 110.	T=14.M
71	LU	178	1+	-50 170.	T=28.4M
72	HF	178	0+	-52 422.	A=27.2
73	TA	178	1+	-50 510.	T=9.3M
74	W	178	0+	-50 420.	T=21.5D
75	RE	178	5-	-45 760.	T=13.2M
71	LU	179	7/2+	-49 100.	T=4.6H
72	HF	179	9/2+	-50 450.	A=13.8
73	TA	179	7/2+	-50 331.	T=600.D
74	W	179	7/2-	-49 230.	T=38.M
75	RE	179	5/2+	-46 540.	T=20.M
71	LU	180	3-	-46 470.	T=5.6M
72	HF	180	0+	-49 766.	A=35.1
73	TA	180	8+	-48 840.	A=0.012

74	W	180	0+	-49 650.	A=0.13	74	W	188	0+	-38 634.	T=69.D
75	RE	180	1-	-45 860.	T=2.45M	75	RE	188	1-	-38 983.	T=16.8H
76	OS	180	0+	-43 960.	T=23.M	76	OS	188	0+	-41 101.	A=13.3
						77	IR	188	2-	-38 268.	T=41.4H
72	HF	181	1/2-	-47 389.	T=42.4D	78	PT	188	0+	-37 728.	T=10.2D
73	TA	181	7/2+	-48 412.	A=99.988	79	AU	188		-32 430.	T=8.8M
74	W	181	9/2+	-48 225.	T=130.D	80	HG	188	0+	-29 520.	T=3.2M
75	RE	181	5/2+	-46 430.	T=19.H						
76	OS	181		-43 400.	T=105.M	74	W	189		-35 440.	T=11.5M
						75	RE	189	5/2+	-37 942.	T=24.H
72	HF	182	0+	-45 900.	T=9.E+6Y	76	OS	189	3/2-	-38 952.	A=16.1
73	TA	182	3-	-46 403.	T=115.D	77	IR	189	3/2+	-38 450.	T=13.3D
74	W	182	0+	-48 208.	A=26.3	78	PT	189		-36 550.	T=10.4H
75	RE	182	7+	-45 348.	T=64.H	79	AU	189		-33 550.	T=28.7M
76	OS	182	0+	-44 250.	T=21.5H	80	HG	189		-29 350.	T=7.7M
77	IR	182	5-	-38 950.	T=15.M						
						75	RE	190		-35 490.	T=3.M
72	HF	183	3/2-	-43 219.	T=64.M	76	OS	190	0+	-38 674.	A=26.4
73	TA	183	7/2+	-45 259.	T=5.0D	77	IR	190	4+	-36 620.	T=12.2D
74	W	183	1/2-	-46 327.	A=14.3	78	PT	190	0+	-37 293.	A=0.013
75	RE	183	5/2+	-45 771.	T=70.D						T=7.E+11Y
76	OS	183	9/2+	-43 370.	T=14.H	79	AU	190	1-	-32 890.	T=42.M
77	IR	183		-39 970.	T=58.M	80	HG	190	0+	-30 890.	T=20.M
73	TA	184	5-	-42 637.	T=8.7H	75	RE	191		-34 460.	T=10.M
74	W	184	0+	-45 667.	A=30.7	76	OS	191	9/2-	-36 362.	T=15.3D
75	RE	184	3-	-44 060.	T=38.D	77	IR	191	3/2+	-36 672.	A=37.4
76	OS	184	0+	-44 158.	A=0.02	78	PT	191	3/2-	-35 672.	T=2.96D
77	IR	184		-39 440.	T=3.0H	79	AU	191	3/2+	-33 770.	T=3.2H
78	PT	184	0+	-36 940.	T=17.3M	80	HG	191		-30 470.	T=56.M
79	AU	184		-30 140.	T=52.S						
80	HG	184	0+	-25 590.	T=30.9S	75	RE	192			T=6.S
						76	OS	192	0+	-35 850.	A=41.0
73	TA	185	7/2+	-41 380.	T=49.M	77	IR	192	4-	-34 799.	T=74.3D
74	W	185	3/2-	-43 345.	T=75.D	78	PT	192	0+	-36 256.	A=0.78
75	RE	185	5/2+	-43 774.	A=37.5	79	AU	192	1-	-32 742.	T=5.0H
76	OS	185	1/2-	-42 759.	T=94.D	80	HG	192	0+	-31 840.	T=5.H
77	IR	185	3/2+	-40 260.	T=14.H	81	TL	192	2-	-25 540.	T=10.M
78	PT	185		-36 460.	T=1.1H						
79	AU	185		-31 490.	T=4.3M	76	OS	193	3/2-	-33 367.	T=30.2H
80	HG	185		-25 930.	T=51.S	77	IR	193	3/2+	-34 499.	A=62.6
						78	PT	193	1/2-	-34 438.	T 620.Y
73	TA	186		-38 580.	T=10.6M	79	AU	193	3/2+	-33 440.	T=17.6H
74	W	186	0+	-42 475.	A=28.6	80	HG	193	3/2-	-31 100.	T=4.H
75	RE	186	1-	-41 881.	T=91.H	81	TL	193	1/2+	-26 900.	T=22.M
76	OS	186	0+	-42 958.	A=1.6						
77	IR	186	6-	-39 127.	T=15.6H	76	OS	194	0+	-32 397.	T=6.Y
78	PT	186	0+	-37 500.	T=2.1H	77	IR	194	1-	-32 494.	T=19.15H
79	AU	186		-31 500.	T=10.7M	78	PT	194	0+	-34 733.	A=32.9
80	HG	186	0+	-27 960.	T=1.4M	79	AU	194	1-	-32 224.	T=39.5H
						80	HG	194	0+	-32 174.	T=1.3Y
74	W	187	3/2-	-39 870.	T=23.9H	81	TL	194	(2-)	-26 670.	T=33.M
75	RE	187	5/2+	-41 181.	T=62.5						
					T=5.E+10Y	77	IR	195	11/2-	-31 851.	T=2.7H
76	OS	187	1/2-	-41 184.	A=1.6	78	PT	195	1/2-	-32 786.	A=33.8
77	IR	187	3/2+	-39 680.	T=11.2H	79	AU	195	3/2+	-32 557.	T=184.D
78	PT	187		-36 780.	T=2.36H	80	HG	195	1/2-	-31 160.	T=9.5H
79	AU	187		-32 750.	T=8.6M	81	TL	195	1/2+	-27 960.	T=1.17H
80	HG	187		-27 910.	T=2.4M						

Table A6 (cont.)

82 PB 195		-23 360.	T=17.M		
77 IR 196		-29 460.	T=53.S		
78 PT 196	0+	-32 635.	A=25.3		
79 AU 196	2-	-31 153.	T=6.18D		
80 HG 196	0+	-31 837.	A=0.15		
81 TL 196	2-	-27 440.	T=1.8H		
82 PB 196	0+	-25 040.	T=37.M		
77 IR 197		-28 410.	T=7.M		
78 PT 197	1/2-	-30 414.	T=20.0H		
79 AU 197	3/2+	-31 161.	A=100.		
80 HG 197	1/2-	-30 746.	T=64.1H		
81 TL 197	1/2+	-28 350.	T=2.84H		
82 PB 197	3/2-	-24 650.			
78 PT 198	0+	-29 906.	A=7.2		
79 AU 198	2-	-29 602.	T=2.696D		
80 HG 198	0+	-30 975.	A=10.1		
81 TL 198	2-	-27 510.	T=5.3H		
82 PB 198	0+	-26 010.	T=2.4H		
83 BI 198		-19 110.	T=11.9M		
78 PT 199	5/2-	-27 406.	T=30.8M		
79 AU 199	3/2+	-29 099.	T=3.139D		
80 HG 199	1/2-	-29 552.	A=16.9		
81 TL 199	1/2+	-28 150.	T=7.42H		
82 PB 199	5/2-	-25 350.	T=1.5H		
83 BI 199	9/2	-20 550.	T=27.M		
78 PT 200	0+	-26 610.	T=11.5H		
79 AU 200	1-	-27 310.	T=48.4M		
80 HG 200	0+	-29 509.	A=23.1		
81 TL 200	2-	-27 055.	T=26.1H		
82 PB 200	0+	-26 350.	T=21.5H		
83 BI 200	7+	-20 350.	T=35.M		
84 PC 200	0+	-16 630.	T=11.5M		
79 AU 201		-26 160.	T=26.M		
80 HG 201	3/2-	-27 662.	A=13.2		
81 TL 201	1/2+	-27 250.	T=73.5H		
82 PB 201	5/2-	-25 450.	T=9.4H		
83 BI 201	9/2-	-21 450.	T=100.M		
84 PO 201	3/2-	-16 420.	T=15.3M		
79 AU 202	1-	-23 850.	T=29.S		
80 HG 202	0+	-27 346.	A=29.7		
81 TL 202	2-	-26 109.	T=12.2D		
82 PB 202	0+	-26 059.	T=3.E+5Y		
83 BI 202	5+	-20 860.	T=1.67H		
84 PC 202	0+	-17 890.	T=45.M		
79 AU 203		-22 770.	T=55.S		
80 HG 203	5/2-	-25 267.	T=46.59D		
81 TL 203	1/2+	-25 758.	A=29.5		
82 PB 203	5/2-	-24 776.	T=52.1H		
83 BI 203	9/2-	-21 590.	T=11.76H		
84 PO 203	5/2-	-17 430.	T=30.M		

80 HG 204	0+	-24 686.	A=6.8
81 TL 204	2-	-24 342.	T=3.78Y
82 PB 204	0+	-25 105.	A=1.4
			T=1.4E+17Y
83 BI 204	6+	-20 710.	T=11.3H
84 PO 204	0+	-18 450.	T=3.52H
80 HG 205	1/2-	-22 282.	T=5.2M
81 TL 205	1/2-	-23 811.	A=70.5
82 PB 205	5/2-	-23 768.	T=1.4E+7Y
83 BI 205	9/2-	-21 064.	T=15.31D
84 PC 205	5/2-	-17 700.	T=1.8H
81 TL 206	0-	-22 244.	T=4.20M
82 PB 206	0+	-23 777.	A=24.1
83 BI 206	6+	-20 125.	T=6.243D
84 PO 206	0+	-18 308.	T=8.8D
85 AT 206		-12 620.	T=31.4M
81 TL 207	1/2+	-21 014.	T=4.77M
82 PB 207	1/2-	-22 446.	A=22.1
83 BI 207	9/2-	-20 041.	T=38.Y
84 PO 207	5/2-	-17 132.	T=5.7H
85 AT 207		-13 290.	T=1.8H
81 TL 208	5+	-16 749.	T=3.07M
82 PB 208	0+	-21 743.	A=52.4
83 BI 208	5+	-18 875.	T=3.68E+5Y
84 PC 208	0+	-17 464.	T=2.898Y
85 AT 208	7+	-12 540.	T=1.63H
81 TL 209	1/2+	-13 632.	T=2.2M
82 PB 209	9/2+	-17 609.	T=3.31H
83 BI 209	9/2-	-18 257.	A=100.
84 PO 209	1/2-	-16 364.	T=102.Y
85 AT 209	9/2-	-12 882.	T=5.42H
81 TL 210	4,5	-9 224.	T=130.M
82 PB 210	0+	-14 720.	T=22.3Y
83 BI 210	1-	-14 783.	T=5.01D
84 PO 210	0+	-15 944.	T=138.40D
85 AT 210	5+	-12 069.	T=8.1H
86 RN 210	0+	-9 723.	T=2.5H
82 PB 211	9/2+	-10 463.	T=36.1M
83 BI 211	9/2-	-11 839.	T=2.13M
84 PO 211	9/2+	-12 429.	T=0.56S
85 AT 211	9/2-	-11 637.	T=7.21H
86 RN 211	1/2-	-8 741.	T=14.6H
82 PB 212	0+	-7 544.	T=10.64H
83 BI 212	1-	-8 117.	T=60.60M
84 PO 212	0+	-10 364.	T=3.04E-7S
85 AT 212	1-	-8 624.	T=0.314S
86 RN 212	0+	-8 648.	T=25.M
82 PB 213		-3 130.	T=10.2M
83 BI 213	9/2-	-5 226.	T=46.M
84 PC 213	9/2+	-6 647.	T=4.E-6S

85	AT	213		- 6 578.	T=1.1E-7S
86	RN	213		- 5 696.	T=2.50E-2S
82	PB	214	0+	- 147.	T=26.8M
83	BI	214	1-	- 1 183.	T=19.8M
84	PO	214	0+	- 4 460.	T=1.64E-4S
85	AT	214		- 3 409.	T=2.E-6S
86	RN	214	0+	- 4 310.	T=2.7E-7S
87	FR	214	1-	- 1 056.	T=5.1E-3S
83	BI	215	9/2-	1 730.	T=7.4M
84	PO	215	9/2+	- 514.	T=1.78E-3S
85	AT	215		- 1 254.	T=E-4S
86	RN	215		- 1 165.	T=2.3E-6S
84	PO	216	0+	1 786.	T=0.15S
85	AT	216	1-	2 260.	T=3.E-4S
86	RN	216	0+	262.	T 4.5E-5S
87	FR	216		2 976.	T=7.0E-7S
85	AT	217		4 398.	T=32.E-3S
86	RN	217	9/2+	3 666.	T=5.4E-4S
87	FR	217		4 318.	T 2.2E-5S
84	PO	218	0+	8 390.	T=3.05M
85	AT	218		8 117.	T=2.S
86	RN	218	0+	5 232.	T=3.5E-2S
87	FR	218		7 013.	T=5.E-3S
85	AT	219		10 540.	T=0.9M
86	RN	219	3/2+	8 856.	T=3.96S
87	FR	219		8 614.	T=0.02S
88	RA	219		9 392.	T=1.E-2S
86	RN	220	0+	10 616.	T=55.6S
87	FR	220		11 483.	T=28.S
88	RA	220	0+	10 279.	T=0.023S
86	RN	221		14 390.	T=25.M
87	FR	221		13 280.	T=4.8M
88	RA	221		12 974.	T=29.S
89	AC	221		14 529.	T=0.05S
86	RN	222	0+	16 402.	T=3.824D
87	FR	222		16 364.	T=15.M
88	RA	222	0+	14 336.	T=38.S
89	AC	222		16 569.	T=5.S
86	RN	223			T=43.M
87	FR	223	3/2+	18 406.	T=22.M
88	RA	223	1/2+	17 257.	T=11.43D
89	AC	223	5/2-	17 821.	T=2.2M
87	FR	224		21 730.	T=2.7M
88	RA	224	0+	18 828.	T=3.64D
89	AC	224		20 231.	T=2.9H
90	TH	224	0+	20 008.	T=1.04S
88	RA	225	5/2-	22 011.	T=14.8D

89	AC	225	3/2-	21 639.	T=10.0D
90	TH	225	3/2+	22 319.	T=8.M
88	RA	226	0+	23 694.	T=1600.Y
89	AC	226		24 327.	T=29.H
90	TH	226	0+	23 212.	T=31.M
91	PA	226		25 980.	T=1.8M
88	RA	227		27 201.	T=41.2M
89	AC	227	3/2-	25 871.	T=21.772Y
90	TH	227	3/2+	25 827.	T=18.72D
91	PA	227	5/2-	26 827.	T=38.3M
88	RA	228	0+	28 962.	T=5.75Y
89	AC	228	3+	28 907.	T=6.13H
90	TH	228	0+	26 770.	T=1.913Y
91	PA	228	3+	28 883.	T=26.H
92	U	228	0+	29 236.	T=9.1M
90	TH	229	5/2+	29 604.	T=7340.Y
91	PA	229	5/2-	29 899.	T=1.4D
92	U	229	3/2+	31 216.	T=58.M
90	TH	230	0+	30 886.	T=7.7E+4Y
91	PA	230	2-	32 190.	T=17.4D
92	U	230	0+	31 628.	T=20.8D
90	TH	231	5/2+	33 829.	T=25.52H
91	PA	231	3/2-	33 443.	T=3.25E+4Y
92	U	231	5/2-	33 800.	T=4.2D
90	TH	232	0+	35 467.	T=1.41E+10Y
					A=100.
91	PA	232	2,3	35 953.	T=1.31D
92	U	232	0+	34 608.	T=72.Y
90	TH	233	1/2+	38 752.	T=22.3M
91	PA	233	3/2-	37 508.	T=27.0D
92	U	233	5/2+	36 937.	T=1.59E+5Y
93	NP	233		38 020.	T=35.M
90	TH	234	0+	40 645.	T=24.10D
91	PA	234	4+	40 382.	T=6.67H
92	U	234	0+	38 168.	T=2.44E+5Y
					A=0.0055
93	NP	234	0+	39 976.	T=4.4D
91	PA	235	3/2-	42 330.	T=24.1M
92	U	235	7/2-	40 934.	A=0.720
					T=7.1E+8Y
93	NP	235	5/2+	41 057.	T=396.D
94	PU	235	5/2+	42 190.	T=24.3M
91	PA	236	1-	45 560.	T=9.1M
92	U	236	0+	42 460.	T=2.4E+7Y
93	NP	236	6-	43 437.	T 5000.Y
94	PU	236	0+	42 900.	T=2.851Y
91	PA	237	3/2-	47 710.	T=8.7M

Table A6 (cont.)

92	U	237	1/2+	45 407.	T=6.75D
93	NP	237	5/2+	44 889.	T=2.14E+6Y
94	PU	237	7/2+	45 113.	T=45.63D
92	U	238	0+	47 335.	T=4.49E+9Y
					A=99.28
93	NP	238	2+	47 481.	T=2.117D
94	PU	238	0+	46 186.	T=87.75Y
95	AM	238		48 490.	T=1.6H
92	U	239	5/2+	50 604.	T=23.54M
93	NP	239	5/2+	49 326.	T=2.35D
94	PU	239	1/2+	48 602.	T=24390.Y
95	AM	239	5/2-	49 406.	T=12.1H
92	U	240	0+	52 742.	T=14.1H
93	NP	240	5+	52 230.	T=65.M
94	PU	240	0+	50 140.	T=6540.Y
95	AM	240	3-	51 540.	T=51.H
96	CM	240	0+	51 721.	T=26.8D
93	NP	241	5/2+	54 330.	T=16.M
94	PU	241	5/2+	52 972.	T=14.8Y
95	AM	241	5/2-	52 951.	T=433.Y
96	CM	241	1/2+	53 723.	T=36.D
94	PU	242	0+	54 742.	T=3.87E+5Y
95	AM	242	1-	55 494.	T=16.02H
96	CM	242	0+	54 827.	T=163.0D
94	PU	243	7/2+	57 777.	T=4.96H
95	AM	243	5/2-	57 189.	T=7370.Y
96	CM	243	5/2+	57 196.	T=28.Y
97	BK	243	3/2-	58 702.	T=4.6H
94	PU	244	0+	59 831.	T=8.3E+7Y
95	AM	244	6-	59 898.	T=10.1H
96	CM	244	0+	58 469.	T=17.9Y
97	BK	244	4-	60 740.	T=4.4H
98	CF	244	0+	61 474.	T=20.M
94	PU	245	9/2-	63 182.	T=10.5H
95	AM	245	5/2+	61 922.	T=2.04H
96	CM	245	7/2+	61 020.	T=8.7E+3Y
97	BK	245	3/2-	61 840.	T=4.98D
98	CF	245	1/2+	63 403.	T=44.M
94	PU	246	0+	65 320.	T=10.85D
95	AM	246	2+	64 940.	T=39.M
96	CM	246	0+	62 641.	T=4.65E+3Y
97	BK	246	2-	64 240.	T=1.8D
98	CF	246	0+	64 121.	T=36.H
95	AM	247	5/2	67 160.	T=22.M
96	CM	247	9/2-	65 556.	T=1.54E+7Y
97	BK	247	3/2-	65 500.	T=1.4E+3Y
98	CF	247	7/2+	66 220.	T=2.5H
96	CM	248	0+	67 417.	T=3.4E+5Y

97	BK	248	8-	68 010.	T 9.Y
98	CF	248	0+	67 264.	T=350.D
99	ES	248		70 320.	T=27.M
100FM		248	0+	71 900.	T=37.S
96	CM	249	1/2+	70 776.	T=64.M
97	BK	249	7/2+	69 868.	T=311.D
98	CF	249	9/2-	69 742.	T=352.Y
99	ES	249	7/2+	71 146.	T=1.7H
100FM		249	7/2+	73 530.	T=2.6M
96	CM	250	0+	73 070.	T=1.1E+4Y
97	BK	250	2-	72 970.	T=3.22H
98	CF	250	0+	71 195.	T=13.1Y
99	ES	250		73 200.	T=8.3H
100FM		250	0+	74 094.	T=30.M
97	BK	251	7/2+	75 280.	T=57.M
98	CF	251	1/2+	74 153.	T=900.Y
99	ES	251	3/2-	74 517.	T=33.H
100FM		251	9/2-	76 010.	T=7.H
98	CF	252	0+	76 059.	T=2.65Y
99	ES	252	7+	77 180.	T=140.D
100FM		252	0+	76 842.	T=23.H
98	CF	253	7/2+	79 337.	T=17.8D
99	ES	253	7/2+	79 038.	T=20.5D
100FM		253	1/2+	79 373.	T=3.0D
102NO		253	9/2-	84 350.	T=1.6M
98	CF	254	0+	81 430.	T=60.D
99	ES	254	7+	82 021.	T=276.D
100FM		254	0+	80 934.	T=3.24H
102NO		254		84 754.	T=56.S
99	ES	255		84 110.	T=39.D
100FM		255	7/2+	83 821.	T=20.1H
101MD		255	7/2-	84 890.	T=27.M
102NO		255	1/2+	86 870.	T=3.2M
99	ES	256		87 280.	T=22.M
100FM		256	0+	85 518.	T=2.63H
101MD		256	0-	87 510.	T=77.M
102NO		256	0+	87 820.	T=3.5S
103LR		256		91 820.	T 35.S
100FM		257	9/2+	88 628.	T=82.D
101MD		257		89 060.	T=5.H
102NO		257		90 249.	T=26.S
103LR		257		92 700.	
104		257			T 4.5S
101MD		258			T=55.D
104		259			T 3.S

Table A7 CUMULATED INDEX TO A-CHAINS. (Courtesy Nuclear Data Project, Oak Ridge National Laboratory.)

Prepared by Nuclear Data Project

April 1973

CUMULATED INDEX TO A-CHAINS

A	Nuclei	Reference	Date
1	H		
2	H		
3	He		*
4	He	NP A109,1	1968*
5		NP 78,5	1966
6	Li	NP 78,19	1966
7	Li	NP 78,36	1966
8	Be	NP 78,54	1966
9	Be	NP 78,79	1966
10	B	NP 78,104	1966
11	B	NP A114,2	1968
12	C	NP A114,36	1968
13	C	NP 152,3	1970
14	N	NP 152,42	1970
15	N	NP 152,93	1970
16	O	NP A166,1	1971
17	O	NP A166,61	1971
18	O	NP A190,1	1972
19	F	NP A190,56	1972
20	Ne	NP A190,105	1972
21	Ne	NP 11,288	1959
21	Ne	NP A105,11	1967
22	Ne	NP 11,295	1959
22	Ne	NP A105,17	1967
23	Na	NP 11,298	1959
23	Na	NP A105,26	1967
24	Mg	NP 11,300	1959
24	Mg	NP A105,40	1967
25	Mg	NP A105,65	1967
26	Mg	NP A105,84	1967
27	Al	NP A105,103	1967
28	Si	NP A105,124	1967
29	Si	NP A105,150	1967
30	Si	NP A105,167	1967
31	P	NP A105,180	1967
32	S	NP A105,196	1967
33	S	NP A105,213	1967
34	S	NP A105,226	1967
35	Cl	NP A105,238	1967
36	S,Ar	NP A105,248	1967
37	Cl	NP A105,261	1967
38	Ar	NP A105,275	1967
39	K	NP A105,290	1967
40	Ar,Ca	NP A105,302	1967
41	K	NP A105,322	1967
42	Ca	NP A105,344	1967
43	Ca	NP A105,357	1967
44	Ca	NP A105,368	1967
45	Sc	B4-237	1970
46	Ca,Ti	B4-269	1970
47	Ti	B4-313	1970
48	Ti	B4-351	1970
49	Ti	B4-397	1970
50	Ti,Cr	B3-5,6-1	1970
51	V	B3-5,6-37	1970
52	Cr	B3-5,6-85	1970
53	Cr	B3-5,6-127	1970
54	Cr,Fe	B3-5,6-161	1970
55	Mn	B3-3,4-1	1970
56	Fe	B3-3,4-43	1970
57	Fe	B3-3,4-103	1970
58	Fe,Ni	B3-3,4-145	1970
59	Co	B2-5-1	1968
60	Ni	B2-5-41	1968
61	Ni	B2-5-81	1968
62	Ni	B2-3-1	1967*
63	Cu	B2-3-31	1967
64	Ni,Zn	B2-3-65	1967
65	Cu	B2-6-1	1968
66	Zn	B2-6-43	1968
67	Zn	B2-6-71	1968
68	Zn	B2-6-93	1968
69	Ga	B2-6-111	1968
70	Zn,Ge	B8-1	1972
71	Ga	B1-6-13	1966*
72	Ge	B1-6-27	1966
73	Ge	B1-6-47	1966
74	Ge,Se	B1-6-59	1966
75	As	B1-6-79	1966
76	Ge,Se	B1-6-103	1966
77	Se	NDS 9,229	1973
78	Se,Kr	B1-4-33	1966*
79	Br	B1-4-49	1966
80	Se,Kr	B1-4-69	1966
81	Br	B1-4-85	1966
82	Se,Kr	B1-4-103	1966
83	Kr	B1-4-125	1966
84	Kr,Sr	B5-109	1971
85	Rb	B5-131	1971
86	Kr,Sr	B5-151	1971
87	Sr	B5-457	1971
88	Sr	A8-4-345	1970
89	Y	A8-4-373	1970
90	Zr	A8-4-407	1970
91	Zr	B8-477	1972
92	Zr,Mo	B7-299	1972
93	Nb	B8-527	1972
94	Zr,Mo	R-661	1960*
95	Mo	B8-29	1972
96	Mo,Ru	B8-599	1972
97	Mo	R-706	1960*
98	Mo,Ru	R-719	1960
99	Ru	R-729	1961
100	Mo,Ru	R-745	1961
101	Ru	R-755	1961*
102	Ru,Pd	R-767	1961*
103	Rh	R-779	1961*
104	Ru,Pd	R-791	1961*
105	Pd	R-805	1961*
106	Pd,Cd	R-820	1961*
107	Ag	B7-1	1972
108	Pd,Cd	B7-33	1972
109	Ag	B6-1	1971
110	Pd,Cd	B5-487	1971
111	Cd	B6-39	1971
112	Cd,Sn	B7-69	1972
113	In	B5-181	1971
114	Cd,Sn	R-933	1960*
115	Sn	R-951	1960*
116	Cd,Sn	R-967	1960*
117	Sn	R-983	1960
118	Sn	R-994	1960*
119	Sn	R-1005	1960
120	Sn,Te	R-1016	1960
121	Sb	B6-75	1971
122	Sn,Te	R-419	1972
123	Sb	B7-363	1972
124	Sn..Xe	R-1064	1960*
125	Te	R-465	1972
126	Te,Xe	NDS 9,195	1973
127	I	B8-77	1972
128	Te,Xe	NDS 9,157	1973
129	Xe	B8-123	1972
130	Te..Ba	R-1149	1961*
131	Xe	R-1158	1961
132	Xe,Ba	R-1181	1961
133	Cs	R-1197	1961*
134	Xe,Ba	R-1211	1961
135	Ba	R-1229	1961
136	Xe..Ce	R-1239	1961*
137	Ba	R-1248	1961
138	Ba,Ce	R-1261	1961*
139	La	R-1271	1961
140	Ce	R-1284	1959*
141	Pr	R-1300	1961*
142	Ce,Nd	B2-1-1	1967*
143	Nd	B2-1-25	1967*
144	Nd,Sm	B2-1-47	1967*
145	Nd	B2-1-181	1967*
146	Nd,Sm	B2-4-1	1967
147	Sm	B2-4-35	1967*
148	Nd,Sm	B2-4-79	1967
149	Sm	R-1401	1962*
150	Nd..Gd	R-1415	1964*
151	Eu	R-1445	1963
152	Sm,Gd	R-1471	1964
153	Eu	R-1503	1963*
154	Sm..Dy	R-1529	1964
155	Gd	R-1555	1963
156	Gd,Dy	R-1578	1964
157	Gd	NDS 9,273	1973
158	Gd,Dy	R-1612	1963*
159	Tb	R-1629	1962*
160	Gd,Dy	R-1642	1964
161	Dy	R-1677	1963
162	Dy,Er	R-1694	1964
163	Dy	B8-295	1972
164	Dy,Er	R-1719	1964*
165	Ho	R-1733	1964*
166	Er	R-1769	1964
167	Er	R-1802	1964
168	Er,Yb	R-1818	1964*
169	Tm	R-1836	1964*
170	Er,Yb	R-1863	1964
171	Yb	R-1877	1964*
172	Tb	R-1897	1965
173	Yb	R-1927	1965
174	Yb	R-1947	1965*
175	Lu	R-1961	1965
176	Hf	R-1980	1965
177	Hf	R-1998	1965
178	Hf	R-2035	1965
179	Hf	R-2055	1965
180	Hf,W	R-2067	1965
181	Ta	R-2083	1965
182	W	B1-1-1	1966
183	W	B1-1-37	1966*
184	W	B1-1-63	1966*
185	Re	B1-1-83	1966*
186	W,Os	B1-2-1	1966*
187	Os	B1-2-23	1966
188	Os	B1-2-53	1966*
189	Os	B1-2-85	1966
190	Os,Pt	B2-223	1963*
191	Ir	R-2237	1963*
192	Os,Pt	NDS 9,195	1973
193	Ir	B8-389	1972
194	Pt	B7-95	1972
195	Pt	B8-431	1972
196	Pt,Hg	B7-395	1972
197	Au	B7-129	1972
198	Pt,Hg	B6-319	1971
199	Hg	B6-355	1971
200	Hg	B6-387	1971
201	Hg	B5-561	1971
202	Hg	B5-581	1971
203	Tl	B5-531	1971
204	Hg,Pb	B5-601	1971
205	Tl	B6-425	1971
206	Pb	B7-161	1972
207	Pb	B5-207	1971
208	Pb	B5-243	1971
209	Bi	B5-287	1971
210	Po	B5-631	1971
211	Po	B5-319	1971
212	Po	B8-165	1972
213	Po	B1-5-1	1966
214	Po	B1-5-7	1966
215	At	B1-5-25	1966
216	Po,Rn	B1-5-29	1966
217	Rn	B1-5-33	1966
218	Rn	B1-5-37	1966
219	Fr	B1-5-41	1966
220	Rn,Ra	B1-5-45	1966
221	Ra	B1-5-49	1966
222	Ra	B1-5-55	1966
223	Ra	B1-5-61	1966
224	Ra,Th	B1-5-75	1966
225	Ac	B1-5-82	1966*
226	Ra,Th	B1-5-91	1966
227	Th	B1-5-97	1966
228	Th	B1-5-107	1966
229	Th	B6-209	1971
230	Th,U	B4-543	1970
231	Pa	B6-225	1971
232	Th,U	B4-561	1970
233	U	B6-257	1971
234	U	B4-581	1970
235	U	B6-287	1971
236	U,Pu	B4-623	1970
237	Np	B6-539	1971
238	U,Pu	B4-635	1970
239	Pu	B6-577	1971
240	Pu	B4-661	1970
241	Am	B4-661	1971
242	Pu,Cm	B4-683	1970
243	Am	B3-2-1	1969
244	Pu,Cm	B3-2-13	1969
245	Cm	B3-2-23	1969
246	Cm	B3-2-37	1969
247	Bk	B3-2-51	1969
248	Cm,Cf	B3-2-57	1969
249	Cf	B3-2-61	1969
250	Cf	B3-2-71	1969
251	Cf	B3-2-77	1969
252	Cf,Fm	B3-2-85	1969
253	Es	B3-2-91	1969
254	Cf,Fm	B3-2-99	1969
255	Fm	B3-2-107	1969
256	Fm	B3-2-113	1969
257	Fm	B3-2-117	1969
258	Fm	B3-2-121	1969
259		B3-2-123	1969
260		B3-2-123	1969
261		B3-2-123	1969

EXPLANATION

The cumulated index gives, for each mass value A, the most recent compilation of experimental information on levels of nuclei with that A-value. For A=20-24, the 1967 compilation only partly supersedes the 1959 compilation.

NUCLEI — The beta-stable member(s) of this A-chain

REFERENCE —
NP = Nuclear Physics
NDS 9,125 = Nuclear Data Sheets, vol.9, p.125
R-779 = Reprint of Nuclear Data Sheets (1959-1965), p.779
B4-269 = Nuclear Data Sheets B4, 269
B1-4-85 = Nuclear Data B1-4-85
A8-4-345 = Nuclear Data Tables A8-4-345

DATE — The year in which the compilation was published. An asterisk(*) indicates that a revision is in progress.

Table A8 SPHERICAL HARMONICS

The spherical harmonics $Y_l^m(\theta, \varphi) \equiv Y_{lm}(\theta, \varphi)$ are the eigenfunctions of the operators L^2 and L_z [Eq. (13.27)]:

$$L^2 Y_{lm} = l(l+1)\hbar^2 Y_{lm},$$

$$L_z Y_{lm} = m\hbar Y_{lm}.$$

They satisfy the *symmetry relation*

$$Y_{l,-m}(\theta, \varphi) = (-1)^m Y_{lm}^*(\theta, \varphi),$$

and the *orthonormality relations*

$$\int_0^{2\pi} d\varphi \int_0^\pi \sin\theta \, d\theta \, Y_{l'm'}^*(\theta, \varphi) Y_{lm}(\theta, \varphi) = \delta_{l'l} \delta_{m'm}.$$

An arbitrary regular function $g(\theta, \varphi)$ can be expanded in spherical harmonics:

$$g(\theta, \varphi) = \sum_{l=0}^\infty \sum_{m=-1}^l A_{lm} Y_{lm}(\theta, \varphi),$$

where the coefficients are

$$A_{lm} = \int d\Omega \, Y_{lm}^*(\theta, \varphi) g(\theta, \varphi).$$

Explicit expressions for the spherical harmonics up to $l = 3$ are given below. The values for negative m follow from the symmetry relation.

Spherical harmonics $Y_{lm}(\theta, \varphi)$

$l = 0$ $Y_{00} = \dfrac{1}{\sqrt{4\pi}}$

$l = 1$
$\begin{cases} Y_{11} = -\sqrt{\dfrac{3}{8\pi}} \sin\theta \, e^{i\varphi} \\[2mm] Y_{10} = \sqrt{\dfrac{3}{4\pi}} \cos\theta \end{cases}$

$l = 2$
$\begin{cases} Y_{22} = \dfrac{1}{4}\sqrt{\dfrac{15}{2\pi}} \sin^2\theta \, e^{2i\varphi} \\[2mm] Y_{21} = -\sqrt{\dfrac{15}{8\pi}} \sin\theta \cos\theta \, e^{i\varphi} \\[2mm] Y_{20} = \sqrt{\dfrac{5}{4\pi}}\left(\dfrac{3}{2}\cos^2\theta - \dfrac{1}{2}\right) \end{cases}$

$l = 3$
$\begin{cases} Y_{33} = -\dfrac{1}{4}\sqrt{\dfrac{35}{4\pi}} \sin^3\theta \, e^{3i\varphi} \\[2mm] Y_{32} = \dfrac{1}{4}\sqrt{\dfrac{105}{2\pi}} \sin^2\theta \cos\theta \, e^{2i\varphi} \\[2mm] Y_{31} = -\dfrac{1}{4}\sqrt{\dfrac{21}{4\pi}} \sin\theta \, (5\cos^2\theta - 1)e^{i\varphi} \\[2mm] Y_{30} = \sqrt{\dfrac{7}{4\pi}}\left(\dfrac{5}{2}\cos^3\theta - \dfrac{3}{2}\cos\theta\right) \end{cases}$

Relations involving the spherical harmonics (also called surface harmonics of the first kind), are given in W. Magnus and F. Oberhettinger, *Formulas and Theorems for the Functions of Mathematical Physics*, Chelsea Publishing Co., New York, 1954, pp. 53–55.

Index

SCIENCE

SCIENCE

SCIENCE